T0331810

The new discipline of nonlinear dynamics, or chaos, has developed explosively in all areas of physics over the last two decades. *Chaos and Complexity in Astrophysics* provides a primer on nonlinear dynamics, and gives researchers and graduate students of astrophysics the investigative tools they need to explore chaotic and complex phenomena. Comprehensive mathematical concepts and techniques are introduced at an appropriate level in the first part of the book, before being applied to stellar, interstellar, galactic and large scale complex phenomena in the Universe. The book demonstrates the application of ideas such as strange attractors, Poincaré sections, fractals, bifurcations, complex spatial patterns, and so on, to specific astrophysical problems. This self-contained text will appeal to a broad audience of astrophysicists and astronomers who wish to learn, and then apply, modern dynamical approaches to the problems they are working on.

ODED REGEV is a professor at the Department of Physics, Technion – Israel Institute of Technology. Following his Ph.D. from Tel Aviv University in 1980, he has held visiting positions in many research institutions and universities in the US and Europe, including University of Florida, Grenoble Observatory, France, Columbia University, New York, Institut d'Astrophysique, Paris and the American Museum of National History. He has been a faculty member at the Technion since 1982, and professor since 1996.

Regev has authored over 60 scientific papers, edited two conference proceedings, and is co-author (with Andrew King) of *Physics with Answers* (Cambridge University Press, 1997). He is also a member of the scientific advisory board of the AIP journal *Chaos*, a member of the American Physical Society and a member of the American Astronomical Society.

CHAOS AND COMPLEXITY IN ASTROPHYSICS

ODED REGEV

Technion – Israel Institute of Technology

CAMBRIDGE
UNIVERSITY PRESS

CAMBRIDGE
UNIVERSITY PRESS

University Printing House, Cambridge CB2 8BS, United Kingdom

Cambridge University Press is part of the University of Cambridge.

It furthers the University's mission by disseminating knowledge in the pursuit of
education, learning and research at the highest international levels of excellence.

www.cambridge.org
Information on this title: www.cambridge.org/9780521855341

© Oded Regev 2006

First published 2006

A catalogue record for this publication is available from the British Library

ISBN 978-0-521-85534-1 Hardback
ISBN 978-1-107-40654-4 Paperback

To Dani, Tamar and Judy

Read not to contradict and confute,
nor to believe and take for granted,
nor to find talk and discourse,
but to weigh and consider.
Francis Bacon, *Of Studies*.

Contents

Preface

The last thing one discovers in composing a work
is what to put first.
Blaise Pascal, *Pensées no. 19.*

In the last two decades or so the astrophysical community – students, teachers and researchers alike – have become aware of a new kind of activity in physics. Some researchers, science historians and philosophers have gone as far as calling it a 'new science' or 'new physics', while others see it as a mere natural extension of 'old' classical mechanics and fluid dynamics. In any case, the subject, variously referred to as *dynamical systems theory, nonlinear dynamics* or simply *chaos*, has undergone an explosive development, causing a lot of excitement in the scientific community and even in the general public. The discoveries look fundamental and there is hope that we will quite soon gain new and basic scientific understanding of the most complex aspects of nature.

The most striking quality of this modern approach to dynamical systems theory is, in my view, its extremely diverse range of applicability. Mechanics, fluid dynamics, chemical kinetics, electronic circuits, biology and even economics, as well as astrophysics, are among the subjects in which chaotic behaviour occurs. At the heart of the theory lies the quest for the *universal* and the *generic*, from which an understanding of complicated and seemingly heterogeneous phenomena can emerge. The ideas of bifurcations, strange attractors, fractal sets and so on, seem to provide the tools for such an unexpected conceptual unification.

My own experience in discussing the subject with astrophysicists suggests that they and their students would like to know more about the new developments in nonlinear dynamics. There is growing interest in the general subject as well as its possible relevance to specific fields of research or study in astrophysics.

The literature on chaos has grown enormously over the years. Books at all levels abound, from the popular to the rigorously mathematical; some of them excellent, some less so. It is not easy for an astrophysicist to pick the right book and to learn the relevant topics from it. Astrophysical applications are dispersed through various journals and a handful of conference proceedings. There seems to be a need for a book presenting the relevant general material and coherently suggesting its possible use in astrophysics. The purpose of this book is to answer this need by providing a useful (one hopes) source of information (and perhaps new ideas) for the entire astrophysical and astronomical community, from senior undergraduates right up to established researchers.

The book is divided into two parts of quite different nature (and size). The first is devoted to a quite comprehensive description of the basic notions in dynamical systems theory and of a few more advanced theoretical topics, that may be important in astrophysics. This part may serve as a self-contained text book for an introductory course in nonlinear dynamics, especially for senior undergraduate students. The second part of the book reviews astrophysical topics, in which work in the context of dynamical systems has already been done or is in progress. Its different chapters could also be used in senior undergraduate and graduate courses in astrophysics, where nonlinear stellar pulsation, convection and hydrodynamical turbulence, the complexity of the interstellar medium, galactic dynamics or large scale structure are discussed. These topics may be taught in a broader than usual way by including, in addition to the standard material, ideas based on dynamical system theory.

Throughout the text only a basic knowledge of ordinary differential equations plus the notion of a partial differential equation are required as the mathematical background. In physics, familiarity with classical analytical mechanics and some fluid dynamics are assumed. Most of the astrophysical applications are explained in a self-contained manner, and assume only quite basic knowledge in astrophysics, which is usually acquired in a typical introductory course.

The first part of the book opens with a short introduction following which, in Chapter 2, a few dynamical systems are presented as typical examples. I have decided to dress some of these celebrated paradigms of chaos (the logistic map, nonlinear oscillators, Hamiltonian non-integrability, transition to turbulence in convection, reaction-diffusion equations) in astrophysical attire. The resulting 'toy models' capture the essentials and thus can serve as a demonstration of the key concepts in nonlinear and chaotic dynamics.

In Chapter 3 these concepts are systematically explored and formulated in rather abstract mathematical language. This chapter is the heart of the first part as it introduces the generic characterisation of chaotic behaviour. In the following chapter strange attractors are defined and discussed and the theoretical tools of Chapter 3 are used in the study of chaotic behaviour and the various types of transitions to it. In Chapter 5 the theoretical diagnostic tools are used in the analysis of time-series data. This type of output is expected from real experiments in the lab, astronomical observations and numerical experiments. Chapters 6 and 7 introduce the basic concepts of Hamiltonian chaos and spatio-temporal pattern theory, respectively. The last four chapters of the first part are particularly valuable for the understanding of the astrophysical applications discussed in the second part of the book.

In the second part of the book we start by discussing several astrophysical Hamiltonian systems. Some recent work in planetary science, the question of stability of the Solar System, binary system dynamics and topics in galactic dynamics are among the subjects discussed. Next, in Chapter 10, we deal with variable astrophysical sources and start with the question of assessing the value of temporally variable astronomical signals as chaos indicators. This is critically examined in view of modern methods, some of which were introduced in Chapter 5. Several examples of theoretical models of irregular astronomical sources, based on nonlinear dynamical systems, follow. Nonlinear stellar pulsation, in which research in this context has been the most fruitful and significant, is given most attention. Chapter 11 describes some attempts at modelling the complexity of extended astrophysical media using dynamical systems and pattern theory, and we include a short discussion of the still controversial subject of a possibly fractal Universe. Finally, in Chapter 12 we discuss a number of fluid dynamical processes, whose understanding is essential to theoretical models of many important astrophysical objects. This chapter is different from all the other chapters in the second part of the book, as we do not include specific classes of astronomical objects as examples. Instead, we elaborate on a number of topics in fluid dynamics (some of which are applicable to astrophysics) in which a dynamical system approach may be fruitful.

My hope is that this book will encourage research in astrophysical problems, using ideas of dynamical systems and pattern theory. The second part may serve as a trigger to such new and hopefully successful approaches, as it contains descriptions of several astrophysical systems in which a dynamical system approach (using the tools introduced in the first part) has already been fruitful. If after reading this book, astrophysicists and their students are less confused about what chaos really is and in what way it is relevant (or irrelevant) to their field of research and interest, this book will have achieved its goal.

Throughout the book some references are given. These include books, review articles and original research papers. The references usually contain details, which are not explained in full in this book. In addition, they may serve as a possible source for broader and deeper study of the topics of the book. All of the references are listed alphabetically at the end of the book. This list is by no means exhaustive and I apologise to the authors of many important and relevant publications for not mentioning them.

Acknowledgements

> You see how lucky you are that I've got so many friends?
> Jackie Mason, *The World According to Me.*

I would like to thank all my teachers, colleagues and students (too numerous to all be mentioned here) with whom I have had the good fortune to do research on topics related to the subject of this book.

I am grateful in particular to those without whom this book would obviously have been impossible: Giora Shaviv, Robert Buchler and Ed Spiegel. They have shown me the way in scientific research and, most importantly, have taught me how to pick out the essentials from the 'sea' of detail. Special thanks are due to Andrew King, who has supported the idea of writing a chaos book for astrophysicists and has helped to bring it into being. Andrew, Attay Kovetz and Mike Shara read parts of the manuscript and their comments (on style and on essence) were very helpful.

Most of this book was written in France and in the United States, where I spent sabbatical leaves. I acknowledge the hospitality of the Institute d'Astrophysique de Paris, the Department of Astrophysics at the American Museum for Natural History and the Astronomy Department of Columbia University, who hosted me during these leaves. But above all it has been the warm welcome and friendship of Claude Bertout, Jean-Pierre Lasota, Mike Shara and Ed Spiegel that have made these leaves not only productive but also very pleasurable.

Over the years of writing this book my research has been supported from several sources, but most of all by continuous grants from the Israel Science Foundation; this support is gratefully acknowledged.

Finally, I would like to thank my close family for their love and support, which was an invaluable contribution towards the completion of this endeavour. My children Dani and Tamar have grown from childhood to adolescence while this book has too been (very slowly) growing. During those years I have also met Judy, who is now my wife. I dedicate this book to them.

Part I

Dynamical systems – general

1

Introduction to Part I

Not chaos-like, together crushed and bruised,
But, as the world harmoniously confused:
Where order in variety we see,
And where, though all things differ, all agree.
Alexander Pope, *Windsor Forest.*

In this part of the book we provide the basic mathematical background for dynamical systems and chaos theory. Some of the ideas introduced here will be applied to various astrophysical systems in the second part of the book. Our discussion here, while not particularly rigorous, will, however, be rather theoretical and abstract. I believe that a reasonable precision in building the basis for further understanding and research is mandatory. Throughout the discussion we continually give specific examples and often return to them in other places in the book. These examples, including some systems that are important by themselves, illustrate the various abstract concepts.

I have made an effort to interest readers, whose background is astronomy and astrophysics, by starting with astrophysical examples. After all, dynamical system theory and chaos have their origins in the studies of the three-body problem and celestial mechanics by Poincaré at the end of the nineteenth century. Fluid turbulence, an important unsolved scientific problem, is now being approached using methods from chaos and dynamical system theory. It has also had many important applications in astrophysics. Readers who are interested more in applications and less in theory and mathematical structure are particularly encouraged to become acquainted with the main concepts and results of this part of the book. Technical details may be skipped, certainly during first reading. When dealing with the second part (applications), the interested reader may return to the relevant material in the first part and study it more deeply.

The contemporary intensive study of chaotic behaviour in *nonlinear* dynamical systems owes its existence to the availability of fast digital computers. Irregular and aperiodic solutions of ordinary differential equations were discovered long ago, but they were, in a way, forgotten. In contrast, enormous analytical progress has been made in the study of *linear* systems. This progress is obviously reflected in the subjects taught in mathematics and the physical sciences at all levels. The bias towards the linear is so strong that a student or a scientist naturally attacks every problem with linear 'weapons' like the various techniques of eigenvalue problems, normal mode analysis, Fourier and other transforms, linear perturbations etc. These techniques often fail when the interest is the studied system's nonlinear behaviour. It may then seem that the only choice left is computer simulation and thus one may think that the 'linear paradise' of powerful analytical methods is lost and what remains is just a 'numerical hell'. I have often encountered the term *nonlinear* being used in astrophysics as a synonym for numerical simulation. This should not be so, as substantial analytical knowledge on nonlinear systems is now available. A lot of information on a nonlinear system may be deduced by analytical and perturbative methods, accompanied by specific numerical computations that are, in general, much easier than brute force direct numerical simulations. In any case, any such knowledge is very useful in devising the right numerical method for a full scale computer simulation and in understanding its result. For the student of modern methods in nonlinear dynamics 'paradise' can not only be regained, but it also reveals new and unexpected beauty. It may be manifested in astounding fractal structures and in fundamental invariances, symmetries and some other more general properties.

One such property, for example, is both very interesting and useful. It is the fact that nonlinear systems have a *generic* behaviour. Some very different-looking systems, describing completely unrelated natural phenomena, fall into classes that have identical behaviour. It seems that though all nonlinear systems differ, all agree in their fundamental properties. A typical example is the finding that transitions between different types of behaviour in a layer of fluid heated from below (transition to convective motions) have exactly the same properties as bifurcations (behaviour changes) of a class of quadratic mappings (like the logistic map obtained in population dynamics). As a result, the essence of very complex phenomena may often be described by a simple model.

We devote the remainder of this chapter to the introduction of the most basic mathematical notion used in the book, that of a *dynamical system*. It may be loosely defined as a set of rules by application of which the state of a physical (or some other, well defined) system can be found, if an initial state is known. Symbolically, we specify the *state* of a system by its being a well defined point, **x**, in some space. The vector notation reflects the fact that several variables may be needed to completely specify the state of a system. The space in which the state **x** lives is naturally called the *state space*.

The previously mentioned set of rules may be thought of as an *evolution operator*, $\mathcal{T}$, acting on a state variable and transforming it into some other state, $\mathbf{x}'$. This is expressed formally by

$$\mathbf{x}' = \mathcal{T}\mathbf{x}$$

The evolution may be continuous in time or performed in discrete steps. In the former case the state variable is a function of time and the continuous time evolution operator (labelled by a subscript t) is understood to carry the system from an initial condition (the state at $t = 0$) to the state at time t. In the latter case the discrete operator (with subscript n) evolves the system from an initial state $\mathbf{x}_0$ through n discrete steps. We can thus formally write for the two cases

$$\mathbf{x}(t) = \mathcal{T}_t \mathbf{x}(0) \tag{1.1}$$

and

$$\mathbf{x}_n = \mathcal{T}_n \mathbf{x}_0 \tag{1.2}$$

respectively.

The continuous case (1.1) can be realised, for example, by a set of two *ordinary differential equations* (ODE) like

$$\frac{du}{dt} = f(u, v) \quad \frac{dv}{dt} = g(u, v) \tag{1.3}$$

where the state variable is two dimensional, $\mathbf{x} = (u, v)$, and where f and g are some well-behaved functions. The evolution operator $\mathcal{T}_t$ symbolises the solution of (1.3), that is, the unique determination of $u(t)$ and $v(t)$ from the initial values $u(0)$ and $v(0)$.

The discrete case (1.2) can be illustrated by an example of an *iterated mapping* (or map) e.g.

$$x_{j+1} = F(x_j) \quad \text{with } j = 0, 1, ... \tag{1.4}$$

where F is some well-defined function. With the help of (1.4) the state variable (in this example it is one-dimensional) can be propagated in discrete steps from an initial value x_0 to its value after n iterations, x_n. The operator $\mathcal{T}_n$ reflects here the repeated (n times) application of F.

Thus, the action of the evolution operator on an initial state (point in state space) transforms the latter continuously, or in discrete steps, into another state. Thus, we may imagine an extra 'axis' (e.g. time t or the iteration number n), along which the state space is being transformed. The state space plus the evolution axis is called the *configuration space*.

So far we have considered only finite-dimensional state spaces, but this is clearly not the case when the dynamical system is represented by a *partial differential equation* (PDE). Such equations can be regarded as the limit ($m \to \infty$) of a set of m ordinary differential equations. Let $\{x^1(t), x^2(t), \ldots, x^m(t)\}$ be the components of the state variable $\mathbf{x}$ satisfying such a set of m ordinary differential equations. The system can be compactly written, using vector notation as

$$\frac{d\mathbf{x}}{dt} = \mathbf{F}(\mathbf{x}, t)$$

or explicitly, using components

$$\frac{dx^1}{dt} = F^1(x^1, x^2, \ldots, x^m, t)$$

$$\frac{dx^2}{dt} = F^2(x^1, x^2, \ldots, x^m, t)$$

$$\vdots$$

$$\frac{dx^m}{dt} = F^m(x^1, x^2, \ldots, x^m, t)$$

The vector state components can also be written in the form $x(k, t) \equiv x^k(t)$, with $k = 1, 2, \ldots, m$. The limit $m \to \infty$ is realised by replacing the discrete set of natural numbers $1 \le k \le m$ by a continuous 'space' variable, say ξ. The state variable is then written as $x(\xi, t)$ and is a continuous function of the space variable, ξ, and time. The evolution operator acts, in this case, on an infinite-dimensional *function space*, whose 'points' (defining states of the system) are functions. A function at some initial time $x(\xi, 0)$, is transformed by the evolution operator into $x(\xi, t)$,

$$x(\xi, t) = T_t(\xi) x(\xi, 0)$$

where we have explicitly stressed that the evolution operator depends now on ξ. In most cases it includes *derivatives* of x with respect to ξ, thus actually it depends on the function values in the *neighbourhood* of ξ as well.

As an example, consider the linear diffusion equation in one space dimension (ξ is a scalar) for the state function $x(\xi, t)$, defined on some interval $a \le \xi \le b$.

$$\frac{\partial x}{\partial t} = \kappa \frac{\partial^2 x}{\partial \xi^2} \tag{1.5}$$

where κ is a constant.

The operator $T_t(\xi)$ symbolises the evolution of the state function from its initial value $x(\xi, 0)$ to its value at time $t - x(\xi, t)$, i.e., the solutions of (1.5). Equation (1.5) has a unique solution, and thus the evolution operator is well defined, if suitable *boundary conditions* at the interval edges, $\xi = a$ and $\xi = b$, are specified.

A discrete version of a partial differential equation is obtained if all the independent variables are discretised. In the case of (1.5) these variables are ξ and t. Thus the function x has to be replaced by an array, $x_{kj} = x(\xi_k, t_j)$ say. This can be done by setting a spatial interval as $\Delta\xi \equiv (b-a)/m$ with some integer m, for example, and any time interval Δt and defining recursively

$$\xi_{k+1} = \xi_k + \Delta\xi \qquad \xi_0 = a$$
$$t_{j+1} = t_j + \Delta t \qquad t_0 = 0$$

As an example we express the derivatives in (1.5) by finite differences. The equation is thus written as

$$x_{k\,j+1} = x_{k\,j} + \alpha(x_{k-1\,j} - 2x_{k\,j} + x_{k+1\,j}) \tag{1.6}$$

for $j = 0, 1, 2, \ldots$ and any $k = 0, 1, \ldots, m$ (so that ξ_k is inside the definition region) and where $\alpha \equiv \kappa \Delta t / (\Delta\xi)^2$.

Equations like (1.6) can be regarded as a mapping from a time step j to the time step $j + 1$. The discrete evolution operator in this case, $T_n(\xi)$ advances the system from an initial state x_{k0} to the state at the nth time step, x_{kn}, through the repeated (n times) application of the map (1.6) and we can write

$$x_{kn} = T_n^{\{k,k\pm1\}} x_{k0} \tag{1.7}$$

In this case the discrete (time) evolution at a particular position, ξ_k, depends explicitly also on the state function values in the neigbourhood of ξ_k. This fact is reflected by writing the evolution operator in (1.7) as dependent on k and the neighbouring indices $k - 1$ and $k + 1$. Such discrete dynamical systems are called *cellular automata*. The question if the discrete system (1.6) can lead to an approximate solution of (1.5) is irrelevant here.

The above discussion covers essentially all the types of dynamical systems that will be discussed in this book in both of its parts.

2

Astrophysical examples

Few things are harder to put up with than the
annoyance of a good example.
Mark Twain, *Pudd'nhead Wilson.*

Rather than starting from abstract mathematical definitions related to dynamical systems and the concepts used to analyse them, I prefer to start from the outset with a few familiar examples. The systems described below are related to the paradigms of deterministic chaos, some of which have indeed been the ones leading to the discovery, definition and understanding of chaotic behaviour. Instead of repeating here the so often quoted examples such as biological population growth, nonlinearly driven electrical oscillations, weather unpredictability, three-body Hamiltonian dynamics and chemical reaction oscillations and patterns, I shall attempt to motivate the reader by trying to find such examples among simplistic models of astrophysical systems. Obviously, the underlying mathematical structure of these will be very similar to the above mentioned paradigms. This only strengthens one of the primary lessons of nonlinear dynamics, namely that this is a generic, universal approach to natural phenomena.

Examples and analogies may sometimes be misleading and decide nothing, but they can make one feel more at home. This was, at least, the view of Sigmund Freud, the father of psychology, whose advice on matters didactic should not be dismissed. Indeed, as stressed before, these examples are the readers' old acquaintances from their astrophysics educational 'home'. In the next chapter, where the basic notions characterising chaotic behaviour will be dealt with in detail, these examples will sometimes be used again for demonstrating abstract concepts.

2.1 Stellar population dynamics

The primary tool of the physical scientist is the infinitesimal calculus. Consequently, the relevant system variables are described as *continuous* functions of time (and often also space). The physical laws are posed as *differential equations*, either *ordinary* (ODE) or *partial* (PDE). In the biological and social sciences, however, the dynamic laws are often formulated as relationships between the variables at *discrete* time intervals whose size relates to typical time scales of change or sampling. Moreover, discretisation naturally appears when differential equations are being solved on digital computers. The basic dynamic relations are, thus, expressed as *difference equations*.

A typical example of this appears in the investigation of population dynamics. May (1976) was the first to recognise the possible complexity arising from a very simple model in this context. We shall choose stars here as the 'organisms' in the population dynamics model.

The number of stars of a particular type, all identical with mass m, all born together as the result of a supernova-triggered star formation event, say, and all ending their lives in a supernova explosion, is denoted by n_i. The subscript i refers to the ith generation and we view this stellar population evolution as a discrete series of events. We further assume that the region where this stellar population lives is well mixed and contains a fixed amount of matter, stars plus gas. Assume now that the number of stars in the next generation, n_{i+1} is proportional to n_i. This is a very reasonable assumption, on average, with the proportionality constant, c_1, say, being the average number of stars formed as the result of a single supernova explosion.

The efficiency of star formation must depend also on gas density, or in the context of our fixed volume zone, on the gas mass. The total amount is constant, say, M and therefore the number n_{i+1} depends on the available gas mass in the ith generation, which is $M - mn_i$. In general, this dependence may take the form of a power law $c_2(M - mn_i)^a$, but to simplify the discussion we take $a = 1$. The constant c_2 reflects some complex properties of star formation. The stellar population dynamical law might therefore look something like

$$n_{i+1} = cn_i(M - mn_i) \tag{2.1}$$

where the two proportionality constants have been combined into one (c). For c, M and m constant, this equation can be rewritten as

$$x_{i+1} = 4rx_i(1 - x_i) \tag{2.2}$$

where $x_i \equiv mn_i/M$ and $r \equiv cM/4$.

This is the well known *logistic map*. It is a typical, well studied nonlinear quadratic map (or mapping) and was the equation that led to the discovery of the period-doubling bifurcation route to chaos with its universal properties, in the celebrated work of Feigenbaum (1978) in the United States. Coullet & Tresser (1978) worked on this problem independently in France and reached similar conclusions.

More sophisticated models of stellar population dynamics can also be made, taking into account the fact that a prescribed fraction of matter is not returned to the interstellar medium, allowing for different stellar types and using a suitable power-law dependence on the amount of gas. Such models could lead to two-dimensional (or multidimensional) mappings, namely two (or more) equations with a suitable number of variables. The simplest model, leading to the mapping (2.2), is however sufficient for our purposes here.

Equation (2.2) can be easily iterated with the help of a computer (or even a programmable calculator), starting from some given $x_0 < 1$ (see below). The result depends, as may be expected, on the value of the constant r. We would like x_k to be non-negative for any k; therefore its value must also always be ≤ 1 (to prevent a negative value in the next generation) and it is also necessary that the constant r be $r < 1$. The logistic equation (2.2) is then a mapping of the interval $[0, 1]$ onto itself. It is qualitatively obvious that a very small value of r (too small star formation efficiency or too little gas in the system) will lead to extinction. What happens for larger values of r is less clear.

Since the logistic difference equation exhibits many of the important properties of nonlinear systems bifurcations, it may be seen as a prototype for one-dimensional (and even multidimensional) nonlinear mappings. Its study and the related concepts will be elaborated on in the next two chapters. Here, we shall just introduce a few simple but important notions and the graphical procedure, which are very useful in the qualitative study of iterated maps and bifurcation theory.

In general, a one-dimensional mapping, like (2.2), can be written in the functional form

$$x_{i+1} = F(x_i) \tag{2.3}$$

The successive iterations give rise to a sequence of x values called *iterates*. The iterates can be followed by plotting $y = F(x)$ and the line $y = x$ on the same graph (see Figure 2.1) and moving successively vertically and horizontally between these two curves. The points at which the line and the curve intersect (i.e., $F(x) = x$) correspond to values of x, for which the iteration sequence values remain fixed. These are referred to as *fixed points* of the map. As is apparent from Figure 2.1 our mapping has two fixed points, x^*, which can be found analytically by solving

$$x^* = 4rx^*(1 - x^*) \tag{2.4}$$

The solutions are: $x^* = 0$ and $x^* = 1 - 1/4r$.

Note that only for $r > 0.25$ do both fixed points lie in the interval $[0, 1]$.

A fixed point of a mapping can be *stable* (or *attracting*), if an iteration starting in its neighbourhood will bring the iterates closer and closer to the point. Conversely, if an iteration of the map starting close to a fixed point results in the iterates moving

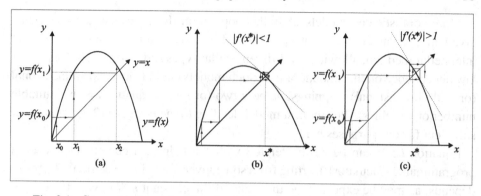

Fig. 2.1. Geometrical construction for finding successive iterates and fixed points of a mapping. The first three iterations are shown in (a). These are followed by spiralling into (shown in (b)), or out of (shown in (c)), the fixed point x^*, depending on its stability.

away from it, it is *unstable* (or *repelling*). Without entering into a complete analysis of the fixed point stability of the logistic map (this will be done in Chapter 4), we just note that the stability here depends on the value of r. More specifically:

(i) $x^* = 0$ is stable for $0 < r < 0.25$ and unstable for $r > 0.25$.
(ii) $x^* = 1 - 1/4r$ is unstable for $0 < r < 0.25$ and stable for $0.25 < r < 0.75$.
(iii) both fixed points are unstable for $r > 0.75$.

It is therefore clear that iterations, starting from any initial x_0, will approach the constant value $x^* = 0$ (extinction) as long as $r < 0.25$, as expected. For any $0.25 < r < 0.75$ the second fixed point will be approached as closely as desired after a sufficient number of iterations. The first fixed point is thus the *attracting point* of the map for $r < 0.25$, while the second one becomes the attracting point for $0.25 < r < 0.75$. At $r = 0.25$ the map changes its qualitative behaviour. Such a point on the parameter axis (here r) is called a *bifurcation* point.

At the value of r at which the second fixed point loses its stability, $r = 0.75$, a more interesting bifurcation occurs. Both fixed points are now unstable and successive iterations will result in the values of x_i alternating repeatedly between two fixed values. This happens only if r is not too far from 0.75. This periodic cycle of the map, which includes just two points, is now attracting; it is called a *limit cycle*.

We can construct the bifurcation diagram of our map by plotting the limiting values of the iterate (for large i) as a function of the *control parameter* r, starting from $r = 0.75$; this is shown in Figure 2.2. We note that the bifurcation at $r = r_1 = 0.75$ is not the last one. Quite the contrary, bifurcation points keep appearing with increasing density. The aforementioned limit cycle loses its stability and a more complicated limit cycle, with four values, appears at some value of r (actually at

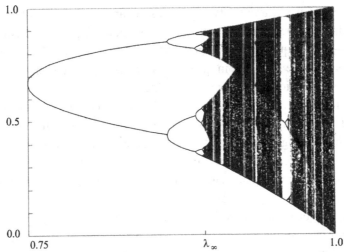

Fig. 2.2. Bifurcation diagram for the logistic map. The vertical axis is the limiting value of the iterate, while the horizontal axis is the parameter r and the diagram is shown for $0.75 \leq r \leq 1$, where the sequence of period-doubling bifurcations leads to chaos. The value $r = \lambda_\infty$ is the accumulation point of the period-doubling bifurcations.

$r = r_2 = 0.862\,37\ldots$). We say that a *2-cycle* is replaced by a *4-cycle*. As r is increased even more a sequence of bifurcations to 2^n cycles is apparent, but the bifurcation points r_n appear to approach a finite limit, $\lim_{n \to \infty} r_n \equiv \lambda_\infty$. Beyond $r = \lambda_\infty$ there is no longer a periodic limit cycle. The iterates have an *aperiodic* or *chaotic* behaviour. They wander indefinitely attaining, in a seemingly random manner, a multitude of values. The *chaotic bands* are interrupted by brief (in r range) appearances of odd-cycle limit cycles. Beyond $r = 1$ there is uniform chaos.

Thus the fate of the stellar population in our simple 'toy model' is by no means simple or trivial. As the star formation efficiency (for a fixed total mass) or the total available mass (for a fixed birth efficiency) increase, the number of stars oscillates, simply at first but with an increased complexity later. Beyond a certain point the oscillations are chaotic and unpredictable.

This simple-looking logistic map will be explored in greater detail in Chapter 4, when it will serve as an example for the investigation of transition to chaotic behaviour in dissipative dynamical systems.

2.2 One-zone model of a nonlinear stellar pulsator

The classical examples of oscillators, which we are usually introduced to in high school, are the spring–mass system and the mathematical pendulum. These are studied assuming that the displacement is sufficiently small that the restoring force is proportional to it. At university level, these examples are often the first ODEs encountered. It also becomes clear that adding velocity-dependent dissipation does not complicate the equations too much and a solution in closed form is still possible. This is so since the ODE in question is *linear*. The situation remains 'under control' even if a periodic forcing term is added.

However if such driven oscillators are nonlinear (usually not mentioned in class) they may, despite their simplicity, exhibit very complicated behaviour. It is sufficient that the restoring force depends also on the cube of the displacement and not only on the first power (see below). It is possible nowadays to buy a multitude of 'chaotic toys' usually based on magnetically driven penduli. Our next example will be a driven, nonlinear 'stellar oscillator'.

We will consider the outer stellar layers, modelled by a one-zone gas layer residing on top of a stellar core of radius, R_c. The value of such models can be demonstrated by the success of the famous one-zone model, proposed by Baker, for elucidating the physical mechanisms responsible for driving stellar pulsation.

Let the outer zone in question have a fixed mass, m, and let it be dynamically characterised by its outer radius, R. The thermal properties of the layer are described by its specific entropy, s. The other state variable, the density, ρ, is expressed by R, m and R_c as: $\rho = m/V$, with the volume V given by $V = 4\pi(R^3 - R_c^3)/3$. We focus here on *adiabatic* motion, i.e., such that s is constant and can be regarded as a parameter. Only R is allowed to vary in time and thus the function $R(t)$ fully describes the motion of the outer zone boundary.

The equation of motion for this single zone can thus be written as

$$\frac{d^2R}{dt^2} = -\frac{GM}{R^2} + 4\pi R^2 \frac{P(\rho; s)}{m} \tag{2.5}$$

where M is the constant stellar mass (assuming $m \ll M$) and s has been explicitly marked as a parameter.

As explained above, ρ can be expressed in terms of R and constants, and therefore P for any R (and s) can be found with the help of an equation of state. Thus this equation can be symbolically rewritten as

$$\frac{d^2R}{dt^2} = f(R; s) \equiv -g(R; s) \tag{2.6}$$

where f is the force per unit mass acting on the stellar layer and depends solely on R (s is a parameter). For most reasonable values of s, one can find the equilibrium radius as a solution of the algebraic equation $f(R) = 0$ for this s. It is usually a unique solution, R_{eq}, the hydrostatic equilibrium solution. As is known from basic stellar structure theory such a solution is stable in most normal conditions (i.e., if the adiabatic exponent γ is not smaller than 4/3).

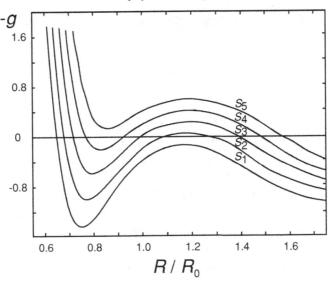

Fig. 2.3. Acceleration of the outer zone radius for different values of the entropy parameter near He$^+$ ionisation. Three equlibrium radii exist for a range of s values. Reprinted, with permission, from Buchler & Regev (1982).

There are however some values of the parameter s for which the thermodynamic state of the outer layer (for some prescribed value of all the other parameters) may be complicated by, say, ionisation–recombination processes. In Figure 2.3 one can see that for some ranges of s values the twisted form of $f(R)$ causes it to cross the line $f = 0$ three times and thus three equilibrium radii (solutions of $f(R) = 0$) are present. This happens because the equation of state undergoes a change due to ionisation (of He$^+$ in this specific model). This is readily understood qualitatively. There is an equilibrium corresponding to a totally ionised gas (the smallest R equilibrium value) and another one corresponding to conditions of full recombination (the largest R equilibrium value). These two are dynamically stable (for both, $\gamma = 5/3$). The middle equilibrium, $R = R_0$ in the figure, corresponds to conditions of partial ionisation with the ionisation–recombination equilibrium causing $\gamma < 4/3$ and thus dynamical instability.

In order to explore our model analytically we first nondimensionalise the equation, scaling the time variable by the dynamical time of the problem, the radius by the value R_{eq} of the equilibrium we wish to study and all the other variables, accordingly, by their typical appropriate values. We then wish to expand $f(R)$ around the equilibrium value ($= 1$ in our units); thus let $x \equiv R - 1$. As long as the parameter s is such that we are far away from any complications arising from ionisation, $f(R)$ is close to being linear in the vicinity of R_{eq} (as may be deduced from Figure 2.3 for $s \ll s_1$, for example). Thus, retaining only the first order in x, (2.6) becomes

$$\frac{d^2x}{dt^2} = -\beta x \tag{2.7}$$

where $\beta > 0$ is a constant.

This is the uninteresting case of a linear oscillator. The outer stellar zone, if disturbed, will harmonically oscillate around its equilibrium. Dissipation processes, not accounted for here, will damp these oscillations. In the stellar model at hand, such processes are connected to heat exchange between the layer and its surroundings, breaking the adiabaticity assumption. However, the inclusion of the heat equation greatly complicates the problem, certainly beyond leaving our model as a simple and telling example. We thus focus here on a simple 'friction-like' dissipation mechanism and crudely model it by adding a $-\delta dx/dt$ term, with $\delta > 0$ a constant, to the right-hand side of the equation. We say more about this below.

Consider now the more interesting case, where s is such that three different hydrostatic equilibrium radii exist. Expansion around the middle (unstable) equilibrium radius must now include higher-order terms (to capture the essence of the problem). Judging from the shape of the s_3 curve in the figure we have chosen the simplest possible analytical form (a cubic polynomial)

$$\frac{d^2x}{dt^2} = \beta x - \alpha x^3 - \delta \frac{dx}{dt} \tag{2.8}$$

with $\alpha > 0$, $\beta \geq 0$ and $\delta > 0$ constant. Actually three solutions exist only for $\beta > 0$, but we allow for the possibility of $\beta = 0$ for the sake of generality (this case will be discussed in Chapter 4). Note that a dissipation term of the above mentioned form has also been included, mainly for the purpose of illustrating its effect below. This equation is known in the literature as the (unforced) Duffing equation and its mathematical properties have been studied extensively in various physical contexts.

The actual form of $f(R)$, even in our one-zone model, is more complicated than just a cubic, but what is essential is its topology. Numerical calculations taking into account ionisation in the equation of state have been performed for this one-zone model by Buchler and Regev (1982) (the source of Figure 2.3). They included also the effects of nonadiabaticity in a consistent way. We shall expound on their (and other) models later in the second part of this book. Here, as we would like to demonstrate the different kinds of behaviour as transparently as possible, we have chosen the simplest analytical form that still captures most of the qualitatively important features of this nonlinear oscillator: cubic nonlinearity and dissipation.

We shall base our discussion on the Duffing system but we would like to introduce here, in passing, an additional particularly elementary nonlinear oscillator, the mathematical pendulum. The forced mathematical pendulum has been an important paradigm of a simple system capable of complex behaviour and has also frequently been used in popular books. In natural units the equation of motion of the (unforced) pendulum can be written as

$$\frac{d^2x}{dt^2} = -\sin x - \delta\frac{dx}{dt} \tag{2.9}$$

where x is the angular displacement of the pendulum from the vertical equilibrium, and we have also included a velocity-dependent dissipation term. It can be shown that the behaviour of the mathematical pendulum is essentially very similar to that of the Duffing system (see e.g., Tabor, 1989).

In dynamical system theory one of the common mathematical characterisations of a system is a set of first-order ODEs. The functions are then called the *state variables* of the system. It is very easy to see that the second-order ODE (2.8) is equivalent to

$$\frac{dx}{dt} = y$$

$$\frac{dy}{dt} = \beta x - \alpha x^3 - \delta y \tag{2.10}$$

Here y (defined by the first equation of the two) is the velocity.

The space spanned by the state variables of a dynamical system is its *state space*. Recall that in classical mechanics it is the space of the generalised coordinates and momenta and it is called *phase space*. Although the term phase space is usually used in Hamiltonian systems we shall use the term *state* and *phase* space interchangeably except when there is a special reason to distinguish between the two (as in parts of Chapter 5).

The evolution of the system can thus be viewed as a motion of the system in its phase space, where the system evolves in time along a trajectory, starting from an initial point, defined by the initial conditions. In our case two such initial conditions are needed (e.g., the displacement and the velocity at $t = 0$) since there are two first order ODEs (or one second order ODE). In this case phase space is two-dimensional (x and y) and we can thus call it the phase *plane*. The system at any given instant is represented by a point in this plane and evolves along a curve (trajectory).

If $\delta = 0$ (no dissipation) the dynamical system (2.10) can be exactly solved with the help of Jacobi elliptical functions. These are mere extensions of the familiar $\propto \sin(\omega t + \phi_0)$ solution ($\omega \equiv \sqrt{\beta}$) valid in the linear case (2.7). The *phase portrait* (plot of the trajectory in phase plane) will be a closed curve, its shape depending on initial conditions. In the case $\delta = 0$, the system is *conservative* and the total mechanical energy is constant. The difference compared to the familiar

elliptical phase trajectories of the linear case stems from the fact that here the force
is derivable from a quartic (double well) rather than quadratic (single well) potential. Indeed we can write for the conservative case

$$\frac{d^2x}{dt^2} = -\frac{dV(x)}{dx} \tag{2.11}$$

with the potential $V(x) = \alpha x^4/4 - \beta x^2/2$.

It is easy to show, using methods from elementary mechanics, that we must have

$$\frac{1}{2}y^2 + V(x) = E \tag{2.12}$$

with E, the energy, being a constant determined by initial conditions.

If $\beta \neq 0$ three equilibrium positions (fixed points of the dynamical system) exist
and they can be calculated by setting the time derivatives in equations (2.10) to
zero. They are easily found to be at $x = 0, \pm\sqrt{\beta/\alpha}$ (and obviously $y = 0$). For
$\beta = 0$ we have only one fixed point. The point $x = 0$ is unstable (it is a local
maximum of the potential) while the other two are stable (potential minima). Some
of the phase trajectories for the case $\alpha = \beta = 1$ are shown in Figure 2.4 along
with the corresponding potential. If $\delta > 0$ the system is no longer conservative
and its phase portrait has to change accordingly. The phase volume of an ensemble

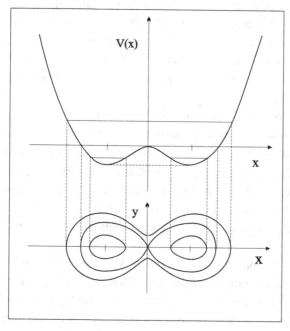

Fig. 2.4. Phase portrait of conservative motion in a symmetrical quartic potential
(shown in the upper part).

of systems, started with initial conditions close to each other, will no longer be conserved. Liouville's theorem of classical mechanics, which guarantees phase volume invariance, is valid only for Hamiltonian (necessarily conservative) systems. In contrast, this volume (here surface) will now shrink as motion progresses, as can be seen from finding the divergence of the phase-space velocity vector, $(\dot{x}, \dot{y})$. Here and henceforth an overdot denotes the time derivative. Using (2.10) one gets

$$\frac{\partial \dot{x}}{\partial x} + \frac{\partial \dot{y}}{\partial y} = -\delta \qquad (2.13)$$

Since $\delta > 0$ the phase volume has to shrink to zero (δ is constant). Indeed, stability analysis of the fixed points (the methods will be explained in the next chapter) shows that the potential minima ($x = \pm\sqrt{\beta/\alpha}$) are now attracting. They are stable *spiral points*. Phase trajectories spiral into them. This is in contrast with the conservative case where these points were *centres*, with the trajectories encircling them. The middle unstable fixed point remains a *saddle point*, as before.

The phase portrait for this case with $\alpha = \beta = 1$ is exhibited schematically in Figure 2.5. Since all neighbouring trajectories converge into a point, the volume in the two-dimensional phase space indeed shrinks to zero. The two points are *attracting*. Which of them will be ultimately approached (exponentially in time) depends on the initial conditions of a trajectory. The stellar outer zone will thus ultimately approach one of its equilibrium radii, oscillating around it with an ever-diminishing amplitude. Which one of the equilibria is chosen depends on the nature of the initial disturbance, imposed from the outside. So far the nonlinearity (the quartic potential) has not complicated the situation too much and we seem to be staying on rather firm ground. Radically new behaviour may, however, appear if the dynamical system (2.8) is driven, say periodically, by an external source. In the

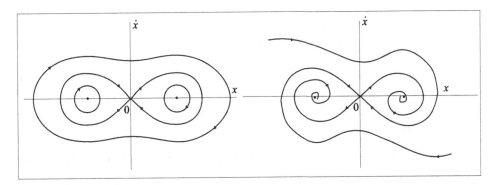

Fig. 2.5. Phase portrait of dissipative motion in a symmetric quartic potential (right part). The conservative case is shown in the left part, for comparison.

astrophysical context this might be caused by pulsation of the underlying core. In its simplest form such a pulsation is sinusoidal. This calls for an additional term in the force function on the right hand side of equation (2.8). We thus get

$$\ddot{x} = \beta x - \alpha x^3 - \delta \dot{x} + \epsilon \cos \omega t \qquad (2.14)$$

where the added forcing term has an amplitude ϵ and angular frequency ω.

This is now a *nonautonomous* differential equation as it explicitly contains the independent variable t. However, a set of three first-order ODEs arising from it, may still be written in an *autonomous* form, analogously to (2.10) if we put $z \equiv \omega t$

$$\dot{x} = y$$
$$\dot{y} = \beta x - \alpha x^3 - \delta y + \epsilon \cos z$$
$$\dot{z} = \omega \qquad (2.15)$$

i.e., without explicit time dependence, but with an additional first-order ODE. The dynamical system thus becomes three-dimensional. In fact, the additional variable, the phase z, increases indefinitely with time, but we must remember that it enters the first two equations only through the $\cos z$ term, whose values are identical for z and $z + 2k\pi$ for any integer k; thus the state variable is better expressed by an angle, $\theta = z$ (mod 2π), say. Fixed points of (2.15) are determined by the conditions $\dot{x} = \dot{y} = 0$ ($\dot{z}$ does not vanish at any point).

This forced Duffing oscillator has been studied numerically for various cases. We report here on some of the interesting results found by Holmes in 1979 and summarised in the book by Guckenheimer and Holmes (1983). For $\alpha = \beta = \omega = 1$, the system (2.15) was integrated numerically with a fixed, not-too-large value of the dissipation parameter δ. Four types of behaviour were found, depending on the value of the forcing amplitude, ϵ.

(i) $\epsilon < \epsilon_0$ – regular motion: trajectories spiralling into one of the fixed points.
(ii) $\epsilon_0 < \epsilon < \epsilon_1$ – transient irregularity: erratic behaviour of the trajectories followed by regular motions, as above.
(iii) $\epsilon_1 < \epsilon < \epsilon_\infty$ – period-doubling sequence: stable limit cycles attract the motion and as ϵ is increased these attracting structures lose their stability, simultaneously with the appearance of more complicated attracting limit cycles; this happens successively at a well defined sequence of ϵ values, whose accumulation point is ϵ_∞.
(iv) $\epsilon > \epsilon_\infty$ – lasting irregular oscillations: no simple attracting objects exist despite volume contraction.

The numerical value of the constants delineating the different types of behaviour depend on δ.

We see that for a fixed ϵ in range (iii) the attracting object is still a rather normal periodic orbit. The outer stellar layer in our model exhibits periodic oscillations, but not just of a sinusoidal form with frequency ω, as could be expected if the oscillator were linear. We note that a periodic attracting orbit, namely a limit cycle, of course satisfies the volume-shrinking condition. The volume of a closed curve in space is zero. In case (iv), i.e., for a forcing amplitude value above the accumulation point of the period-doubling bifurcations ϵ_∞, the behaviour is truly complex. Attracting fixed points or closed curves, however distorted, no longer exist. Still there must exist an attracting geometrical object of volume zero since $\nabla \cdot (\dot{x}, \dot{y}, \dot{z}) = -\delta$, and is negative. The object's dimension must therefore be less than 3 (the dimension of phase space). What can it be?

It turns out that the object (such geometrical structures will later be called *attractors*) is extremely complicated and does not have the form of any normal geometrical object of dimension 0 (point), 1 (curve) or 2 (surface). There is no escaping the conclusion that this unusual attractor's dimension is not an integer number at all. It must have a *fractal* dimension. We will say more about fractals in the next chapter. Such a chaotic, fractal attractor is called *strange*.

Results of numerical integrations of the forced Duffing oscillator (2.15) with $\alpha = \beta = 1$ for $\epsilon > \epsilon_\infty$ are shown in Figure 2.6, and they illustrate the behaviour described above. The oscillations of $x(t)$ are shown in the figure, but the other state variables have similar behaviour. The case of irregular oscillations indicates that the phase space trajectory must be very complex for these values of the parameter ϵ (see Chapter 4). Repeated numerical calculations in the irregular case, reveal one of the most characteristic features of chaotic solutions, *sensitivity to initial conditions* (SIC). Two calculations starting from arbitrarily close initial states give, after some time, completely different trajectories, diverging as time progresses. This is the reason for the *unpredictability* of chaotic behaviour. Although the system is

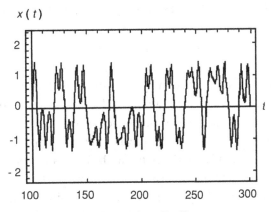

Fig. 2.6. Irregular oscillations of the driven Duffing system at a relatively large driving amplitude ($\epsilon = 0.3$) value. Here $\delta = 0.2$, $\omega = 1$ and $\alpha = \beta = 1$.

deterministic (a set of ODEs), the SIC property makes the prediction impossible, as in practice we can only know the initial condition to finite accuracy. In contrast to this, *linear* dynamical systems do not have SIC.

One-zone models of stellar pulsations, giving rise to complex aperiodic behaviour, had already been proposed by Baker, Moore and Spiegel in 1966, well before chaos became popular and ubiquitous. The idea was prompted by the mathematical similarity to the one-element model of thermal convection which will be one of our examples later. We shall discuss this idea along with its more modern versions in the second part of the book.

We conclude this section with the following remarks.

(i) The seemingly random and complex oscillations in our model (appropriate perhaps for an irregular variable star) result from a strikingly simple deterministic system.

(ii) The transition to chaos proceeds through a sequence of bifurcations, remarkably similar to the bifurcations of the logistic map.

(iii) Chaotic oscillations have SIC, causing unpredictability.

(iv) Fractal geometry is needed to describe chaotic motion.

(v) One may hope to find chaos and fractal attractors in dissipative ODEs only if the phase space has at least three dimensions. Strange attractors cannot live in a two-dimensional phase space.

2.3 Stellar orbits in a model galactic potential

The eighteenth century saw intensive analytical efforts to formalise, generalise and extend Newton's 'Principia', the great scientific masterpiece published in 1687. These efforts culminated in what is now called the *Hamiltonian* or *canonical theory* of dynamics. The first two sections of Chapter 6 are devoted to a review of this theory.

It became apparent soon, however, that the canonical formalism, including its powerful and general methods, could not provide a full analytical understanding of even some very simple looking problems. In his celebrated work on celestial mechanics (near the beginning of this century), the great French mathematician Poincaré consolidated this conclusion by demonstrating that the difficulties, which had been encountered in some applications of the theory, are fundamental.

As anyone who studies elementary mechanics knows, the gravitational two-body problem can be reduced exactly to an equivalent one-body problem. The latter can be solved in general, the only difficulty being the evaluation of integrals. Poincaré showed that in the gravitational many-body problem such a procedure cannot be generalised in an analytical way. The presence of a third body, even if treated as a perturbation, may already give rise to profound difficulties. The simple elliptic orbits known from the bound two-body problem could not even serve as a starting point for the attempt to solve generally the three-body problem. In addition, Poincaré realised that the trajectories in such a case may be aperiodic (certainly not periodic and not even quasiperiodic) and have an extremely complicated and intricate geometrical structure. As there were no computers in his time, Poincaré could not go far beyond the above-mentioned fundamental observations, which nevertheless looked rather depressing for astronomers.

We now introduce a beautiful example, from the astronomical literature of the 1960s, in which such surprisingly irregular behaviour was discovered in a rather standard dynamical system. It is the work of Hénon and Heiles (1964), who numerically investigated a model for the orbits of a point mass (a star) in the gravitational potential of an axially symmetric galaxy. This standard approach circumvents the difficulty of computing a full n-body problem, by approximating the gravitational effect of all the other stars by a single time-independent function of position (the mean potential).

The first steps in the solution of this problem are analytical and rather elementary. Assume that the cylindrical coordinates are r, z and ϕ and the mass of the star is m. The Hamiltonian (in this case the *energy* of the system) can be written as the sum of the kinetic and gravitational potential energies of the system (one star)

$$H(p_r, p_z, p_\phi, r, z, \phi) = \frac{1}{2m}(p_r^2 + p_z^2) + \frac{1}{2mr^2}p_\phi^2 + V_g(r, z) \qquad (2.16)$$

where $p_r = m\dot{r}$ and $p_z = m\dot{z}$ are the linear momenta in the r and z directions respectively, $p_\phi = mr\dot{\phi}$ is the angular momentum around the symmetry axis and V_g is the axially symmetric potential energy function.

As is always the case in axially symmetric potentials the Hamiltonian does not depend explicitly on the coordinate ϕ. Thus, the corresponding momentum, p_ϕ is an integral of the motion. In more elementary terms – the angular momentum around the z-axis is conserved since there are no torques to change it. We thus have two conserved dynamical quantities (called integrals or constants of motion), the angular momentum, l and the total energy, E. They are given by

$$p_\phi = l \quad \text{and} \quad \frac{1}{2m}(p_r^2 + p_z^2) + \frac{1}{2mr^2}l^2 + V_g(r, z) = E$$

The common practice in such problems is to redefine the potential energy, by including in it the centrifugal term, $l^2/(2mr^2)$, thus

$$V(r, z) \equiv V_g(r, z) + \left(\frac{l^2}{2m}\right)\frac{1}{r^2}$$

In this way the problem is reduced to two spatial dimensions, with the angular motion, given by

$$\dot{\phi} = \frac{l}{r}$$

effectively removed (it can be solved later, when $r(t)$ is known). This reduction is possible due to the existence of the integral of motion l *and* the fact that the motion in ϕ can be trivially separated out.

In general, a one-particle in an external potential system, as the one studied here, has a 6-dimensional phase space (this term can be used here with its full glory, as the system is Hamiltonian). Because of the symmetry, an integral of motion exists, reducing by one the dimension of the portion of phase space in which the system can move. Such an integral is sometimes called an *isolating* integral. It isolates, in some sense, the phase trajectory of the system from inaccessible portions of phase space. The additional reduction, resulting from ϕ being separable, allows us to focus on a system with just four phase-space variables.

Following Hénon and Heiles we write the reduced Hamiltonian as

$$H(p_x, p_y, x, y) = \frac{1}{2m}(p_x^2 + p_y^2) + V(x, y) \tag{2.17}$$

where x and y, the usual Cartesian coordinates have been written instead of r and z for simplicity.

As a result of the energy conservation, we have an additional isolating integral. The fixed energy is also such an integral since it is *independent* of the previously defined isolating integral (angular momentum). Since the Hamiltonians (2.17) and (2.16) are equal, the energy is also an isolating integral for the reduced problem. Thus we know, at least in principle that the dimension of the portion of phase space in which the system is constrained to move, is three.

If now the potential in (2.17) can be written as the sum of an x-dependent part and a y-dependent one, the problem separates, is trivial and uninteresting and most importantly, totally inadequate for modelling a galactic potential. Thus Hénon and Heiles took a non-separable potential $V(x, y)$. They addressed the problem of existence of an additional *third isolating integral*. If such an integral exists one can be sure that the reduced system's motion is in a two-dimensional portion of phase space (on a surface).

Such motion is very pleasing (for some) as it consists of a closed, periodic trajectory in any normal case, i.e., when the motion is *bounded*. The technical details of finding such a periodic orbit are not very important to this discussion. The principle is very simple. If two independent integrals of motion exist for the system described by (2.17), they constitute constraints on the four generalised coordinates (two Cartesian coordinates and their conjugate momenta). Thus two generalised coordinates, x and p_x say, can be eliminated (expressed in terms of y and p_y). What is left is a one-degree of freedom system that can be solved, in principle, by quadrature. The result is a periodic orbit, whose nature must be found by solving in turn all the equations on the way back to the original problem.

If a third isolating integral does not exist, the motion may be three-dimensional in phase space. That is, it is possible in principle that the orbits may be very complicated, e.g., covering all points in some spatial region in the $x-y$ plane and coming arbitrarily close to any given point in this region. Hénon and Heiles called such motions *ergodic*. Obviously, phase trajectories cannot intersect. Ergodicity is thus realised in this case on a three-dimensional region of phase space (one of the momenta is an additional independent phase variable).

The above findings have important observational implications. As it is impossible to determine observationally the full orbit of even the closest star (at least in the present state of medicine!) statistical methods must be used. Many stars can be observed (position and velocity) and the nature of their statistical distribution compared to theory. As is known from statistical mechanics, this distribution depends on the constants of motion. Studies of stars in the solar neighbourhood have indicated that a third integral of motion most probably does exist. The astronomers called it simply the 'third isolating integral' and the question of its nature and conditions for existence was the primary issue of celestial mechanics in the 1960s. We shall discuss some aspects of this problem together with several later developments in the second part of this book.

Hénon and Heiles had a very fruitful insight. They decided to investigate a very simple (but not trivial) potential in the Hamiltonian (2.17). They even studied a related area-preserving mapping, as this was a much easier task given the limited power of electronic computers at that time. The potential they decided to study (we shall call it the HH potential) contains a harmonic part (with equal frequencies for both degrees of freedom) and a nonlinear part. The harmonic part gives rise to linear oscillations. In two dimensions the orbit traces an ellipse. A uniform sphere, for example, gives rise to such a harmonic potential at any point inside itself. The HH potential also includes a third-order (in the displacement) term. Here Hénon and Heiles chose a combination of $\propto x^2 y$ and $\propto y^3$ dependence. In their words 'it seems probable that this potential is a typical representative of the general case, and nothing would be fundamentally changed by the addition of higher order terms'.

The HH potential can be written in the form

$$V(x, y) = \frac{1}{2} m \omega^2 \left[x^2 + y^2 + \frac{2}{\Lambda} \left(x^2 y - \frac{y^3}{3} \right) \right] \tag{2.18}$$

where ω is the harmonic angular frequency and Λ is a constant, having the dimension of length. As required, V has the dimensions of energy. Choosing the mass, length and time units as m, L and $1/\omega$ respectively, the energy unit becomes

$E_0 = m\omega^2 L^2$ and the potential in these units is

$$V(x, y) = \frac{1}{2}\left(x^2 + y^3\right) + \eta\left(x^2 y - \frac{y^2}{3}\right) \tag{2.19}$$

where the nondimensional constant $\eta \equiv L/\Lambda$ is a measure of the nonlinearity, because it is defined as the ratio of a typical length scale of the problem to the scale on which the cubic term is important (the *nonlinearity scale*).

Hénon and Heiles chose $\eta = 1$ as they wanted to explore the nonlinear regime. They then proceeded to numerically solve the Hamiltonian system resulting from (2.17) with (2.19). What happened next reminds one of the biblical story of Saul, who went to look for his father's lost herd, but ended up as King. It seems, however, that the deep and beautiful results of Hénon and Heiles were not accidental. They understood that it may be possible to ignore complicated details in order to see the essence.

The explicit form of the HH dynamical system consists of the appropriate Hamilton equations of motion,

$$\dot{x} = \frac{\partial H}{\partial p_x} = p_x$$

$$\dot{y} = \frac{\partial H}{\partial p_y} = p_y$$

$$\dot{p}_x = -\frac{\partial H}{\partial x} = -x - 2xy$$

$$\dot{p}_y = -\frac{\partial H}{\partial y} = -y - x^2 + y^2 \tag{2.20}$$

This ODE system can also be written in the form of the following two second-order equations

$$\ddot{x} = -x - 2xy$$
$$\ddot{y} = -y - x^2 + y^2 \tag{2.21}$$

Hénon and Heiles numerically integrated equations (2.21) for several values of the energy parameter E. Because $H = E$ this is obviously equivalent to fixing E and varying the parameter η, the strength of the nonlinearity. For each parameter value a number of orbits were computed. A particular orbit was determined by fixing the values of three parameters (e.g., the initial conditions $x(0)$, $y(0)$ and $\dot{y}(0)$). The fourth initial condition can be found (to within a sign) from the value of the energy and the above three initial conditions. The orbit resulting from such a calculation must be confined to a three-dimensional *energy shell*. Thus the motion can be plotted in a three-dimensional space, x–y–$\dot{y}$, say. Still, the visualisation of such an orbit on paper remains difficult. Thus Hénon and Heiles used the *surface of section* method to characterise their results.

This method will be dealt with in detail in Chapter 3 and applied in various places in the book. Here we just briefly describe how it was used by Hénon and Heiles. They looked at a two-dimensional surface (actually a plane) in phase space, defined by the equation $x = 0$, that is, this surface is the y–$\dot{y}$ plane. In the course of the numerical integration the positions where a given orbit cut this plane were recorded. To remove the ambiguity of sign in the determination of the fourth variable $\dot{x}$, Hénon and Heiles recorded just those points at which the orbit crossed the $x = 0$ in the 'upward' direction, that is, when $\dot{x} > 0$. Except for this purpose the fourth variable, $\dot{x}$ was redundant. Actually, as this variable was also obtained in the output of the calculation, it could be used to check the accuracy of the computation.

The successive values of the crossing points can be regarded as a two-dimensional discrete mapping. It is called the *Poincaré map* and its properties will be discussed in detail in Chapter 3.

For a rather low value of the nondimensional energy parameter ($E = 1/12$) an expected result emerged. The points on the surface of section were seen to lie, for each orbit, on some well-defined curve. This fact indicates that in this case the orbits lie on two-dimensional phase surfaces, some of which are nested in a complicated manner. Thus the motion is regular and simple – periodic or quasiperiodic (having several incommensurate frequencies). Hénon and Heiles concluded that this result strengthens the conjecture that a third isolating integral exists in the problem. They also found that all the surface of section (up to some last curve, the outermost in Figure 2.7) could be filled by curves resulting from the set of all possible orbits having the prescribed energy. In addition they found the fixed points of the Poincaré map. Stable points, corresponding to a stable periodic orbit of the system, were located near the middle of the nested, closed curves. Unstable ones (unstable periodic orbits) appeared at curve intersections.

When the value of the parameter was increased to $E = 1/8$, still below any possibility of escape from the potential, a surprising result was obtained. The corresponding surface of section is shown in the lower part of Figure 2.7. There are still some closed curves (around the stable fixed points), but they no longer fill the whole surface. The gaps between the periodic orbits are full of isolated points, indicating ergodic three-dimensional motion. The fact that all such points shown in the figure belong to a single trajectory (!) strengthens this conclusion. Hénon and Heiles also pointed out that the five little loops around the large closed curve, in the right portion of the figure, belong in fact to one orbit. Such chains of *islands* (as these features are called) were found also in other parts of the section, all associated with a stable periodic orbit passing through the middle of each island.

The surprise was even greater when the energy parameter was increased to $E = 1/6$. This energy value allows the particle to escape from the potential at

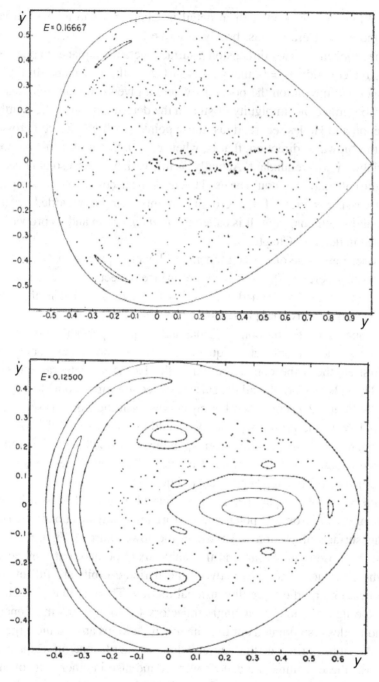

Fig. 2.7. Surface of section plots for the Hénon–Heiles system. The two panels correspond to different values of the energy E shown in the figure (see text). Reprinted, with permission, from Hénon & Heiles (1964).

a single point, $y = 1$; $\dot{y} = 0$. In this case, as shown in the upper part of Figure 2.7, the crossing points of a single orbit wandered around almost all of the available area. Hénon and Heiles concluded that in this case almost all orbits in the domain are ergodic. The tiny islands covering a small area of the section represent well-behaved orbits but they correspond only to very special conditions. The probability of picking them in some statistical distribution of star orbits is negligible.

Hénon and Heiles' conclusion was that above some value of the energy the third integral exists for only a limited set of initial conditions. They estimated the measure of this set by calculating the relative area of the surface of section covered by closed curves, as a function of the energy. They found that the full area is covered for $0 < E < E_l$ with $E_l \approx 0.11$. Thus, up to this value of energy the motion is regular for all initial conditions. Above this value the relative area seemed to decrease linearly with E, becoming close to zero for $E \approx 0.16$.

The HH system has been investigated in a huge number of works since its introduction, and a lot has been learned about regular and irregular motion in Hamiltonian systems from these studies. Hénon and Heiles' pioneering work could perhaps not directly predict the properties of the stellar distribution in some particular galaxy; it has, however, remained to this day a fundamental contribution to the theory of dynamical systems.

2.4 One-element model of thermal convection

Thermal convection is a well known phenomenon. It has been studied extensively in the laboratory and a considerable effort has also been made on the theoretical side; consequently a large amount of literature is available on this subject. We shall discuss it only very briefly in the second part of this book. The theoretical difficulty with the study of convection stems from the fact that we deal here with fluid dynamics, described by nonlinear *partial* differential equations (PDE). Analytical solutions to the equations of fluid dynamics exist only for the simplest cases. It is well known that hydrodynamical flows exhibit instabilities and can become incredibly complex. Such *turbulent* flows have long been the bugbear of physicists, meteorologists, geophysicists and mathematicians.

Students of stellar structure encounter thermal convection as a possible heat transport mechanism in a star. Whenever the temperature gradient in a star exceeds a certain value, a convective flow starts and readily becomes turbulent. Some average properties of such turbulent convection can be estimated using various practical prescriptions (e.g., the mixing length theory), but the phenomenon still defies satisfactory physical modelling and understanding. In the 1960s two simple models were proposed in an effort to unfold the physical ingredients of turbulent thermal convection. We shall use one of these models as our present example. The second one, the Lorentz model, famous for its role as a paradigm of dissipative chaos and its meteorological consequences, will be discussed in the second part of the book. In both cases the hydrodynamical PDEs have been replaced by a third-order dissipative ODE system, mathematically belonging to the same family as the driven oscillator of Section 2.2.

In 1966 Moore and Spiegel constructed a thermally driven model oscillator in an astrophysical context. They wished to clarify the role of thermal dissipation in convective *overstability*. This form of instability (growing oscillations) had earlier been shown to occur in convectively unstable rotating fluids and Moore and Spiegel wanted to investigate it in a compressible fluid with thermal dissipation. They focused on one small mass element of the fluid (a 'parcel'), immersed in a horizontally stratified fluid in a constant gravitational field.

That such a 'parcel' approach can be useful is demonstrated by the heuristic derivation of the Schwarzschild criterion for convective instability. As we have seen in the previous example, Hénon and Heiles discovered fundamental properties of irregular motion in Hamiltonian systems by studying a simple astrophysical model. This time it was Moore and Spiegel who found and studied aperiodic motion in a dissipative system by considering another astrophysical toy model.

Let a small fluid element having a fixed mass M be allowed to exchange heat with its surroundings and adjust its pressure instantaneously to the ambient pressure. Let it also be acted upon by some restoring force $Mr(z)$, where z is the vertical coordinate. If the volume of the mass element as a function of time is $V(t)$, its equation of motion can be written as

$$M\frac{d^2z}{dt^2} = -g[M - \rho_a(z)V(t)] + Mr(z) \tag{2.22}$$

where $\rho_a(z)$ is the density of the ambient fluid. This equation includes the buoyancy force in addition to the the element's weight and the restoring force.

Let $z = 0$ be the equilibrium position of the element, where its density is identical to the ambient density, and let also the restoring force vanish there. We consider only very small displacements from this equilibrium so that the restoring force will be approximated by its linear term, $r(z) = -\omega^2 z$, where ω is a constant. The smallness of z is understood in the sense of its being much smaller than the scale height of the ambient density distribution. Thus the element's density variations will also be small. Since the mass of the element is constant we have

$$\rho(t)V(t) = M = \rho_0 V_0$$

where $\rho_0 = \rho_a(0)$ and is also equal to the element's density at its equilibrium position, and V_0 is the element's volume at that point.

Writing now $\Delta\rho$ and ΔV for the variations in the element's density and volume respectively, and keeping only the linear terms in these variations, the last equation gives

$$V_0\Delta\rho + \rho_0\Delta V = 0$$

or

$$V(t) - V_0 = -\frac{M}{\rho_0^2}[\rho(t) - \rho_0] \tag{2.23}$$

This is reminiscent of the Boussinesq approximation in fluid dynamics (which will be discussed in Chapter 12). In this spirit after the substitution of $V(t)$ from (2.23) into the equation of motion (2.22), $\rho_a(z)$ is replaced by ρ_0 when it multiplies a small density difference. The equation of motion thus becomes

$$\rho_0\frac{d^2z}{dt^2} = g[\rho_a(z) - \rho_0] - g[\rho(t) - \rho_0] - \rho_0\omega^2 z \tag{2.24}$$

The Boussinesq approximation also allows one to express the density variations in (2.24) in the ambient medium and the element by the corresponding temperature variations using the Boussinesq equation of state

$$\Delta\rho = -\alpha\rho_0\Delta T$$

where α is a constant.

The equation of motion then becomes

$$\frac{d^2z}{dt^2} = g\alpha[T(t) - T_a(z)] - \omega^2 z \tag{2.25}$$

Now the cooling law of the element has to be specified. Assuming that the element is small enough to be optically thin, Moore and Spiegel used Newton's cooling law, i.e., $dT(t)/dt = -q[T(t) - T_a(z)]$, so that the cooling rate is proportional to the temperature difference between the element and the ambient medium; q is the inverse of the characteristic radiative cooling time. Now let $\Theta \equiv T(t) - T_a(z)$. Differentiating Θ with respect to time and using the cooling law one gets

$$\frac{d\Theta}{dt} = \frac{dT}{dt} - \frac{dT_a}{dz}\frac{dz}{dt} = -q\Theta + \beta(z)\frac{dz}{dt} \tag{2.26}$$

where $\beta(z)$ is the gradient of the ambient temperature profile. As the simplest non-linear nontrivial temperature profile gradient of the ambient medium, Moore and Spiegel chose $\beta(z) = \beta_0[1 - (z/L)^2]$, where L is an arbitrary length and β_0 is a constant.

It is advantageous to write equations (2.25) and (2.26) in a nondimensional form. Choosing q^{-1} as the time unit and L as the unit of length, and using the assumed ambient temperature gradient, the equations can be cast in the following simple form

$$\ddot{z} = -\sigma z + r y$$
$$\dot{y} = -y + (1 - z^2)\dot{z} \qquad (2.27)$$

where $y(t) \equiv \Theta/(\beta_0 L)$ is the nondimensional temperature difference. The terms r and σ are nondimensional parameters, whose physical meaning is readily understood:

$$r \equiv \frac{g\alpha\beta_0}{q^2}$$

and is therefore the square of the ratio of thermal relaxation time to the free fall time. Similarly,

$$\sigma \equiv \left(\frac{\omega}{q}\right)^2$$

is the square of the ratio of thermal relaxation time to the free oscillation time.

As the first equation in (2.27) is a second-order ODE, the two ODEs constitute a dynamical system with three degrees of freedom. It can be looked upon as a driven oscillator, similar in principle to the oscillator of Section 2.2. We shall name this system the Moore and Spiegel (MS) oscillator in this book. In contrast to the nonlinear Duffing system, this oscillator has a linear restoring force and the nonlinearity enters through the driving term. Physically, the driving term has a thermal origin and it is given by a separate ODE and not in the form of an external force.

Moore and Spiegel numerically studied their oscillator for different values of the parameters r and σ. They found that for $r \gg \sigma$ the oscillations are periodic, i.e., the attractor is a limit cycle. Physically this corresponds to thermal relaxation and oscillation times that are both long compared with free fall time (modified by buoyancy). In this regime, the temporal behaviour took the form of *relaxation oscillations*, i.e., decaying oscillations around an (unstable) equilibrium point followed by a "jump" to the region of the other point with concomitant oscillations around that point. This behaviour was repeated in time. As σ was increased the duration of the oscillations around one of the points shortened and the jump transitions were more frequent.

As σ was increased further the periodic behaviour was lost and the oscillations looked irregular. This happened for some range of σ values; simple periodic oscillations began to appear again above some fixed value of σ. Moore and Spiegel were able to delineate a rather narrow region in the r–σ plane for which aperiodic oscillations were found to occur. A schematic presentation of this is shown in Figure 2.8 and an explicit example of the behaviour in the aperiodic regime, as reflected by the plot of the function $y(t)$, is shown in Figure 2.9.

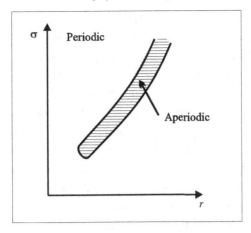

Fig. 2.8. Schematic view of the regions of aperiodic motions in parameter space of the Moore and Spiegel thermal oscillator.

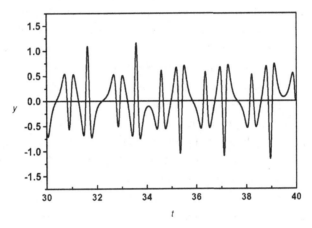

Fig. 2.9. The nondimensional temperature excess (y) as a function of time in the Moore and Spiegel oscillator (2.27) in the chaotic region. In this calculation $r = 100$ and $\sigma = 27$.

In this example too, the aperiodic (chaotic) regime is characterised by the presence of a strange fractal attractor and by unpredictability due to SIC. The MS oscillator demonstrates that a detailed study of a simple instability mechanism can yield a surprising range of phenomena, some of them characterised by complex aperiodic motions, whose properties are quite similar to the ones summarised in the concluding remarks of Section 2.2

2.5 Patterns in a thermally unstable medium

The previous examples were models of systems exhibiting a complex behaviour during their evolution. In the case of mappings such evolution proceeds in discrete steps, while in ODE flows the evolution is continuous in some variable (generally time). In both cases we usually speak of a *temporal* evolution and, possibly, complexity. Another manifestation of complexity in nature is that of form, or *spatial* structure. In this case we are interested in spatial patterns apparent in the system and we try to understand their origin and behaviour.

Systems exhibiting spatial patterns cannot, by definition, be homogeneous. They must consist of some spatially varying characteristic fields defining the pattern. A particularly simple example is a system composed of just two different phases, each one being locally uniform. The shape of the boundaries separating such different regions define the pattern in this case. Pattern-forming systems have, in the past two decades or so, been the subject of many experimental, theoretical and mathematical studies. The basic theoretical results will be summarised in Chapter 7 and some applications, relevant to astrophysics, will be described in the second part of the book. Here we shall consider, as an example, one of the simplest cases – a system having only one space dimension and composed of just two distinct components.

The system we shall discuss can serve as a toy model of thermally unstable astrophysical plasmas (occurring in the interstellar medium, coronal gas, intergalactic gas etc.). It has been proposed as such by Elphick, Regev and Spiegel (1991a) and was also studied by Elphick, Regev and Shaviv (1992). In this section we describe the basics of the approach by the above authors (henceforth referred to as ERSS). In order to make the example as clear and simple as possible we forget about the enormous complexity related to the many non-trivial physical processes in such plasmas and focus on just two – thermal bistability and conduction. In addition, as stressed before, the model is in the simplest, one-dimensional, geometry.

We consider gas heated by some constant external source and cooled by radiative processes in such a way that the change of the heat content of a fluid element can be expressed in the equation

$$\rho T \left(\frac{\partial s}{\partial t} + v \frac{\partial s}{\partial x} \right) = -\rho L(\rho, T) + \frac{\partial}{\partial x} \left(\kappa \frac{\partial T}{\partial x} \right) \tag{2.28}$$

The system is one-dimensional, so there is only an x dependence of the thermodynamic variables ρ (density), T (temperature) and s (specific entropy). L is the cooling function – the radiative heat loss minus the heat gain per unit mass, and is assumed to be known as a function of T and ρ. The coefficient of thermal diffusivity is κ (also assumed to be known as a function of the thermodynamic variables) and v is the velocity – $v(x, t)$. The left-hand side of the equation is the heat gain per unit volume and the right-hand side specifies the physical processes – local radiative cooling and heat transfer by conduction or some other diffusive process.

In many relevant astrophysical systems the dynamical time (the time for the system to be crossed by a sound wave) is much shorter than its thermal time (the time to change the internal energy significantly). Thus equation (2.28) is satisfied essentially at constant pressure, fixed by the external pressure on the system. In addition, assuming the perfect gas equation of state, ρ can be expressed in terms of T and the constant pressure, p_0, say, as $\rho = (p_0\mu/\mathcal{R})/T$, where $\mathcal{R}$ is the gas constant and μ the mean molecular weight. Thus (2.28) becomes

$$c_p\left(\frac{\partial T}{\partial t} + v\frac{\partial T}{\partial x}\right) = -L(T; p_0) + \frac{1}{\rho}\frac{\partial}{\partial x}\left(\kappa\frac{\partial T}{\partial x}\right) \tag{2.29}$$

where c_p is the specific heat (per unit mass) at constant pressure and p_0 has been explicitly marked as a fixed parameter.

Now $T = T_0$ = const is a steady uniform solution of equation (2.29) provided that it is a zero of the function $L(T)$ for a given value of the pressure p_0. In his pioneering work on thermal instability in 1965, Field has shown that such a uniform state is unstable to infinitesimal perturbations if at $T = T_0$

$$\left(\frac{\partial L}{\partial T}\right)_p < 0$$

and it is stable otherwise. This condition can be readily understood if we interpret it as stating that L *decreases* with T, consequently a small increase in T will cause L (the cooling rate) to decrease, increasing T even more. This is the essence of the isobaric thermal instability and it depends, as we see, on the slope of $L(T)$ for some fixed pressure. In a number of interesting regimes $L(T) = 0$ has *three* roots. Thus there are three different steady uniform states. In such situations the middle state is usually unstable, while the other two are thermally stable. The condition for this is that the function $L(T)$ has a characteristic shape, cutting the $L = 0$ line at three points with a positive slope at the outer two zeros and negative slope at the middle zero. In such a case one talks of *bistability*.

We shall explore the consequences of this topology of the cooling function later on. Before that we would like to simplify as much as possible the heat equation describing the thermal energy balance of the system. The variable v in our equation is the fluid velocity at point x at time t. This must come from the equation of motion of the fluid, supplemented by the continuity equation. Our system is thus quite complicated.

A significant simplification can be achieved, however, owing to the fact that in a one-dimensional fluid system a Lagrangian description (labelling fluid elements and following their motion, see the first section of Chapter 12 for details) is particularly useful. Let x_0 be the location of some fixed reference fluid particle. Then

$$\sigma(x, t) = \int_{x_0}^{x} \rho(x', t)dx'$$

is the mass per unit area from the reference point to the point x at time t, and it can serve as a good Lagrangian variable, labelling fluid elements. The independent variables thus become σ and t and the time derivative (for σ fixed) is the Lagrangian, material derivative (following the fluid element). This derivative is sometimes written as D/Dt and is connected with the Eulerian time and space derivatives by

$$\frac{D}{Dt} \equiv \frac{\partial}{\partial t} + v \frac{\partial}{\partial x}$$

We shall momentarily use this notation here but remember that its meaning is partial derivative with respect to t with σ fixed. After the transformation of independent variables from (x, t) into (σ, t) we shall return to the standard notation.

In the Eulerian framework mass conservation is expressed by the following continuity equation

$$\frac{\partial \rho}{\partial t} + v \frac{\partial \rho}{\partial x} = -\rho \frac{\partial v}{\partial x}$$

This yields in Lagrangian notation

$$\frac{D}{Dt} \rho = -\rho^2 \frac{\partial v}{\partial \sigma} \tag{2.30}$$

and (2.29) transforms into

$$c_p \frac{D}{Dt} T = -L(T; p_0) + \frac{\partial}{\partial \sigma} \left[\frac{p_0 \mu}{\mathcal{R}} \kappa(T; p_0) \frac{1}{T} \frac{\partial T}{\partial \sigma} \right] \tag{2.31}$$

where we have used the relation $\rho = p_0 \mu / (\mathcal{R} T)$ resulting from the equation of state.

Equation (2.31) is now a PDE for $T(\sigma, t)$, free of any additional functions, and after it is solved it will be possible to get $\rho(\sigma, t)$ using the solution, $T(\sigma, t)$, and the equation of state, and from this, with the help of equation (2.30) the velocity $v(\sigma, t)$. Now since in Lagrangian description $Dx(\sigma, t)/Dt = v(\sigma, t)$, we can transform back to the space variable x by integration, if necessary.

Returning now to the Lagrangian heat equation (2.31), we assume a power law dependence of the heat conductivity on temperature: $\kappa = \kappa_0 T^\alpha$, with κ_0 and α being constant. This general formula is valid in a variety of circumstances. In addition, we wish to write equation (2.31) in a nondimensional form. The temperature T can be scaled by its typical value, $\tilde{T}$, say, and the pressure by an appropriate $\tilde{p}$. Thus the parameter of the equation becomes nondimensional $\eta \equiv p_0/\tilde{p}$ and the density is scaled by $\tilde{\rho} = \tilde{p}\mu / (\mathcal{R}\tilde{T})$. Let $L_0 = L(\tilde{T}, \tilde{p})$ serve as the unit of the cooling function and $\tilde{\kappa} = \kappa_0 \tilde{T}^\alpha$ as the unit of the heat conduction coefficient. In addition let t_0 be the unit of time and l_0 the length unit (thus $\tilde{\rho} l_0$ is the unit of the Lagrangian variable σ).

It is advantageous to choose characteristic, physically meaningful, time and length scales as the time and length units in the equation. Thus we use the characteristic *cooling time* $t_0 = c_p \tilde{T}/L_0$ and conduction length $l_0 = (\eta \tilde{\kappa}/\tilde{\rho} L_0)^{1/2}$ and get the following nondimensional equation

$$\frac{\partial T}{\partial t} = -L(T; \eta) + \frac{\partial}{\partial \sigma}\left[T^{\alpha-1} \frac{\partial T}{\partial \sigma} \right] \tag{2.32}$$

The length unit defined above (heat conduction length in a cooling time) is known in the literature as the *Field length* and its meaning will be apparent soon. Note that here and henceforth the Lagrangian derivative is written as a partial derivative because the independent variables are already σ and t.

A further simplification of our equation is achieved by the transformation $Z \equiv T^\alpha$, leading to the following form of the heat equation

$$\frac{\partial Z}{\partial t} = Z^\lambda \left[G(Z; \eta) + \frac{\partial^2 Z}{\partial \sigma^2} \right] \tag{2.33}$$

where $\lambda \equiv (\alpha - 1)/\alpha$ and the definition of the new function G introduced here is

$$G(Z; \eta) \equiv -\alpha L(T, \eta) \tag{2.34}$$

with the variable T being expressed in terms of Z as $T = Z^{1/\alpha}$.

It is useful to write $G(Z)$ as a derivative of a 'potential' $V(Z)$. This is always possible (if G is not pathological). We thus have

$$G(Z; \eta) = -\frac{dV(Z; \eta)}{dZ}$$

where V is defined, as usual, up to an additive constant.

With this definition, equation (2.33) becomes

$$\partial_t Z = \left(-\frac{dV}{dZ} + \partial_\sigma^2 Z \right) Z^\lambda \tag{2.35}$$

where we remember that the potential is a function (actually a *functional*) of $Z(\sigma, t)$ and contains also a parameter, η. This PDE (2.35) is an example of a spatio-temporal *gradient system*. Such systems, as we shall see in Chapter 7, are endowed with some general mathematical properties.

The roots of $G(Z) = 0$ are obviously the extrema of the potential $V(Z)$. Since the negative slope (derivative) of $L(T)$ corresponds to a positive derivative of $G(Z)$, and thus to a negative second derivative of $V(Z)$, the potential is maximal at an unstable uniform solution. Similarly, at the stable solutions the potential is minimal. This is, of course, consistent with the usual properties of a potential function. If the cooling function is such that there is bistability we have again (like in the example in Section 2.2) a double-well potential (two minima with a maximum in-between). A thermal isobaric bistability usually only happens for pressures in some range, i.e., $\eta_2 > \eta > \eta_1$, where the potential $V(Z)$ has two, generally unequal, minima. There also exists a value of η in this range for which the potential is symmetric, that is, both minima are of the same depth. Outside the above range of η there is only one minimum (stable uniform solution for Z).

The physics responsible for an appropriate 'turn' in the cooling function (for some pressure range), so as to give a double-well potential, involves relevant atomic processes – the competition between collisional de-excitation and radiative energy loss. For a detailed description of the cooling function and the atomic physics involved in astrophysical conditions the reader is referred to Lepp *et al.* (1985).

The special value of the pressure parameter η_Z, for which V has two *equally deep* minima, was singled out by Zeldovich and Pikelner (1969). They obtained this by using an idea similar to the Maxwell construction in thermodynamics and proposed that, at this value of the pressure, the two stable phases can coexist with a stationary *front* between them.

This result, together with additional properties of fronts in the general case, was found again by ERSS. They used ideas from dynamical systems and pattern formation theory and this will be discussed in more detail in the second part of this book. Although we are not yet equipped with all the concepts and techniques needed for such a treatment, we can attempt to understand some of the results using a mechanical analogy. Let us try a solution of the type $Z(\sigma, t) = H(\sigma - ct)$, with c constant. Clearly, this is a uniformly translating solution with velocity c in the σ-space. Defining $\chi \equiv \sigma - ct$ and substituting this solution into equation (2.35) gives the following ODE for $H(\chi)$

$$-cH^{-\lambda}H' = -\frac{dV}{dH} + H''$$

where the primes denote derivatives with respect to χ. The equation follows from the fact that $\partial_t = -cd/d\chi$ and $\partial_x = d/d\chi$. If we now define $U \equiv -V$ the above equation is

$$H'' = -\frac{dU}{dH} - \gamma(H)H' \qquad (2.36)$$

where $\gamma(H) \equiv cH^{-\lambda}$.

Equation (2.36) can be looked upon as an equation of motion (Newton's second law) of a 'particle' of unit mass, whose position is $H(\chi)$ at time χ. The 'force' acting on the 'particle' consists of a conservative part (derived from the 'mechanical potential' $U(H)$) and a dissipative term. It can easily be shown that if $\gamma > 0$ 'mechanical energy' is lost by the system. Since H is always positive (remember that it corresponds to the temperature in the original system) this means that energy

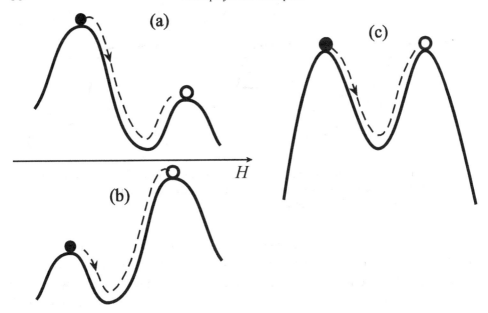

Fig. 2.10. $U(H)$ of the equivalent mechanical problem for three cases. (a) the particle starts from a higher maximum of U, (b) the particle starts from a lower maximum of U, (c) the potential U is symmetric.

is lost for $c > 0$. Conversely, $c < 0$ implies 'negative dissipation', i.e., 'mechanical energy' gain.

In Figure 2.10 we show three situations of a mechanical gedanken experiment, where the potential energy has a double-hump shape (remember that $U = -V$). A particle, located at the potential energy maximum on the left side, starts moving (by an infinitesimal push, i.e., with negligible velocity) towards the other maximum.

We now examine under what conditions the particle can be made to just reach (i.e., with negligible velocity) the other potential energy maximum. Since the particle starts from a maximum by an infinitesimal push we find the following conditions for the three cases.

(i)   Starting from the *higher* maximum – some *positive* dissipation ($c > 0$) is required to just reach, but not overshoot, the other maximum.
(ii) Starting from the *lower* maximum – some *negative* dissipation ($c < 0$) is required to just reach the other maximum.
(iii) In the *symmetric* potential case zero dissipation ($c = 0$) is required.

These observations are rather elementary in the context of Newtonian mechanics. What is new and interesting here is the fact that the above-mentioned special solutions of the ODE 'equation of motion', i.e., those starting at the potential maximum with zero (actually infinitesimal) velocity and reaching the other maximum

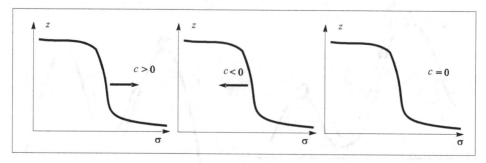

Fig. 2.11. Fronts connecting the two stable uniform solutions of the PDE corresponding to the three cases of the previous figure. The direction of front motion is towards the less stable uniform solution.

again with zero velocity, have a special meaning in the original PDE. The term $H(\chi)$ starts at an H value corresponding to one of the stable minima of V (the maxima of U) and ends at the other one. Thus H is connecting the two stable uniform solutions of the PDE when χ goes from $-\infty$ to ∞. The corresponding $Z(\sigma, t)$ is thus a *front* solution of the PDE (2.35).

If the potential V is symmetric, $c = 0$ and this solution is stationary $- Z(\sigma)$; it is a motionless front. In the two asymmetric cases c is such that the front moves in the direction of the 'less stable' uniform solution of the PDE. Less stable means here the one corresponding to the shallower minimum of V. The fronts corresponding to the three cases of the previous figure are sketched in Figure 2.11.

We have thus achieved a significant understanding of the nature of front solutions, connecting the two stable uniform solutions of our bistable PDE. This was done by studying an associated ODE (we used a mechanical analogy here). This technique of looking for uniformly translating solutions and finding *localised structures* will be explained in detail in Chapter 7 and applied in the second part of the book. In this example we have just outlined the first steps of an approach, based on spatio-temporal pattern theory, to a bistable PDE system. The fronts (stationary or moving) we have found are very special solutions of the PDE, corresponding to certain boundary conditions at $x = \pm\infty$ and some specified initial conditions. They seem, however, to play a fundamental role towards further understanding of spatio-temporal patterns in bistable systems of this kind.

3

Mathematical properties of dynamical systems

Mathematics may be defined as the subject in which we
never know what we are talking about, nor whether
what we are saying is true.
Bertrand Russell, *Mysticism and Logic*.

A *dynamical system* has already been loosely defined in Chapter 1 as a set of rules
by the application of which the state of a physical (or other well defined) system
can be found, if some initial state is known. The origin of this name is in Newtonian
mechanics, where a system of particles is described by ODE in time. If the initial
state is specified, the ODE can be solved (at least in principle) to give the particles'
state at any other later time. It should be stressed that the above mentioned set of
rules can define a meaningful dynamical system only if it implies a *unique* evolution. Dynamical systems are *deterministic*. This is obvious in mechanics, since a
particle cannot be at two different positions at the same time.

Mathematically, a dynamical system is a set of equations – algebraic, differential, integral or combination thereof, the solution of which provides the above
mentioned rule of evolution with time (or some other independent variable). The
equations may contain given parameters and the study of the dependence of the
systems' behaviour on these parameters is often very important and interesting.

The notion of a dynamical system is used nowadays almost always as a synonym for a *nonlinear system*, i.e., one containing nonlinear evolution equations.
In this book we shall deal mainly with such nonlinear dynamical systems; linear
behaviour, when encountered, will be an exception. An overwhelming majority of
systems taught and studied in the exact sciences are linear. Thus our scientific education and practice may leave the impression that nonlinear systems, idiosyncratic
and odd, are also rare in nature. The converse, however, is true. It is as if a zoologist,
studying only elephants, described all the other animals as 'non-elephant'.

One prototype of a dynamical system is the iterated *map* (or mapping). The
evolution rules are given as algebraic equations, in which the values of the variables

40

at some discrete step are expressed as a function of their values at the previous step. The logistic equation (2.2), mentioned in the previous chapter, is an example of such a map. A map need not be one-dimensional. It is, of course, possible that the dynamical system's state is described by several *state variables*. If we denote the n (say) state variables by the n-tuple $(x^1, x^2, \ldots, x^n)$, we may, using vector notation, write the dynamical system as

$$\mathbf{x}_{i+1} = \mathbf{F}(\mathbf{x}_i) \tag{3.1}$$

The n-tuple is written as the vector $\mathbf{x}$, whose value at the $(i+1)$th step is expressed as a (vector) function $\mathbf{F}$ of its value at the ith step.

It is also possible that a dynamical system is such that the value of $\mathbf{x}$ depends not only on its value at the previous step, but also on the values of some of its components at even earlier steps, or even the iteration index, i, itself. Then one must write

$$\mathbf{x}_{i+1} = \mathbf{F}(\mathbf{x}_i, \mathbf{x}_{i-1}, \ldots) \tag{3.2}$$

and now there is need for starting values not only at one step, but at several successive ones. It is, however, generally possible to bring such a system to the generic form (3.1), by redefining variables.

Another prototype of a dynamical system is a set of ODEs. This is the form of the dynamical system describing particle dynamics. In a manner similar to the one used to describe the driven stellar oscillator in the previous chapter, the relevant equations can be reduced to a set of first order ODEs like (2.15). Generalising to an n-dimensional system we again use vector notation. The term $\mathbf{x}(t)$ is the n-tuple of state variables, as a function of time, and the dynamical system is written as

$$\dot{\mathbf{x}} = \mathbf{F}(\mathbf{x}, t) \tag{3.3}$$

where the overdot denotes, as usual, the time derivative.

The vector function $\mathbf{F}$ contains the essence of the dynamical system. If it does not depend on time explicitly, i.e., if we have $\mathbf{F} = \mathbf{F}(\mathbf{x})$, the system is said to be *autonomous*. As was the case in the example leading to (2.15), nonautonomous systems can usually be transformed to autonomous ones, but of a higher dimension.

The n-dimensional space where the vector $\mathbf{x}$ lives is the *phase space* (or, more generally, *state space*) of the dynamical system. As the system evolves in time the point defined by the vector $\mathbf{x}$ moves in phase space, starting from its initial position defined by the initial conditions. The system (3.3) contains, thus, the information on the time evolution of any point (which can be viewed as an initial condition) in phase space. It therefore induces a continuous transformation of phase space into itself which is sometimes symbolically written as

$$\mathbf{x}' = \mathbf{\Phi}_t(\mathbf{x}) \tag{3.4}$$

expressing the fact that a point $\mathbf{x}$ is mapped to a point $\mathbf{x}'$ and that the mapping is time-dependent. A specific initial condition (a point in phase space) results in one *trajectory* or *orbit* in phase space. The collection of all such trajectories constitutes what is called a *flow* in the n-dimensional phase space, and is usually identified with the mapping $\boldsymbol{\Phi}_t(\mathbf{x})$. In this book the term flow will frequently be used to also denote the ODE dynamical system as such. A graphical representation of relevant important trajectories is referred to as a *phase portrait* of the system or flow. In the case of discrete mappings like (3.1) one also talks of the state space and trajectories, or orbits, which are in this case infinite sequences of points.

In this chapter the basic mathematical properties of maps and flows will be discussed. In addition, we shall define several mathematical concepts needed for the exploration of these properties (e.g., invariant manifolds, bifurcations, fractal sets). Throughout the chapter we shall deal only with maps and flows (algebraic and ordinary differential equations) since they are the most important, and other possibilities are often generalisations of them (for example a PDE can be viewed as an infinite-dimensional ODE system). The presentation here will be quite formal but still far from absolute mathematical rigour. The chapter should thus be perceived not as a formal mathematical study, but (consistently with the purpose of this book) as an introduction to dynamical systems for astrophysicists.

3.1 Local geometrical properties

'Excellent!' I cried. 'Elementary,' said he.
Sir Arthur Conan Doyle, *The Crooked Man*.

We shall be exploring here some basic *local* geometrical properties of flows of the type (3.3). Analogous properties of maps will also be mentioned as we proceed. As we are dealing with systems of differential equations which are usually nonlinear it is difficult, if not impossible, to find general analytic solutions that are globally valid. However considerable insight can be achieved by examining the local geometrical properties of the system, that is, the structure of orbits in phase space in the neighbourhood of some special solutions. This can usually be done without actually integrating the equations.

Phase volume – conservation, dissipation and attraction

One geometrical property of a flow in phase space is of considerable importance, and we start by formally defining this and explaining its meaning. Since $\mathbf{x}(t)$ is the position of the dynamical system (3.3) in phase space as a function of time (its trajectory), $\dot{\mathbf{x}}(t)$ is the velocity along this trajectory. Generalising from the two-dimensional example given by equation (2.10) to n dimensions, one can form the divergence of the phase space velocity vector field (see 2.13)

$$D \equiv \nabla \cdot \dot{\mathbf{x}} = \nabla \cdot \mathbf{F}(\mathbf{x}) \tag{3.5}$$

where the equality follows from the dynamical system definition (3.3). Thus, D measures the rate of change of phase space volume. If we consider a small ball in phase space around a given point, D will give the relative rate at which the volume of this ball changes as a result of the flow. If $D < 0$ the volume shrinks and neighbouring trajectories converge. The opposite is true if $D > 0$. Note that D measures, in general, a local property of the flow, $D = D(\mathbf{x}, t)$. In an autonomous system it is not an explicit function of time, but may still be a function of position.

It may happen that D is a constant, namely that an explicit calculation of $D = \sum_{k=1}^{n} \partial \dot{x}^k / \partial x^k$ yields a number (independent of x^j and t). In such systems the relative rate of phase volume change is equal in all phase space. Such was the case in the Duffing oscillator of (2.15). It had $D = -\delta$, and with $\delta > 0$ this ensured the shrinking of phase volume and the tendency of all trajectories to converge on an object of zero volume in phase space. We have seen that, in this case, stable fixed points, limit cycles, or sometimes a more complicated structure attracted the flow.

In general, when $D < 0$ everywhere (even if it is not a constant) the dynamical system is said to be *dissipative*. The flow causes phase-space volumes occupied by neighbouring states of the system to shrink and approach a zero-volume object. This phenomenon is called *attraction*. In contrast, when $D > 0$, phase volumes expand and the solutions diverge for large times. This last observation is usually not very useful in understanding the flow without fully solving the equation, but the diagnosis in itself is often important.

Of particular importance are systems for which $D = 0$ exactly. These are the *conservative* or *volume-preserving* systems. Liouville's theorem guarantees that Hamiltonian dynamical systems, like the ones describing the dynamics of particles in classical mechanics, are volume-preserving. Indeed, for an n_p-particle system in a three-dimensional physical space, there are $N = 3n_\mathrm{p}$ degrees of freedom and phase space is spanned by the $2N = 6n_\mathrm{p}$ generalised coordinates q_j and momenta p_j ($j = 1, \ldots, N$). Hamilton's equations of motion

$$\dot{q}_j = \frac{\partial H}{\partial p_j}$$

$$\dot{p}_j = -\frac{\partial H}{\partial q_j} \tag{3.6}$$

with $H(q_k, p_k)$ the Hamiltonian, yield

$$D = \sum_j \left(\frac{\partial^2 H}{\partial p_j \partial q_j} - \frac{\partial^2 H}{\partial q_j \partial p_j} \right) = 0$$

For maps there exists an analogous property. Consider for a moment the function $\mathbf{F}$ defining the mapping (3.1) as an n-dimensional vector function of an n-dimensional vector variable $\mathbf{x}$. We shall not discuss here the more complicated case of (3.2), but remember that it can usually be transformed to the form of (3.1). This function can be viewed as a transformation of the n-dimensional vector space into itself. This time, however, it is a discrete transformation. Each application of $\mathbf{F}$ results in a given finite change, in contrast to the phase-space flow inducing a continuous change as t increases. A set of points $\mathbf{S}_i$ will be transformed by the map into another set $\mathbf{S}_{i+1}$, where every point $\mathbf{x}_i \in \mathbf{S}_i$ goes to a point $\mathbf{x}_{i+1} \in \mathbf{S}_{i+1}$. The crucial mathematical construction, regarding the comparison of the volumes (or, more precisely, when general sets of points are concerned, *measures*) of the sets $\mathbf{S}_i$ and $\mathbf{S}_{i+1}$, is the Jacobian of the transformation $\mathbf{F}$. Thus one must form the matrix

$$\mathcal{J} = \begin{pmatrix} \frac{\partial F^1}{\partial x^1} & \cdots & \frac{\partial F^1}{\partial x^n} \\ \vdots & \ddots & \vdots \\ \frac{\partial F^n}{\partial x^1} & \cdots & \frac{\partial F^n}{\partial x^n} \end{pmatrix} \tag{3.7}$$

and calculate its determinant

$$J \equiv \det \mathcal{J} \tag{3.8}$$

The absolute value of J gives the ratio of the measures of $\mathbf{S}_{i+1}$ and $\mathbf{S}_i$. This is an elementary result from calculus. If $|J| = 1$ the map is *volume-preserving*; $|J| < 1$ ($|J| > 1$) indicates that the map reduces (increases) volume (measure).

Analogously to the case of flows, if $|J| < 1$ in some neighbourhood, the repeated application of the map to all points in that neighbourhood results in the iterates approaching an *attracting* geometrical object of measure zero (a stable fixed point, periodic cycle, or a more complicated structure).

In the course of this book we shall seldom, if at all, meet mappings in more than two dimensions. In the case of two dimensions one talks of *area*-preserving (or non-preserving) maps. The logistic map (2.2) is one-dimensional. Simple calculation gives the value of its Jacobian determinant, degenerating here into just one derivative $|J| = |4r - 8rx| = 4r|1 - 2x|$. This indicates, for example, that a sufficiently close neighbourhood of $x = 0$ shrinks as long as $r < 0.25$, and expands for $r > 0.25$.

As an another example consider the two-dimensional map acting on points (x, y) of the plane

$$x_{i+1} = y_i$$
$$y_{i+1} = -x_i \tag{3.9}$$

This is clearly a rotation of the plane by $90°$ and should preserve area. Indeed, as can easily be verified, $J = 1$.

Attracting sets

In order to precisely define an *attracting set* we have first to recall the concept of a cluster point, usually introduced in differential calculus.

We define the notion of a cluster point for ODE flows (i) and discrete maps (ii) in the following way.

(i) Let $x(t)$ be an orbit of a flow (or, in general, a continuous function). The point x_∞ is a *cluster point of* $x(t)$ *as* $t \to \infty$ if for every $\epsilon > 0$, however small, and every $T > 0$, however large, $t_\epsilon > T$ exists such that

$$|x(t_\epsilon) - x_\infty| < \epsilon \tag{3.10}$$

(ii) Let $\{x_n\}$, where $n = 0, 1, 2, \ldots$, be an orbit of a mapping, that is, the set of iterates of this mapping (or, in general, any discrete sequence). If there exists an infinite *subsequence* $\{x_{n_k}\} \subseteq \{x_n\}$, which approaches some point x_∞ as $k \to \infty$, x_∞ is called a *cluster point of* x_n *as* $n \to \infty$.

Cluster points are sometimes called *accumulation* or *limit points*.

The set of *all cluster points of a given orbit* of a map or a flow is called the *ω-set of the orbit*. This is easy to remember as ω, the last letter in the Greek alphabet, symbolises here the set of 'end points' (in the above sense) of an orbit. A stable periodic orbit of a map or a flow is an example of an ω-set of all trajectories approaching this orbit.

The set of *all cluster points of all orbits* of a map or a flow is called the *attracting set of the dynamical system* (map or flow). Obviously, the attracting set of a dynamical system is the union of all the ω-sets corresponding to all orbits of this system.

Fixed points and their stability

An important geometrical concept for flows and maps is the notion of a *fixed* point. Fixed points of a flow (3.3) in phase space are given by the solutions of the equation $F(x) = 0$. In the context of ODEs these fixed points are thus also the *equilibrium solutions* of these equations. When the system is placed in a state represented by a fixed point it does not evolve, and stays there indefinitely, since the velocity vector $\dot{\mathbf{x}}$ is, by definition, zero. The system is thus at an equilibrium state and the constant value of $\mathbf{x} = \mathbf{x}_e$, say, is an equilibrium solution. For example, the fixed points of the unforced Duffing oscillator (2.10) can easily be found by solving

$$y = 0$$
$$\beta x - \alpha x^3 - \delta y = 0 \qquad (3.11)$$

for x and y. The notation here is such that x and y are the two components (phase space is two-dimensional) of $\mathbf{x}$. There are three fixed points in phase plane as (3.11) allows three different solutions. These are the origin $\mathbf{x}_{e0} = (0, 0)$ along with $\mathbf{x}_{e1} = (+\sqrt{\beta/\alpha}, 0)$ and $\mathbf{x}_{e2} = (-\sqrt{\beta/\alpha}, 0)$, the last two existing provided that α and β have the same sign.

In the case of maps like (3.1) the notion of a fixed point is, of course, also meaningful. The function $\mathbf{F}$ in (3.1) when applied to a point provides the next iterate. Thus a fixed point, one that further iterations will not change, is in the case of maps the solution of the equation $\mathbf{x} = \mathbf{F}(\mathbf{x})$. The difference between this prescription for finding the fixed points of a map, and that appropriate for flows, $\mathbf{0} = \mathbf{F}(\mathbf{x})$, is self-evident.

The fixed points of the logistic map were found in the previous chapter with the help of (2.4). As another example consider the rotation map (3.9). Clearly, the only fixed point of this map is the origin, $(x = 0, y = 0)$, as is intuitively obvious since in each rotation the origin remains (literally) fixed. Everyday experience is enough to understand that fixed points can vary in their nature. A particle resting at the bottom of a potential well is in a *stable* equilibrium. Such a particle can also be in an equilibrium state when it is at the top (maximum) of the potential curve, but such an equilibrium is *unstable*. At a next level of abstraction we may say that a fixed point of a flow is stable if any sufficiently small displacement away from the point will not cause the flow to push the system far away from the fixed point. Conversely, if the flow acts so as to increase the perturbation the fixed point is unstable. The concept is also similar for maps, the only difference being that the evolution is not a flow but discrete iterations.

We now turn to the precise mathematical definition of stability of fixed points. Goethe once wrote: 'Mathematicians are like Frenchmen: whenever you say something to them, they translate it into their own language and all at once it is something entirely different'. I hope that at least some of the readers are not discouraged by Goethe's maxim (translated from German) so as to stop reading the book at this point. Those who decide to continue may be glad to hear that we will usually be no more 'mathematical' than here at the outset, where we encounter a couple of old, small 'friends' – ϵ and δ.

A fixed point $\mathbf{x}_e$ of the *flow* defined by (3.3) is *stable* if for every $\epsilon > 0$ there exists a $\delta(\epsilon) > 0$ such that

$$|\mathbf{x}(t) - \mathbf{x}_e| < \epsilon \qquad (3.12)$$

for every $t > 0$, if only $\mathbf{x}_0 \equiv \mathbf{x}(0)$ (the initial value) is such that

$$|\mathbf{x}_0 - \mathbf{x}_e| < \delta \qquad (3.13)$$

The fixed point is called *asymptotically stable* if in addition to being stable, in the above sense, $\mathbf{x}(t) \rightarrow \mathbf{x}_e$ as $t \rightarrow \infty$ for any initial value $\mathbf{x}_0$ satisfying (3.13).

The definition of stability of a fixed point for maps like (3.1) is analogous to the above definition for flows. Instead of depending on the continuous parameter t the dynamical system is advanced in discrete steps by iterating (3.1) with increasing index i. Thus we have in the definition of stability instead of (3.12) the requirement that

$$|\mathbf{x}_i - \mathbf{x}_e| < \epsilon$$

for all i, if only the initial value $\mathbf{x}_0$ is such that (3.13) is satisfied.

Any $\mathbf{x}_i$ is, of course, obtained by the repeated i times application of $\mathbf{F}$ to the initial value - $\mathbf{x}_i = \mathbf{F}^i(\mathbf{x}_0)$. $\mathbf{F}^i$ denotes a function obtained by i successive applications of the function $\mathbf{F}$.

Asymptotic stability of a map fixed point means stability plus the requirement $\mathbf{x}_i \rightarrow \mathbf{x}_e$ as $i \rightarrow \infty$.

Linear stability analysis

A practical way of examining the stability of an equilibrium solution (fixed point) of a dynamical system is to observe the evolution of a point *very close* to the fixed point. Since we shall neglect, in the following discussion, terms that are of higher order than the first (in the displacement from the fixed point), the procedure is called *linear stability analysis*. The results of linear stability analysis are not always identical with the properties of the original nonlinear system. In many cases, however, stability properties of the linearised system at a fixed point do coincide with the properties of the original system, provided that the initial perturbation is sufficiently small. A general method for testing nonlinear stability, which is not always practically possible however, will be mentioned later on in this section.

To effect the procedure for a flow defined by the dynamical system (3.3) we perform a small displacement $\delta\mathbf{x}$ from the fixed point $\mathbf{x}_e$ and examine the evolution of the system under the action of (3.3). Substituting $\mathbf{x} = \mathbf{x}_e + \delta\mathbf{x}$ into (3.3) we get

$$\dot{\delta\mathbf{x}} = \mathbf{F}(\mathbf{x}_e + \delta\mathbf{x}) \qquad (3.14)$$

since $\mathbf{x}_e$ is constant.

Now $\mathbf{F}$ can be Taylor expanded around the fixed point, and since $\delta\mathbf{x}$ is very small we retain only first-order terms. This procedure is called *linearisation* and we thus obtain a linear system near the fixed point, related to the nonlinear system (3.3). Written for each component of $\delta\mathbf{x}$ it reads

$$\dot{\delta x}^k = \sum_{j=1}^{n} \frac{\partial F^k}{\partial x^j} \delta x^j$$

where the partial derivatives are all evaluated at the fixed point $\mathbf{x}_e$. In matrix notation we have

$$\dot{\delta\mathbf{x}} = \mathcal{J}\delta\mathbf{x} \tag{3.15}$$

where $\mathcal{J}$ is the Jacobian matrix, the same as before in (3.7). This matrix is sometimes called the *stability matrix*.

Now the linear system (3.15) can be solved by the standard *normal mode* analysis. First one performs the substitution

$$\delta\mathbf{x}(t) = \mathbf{D}\exp(st) \tag{3.16}$$

with $\mathbf{D}$ and s arbitrary constants. It is then obvious that this is a possible solution of the linear system provided

$$\mathcal{J}\mathbf{D} = s\mathbf{D} \tag{3.17}$$

Therefore s must be an eigenvalue of the matrix $\mathcal{J}$ and $\mathbf{D}$ is the corresponding eigenvector. The general solution is a linear combination of such normal modes

$$\delta\mathbf{x}(t) = \sum_{j=1}^{n} c_j\mathbf{D}_j\exp(s_jt) \tag{3.18}$$

where s_j are the eigenvalues, $\mathbf{D}_j$ the corresponding eigenvectors and c_j are constants determined by the initial value of the perturbation, $\delta\mathbf{x}(0)$. Note that since $\delta\mathbf{x}(0)$ and $\mathcal{J}$ are real, s_j are either real or include complex conjugate pairs. Thus one can write $s_j = \sigma_j + i\omega_j$ distinguishing between the real and imaginary parts of the eigenvalues. ω_j is called the *frequency* of the jth mode and σ_j its *growth rate*. If $\omega_j = 0$ the jth mode is called *stationary*, and otherwise it is *oscillatory*.

A degenerate case, where not all the eigenvectors are linearly independent (multiple eigenvalues) requires special attention (see below, where a detailed classification of fixed points in two dimensions is given).

For the determination of linear stability only the values of σ_j are important. Clearly, if even one s_j is such that $\sigma_j = \mathrm{Re}(s_j) > 0$, the solution will grow exponentially in time and the fixed point will be unstable under the linear system (3.15), i.e., the point is said to be *linearly unstable*. Unstable oscillatory modes exhibit growing oscillations (this situation is sometimes called *overstability*, in the astrophysical literature). If for all modes $\sigma_j \leq 0$, the fixed point is *linearly stable*. The linearly stable case is subdivided into two. If $\sigma_j < 0$ for all modes, we say that the fixed point is linearly asymptotically stable (each such mode undergoes exponential decay or has an exponentially decaying oscillatory behaviour). In the case $\sigma_j = 0$ the jth mode is *neutral* and such a situation is called *marginal*.

It might be thought that marginal situations are very rare, and indeed it is so when a given particular physical system is considered. However, marginality plays a very important role in understanding systems when they become unstable. Moreover, in conservative ODE systems one cannot even hope for asymptotic stability of stable fixed points. In any case, close to marginality perturbational analysis of the system is possible and analytical properties of the instability can be found. In addition, several other very useful analytical techniques can be used to better understand the properties of the nonlinear system.

Often one is interested in dynamical systems that depend on a parameter and as this parameter varies a normal mode becomes unstable. At the marginal state the parameter is said to have a *critical* value. A critical value of the parameter signifies a change of the fixed point's stability. The dynamical system changes its qualitative behaviour – a *bifurcation* is said to occur. We shall discuss these issues in detail in Section 3.4 below.

The prescription for performing a linear stability analysis in maps is similar to that described above. The small perturbation is now defined to be the difference between some close (to $\mathbf{x}_e$) point taken to be the mth, say, iterate and $\mathbf{x}_e$ itself. Thus $\delta\mathbf{x}_m = \mathbf{x}_m - \mathbf{x}_e$ and we are interested in checking how this perturbation develops under repeated iteration. The substitution of $\mathbf{x}_m = \mathbf{x}_e + \delta\mathbf{x}_m$ into the map, defined by $\mathbf{x}_{i+1} = \mathbf{F}(\mathbf{x}_i)$, yields, on the one hand

$$\mathbf{x}_{m+1} = \mathbf{F}(\mathbf{x}_m) = \mathbf{F}(\mathbf{x}_e + \delta\mathbf{x}_m) = \mathbf{F}(\mathbf{x}_e) + \mathcal{J}\delta\mathbf{x}_m$$

where the last equality is correct to first-order in $\delta\mathbf{x}_m$ and the derivatives in $\mathcal{J}$ are evaluated at $\mathbf{x}_m$. On the other hand

$$\mathbf{x}_{m+1} = \mathbf{x}_e + \delta\mathbf{x}_{m+1}$$

Thus the linearised system, analogous to (3.15) is

$$\delta\mathbf{x}_{m+1} = \mathcal{J}\delta\mathbf{x}_m \tag{3.19}$$

because $\mathbf{x}_e = \mathbf{F}(\mathbf{x}_e)$.

Equation (3.19) admits solutions of the form $\delta\mathbf{x}_m = q^m\mathbf{D}$, provided that the constants q and $\mathbf{D}$ satisfy the characteristic equation analogous to (3.17)

$$\mathcal{J}\mathbf{D} = q\mathbf{D} \tag{3.20}$$

The solution of the linear system is thus

$$\delta \mathbf{x}_m = \sum_{j=1}^{n} c_j \mathbf{D}_j q_j^m \tag{3.21}$$

where q_j are the eigenvalues of the Jacobian matrix, $\mathbf{D}_j$ the corresponding eigenvectors and the constants c_j depend on the initial value. Because of the above form of the solution the stability properties depend on the q_j. If at least one q_j is such that $|q_j| > 1$ the fixed point is linearly unstable. Stability is guaranteed if all perturbations remain bounded, that is, $|q_j| \leq 1$ for all j. If $|q_j| < 1$ the stability is asymptotic. Again, this is linear stability analysis, but it can be shown that linear stability is sufficient for nonlinear stability except for some very special cases.

We now illustrate the above ideas by performing linear stability analysis of fixed points in some examples of flows and maps in two dimensions. The one-dimensional case is trivial. Two-dimensional flows can be analysed in a general way too, yielding a systematic classification of such points. The procedure is described in considerable detail below and some examples are then given. We note that two-dimensional systems have many applications and some important mechanical, electronic, chemical and biological phenomena can be modelled by such systems. Multidimensional cases are technically more difficult, although the principles are essentially the same.

Fixed point classification in two dimensions and examples

Consider a two-dimensional autonomous dynamical system

$$\begin{aligned}
\dot{x} &= F^x(x, y) \\
\dot{y} &= F^y(x, y)
\end{aligned} \tag{3.22}$$

generating a flow whose trajectories lie in the x–y phase plane.

A number of powerful general theorems on systems of this kind exist and these can be found in the mathematical literature (see e.g., the first chapter in Guckenheimer & Holmes, 1983; and references therein). Here we shall discuss the stability analysis of fixed points of (3.22) and the classification of such points in various cases. As we have seen above, the key to such analysis is the Jacobian matrix $\mathcal{J}$ (defined in 3.7), evaluated at the fixed point $\mathbf{x}_e = (x_e, y_e)$, say, i.e.

$$\mathcal{J}(x_e, y_e) = \begin{pmatrix} a & b \\ c & d \end{pmatrix} \tag{3.23}$$

where we use the notation (x, y) for a point in the phase plane and the constants a, b, c, d are a shorthand for the partial derivatives

$$a \equiv \frac{\partial F^x}{\partial x}(x_e, y_e)$$

$$b \equiv \frac{\partial F^x}{\partial y}(x_e, y_e)$$

$$c \equiv \frac{\partial F^y}{\partial x}(x_e, y_e)$$

$$d \equiv \frac{\partial F^y}{\partial y}(x_e, y_e) \tag{3.24}$$

To find the eigenvalues, s, of $\mathcal{J}(x_e, y_e)$ one has to solve (for s) the characteristic equation,

$$\det(\mathcal{J} - s\mathcal{I}) = \begin{vmatrix} a - s & b \\ c & d - s \end{vmatrix} = 0 \tag{3.25}$$

where $\mathcal{I}$ is the unit matrix.

A quadratic equation for s, $s^2 - (a + d)s + ad - bc = 0$, is obtained with the solutions : $s_{1/2} = [p \pm \sqrt{p^2 - 4q}]/2$, where $p \equiv a + d = \text{trace } \mathcal{J}$ and $q \equiv ad - bc = \det \mathcal{J}$.

The stability properties of a fixed point depend on the nature of s. The possible different cases are covered systematically as follows:

(i) s_1 and s_2 are real, distinct and both have the same sign.
    This happens if $p^2 > 4q$ and $q > 0$.
    If $p < 0$ the fixed point is stable and is called a *stable node*. The local trajectories flow into the fixed point from two directions determined by the corresponding eigenvectors.
    If $p > 0$ the fixed point is unstable and is called in this case an *unstable node*. The trajectories look similar to the previous case, but the flow direction is directed *out* of the fixed point in both directions.

(ii) s_1 and s_2 are real, distinct and have different signs.
    The conditions for this case are $p^2 > 4q$ and $q < 0$.
    One direction grows exponentially (thus the point is unstable) and the other decays exponentially, this is a *saddle* point.

(iii) s_1 and s_2 are a complex conjugate pair.
    Two complex conjugate roots exist if $p^2 < 4q$ and $p \neq 0$.
    The condition $p < 0$ guarantees the negativity of the real part and thus stability. The fixed point is termed a *stable spiral* point (sometimes called *stable focus*). Phase trajectories spiral into the fixed point, the pitch of the spiral being determined by the relative size of the real and imaginary parts of s.
    If $p > 0$ the fixed point is an *unstable spiral* or *focus*. Phase trajectories are similar to the previous case but the flow direction is reversed – they spiral *out* of the point.

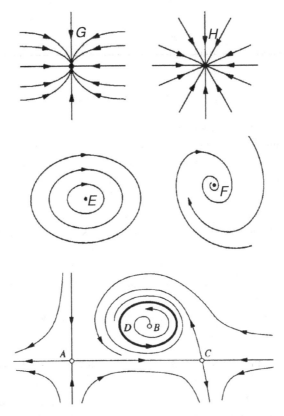

Fig. 3.1. Schematic phase portraits near fixed points of two-dimensional systems. Points A, B, C are unstable and E, F, G, H are stable. Can you identify the class to which each point belongs? D indicates a *limit cycle* (to be discussed later).

In all the above cases the real part of the eigenvalue was not zero. These are the generic cases where the fixed point is called *hyperbolic* (see in the next subsection for a more general definition, in terms of invariant subspaces). Hyperbolic fixed points are important because their stability and nature in the linearised system also carries over to the nonlinear case for a sufficiently small, well behaved perturbation. In addition, in the cases described above, no two eigenvalues were equal, i.e., they were not *degenerate*. We turn now to the discussion of non-hyperbolic and degenerate cases.

(iv) s_1 and s_2 are purely imaginary (have zero real part).

This happens for $p^2 < 4q$ and $p = 0$, thus necessarily $q > 0$.

The fixed point in this case is stable (but not asymptotically stable) and is called a *centre* or *vortex* and is an *elliptic* fixed point. Phase trajectories are ellipses encircling the fixed point.

(v) $s_1 = s_2 \equiv s \neq 0$.

This happens when $p^2 = 4q$ and $q > 0$.

In this degenerate case the solution can no longer be written in the form (3.18) but rather will be (for two dimensions)

$$\delta \mathbf{x}(t) = [c_1 \mathbf{D}_1 + c_2(\mathbf{D}_2 + \mathbf{D}_1 t)] \exp(st)$$

where $\mathbf{D}_j$ are the eigenvectors.

If $\mathbf{D}_2$ is a null eigenvector (this happens if $b = c = 0$ and $a = d \neq 0$), the trajectories will leave or approach the fixed point at the same rate in all directions and the point is then called a *star* or *proper node*. For $p < 0$ it is stable and if $p > 0$ the direction of the flow is reversed and it is unstable.

If $\mathbf{D}_2$ is not a null eigenvector the point is usually called a *degenerate* or *inflected node*. The fixed point is no longer a symmetric sink or source. Again $p > 0$ implies instability, i.e., outflow of trajectories, while $p < 0$ implies stability.

(vi) $s_1 = 0, s_2 \neq 0$. This case is realised when $p^2 > q$ and $q = 0$.

It is a limiting case of (i) and the trajectories are parallel lines (parallel to $\mathbf{D}_2$) approaching (for $p < 0$), or leaving (for $p > 0$) a line that passes through the fixed point and is in the direction of the $\mathbf{D}_1$ eigenvector. Such a point is usually referred to as an *improper node*.

Some representative phase portraits for the different cases are shown schematically in Figure 3.1. It is also important to remark that different books may vary in the exact nomenclature of fixed point stability classes, especially in the last two cases. Turning now to specific examples, we start with the unforced Duffing oscillator (2.10), which we have used in the simplistic model of oscillations of outer stellar layers. We rewrite the system here with the choice $\alpha = \beta = 1$, noting again that such a situation can always be achieved by proper rescaling of the variables, provided that α and β are positive.

$$\begin{aligned} \dot{x} &= y \\ \dot{y} &= x - x^3 - \delta y \end{aligned} \tag{3.26}$$

where we chose $\delta \geq 0$.

The three fixed points in the phase plane (x, y) have already been found in Chapter 2. They are $\mathbf{x}_{e0} = (0, 0)$, $\mathbf{x}_{e1} = (1, 0)$ and $\mathbf{x}_{e2} = (-1, 0)$. Using the definition (3.24) and the nomenclature introduced above, we find the four components of the stability matrix at any of the fixed points (x_e, y_e). This gives $a = 0, b = 1$, $c = 1 - 2x_e^2$ and $d = -\delta$. The quantities $p = a + d$ and $q = ad - bc$, crucial to the determination of the stability and nature of the fixed points, are easily calculated to be: $p = -\delta$ and $q = 3x_e^2 - 1$. The middle fixed point thus differs from the other two. For this point $p = -\delta, q = -1$, while for the other two $p = -\delta, q = 2$ (for both).

First examine the case $\delta = 0$. For this case in the middle fixed point $p^2 - 4q = 4 > 0$ and $q < 0$. The stability eigenvalues can be calculated as well and they are $s_{1/2} = \pm 1$. The origin is thus an unstable saddle point (case (ii) above). The two other fixed points have $p^2 - 4q = -8$ and $p = 0$, thus the eigenvalues are a pair of complex conjugate purely imaginary numbers,

$$s_{1/2} = \pm i \sqrt{2}$$

The fixed points are stable centres (case (iv) above).

These results are no surprise. Since $\delta = 0$, the motion is conservative and, as has already been shown in the previous chapter, the system is equivalent to the motion of a particle in a symmetric quartic potential (2.11). The energy is a constant of motion (2.12) and the phase portrait is displayed in Figure 2.4. Similarly, the nonlinear pendulum system (2.9) in its conservative case ($\delta = 0$) has stable centres at $x = 2n\pi$ and unstable saddle points at $x = (2n - 1)\pi$, for all integers n.

If $\delta > 0$ the motion is dissipative (phase volume shrinks). In this case we still have $p^2 - 4q = \delta^2 + 4 > 0$ and $q < 0$ for the middle point. The stability eigenvalues for this point are real, distinct and have opposite signs. It thus remains an unstable saddle point. At the other two fixed points $p^2 - 4q = \delta^2 - 8$ and $p = -\delta < 0$. The stability eigenvalues are

$$s_{1/2} = -\delta \pm \sqrt{\delta^2 - 8} \tag{3.27}$$

These points are therefore stable. Their nature, however, depends on the magnitude of the dissipation parameter δ. If δ is not too large ($\delta < \sqrt{8}$ in these units), $p^2 < 4q$ and we have stable spiral points (case (iv) above). A larger δ yields two negative eigenvalues, i.e., the two outer fixed points are stable nodes (case (i)). This behaviour is fairly obvious as large dissipation will cause a decay in one of the potential minima, without even a single oscillation (called overdamping in the linear oscillator). Sufficiently small dissipation allows decaying oscillations (spiral motion in phase plane) before the system approaches the potential minimum as closely as desired. The phase portrait for the dissipative case when the potential minima are stable spiral points was sketched in Figure 2.5. The reader is encouraged to repeat these considerations for the dissipative pendulum system, i.e., (2.9) with $\delta > 0$.

For the next example we introduce a famous two-dimensional dissipative map introduced by Hénon (1976), the French astronomer, whose interest in nonlinear dynamics and chaos problems arose from his work on stellar orbits in a galactic potential, mentioned in Chapter 2.

$$
\begin{aligned}
x_{i+1} &= 1 - \alpha x_i^2 + y_i \\
y_{i+1} &= \beta x_i
\end{aligned}
\tag{3.28}
$$

with constant α and $\beta > 0$. The fixed points of this map can be found rather easily by putting $x_{i+1} = x_i \equiv X$ and $y_{i+1} = y_i \equiv Y$ in (3.28). This gives a quadratic equation for X

$$\alpha X^2 + (1 - \beta)X - 1 = 0$$

and Y is trivially expressed as $Y = \beta X$. The solutions for X are

$$X_{1/2} = \frac{1}{2\alpha}\left[(\beta - 1) \pm \sqrt{(1 - \beta^2) + 4\alpha}\right] \tag{3.29}$$

For all values of α, such that $\alpha > \alpha_0 \equiv -(1 - \beta^2)/4$ and $\alpha \neq 0$, two distinct fixed points exist. Their stability depends on the eigenvalues of the Jacobian matrix $\mathcal{J}$, which in this case is

$$\mathcal{J} = \begin{pmatrix} -2\alpha X_j & 1 \\ \beta & 0 \end{pmatrix} \tag{3.30}$$

For each root X_j ($j = 1, 2$) there are two eigenvalues

$$q_1 = -\alpha X_j + \sqrt{\alpha^2 X_j^2 + \beta}$$

$$q_2 = -\alpha X_j - \sqrt{\alpha^2 X_j^2 + \beta} \tag{3.31}$$

resulting from the solution of the characteristic equation

$$\det(\mathcal{J} - q\mathcal{I}) = 0$$

with $\mathcal{I}$ being the unit matrix. Let us numerically check one case, $\alpha = 0.1$ and $\beta = 0.5$. We get the fixed points $X_1 = 1.531$, $Y_1 = 0.766$ and $X_2 = -6.531$, $Y_2 = -3.266$. The eigenvalues corresponding to the first fixed point are $q_1 = 0.570$ and $q_2 = -0.877$. This point is thus stable. The eigenvalues corresponding to the second point are $q_1 = 1.616$ and $q_2 = -0.309$. Since $|q_1| > 1$ this point is unstable. The other eigenvalue, corresponding to this point, is smaller than unity (in its absolute value), so the second point is a saddle point.

The last example is a flow in which results from linear stability analysis, regarding the nature of a fixed point, do not carry over to the nonlinear case. The system is

$$\dot{x} = -y + x(x^2 + y^2)$$
$$\dot{y} = x + y(x^2 + y^2) \tag{3.32}$$

As it happens these equations can be integrated exactly by the introduction of the variable $\rho \equiv x^2 + y^2$. This definition implies

$$\frac{d\rho}{dt} = 2(x\dot{x} + y\dot{y})$$

the right-hand side of which can be written in term of ρ by multiplying the first of the equations (3.32) by x, the second one by y, and adding these products. The following differential equation for $\rho(t)$ results

$$\frac{d\rho}{dt} = 2\rho^2$$

whose exact solution is $\rho(t) = \rho(0)/[1 - 2\rho(0)t]$.

The term $\rho(t)$ is the square of the distance from the origin, which is clearly a fixed point. The fixed point is unstable since for any $\rho(0)$, however small, the trajectory carries the system far out of the fixed point. The solution even actually diverges in finite time. In contrast to this, linear stability analysis of the fixed point $(0, 0)$ for the system (3.32) gives the stability matrix

$$\mathcal{J}(0, 0) = \begin{pmatrix} 0 & -1 \\ 1 & 0 \end{pmatrix} \tag{3.33}$$

Thus, again using the notation introduced above, $p = 0$ and $q = 1$, and the fixed point is classified as a linearly stable centre. As mentioned before, the linear analysis may be misleading only in special circumstances. One has to be careful, for example, in the case of linearly stable elliptic points, as is the case here.

Liapunov method for nonlinear stability

The Russian mathematician Liapunov provided a powerful method for proving the stability (or instability) of fixed points of nonlinear flows. This is important since, as we have seen, nonlinear stability properties are not always identical to the linear ones. The problem is, as will shortly be apparent, that it is usually very difficult (and often impossible) to construct the mathematical object, the *Liapunov function* or *functional*, necessary for the application of the method.

The essence of the Liapunov method can be demonstrated by examining a mechanical system of a unit mass particle, moving in a potential $V(\mathbf{r})$ and subject to a velocity-proportional dissipation. The equation of motion is

$$\ddot{\mathbf{r}} = -\nabla V - \gamma \dot{\mathbf{r}} \tag{3.34}$$

where $\mathbf{r}(t)$ is the particle's position vector and γ is a positive constant. We assume that the potential is such that (3.34) is a nonlinear system. We also assume that the potential is zero at the origin and $V(\mathbf{r}) > 0$ in some *finite* region around it. Both the choice of the value 0 for the potential and the origin as its minimal point are for convenience.

The origin is thus a fixed point of the system (in the phase space spanned by the position and velocity components). We know that this fixed point is linearly stable (from studying mechanics or by performing linear stability analysis). Can we be sure that the fixed point is stable also for the *nonlinear* system?

In this system the answer is in the affirmative. We know that there exists a function

$$H(\mathbf{r}, \dot{\mathbf{r}}) \equiv \frac{1}{2}\dot{\mathbf{r}} \cdot \dot{\mathbf{r}} + V(\mathbf{r}) \tag{3.35}$$

the mechanical energy of the system, such that for $\gamma = 0$ it has a constant value and for $\gamma > 0$ it *decreases* with time for all points in a finite region of phase space around the origin. At the origin the time derivative of the energy is zero. This is so because the differentiation of (3.35) with respect to time gives

$$\frac{dH}{dt} = \dot{\mathbf{r}} \cdot (\ddot{\mathbf{r}} + \nabla V) = -\gamma \dot{\mathbf{r}} \cdot \dot{\mathbf{r}} \tag{3.36}$$

where the second equality follows from the equation of motion (3.34).

The motion must therefore be such that the system approaches the origin in phase space as $t \to \infty$ for any initial condition in the finite region where $V > 0$. Thus the origin is an asymptotically stable fixed point of the nonlinear system.

The energy, or the Hamiltonian, of the mechanical system serves as what is called the Liapunov functional. In a general dynamical system, however, its choice is often not obvious. We give below the Liapunov theorem, without proof, for determining the stability of a fixed point of a dynamical system at the origin $\mathbf{x}_e = \mathbf{0}$. This should not limit the generality, as any fixed point can be made to be the origin of phase coordinates by a suitable translation transformation. Before quoting the theorem we remind the reader that a function is *positive definite* if it is everywhere positive except at the origin, where its value is zero. If instead of being positive, as above, it is just non-negative, it is *positive semidefinite*.

The essence of the theorem is the following.

(i) If $\mathbf{0}$ is an equilibrium point of the system $\dot{\mathbf{x}} = \mathbf{F}(\mathbf{x})$ and there exists a well-behaved positive definite function $H(\mathbf{x})$, such that $-\mathbf{F} \cdot \nabla H$ is positive definite in some domain around $\mathbf{0}$, then the equilibrium point is *asymptotically* stable.

(ii) If $-\mathbf{F} \cdot \nabla H$ is just positive semidefinite in such a domain then the equilibrium point is only stable.

(iii) If $\mathbf{F} \cdot \nabla H$ is positive definite in that domain the point is unstable.

Observing that $dH/dt = (\nabla H) \cdot \dot{\mathbf{x}} = \mathbf{F} \cdot \nabla H$ it is easy to see that the meaning of the Liapunov theorem is exactly the same as explained above in the examination of the energy in a mechanical system.

Invariant subspaces of a fixed point

We introduce now the notion of invariant subspaces of a fixed point in linear and nonlinear dynamical systems. We start with a nonlinear ODE system in its general form

$$\dot{\mathbf{x}} = \mathbf{F}(\mathbf{x}) \tag{3.37}$$

where $\mathbf{x}$ is a vector living in the n-dimensional Euclidean space $\mathbf{R}^n$. We shall refer below to (3.37) as the *nonlinear system*.

Let $\mathbf{x}_e$ be a fixed point of the nonlinear system. Linearisation around this fixed point yields the system

$$\dot{\mathbf{y}} = \mathcal{J}\mathbf{y} \tag{3.38}$$

This system is identical to the one we have already seen in (3.15), but now we call the small displacement from the fixed point $\mathbf{y}$ (instead of $\delta\mathbf{x}$) for convenience. $\mathcal{J}$ is the Jacobian matrix defined in (3.7) evaluated at $\mathbf{x}_e$. System (3.38) will be called here the *linearised* or simply *linear* system.

In the course of linear stability analysis, only system (3.38) is used and a characteristic equation is solved yielding the set of eigenvalues, s_j, of the stability matrix $\mathcal{J}$. The corresponding eigenvectors, $\mathbf{D}_j$, can also be found generally. We assume here that all the eigenvalues are distinct and thus we have n (the dimension of the vectors in the original nonlinear system) eigenvalues and the same number of independent eigenvectors.

Assume now that n_s of the eigenvalues have negative real parts and therefore the corresponding modes are asymptotically stable, n_u modes are unstable (positive real parts of the eigenvalues) and the remaining n_c modes are marginal (their eigenvalues have zero real parts), where obviously $n_s + n_u + n_c = n$. In considering the corresponding eigenvectors we construct their generalized counterparts. This means that for a real eigenvalue we take its real eigenvector as is, and for the pairs of complex conjugate eigenvalues we take the real and imaginary parts of their complex conjugate eigenvectors. In this way three sets of real, distinct and linearly independent eigenvectors are obtained. The three sets correspond to decaying, growing and neutral modes and contain n_s, n_u and n_c eigenvectors respectively.

These sets of eigenvectors span three linear subspaces E^s, E^u and E^c of the full vector space $\mathbf{R}^n$. Since all the eigenvectors in the three sets are linearly independent (eigenvectors of a linear system), the direct sum of the three subspaces is the whole space $\mathbf{R}^n$ of the dynamical system.

E^s, E^u and E^c all have a very important property, they are *invariant subspaces* of the linearised system corresponding to the nonlinear system. This means that any point in one of the invariant subspaces is carried by the flow generated by the *linear* system on a trajectory lying entirely in the same subspace. A more general definition of an invariant set will be given in Chapter 4.

All solutions of the linear system, that are initiated in E^s, approach the fixed point as $t \to \infty$ and stay in E^s while doing so. Consequently E^s is called the the *linear stable subspace* of the fixed point in the linearised system. Similarly, any such solution initiated in E^u runs away from the fixed point, or alternatively, approaches the fixed point as $t \to -\infty$, while the entire trajectory remains in E^u. Thus E^u is the *linear unstable subspace* of the same fixed point.

The third subspace, E^c, is called the *linear centre subspace*. Trajectories of the linearised system initiated in E^c will stay in E^c. They will not approach the fixed point exponentially, neither will they escape away from it as $t \to \infty$. This is so because the real part of the relevant eigenvalues is zero.

The linear subspaces defined above are Euclidean, or flat. Such subspaces belong to the family of more general subspaces called *manifolds*. A rigorous definition of this concept is beyond the scope of this book. We shall just state here that a differentiable manifold is a geometrical object that can be *locally* approximated by a part of a smooth, flat (or Euclidean) space of a certain dimension. Differential calculus (and equations) can be defined on such differentiable manifolds and not only on the 'normal' Euclidean spaces (like R^n or the E^α). For example, the surface of a three-dimensional sphere (or infinite cylinder, or a torus etc.) is a two-dimensional differentiable manifold. As dwellers on a spherical planet we need not be convinced, I hope, that *locally* the surface of a sphere can be well represented by a plane.

Considering now the original nonlinear system, we define the *invariant manifolds* of the fixed point.

The *stable manifold* of a fixed point is the set of all points in phase space, that are brought by the nonlinear flow (when regarded as initial conditions) as close as desired to the fixed point for $t \to \infty$. The corresponding *unstable* and *centre* manifolds of the nonlinear system are defined in an obvious, analogous way. Denoting these invariant manifolds by W^s, W^u and W^c respectively, we shall discuss next the relationship between them and the corresponding subspaces of the linear system (E^s, E^u and E^c), noting that in what follows we shall use the term 'subspace' for the linear invariant subspace and 'manifold' for invariant manifolds of the nonlinear system.

Clearly, when the initial condition is very close to the fixed point, one may approximate the nonlinear system quite well by the linearised one. The corresponding manifolds are thus close to each other near the fixed point and they actually coincide at the fixed point. It is thus intuitively obvious that the stable, unstable and centre subspaces of the linearised system are *tangent* to the corresponding manifolds of the nonlinear system. The invariant manifolds of a fixed point in the nonlinear system are generally not flat. The flat subspaces E^α are tangent to the corresponding W^α for $\alpha = s,\ u,\ c$, at the fixed point.

Mathematical theorems guarantee the existence of the invariant manifolds in a nonlinear system, when the corresponding linearised system has such subspaces. The intuitive conjecture, about their being tangent to the manifolds of the linearised system, can also be proven. The main problem is that the invariant manifolds in the nonlinear system can actually be found (and thus proven to exist) only in the neighbourhood of the fixed point. Thus they guarantee the existence of invariant manifolds only *locally* (in this sense).

It is obviously possible that a fixed point does not have all three linear invariant subspaces (e.g., if all stability eigenvalues have a negative real part) and the corresponding invariant manifold may also be absent.

Fixed points which have only stable and unstable manifolds (the centre manifold does not exist) are called *hyperbolic*. The notion of *hyperbolicity* can also be extended to more general invariant sets of dynamical systems. Both hyperbolicity (absence of the centre manifold) and the existence of the centre manifold are very important in the unfolding of essential dynamical features of a system. On the one hand, as it will be shown in Chapter 4, chaotic dynamics is associated with invariant hyperbolic sets. On the other hand, the existence and nontriviality of the centre manifolds allow one to achieve significant progress in understanding the properties of the solutions of the nonlinear system in the vicinity of a neutral fixed point. This idea will be explained in detail and used in Section 3.4 below and in Chapter 7.

A fixed point of a discrete linearised (around that point) map may also have three invariant (stable, unstable and centre) linear subspaces. The corresponding invariant manifolds can also be defined. The definitions of the invariant linear subspaces and the corresponding manifolds of the original nonlinear map are analogous to those given above for flows. Denoting, as above, the small perturbation by $\mathbf{y}_m \equiv \delta\mathbf{x}_m$, the linearised system (3.19) is written as

$$\mathbf{y}_{m+1} = \mathcal{J}\mathbf{y}_m \tag{3.39}$$

Stability properties, in the case of mappings, depend on the comparison of $|q_j|$, the absolute value of the stability matrix eigenvalues, to one. In addition, asymptotic behaviour occurs for $m \to \infty$ as the evolution is discrete. The rest of the discussion is essentially the same as in the case for flows.

In Figure 3.2 we illustrate the concept of invariant subspaces of a fixed point by schematically showing the structure of representative trajectories (the phase portrait) near typical fixed points of some low-dimensional flows. In each case the stable ($\boldsymbol{E}^s$) and unstable ($\boldsymbol{E}^u$) invariant subspaces are indicated on the figure. The invariant manifolds are usually more complex than these linear subspaces which are tangent to them at the fixed point. The different types of fixed points in the four cases should be evident from the nature of the trajectories.

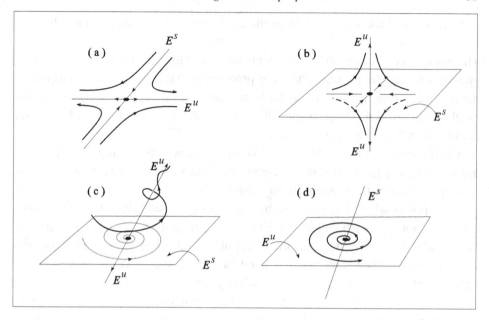

Fig. 3.2. Stable and unstable subspaces of different-type fixed points in ODE flows. In each case representative trajectories are shown.

Periodic solutions

Dynamical systems may have periodic solutions, characterised by a return of the system to a state (phase point), which is one of its past states, and thus the evolution repeats itself.

In the case of flows, periodic solutions are closed trajectories in phase space. For example, a conservative system of the type (2.11) with a quartic potential, e.g., $V(x) = x^4/4 - x^2/2$, will exhibit closed phase trajectories for a given energy (see Figure 2.4). The harmonic oscillator is obviously also an example, the orbit being an ellipse. These two-dimensional conservative examples are rather simple and boring. A more interesting case can be found by examining the following dissipative dynamical system:

$$\dot{x} = x + y - x(x^2 + y^2)$$
$$\dot{y} = -x + y - y(x^2 + y^2) \tag{3.40}$$

Linear stability analysis of the fixed point $x = 0$, $y = 0$ (the origin) classifies it as an unstable spiral point. No other fixed points exist, so it might seem that the phase trajectories should always diverge. However, the rate of phase volume change, D, is

$$D = \frac{\partial \dot{x}}{\partial x} + \frac{\partial \dot{y}}{\partial y} = 1 - 3x^2 - y^2 + 1 - x^2 - 3y^2 = 2 - 4(x^2 + y^2) \tag{3.41}$$

Thus for all points for which $x^2 + y^2 > 1/2$, i.e., far enough from the origin, $D < 0$ and there is contraction. Phase plane structure must therefore be such as to allow trajectories to spiral out from the origin but spiral *in* from far outside. Such a situation is possible in two dimensions only if a closed curve exists and it attracts neighbouring trajectories. This closed curve is itself an invariant trajectory – a periodic orbit.

Indeed, such is the case here, and this claim can be exactly proven since (3.40) is soluble analytically. Multiplying the first equation by x and the second by y, adding them and defining $\rho(t) \equiv x^2(t) + y^2(t)$ in the resulting equation gives

$$\frac{d\rho}{dt} + 2\rho^2 - 2\rho = 0$$

The exact solution of the last equation is

$$\rho(t) = \frac{A \exp(2t)}{A \exp(2t) + B}$$

with A and B constant, as can be verified by a direct substitution; A and B depend on the initial condition. From the solution it is apparent that $\lim_{t \to \infty} \rho(t) = 1$, i.e., the trajectory tends to the unit circle irrespective of the values of A and B if only $A \neq 0$, that is, for all initial points save the origin. The limiting periodic trajectory is called a *limit cycle*. In this case it is attracting.

Another example of a two-dimensional dissipative system exhibiting a more general limit cycle is the famous Van der Pol oscillator. It is an ODE system describing a certain electronic circuit, comprising a linear oscillator with a nonlinear dissipation term, which changes sign at a particular value of the state variable. In a normalised form the second-order ODE describing the Van der Pol oscillator reads

$$\ddot{x} - \lambda(1 - x^2)\dot{x} + \omega^2 x = 0 \tag{3.42}$$

where λ and ω are constants. The origin of the $(x, y \equiv \dot{x})$ space is an unstable spiral fixed point for $\lambda^2 < 4\omega^2$ (which we assume here to hold), as can easily be seen using the standard technique.

It is also obvious that for a sufficiently large x the dissipation term will be positive, causing the trajectory to move inwards. An exact solution to (3.42) is not known, but the above qualitative analysis suggests the existence of a limit cycle. The exact shape of this attracting, stable limit cycle must, in general, be calculated numerically. For a small λ, however, equation (3.42) can be solved approximately using perturbation techniques.

This is the first time in this book that perturbation theory is mentioned and/or used and we shall take this opportunity to briefly introduce the subject. A physical problem is usually formulated in an appropriate set of mathematical equations (differential equations, in most cases). Unfortunately, the instances in which an exact analytical solution to these equations is available are quite rare. The approach must then either involve numerical calculations or approximation techniques. The latter can be made systematic and effective if it is possible to identify a simpler soluble problem, which is close (in a well-defined sense) to the original one. The collection of methods for the systematic analysis of the global behaviour of the solutions is then called *perturbation theory*. The idea behind perturbation theory is to identify in the equations a small parameter (usually called ϵ), such that for $\epsilon = 0$ the solution is known.

We shall illustrate here one technique, belonging to what is called *singular* perturbation theory (employed when a straightforward expansion of the solution in ϵ is not *uniformly* valid) (for a thorough discussion of these concepts see Bender & Orszag, 1999 and Nayfeh, 1973), with the help of the Van der Pol oscillator in the weakly nonlinear case, i.e., when the nonlinear term is small $\lambda \ll 1$ (remember that the equation is nondimensional, i.e the other terms are of order 1). The perturbative technique, which we shall employ here, is known as the *multiple scales* method. This powerful method will be used several times in this book and we refer the reader to Nayfeh (1973) and Kevorkian & Cole (1981) for a detailed account. Applying the technique to the Van der Pol oscillator in the weakly nonlinear case, we expect to find the approximate analytic form of the limit cycle of the Van der Pol oscillator in this limit.

We first simplify the dynamical system (3.42) by setting $\omega = 1$ (this can be done by rescaling the time) and by defining $\epsilon = \lambda/\omega \ll 1$, thus

$$\frac{d^2}{dt^2}x = -x + \epsilon(1 - x^2)\frac{d}{dt}x \tag{3.43}$$

Because of the small size of ϵ two very different time scales are present in the problem. Using the usual prescription for defining multiple-scale time variables (subsequently treated as independent variables),

$$t_n \equiv \epsilon^n t$$

we choose just two of them , i.e., $n = 0, 1$, and thus distinguish between the physical fast (oscillations) and slow (dissipation) time scales.

The function x is then expanded in powers of ϵ

$$x(t) = x_0 + \epsilon x_1 + \epsilon^2 x_2 + \cdots \tag{3.44}$$

where the expansion terms x_i are considered to be functions of both *independent* time variables t_0 and t_1. From the definition of t_n it follows, using the chain rule, that

$$\frac{d}{dt} = \frac{\partial}{\partial t_0} + \epsilon\frac{\partial}{\partial t_1}$$

and

$$\frac{d^2}{dt^2} = \left(\frac{\partial}{\partial t_0} + \epsilon\frac{\partial}{\partial t_1}\right)\left(\frac{\partial}{\partial t_0} + \epsilon\frac{\partial}{\partial t_1}\right)$$

Substituting the above expansions for the functions and the operators d/dt and d^2/dt^2 into the differential system (3.43) and collecting terms of the same order in ϵ we obtain the following set of equations:

(i) *Order ϵ^0:*

$$\frac{\partial^2 x_0}{\partial t_0^2} + x_0 = 0 \tag{3.45}$$

(ii) *Order ϵ:*

$$\frac{\partial^2 x_1}{\partial t_0^2} + x_1 = (1 - x_0^2)\frac{\partial x_0}{\partial t_0} - 2\frac{\partial^2 x_0}{\partial t_1 \partial t_0} \tag{3.46}$$

We stop at this point, deciding to be satisfied with an approximate solution up to first order in ϵ. This is consistent with introducing just two time scales. Consequently, only the first two terms are kept in the function expansion and, to simplify the notation, we put $t \equiv t_0$ and $\tau \equiv t_1$ for the fast and slow time scales respectively.

The zeroth order equation (3.45) is a linear homogeneous equation and it can be solved trivially. We write its general solution as

$$x_0(t, \tau) = A(\tau)e^{it} + A^*(\tau)e^{-it} \tag{3.47}$$

The slow time scale dependence is explicitly shown in the complex amplitudes, which instead of being constants (as in the corresponding ODE) are slowly varying here. The method thus allows as to separate the dependence of the solution on the two different time scales. The complex notation is convenient, but the calculations can also be done when the solution is written in its real form.

The equation of order ϵ (3.46) in the new notation is

$$\frac{\partial^2 x_1}{\partial t^2} + x_1 = \left(1 - x_0^2\right)\frac{\partial x_0}{\partial t} - 2\frac{\partial^2 x_0}{\partial \tau \partial t} \tag{3.48}$$

As is typical in the method of multiple scales, equation (3.48) is inhomogeneous, but the term on the right-hand side of this equation depends only on the previous (here zeroth) order solution $x_0(t, \tau)$. This can be substituted from the solution (3.47) giving the following form of the inhomogeneous term, which we denote by $f(t, \tau)$:

$$f(t, \tau) = i\left(A - A|A|^2 - 2A'\right)e^{it} - iA^3 e^{3it} -$$
$$- i\left[A^* - A^*|A|^2 - 2(A^*)'\right]e^{-it} + i(A^*)^3 e^{-3it} \tag{3.49}$$

where primes denote derivatives with respect to τ. Note that the second line in this equation is the complex conjugate of the first (an advantage of the complex notation). Thus $f(t, \tau)$ is real.

The differential operators on the left-hand sides of both equations, (3.46) and (3.45), are identical. Thus f contains resonant driving terms and the solution of the inhomogeneous equation (3.48) includes, *secular* (i.e., linearly growing) terms. These terms must be absent, however, if we wish the expansion (3.44) to be uniformly valid. If $x_1(t, \tau)$ is allowed to grow beyond any bound with t, the expansion breaks down. We need not worry about any other conditions in this relatively simple problem. The elimination of the terms including $e^{\pm it}$ on the right-hand side of the inhomogeneous equation is sufficient to guarantee the existence of a bounded solution to the inhomogeneous problem.

Thus we require that the coefficient of e^{it} in f should vanish identically (the complex conjugate term with e^{-it} will then also vanish). This immediately gives a (complex) ODE for A, with τ as the independent variable. This last step (elimination of resonances to prevent breakdown of the perturbation expansion of the solution) is usually referred to as 'imposing solvability conditions'. It is the essence of the method of multiple scales. In this case the step is not too difficult – we simply require that the terms that can give rise to secular behaviour should be identically equal to zero.

We thus get

$$\frac{\mathrm{d}A}{\mathrm{d}\tau} = \frac{1}{2}(A - A|A|^2)$$

This complex equation is equivalent to two real ones. Its solution furnishes the slowly varying amplitudes in (3.47).

Writing now $A \equiv a e^{i\theta}$, with a and θ being real functions and separating the real and imaginary parts of the above equation for A, we get the following two real ODEs

$$\frac{\mathrm{d}}{\mathrm{d}\tau}a = \frac{1}{2}(r - r^3)$$
$$\frac{\mathrm{d}}{\mathrm{d}\tau}\theta = 0$$

The first equation is exactly soluble and the second one is trivial. We get

$$a(\tau) = \left[1 + \left(\frac{1}{a_0^2} - 1\right)e^{-\tau}\right]^{-1/2}$$

and

$$\theta(\tau) = \theta_0$$

where θ_0 and a_0 are the initial values of θ and a. Since $a \to 1$ as $\tau \to \infty$ we deduce that, irrespective of the initial condition, the amplitude of the fast oscillation in (3.47) approaches a constant value. The solution tends to a limit cycle. For example, if we assume the following explicit initial conditions for the original oscillator: $x(0) = \xi_0$ and $\dot{x}(0) = 0$, we get from (3.47), in lowest order in ϵ

$$A(0) = A^*(0), \quad \text{and} \quad A(0) + A^*(0) = \xi_0$$

Thus $A(0) = \xi_0/2$ (the amplitude is real) and therefore $a_0 = \xi_0/2$ and $\theta_0 = 0$. The lowest order solution (3.47) becomes

$$x(t) = 2\left[1 + \left(\frac{4}{\xi_0^2} - 1\right)e^{-\epsilon t}\right]^{-1/2} \cos t$$

The limit cycle $x(t) = 2\cos t$ and $\dot{x}(t) = -2\sin t$ is a circle of radius 2 and is being approached rather slowly (exponentially fast, but slow in time!) as $t \to \infty$. This happens for all the starting points, excluding the origin and the circle itself.

We conclude this derivation by stressing that its result (a stable limit cycle in the form of a circle) is only approximate, valid up to order ϵ. The Van der Pol oscillator is known, however, to also have a limit cycle when the dissipation term in the original equation (3.42) is not small. Numerical studies show that this limit cycle is a distorted closed curve and not a circle. The form of the limit cycle, numerically calculated for a few values of ϵ in (3.42), is shown in Figure 3.3. It is apparent that a circle is a fairly good approximation even if $\epsilon = 0.2$.

We shall now return to the general discussion of periodic solutions, but first remark that other perturbation techniques can also be used in the analysis of the Van der Pol oscillator (see Nayfeh, 1973).

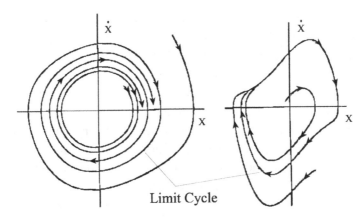

Fig. 3.3. Stable limit cycles of the Van der Pol equation for $\epsilon = 0.2$ (left) and $\epsilon = 2$ (right).

A limit cycle solution of a dynamical system can also be unstable. This occurs if trajectories, when slightly perturbed from the limit cycle (an invariant orbit), progressively depart from it. The complete mathematical stability theory of periodic solutions is beyond the scope of this book. In what follows we concentrate on *local* stability properties, which can be formulated analogously to those of fixed points.

To define local stability of a limit cycle consider $\mathbf{X}(t)$ to be a periodic solution of an ODE system, with period T,

$$\frac{d\mathbf{x}}{dt} = \mathbf{F}(\mathbf{x}) \tag{3.50}$$

that is, $\mathbf{X}(t+T) = \mathbf{X}(t)$ for all t. $\mathbf{X}$ is represented by a closed curve in phase space. It is said to be *orbitally stable*, or *Poincaré stable*, if all neighbouring solution orbits, $\mathbf{x}(t)$, always remain close in phase space to the orbit $\mathbf{X}$.

A more stringent type of stability is that analogous to the definition of a stable fixed point (see 3.12, 3.13). The requirement here is that $|\mathbf{x}(t) - \mathbf{X}(t)|$ remains sufficiently small for *every* t. This defines the *uniform* or *Liapunov stability* of a periodic solution $\mathbf{X}$.

Stability in the sense of Liapunov implies stability in the sense of Poincaré. In the latter case the requirement is that any initially close orbit $\mathbf{x}(t)$ remains close to the periodic orbit $\mathbf{X}(t)$, namely within a tube of radius ϵ, say, around $\mathbf{X}(t)$. Stability in the sense of Liapunov requires, in addition to $\mathbf{x}(t)$ staying in the tube, that there is also no significant 'shear' between the orbits.

The method for finding if a periodic solution $\mathbf{X}(t)$ of (3.50) is orbitally stable is based on *Floquet theory*, whose essential idea is based on reducing the problem to one of a discrete mapping. This is done in the following way. Let $\mathbf{x}(t)$ be an orbit starting close to the periodic orbit $\mathbf{X}(t)$, so that $\delta\mathbf{x}(t) \equiv \mathbf{x}(t) - \mathbf{X}(t)$ is a perturbation whose evolution is governed by the linearised system of (3.50) (analogous to 3.15). One can now construct a linear mapping by starting from $\delta\mathbf{x}(0)$ and advancing in the nth iteration $\delta\mathbf{x}(t)$ from $t = (n-1)T$ to $t = nT$, where T is the period. Stability of the periodic solution in question is then examined by finding the eigenvalues of the mapping (compare with equations 3.19–3.21). For a complete exposition of Floquet theory the reader is referred to e.g., Coddington & Levinson (1955). Periodic solutions give rise, as fixed points do, to invariant (stable, unstable and centre) manifolds. The definitions of these invariant manifolds are analogous to the fixed point case. The invariant manifolds are related to Floquet theory as the fixed point invariant manifolds are related to linear stability analysis.

Before discussing a few additional topics related to periodic orbits and to general behaviour of dynamical systems, we define some special types of orbits, that will later be seen to play an important role in the context of chaotic behaviour. Consider the case when the stability matrix of a fixed point has at least one eigenvalue with a positive real part, and at least one eigenvalue with a negative real part. It is possible then that a trajectory, starting very close to this fixed point, will leave along the unstable manifold but ultimately return back to the close vicinity of the point. Such orbits, essentially connecting a fixed point to itself, are called *homoclinic orbits*. They are defined as orbits approaching a fixed point for both $t \to \infty$ *and* $t \to -\infty$. This is possible only if the stable and unstable manifolds intersect somewhere, in addition to their trivial intersection at the fixed point itself. Assume now that two fixed points, with the above property, i.e., having stable and unstable manifolds, exist. Trajectories connecting the two fixed points are called *heteroclinic* orbits, or half-orbits. A pair of two distinct half-orbits is a closed heteroclinic orbit. Closed homoclinic or heteroclinic orbits are a limiting case of periodic orbits (the period is infinite). We have already seen an example of a heteroclinic orbit. It was a special solution of the ODE resulting from the examination of the uniformly translating solution of a bistable PDE (2.36). As invariant manifold intersections are an example of a *global* property of a dynamical system, we shall discuss this topic in the next section.

If phase space is two-dimensional, a limit cycle is necessarily a closed curve in the plane. The only attracting sets possible are points (stable fixed points), closed curves (stable limit cycles), or generalised limit cycles with fixed saddle points on them (homoclinic or heteroclinic orbits). This statement follows from the famous Poincaré–Bendixson theorem, which we repeat here without proof (see Guckenheimer & Holmes, 1983).

If a solution of a system (3.50) in two dimensions (with well-behaved **F**) remains for all t inside a closed bounded region, then the orbit is either a closed path or a point, or approaches a closed path (limit cycle, homoclinic or heteroclinic orbit), or a point (fixed point) as $t \to \infty$. This is a fundamental result. It is very useful to remember it when one attempts to explain a particular property by using a two-dimensional ODE flow as a model.

In more than two dimensions the situation may be more complicated. This can easily be understood by noting that in two dimensions an orbit starting inside (or outside) a limit cycle can never cross it. In contrast, already in three-dimensional phase space, a closed curve (periodic solution) does not divide the space into two parts. In addition the closed orbit may itself be quite complex as it is a curve in space. For example, if the solution is a multiple periodic function $\mathbf{x}(t) = \mathbf{p}(\omega_1 t, \omega_2 t)$, having periods $2\pi/\omega_i$, $(i = 1, 2)$ in its two arguments, then the trajectory winds on a torus. When the two frequencies are commensurate (i.e., $\omega_1/\omega_2 = $ rational number) the solution is periodic and the orbit is a closed curve wound on the torus (see Figure 3.4). If the frequencies are incommensurate the orbit will densely cover the torus. In the latter case, if this *quasi-periodic* (see

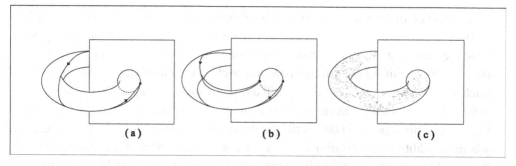

(a) **(b)** **(c)**

Fig. 3.4. Orbits on a torus. (a) and (b) – commensurate frequencies, periodic solution; (c) – incommensurate frequencies, quasi-periodic solution.

below in Chapter 4) orbit is stable, the entire surface of the torus is attracting the flow around it. Thus we see that already in three dimensions there exist attracting invariant sets of a more complicated nature than just a point or a closed curve. Truly complex geometrical objects of non-integral dimension may also be invariant and attracting in three-dimensional phase space. We shall see that they give rise to chaotic solutions, which are ruled out in two dimensions by the Poincaré–Bendixson theorem.

These observations carry over into $n > 3$ dimensions as well, where the tori may be multidimensional. The geometrical analysis of the orbits is then more involved, and the idea of Poincaré sections is very useful in examining the global behaviour (see the next section).

The discussion in this section has been limited so far to dynamical ODE systems represented by flows in phase space. However, as we have already seen at the beginning, the logistic map may possess limit cycles too. In general, if for a given discrete mapping $\mathbf{x}_{i+1} = \mathbf{F}(\mathbf{x}_i)$ there exists a positive integer, p, such that $\mathbf{x}_{n+p} = \mathbf{x}_n$, for all n, then the mapping is said to be *periodic with period p* or *possess a p-cycle*, provided p is the smallest integer having this property.

In this case the sequence of (all distinct) points $C = \{\mathbf{z}_0, \mathbf{z}_1, \ldots, \mathbf{z}_{p-1}\}$ where $\mathbf{z}_{r+1} = \mathbf{F}(\mathbf{z}_r)$, for $r = 0, 1, \ldots, p - 2$, is an *invariant set* under the operation of the map $\mathbf{F}$, in the same way as a fixed point of this map is. This is so since

$$\mathbf{F}(C) = \{\mathbf{F}(\mathbf{z}_0), \mathbf{F}(\mathbf{z}_1), \ldots, \mathbf{F}(\mathbf{z}_{p-1})\} = \{\mathbf{z}_1, \mathbf{z}_2, \ldots, \mathbf{z}_{p-1}, \mathbf{z}_0\} = C$$

because $\mathbf{F}(\mathbf{z}_{p-1}) = \mathbf{z}_p = \mathbf{z}_0$ by definition.

We turn now to the definition of stability of p-cycles in discrete mappings. The construction of a p-generation mapping is very useful for this purpose. If a given discrete map, defined by the function $\mathbf{F}$, is composed with itself p-times (p is any integer > 2), the resulting map $\mathbf{G} = \mathbf{F}^p$ is called the *p-generation map* of the map $\mathbf{F}$. It is easy to understand that all the points belonging to a p-cycle of $\mathbf{F}$ are fixed points of **G**. The set of all fixed points of **G** contains additional members (e.g., the fixed points of **F**), but it is easy to identify those of a particular p-cycle of **F**.

A p-cycle of a mapping **F** is defined as stable if and only if the corresponding fixed points of the p-generation map **G** are stable. In this way the issue of stability of a periodic orbit of a map is reduced to the determination of the stability of fixed points (of another map).

For example, consider the logistic map (2.2) with $r = a/4$, chosen for simplicity. The second generation map for this value of r is

$$G(x) = F^2(x) = a^2x(1-x)[1-ax(1-x)]$$

The fixed points of **G** can be found by solving $G(x) = x$, which here is equivalent to

$$ax(x - 1 + a^{-1})[a^2x^2 - a(a+1)x + a + 1] = 0$$

If $a \leq 3$ ($r \leq 0.75$) this equation has only two roots, the two fixed points of **F**. If $a > 3$ two additional real roots,

$$X_{1,2} = \frac{a + 1 \pm \sqrt{(a+1)(a-3)}}{2a}$$

exist. Therefore $\{X_1, X_2\}$ is a two-cycle of the logistic map. Linear stability analysis shows that the fixed points $X_{1,2}$ of **G** are stable for $3 < a < 1 + \sqrt{6}$ and thus $\{X_1, X_2\}$ is a stable two-cycle of **F**.

Homoclinic and heteroclinic orbits can be defined also for maps, in an analogous way to that used in flows. We thus see that in many aspects ODE systems and maps are similar. Fundamental differences between discrete maps and flows are found, however, if one tries to list all the possible fates of a point in phase space, taken as the dynamical system's initial condition.

When starting an iteration of a map from some given point x_0 it may happen that although x_0 is not itself a fixed point, a fixed point is reached after a *finite* (say $m > 0$) number of iterations. Likewise, a p-cycle of the map may also be reached after m iterations. Such behaviour in flows, i.e., reaching a fixed point or a limit cycle in a finite (but $\neq 0$) time, is excluded by smoothness requirements. Note also that in the case of mappings there are no analogues to the quasi-periodic solutions of ODE flows. This is because difference equations (maps) have solutions of only integral period and any two natural numbers are obviously commensurate.

Considering only bounded dynamical systems we may summarise the possible ultimate fate of an initial point x_0 in phase space, in the following way.

(i) In discrete maps the possibilities are:
    (a) x_0 is a fixed point, or belongs to a p-cycle,
    (b) x_0 *eventually* (in a finite number of iterations – $m > 0$) reaches a fixed point, or a periodic p-cycle,

   (c) x_0 *asymptotically* (as the number of iterations – $m \rightarrow \infty$) approaches a fixed point, or a periodic p-cycle,

   (d) none of the above happens and the orbit is aperiodic.

 (ii) In smooth flows with phase-space dimension $n = 2$ the possibilities are given by the Poincaré–Bendixson theorem:

   (a) x_0 is a fixed point or lies on a closed curve (periodic orbit),

   (b) x_0 approaches, as $t \rightarrow \infty$, a point or a closed curve.

(iii) If the dimension of the flow is $n > 3$, two *additional* possibilities exist:

   (a) x_0 lies on, or approaches, as $t \rightarrow \infty$, a generalised surface of integral dimension and the orbit covers this surface densely,

   (b) none of the above happens and the orbit is aperiodic.

It is worth stressing again that maps of any dimension can exhibit aperiodic behaviour (e.g., the one-dimensional logistic map), while this possibility is ruled out for flows of dimension less than three.

3.2 Global attributes of dynamical systems

> Far away is close at hand
> Close joined is far away,
> Love shall come at your command
> Yet will not stay.
> Robert Graves, *Song of Contrariety.*

This chapter has so far provided a rather detailed account of the properties of dynamical systems. It was shown that aperiodic behaviour is possible in multidimensional ($n > 2$) flows as well as in discrete mappings of any dimension. We have, however, stopped short of discussing these complex solutions in detail. They were defined ad hoc to serve as 'anything else' when all possible regular solutions of a bounded dynamical system were considered.

The understanding of more features of the aperiodic solutions, rather than what they are not, depends on the ability to analyse *global* properties of dynamical systems. The fact that the trajectories of these solutions, although bounded, do not stay in the close neighbourhood of any normal geometric object in phase space, hints that local analysis is insufficient. In addition, we know that such aperiodic solutions are absent in linear systems. Essential nonlinearity is required to bring about chaotic behaviour, and linearisation of the system around a point, a curve, or even a multidimensional surface, annihilates aperiodic solutions.

We have already mentioned a situation that cannot occur in a linear (or linearised) dynamical system. The linear stable and unstable invariant subspaces (of a fixed point, for example) intersect only at the invariant object itself, where they are tangent to the corresponding invariant manifolds of the nonlinear system. In contrast, the nonlinear stable and unstable manifolds may be significantly curved and intersect at an additional location. *Homoclinic* and *heteroclinic* orbits are present as a result of such intersections, but their existence per se is not sufficient for chaos as they may also occur in two-dimensional flows, where aperiodic behaviour is forbidden. We shall discuss invariant manifold intersections in the first part of this section.

Since chaotic behaviour may exist in mappings of any dimension it is very useful to study the global properties of two-dimensional maps (the maximal dimension in which the results can be visualised in a reasonable way). The *surface of section* technique, introduced by Poincaré and resulting in a two-dimensional map bearing his name, is thus of primary importance in the investigation of chaotic behaviour in multidimensional flows. We have already seen this technique used by Hénon and Heiles in the study of their system's behaviour (Section 2.3). The Poincaré map and its properties are the subject of the second part of this section.

If one were asked to diagnose chaotic behaviour, the best bet would probably be to test the system for sensitivity to initial conditions (SIC). Such a test involves the comparison of the ultimate fates of two, initially very close, points. The inherent unpredictability exhibited by the simple deterministic dissipative systems, discussed in Chapter 2, the forced Duffing system (Section 2.3) and the MS oscillator (Section 2.4), is due to SIC. A forced pendulum, that is equation (2.9) with a harmonic forcing term, say, added to it clearly has this property as well. In the last part of this section we shall define and explain a quantitative indicator of SIC – the *Liapunov exponents*.

Intersections of invariant manifolds

Invariant manifolds of a fixed point or a periodic orbit in nonlinear dynamical systems were defined in the previous section, where it was also noted that these manifolds can intersect. For the sake of conciseness and simplicity we shall limit our discussion on this topic here to two-dimensional systems. A particular three-dimensional flow could be so complicated as to allow mathematicians to write a full book on the subject.

Two-dimensional flows may have intersecting fixed point manifolds, but their structure cannot be rich enough to allow for irregular behaviour. Nevertheless, phase portraits of typical cases can be useful in defining some basic concepts. Two-dimensional maps are also simple enough and, in addition, may have chaotic solutions. Moreover, as we may already guess, essential features of a multidimensional flow can be captured by its two-dimensional surface of section (i.e., a mapping).

Consider first the two-dimensional dynamical system defined by the second-order ODE

$$\frac{dx}{dt} = y$$

$$\frac{dy}{dt} = -x + \alpha_3 x^3 - \alpha_5 x^5 - \delta_1 y - \delta_3 y^3 \tag{3.51}$$

where all the Greek letters denote constants. This system can be perceived as a generalisation of the unforced Duffing system (3.15) in a similar way to the latter being a generalisation of a harmonic oscillator. A quintic term is added to the restoring force and a cubic term to the dissipation. This system is therefore 'more nonlinear' and thus richer behaviour can be expected. It has been proposed for modelling the rolling motion of a ship, but I hope that its study here will not cause any seasickness! The system has mostly been explored by Nayfeh and his collaborators (see the book by Nayfeh and Balachandran, 1995).

It is not difficult to see that (the reader is encouraged to work it out) for appropriate values of the parameters the system has five fixed points of alternating stability on the $y = 0$ line in phase space. Schematic phase portraits of the system for the conservative case ($\delta_1 = \delta_3 = 0$) and the nonlinear dissipative case (both $\delta_i > 0$) are shown in Figure 3.5. The nomenclature used in the figure caption is based on the classification of fixed points given in Section 3.1. We concentrate mainly on the saddle points, U_1 and U_2, since they are more interesting than the stable fixed points, S_1, S_2 and S_3. Note also that the phase portrait is symmetric and it is sufficient to consider in detail just one point in each of the symmetric pairs, (U_1, U_2) and (S_1, S_3). The following is apparent from the figure.

(i) In the conservative case:
   (a) the closed curve marked as H_1 is a homoclinic orbit of U_1,
   (b) the curves marked as H_2 and H_3 are heteroclinic orbits between U_1 and U_2,
   (c) the stable manifold of U_1 consists of H_1 and H_3,
   (d) the unstable manifold of U_1 consists of H_1 and H_2,
   (e) all the stable fixed points have neither stable nor unstable manifolds.
(ii) In the dissipative case:
   (a) all the homoclinic and heteroclinic orbits were 'broken up' and are absent,
   (b) the stable manifold of U_1 is an infinite curve including the 'fragments' of H_1 and H_3 (the stable manifold in the conservative case),
   (c) the unstable manifold of U_1 is composed of the two curves that originate from it and then spiral asymptotically into the stable points S_1 and S_2 – it includes the 'fragments' of H_1 and H_2 (the unstable manifold in the conservative case),
   (d) the stable fixed points have two-dimensional stable manifolds. Their union is the whole plane minus the stable manifolds of both saddle points. The stable manifolds of the saddles separate between the stable manifolds of the different stable fixed points.

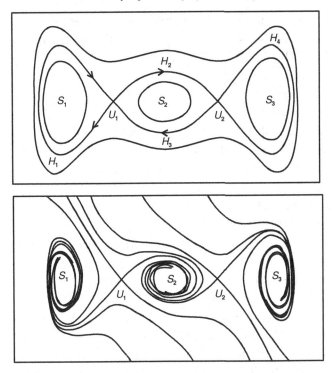

Fig. 3.5. Schematic phase portraits of the system (3.51). The conservative (dissipative) case is depicted in the upper (lower) panel. Unstable fixed points are denoted by U_j and are hyperbolic saddles, S_j are stable fixed points. They are all centres (elliptic) in the conservative case and spiral points (hyperbolic) in the dissipative case. For more details see text.

With the help of the phase portrait of this relatively simple two-dimensional case we identify some of the typical global features of phase-space structure. We clearly see that globally, the invariant manifolds can intersect in different ways. In case (i) the stable and unstable manifolds of the same fixed point (U_1) contain a common portion (the homoclinic orbit H_1). This is an example of a *non-transversal intersection* of the stable and unstable manifold. These manifolds obviously also intersect at the saddle fixed point itself in a *transversal intersection*. In a two-dimensional flow such transversal intersections can occur only at a fixed point.

In general, curves (or surfaces, if phase space is multidimensional) separating regions of qualitatively different behaviour are called *separatrices*. In the conservative case, the homoclinic orbits H_1 and $H_{23} \equiv H_2 \cup H_3$ delimit the regions where oscillations around a particular fixed point occur. In the dissipative case, the stable manifolds of the saddles separate between the stable manifolds of the stable spiral points. The essential global difference seen here between the conservative and dissipative similar systems is a special case of an additional general feature, the *breakup of homoclinic* orbits. In our two-dimensional system it occurs when

non-zero dissipation is introduced. The fixed points change their stability properties (a local property) but there are also global changes in the invariant manifolds
and phase space. All possible motions remain regular however, in agreement with
the Poincaré–Bendixson theorem.

We turn now to the study of invariant manifolds intersections in two-dimensional
maps, defined as usual by the discrete equation $\mathbf{x}_{i+1} = \mathbf{F}(\mathbf{x}_i)$. For simplicity we
consider just the case of fixed points, remembering however that periodic orbits
can be made to be fixed points of a suitable (multi-generation) mapping. The global
structure of a map's phase plane is expected to be, at least in some cases, much
more complicated than that of a two-dimensional flow. The Poincaré–Bendixson
or, for that matter, any other theorem does not guarantee regularity of solutions in
two-dimensional discrete maps.

To demonstrate this we consider an unstable saddle-type fixed point U, that
is, one for which the stability matrix eigenvalues q_1 and q_2 satisfy $|q_1| < 1$ and
$|q_2| > 1$. The stable and unstable invariant manifolds of U obviously intersect
transversally at the fixed point. This observation is trivial. However, the important
point is that in two-dimensional discrete maps additional transversal intersections
of $W^s(U)$ with $W^u(U)$ cannot be ruled out in the same way as they are in two-
dimensional flows. Assume that $\boldsymbol{\xi}_0$ is an intersection point (but not the saddle point
itself) of $W^s(U)$ and $W^u(U)$. Since both manifolds are invariant sets of the map-
ping, $\boldsymbol{\xi}_1 = \mathbf{F}(\boldsymbol{\xi}_0)$ must also belong to both manifolds and therefore be an additional
intersection point. In fact all the points in the infinite series of iterates of $\boldsymbol{\xi}_0$ are intersections of $W^s(U)$ and $W^u(U)$! Thus if the stable and unstable manifolds cross

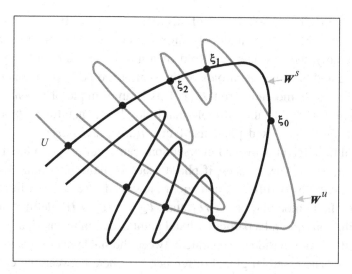

Fig. 3.6. Schematic illustration of the homoclinic tangle of the unstable saddle
point U of a two-dimensional discrete map.

once, they must cross an infinite number of times. Since $\boldsymbol{\xi}_0$ is a transversal intersection $W^s(U) \neq W^u(U)$. This situation is shown schematically in Figure 3.6.

The orbit depicted in the figure is very far from the simple-looking homoclinic orbits we have discussed in two-dimensional flows. It is called the *homoclinic tangle* and reflects an inherent complexity in the dynamics. Such complexity is absent in two-dimensional flows, but as we shall see it is already possible in three-dimensional ones. Heteroclinic tangles can also occur in two-dimensional discrete maps having two saddle points. We shall discuss homoclinic and heteroclinic tangles in more detail in Chapters 4 and 6, where their role in bringing chaos into the dynamics will be stressed.

Surfaces of section and the Poincaré map

It is obvious and it has already been stated several times in this book that the analysis of multidimensional flows must necessarily be rather abstract, since the practical visualisation of trajectories in more than two dimensions is difficult, if not impossible. Chaotic behaviour can be expected in flows only if the dynamical system's phase space is at least three-dimensional, and two-dimensional paper can only picture projections of trajectories. Such simple projections may often be misleading since they include spurious trajectory crossings.

The above remarks should be sufficiently convincing that any idea providing a way of faithfully describing complicated multidimensional motions in two dimensions would be very valuable. Such an idea was provided by Poincaré and Birkhoff and is based on the concept of *surface of section* (or *Poincaré section*) in multidimensional flows. It facilitates a reduction of the flow to a discrete mapping and the resulting technique is applicable to both conservative and dissipative systems. The idea was originally proposed to study analytically the qualitative behaviour of dynamical systems. Nowadays it is widely utilised in numerical calculations with computer graphics.

Consider a general n-dimensional flow defined as usual by $\dot{\mathbf{x}} = \mathbf{F}(\mathbf{x})$. A surface of section of this flow is any generalised (or hyper-) surface (a surface of dimension less than n) in phase space that is transversal to the flow. This means that the unit normal to the surface $\hat{\mathbf{n}}(\mathbf{x})$ satisfies the condition $\hat{\mathbf{n}} \cdot \mathbf{F} \neq 0$ on the surface. These conditions can be phrased in simple terms as a requirement that the flow trajectories cross the surface at a nonzero angle (to the tangent plane). We shall limit ourselves in this book to only two-dimensional surfaces (usually planes) of section. Such a plane of section is obviously not unique, but an appropriate choice yields sections that are both meaningful and easily analysed.

After a surface of section has been specified, one can follow a given orbit (obtained, for example, by numerically integrating the relevant ODE on a computer) and mark its successive crossing points, whenever the orbit crosses the surface in one of the two possible directions. The successive points thus obtained form a sequence $\{\boldsymbol{\xi}_1, \boldsymbol{\xi}_2, \dots\}$. If the section is planar this sequence is just an ordered collection of points in the plane (each having two coordinates). The sequence can be perceived as a two-dimensional discrete mapping

$$\boldsymbol{\xi}_{n+1} = \mathbf{P}(\boldsymbol{\xi}_n) \tag{3.52}$$

The function $\mathbf{P}$, advancing the iterates, is defined by the flow and plane of section. Mapping (3.52), obtained from the flow in the way explained above, is called the *Poincaré* map. Since each point in the sequence is uniquely specified by the previous point and the flow and vice versa, the Poincaré map is *invertible*. Only in exceptional cases is the time interval between the successive crossings defining the Poincaré map constant. The Poincaré map we have defined here is a two-dimensional mapping. Remarkably, however, it is endowed with many of the essential geometrical properties of its parent multidimensional flow. Volume-conserving flows give rise to area-conserving Poincaré maps and dissipative flows have area-contracting maps. In addition, many essential properties of attracting sets of the flows are reflected in suitably chosen Poincaré maps.

If Figure 3.7 the idea behind the Poincaré map is illustrated. The figure is also helpful in understanding the following features of the Poincaré map which result from the corresponding properties of the flow.

(i) Fixed point of the map – periodic orbit (or fixed point) of the flow.
(ii) Periodic orbit of the map (several points) – multiperiodic (or periodic) orbit of the flow.

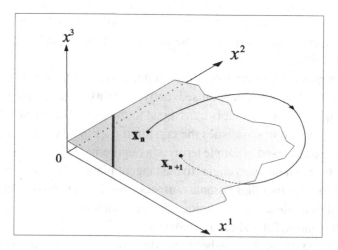

Fig. 3.7. A sketch of the idea behind a Poincaré map.

(iii) Closed curve or curves in the map – quasiperiodic (incommensurate frequencies) orbit of the flow.

(iv) Complicated features in the map – aperiodic behaviour in the flow.

We have already seen an example of a Poincaré map of a Hamiltonian system in our discussion of the Hénon and Heiles model in Chapter 2. All the possibilities listed above appear in the HH system and the aperiodic behaviour is reflected by wildly scattered area-filling points in the Poincaré map. In a dissipative system such behaviour is impossible even if its orbits are aperiodic. Since dissipative systems must have area-contracting Poincaré maps, the attracting sets should be reflected by one-dimensional structures in the surface of section. Aperiodic behaviour may then be reflected by curves or segments arranged in a complicated way on the surface. In fact, as it will be shown in the next chapter, such attracting sets are incredibly complex.

The dissipative case can be well demonstrated using, for example, the forced Duffing oscillator introduced in Chapter 2 (2.15). As we said there, different types of behaviour are possible, depending on the values of the parameters. In Figure 3.8 a Poincaré map for the system is shown to illustrate its typical features in the chaotic regime. Because the attracting set of a multidimensional dissipative flow is reflected by a one-dimensional, albeit very intricate, structure in the Poincaré map, it is often useful to reduce the system further, to a one-dimensional mapping. The points on a curve can be labelled by just one coordinate ζ, say, and since the relevant one-dimensional structure is invariant, the Poincaré map can be reduced to $\zeta_{k+1} = P(\zeta_k)$. This one-dimensional mapping is called the *first return* map. The analysis of first return maps may reflect chaotic behaviour. For example, if the first return map of some system has a form similar to the logistic map, aperiodic behaviour can be expected. In general, first return maps in which the function $P(\zeta)$ has a 'hump' or a 'spike' in the relevant range of its variable, indicate that the underlying dynamical system can exhibit chaotic behaviour.

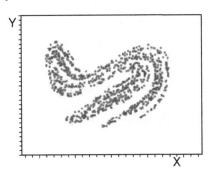

Fig. 3.8. A Poincaré map (a suitable plane section of phase space) for the forced Duffing oscillator in the chaotic regime. This calculation was not continued for long and neither is the resolution good enough, but the complex structure of folded layers of points is nevertheless apparent.

While being a powerful tool in the study of multidimensional dynamical systems, it has to be remembered that the interpretation of Poincaré maps is largely qualitative and cannot be a definitive indication of chaos. The quantitative characteristic 'measures' of the dynamics, discussed next, are the generally accepted indicators of this sort.

Liapunov exponents

The great sensitivity to initial conditions (SIC) has already been mentioned as an important characteristic of chaotic motion. Neighbouring trajectories in a chaotic flow are found to diverge *exponentially*, while in a regular flow they might diverge only linearly, if at all. Since in dissipative systems the phase volume must shrink and in Hamiltonian ones it is preserved (Liouville's theorem), the only way to separate neighbouring trajectories is by greatly distorting volume elements. Trajectories may diverge in some directions, but they must converge in other directions so as not to increase the volume element. Moreover, as we are almost always interested in bound motions, exponential divergence of trajectories cannot continue forever without being accompanied by some sort of 'folding' of the volume element. If we are to quantify this divergence of trajectories, we must take these considerations into account. Understanding that trajectories can exponentially diverge only locally, the quantitative global measure of the orbit divergence must be defined as some sort of average.

Consider a flow resulting from some autonomous n-dimensional differential system,

$$\frac{d\mathbf{x}}{dt} = \mathbf{F}(\mathbf{x}) \tag{3.53}$$

We first pick a solution (trajectory) of (3.53) passing through the point $\mathbf{x}_0$ at time t_0. Looking now at a point close to $\mathbf{x}_0$ (a small distance d_0 from it) we examine the trajectory passing through it at time t_0 and define $d(t)$ to be the distance, as a function of time, of this new trajectory from the original one. If the trajectories diverge exponentially (locally)

$$d(t) = d_0 \exp(\lambda t) \tag{3.54}$$

with $\lambda > 0$ at least for t close enough to t_0. Clearly, λ is a quantitative measure for trajectory divergence, SIC, and therefore an indicator of chaos. However, for the definition to be meaningful some limiting process has to be defined, since we do not want λ to depend on d_0, for example.

The formal procedure is thus the following. Let $\mathbf{X}(t)$ be a trajectory passing through a point $\mathbf{x}_0$ at time t_0 so that $\mathbf{x}_0 = \mathbf{X}(t_0)$. Define now $\mathbf{x}(t) = \mathbf{X}(t) + \delta\mathbf{x}(t)$ to be a neighbouring trajectory, passing at t_0 through a point close to $\mathbf{x}_0$.

Define $d(t)$ to be the usual Euclidean norm (length) of $\delta\mathbf{x}$, i.e.,

$$d(t) \equiv |\delta\mathbf{x}| = \sqrt{\sum_{j=1}^{n} (\delta x_j)^2} \tag{3.55}$$

and $d_0 \equiv d(0)$. The rate of exponential divergence is defined as

$$\lambda = \lim_{t\to\infty,\, d_0\to 0} \left(\frac{1}{t}\right) \ln\left(\frac{d(t)}{d_0}\right) \tag{3.56}$$

We shall proceed to show that there are, in general, n such quantities λ_i, $i = 1, \ldots, n$ for an n-dimensional system, depending on the *direction* of the perturbed trajectory from the original one. These λ_i are called the *Liapunov characteristic exponents*. It is sufficient that the largest Liapunov exponent is positive to diagnose the orbit as chaotic. Since such a positive exponent is a measure of the exponential divergence of neighbouring trajectories in a flow, its inverse should indicate the e-folding time for such a divergence. This quantity is sometimes referred to as the *Liapunov time*.

Linearisation of the system (3.53) around the original orbit $\mathbf{X}(t)$ gives

$$\frac{d}{dt}\delta\mathbf{x} = \mathcal{J}\delta\mathbf{x} \tag{3.57}$$

where $\mathcal{J}$ is the usual stability matrix evaluated for the unperturbed flow.

It is clear that the growth rates in the different directions of the eigenvectors of $\mathcal{J}$ can be different. This yields, in principle, n exponents.

As time progresses, a small volume element will be stretched most in the direction of the largest exponent. In practice we would like to be able to calculate the largest exponent and thus (3.56) should yield just this exponent. The calculation of Liapunov exponents in multidimensional flows is sometimes not an easy task (see below) and special numerical techniques have been developed for this purpose. In any case, the identity of the largest Liapunov exponent should be the same for all starting points which are in the same *basin of attraction* (see Chapter 4).

Before proceeding to such technical details, it is useful to remark that the concept of Liapunov exponents is also meaningful for mappings. The ideas are best demonstrated in the case of *one*-dimensional maps. For such a map, $x_{i+1} = F(x_i)$, with $i = 0, 1 \ldots$, say, it is clear that $\delta x_{i+1} = F'(x_i)\delta x_i$, i.e., a small perturbation is propagated by the mapping as a product with the derivative. Thus for any natural N

$$\delta x_N = F'(x_{N-1})F'(x_{N-2})\ldots F'(x_0)\delta x_0 = \left[\prod_{j=0}^{N-1} F'(x_j)\right]\delta x_0 \tag{3.58}$$

The Liapunov exponent can thus be defined, analogously to the case of flows, as

$$\lambda \equiv \lim_{N\to\infty} \frac{1}{N} \ln \left| \prod_{j=0}^{N-1} [F'(x_j)] \right| = \lim_{N\to\infty} \frac{1}{N} \sum_{j=0}^{N-1} \ln|F'(x_j)| \tag{3.59}$$

So in the case of mappings the condition for SIC and thus chaos is also $\lambda > 0$. In this case of a one-dimensional map there is only one Liapunov exponent. The Liapunov exponent is independent of the initial point, apart from maybe a set of initial conditions of measure zero, as it was for flows. The idea of Liapunov exponents for multidimensional mappings is also useful. We shall omit here their rigorous definitions, remarking only that in an n-dimensional map there should be n exponents and from an expression analogous to (3.58), i.e.

$$\delta\mathbf{x}_N = \left[\prod_{j=0}^{N-1} \mathcal{J}(\mathbf{x}_j) \right] \delta\mathbf{x}_0 \tag{3.60}$$

where $\mathcal{J}$ is the Jacobian matrix of the linearised map, we find the expression for the Liapunov exponents

$$\lambda_k = \lim_{N\to\infty} \frac{1}{N} \ln \left| \prod_{j=0}^{N-1} [\mu_k(\mathbf{x}_j)] \right| \tag{3.61}$$

Here μ_k are the eigenvalues of the $n \times n$ matrix $\mathcal{M}$, defined as the product of the Jacobian matrices

$$\mathcal{M} = \mathcal{J}(\mathbf{x}_0)\mathcal{J}(\mathbf{x}_1)\ldots\mathcal{J}(\mathbf{x}_{N-1})$$

Since in general there are n such eigenvalues, we have n Liapunov exponents and the largest one is the most important.

Except for a few very simple cases the value of the (largest) Liapunov exponent must be calculated numerically. For example, consider the one-dimensional map defined on the $[0, 1]$ interval by

$$x_{i+1} = 2x_i \quad (\text{mod } 1) \tag{3.62}$$

called the *Bernoulli* or *shift* map. Clearly, for this map $|F'| = 2$ everywhere, except for $x = 0.5$. Thus $\lambda = \ln 2$. The Liapunov exponent is positive and hence the map is chaotic. The occurrence of chaos in the Bernoulli map can be rather easily understood when one expresses x in its *binary* form, i.e., as a sequence of 0s and 1s after the decimal point. Multiplications by 2, will result in moving the digits of the binary decimal fraction one place to the left and modulo 1 means that we have to replace 1 by 0 in the position just before the decimal point, should 1 be there. In this way more and more distant digits of the original number are brought to the front. It is well known that rational numbers can be expressed as finite decimal fractions or decimal fractions with an ultimately repeating sequence of digits (periodic sequence). Thus a start from a rational number in the Bernoulli map will end at the fixed point 0 or on a periodic limit cycle. If the initial number is irrational (infinite aperiodic sequence) the iterations will continue forever in an aperiodic manner with information about the starting point being progressively lost as the iteration continues. The positive value of the Liapunov exponent thus reflects aperiodic behaviour with SIC, namely chaos, for all initial conditions, except for a set of measure zero (the rational numbers).

In Chapter 4 we shall analytically calculate the Liapunov exponents of a well known two-dimensional mapping (the horseshoe map), where this map will be shown to play an important role in the understanding of chaotic dynamics. The Liapunov exponent for the familiar logistic map (2.2) must be calculated numerically and as can been seen in Figure 3.9 it becomes positive for those values of the parameter r for which chaotic behaviour was described in Chapter 2.

We end this first discussion of Liapunov exponents by outlining a numerical method for the computation of the largest exponent for a flow like (3.53). It is impossible to follow $d(t)$ (see definition 3.55) for a long time on the computer because it grows exponentially. Instead, one divides the reference orbit into segments, corresponding to a fixed time interval Δt. Let the initial point be $\mathbf{x}_{k-1}$, say, with an initial perturbation $\delta\mathbf{x}_{k-1}(0)$, whose norm $d(0)$ is normalised to unity. The perturbed orbit is integrated for time Δt yielding $\delta\mathbf{x}_k = \delta\mathbf{x}_{k-1}(\Delta t)$, the norm of which we denote by d_k. This, in turn, is normalised to unity, called $d(0)$, and the procedure is repeated for the next segment. In this way the set $\{d_1, d_2, \ldots, d_N\}$ is

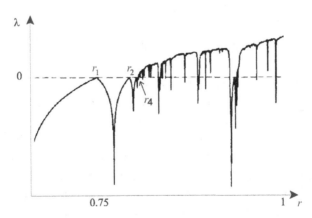

Fig. 3.9. The Liapunov exponent the logistic map as a function of r.

formed and by analogy to (3.56) we define

$$\lambda_N = \frac{1}{N \Delta t} \sum_{j=1}^{N} \ln d_j \tag{3.63}$$

It can be shown that the largest Liapunov exponent is obtained by taking in (3.63) the limit $\lambda = \lim_{N \to \infty} \lambda_N$. For sufficiently small Δt the limit exists and is independent of Δt. The considerations in the choice of a suitable Δt, further details on the numerical procedure and additional references on this subject can be found in Wolf *et al.* (1985).

In the last section of this chapter we shall see that the Liapunov exponents of a flow are connected with the dimension of attracting sets of this flow. The implications of Liapunov exponents for chaotic dynamics and some examples will be given in the following chapters of this part of the book. The concept will also frequently appear in the second part.

3.3 Elements of bifurcation theory

> Two roads diverged in a wood, and I —
> I took the one less traveled by,
> And that has made all the difference.
> Robert Frost, *The Road Not Taken.*

The example of the logistic map (2.2) in Chapter 2 provided us with a first glimpse of the *bifurcation* phenomenon. The route to chaotic behaviour, as a parameter was varied, consisted of a series of bifurcations, with the dynamical system solutions changing their nature. Similar behaviour may be expected in a variety of dynamical systems. Indeed, most systems we have seen so far have bifurcating solutions, when some parameter of the system is varied.

The first ideas of bifurcation theory arose very long ago, together with the development of algebra. The concept of a bifurcation can be very simply illustrated by considering an *algebraic* equation, e.g., $x^2 - 2bx + 1 = 0$, where b is regarded as a parameter and we want to find the solution(s) as b is varied. The solution is $x_{1,2} = b \pm \sqrt{b^2 - 1}$, and it is obvious that for $b < 1$, no real solutions exist, for $b = 1$, one solution $x_1 = b$ exists, and for $b > 1$, there are two distinct real solutions and they change monotonically with b. Thus we see that at a specific value of the parameter ($b = 1$) a *qualitative* change in the nature of the solution occurs – we pass from zero through one to two solutions. Considering the roots as a function of b one notices the appearance of two branches of solutions at $b = 1$. The solutions are said to bifurcate at this point.

Bifurcations in dynamical systems are less trivial than this. Most readers of this book are probably aware of the famous problem of finding the figures of equilibrium for uniformly rotating incompressible, self-gravitating fluid bodies. Many of the most prominent mathematicians of the last three centuries worked on this problem starting with Newton and D'Alembert and continuing with Maclaurin, Jacobi, Dirichlet, Riemann, Poincaré and others. It has been found that as the rotation becomes more and more important (a parameter reflecting its value relative to gravity grows) oblate Maclaurin spheroids become unstable, and ellipsoidal (Jacobi) figures of equilibrium appear, only to bifurcate again into the pear-shaped figures found by Poincaré. Poincaré is the one who coined the word *bifurcation* in this sense.

Rotating figures of equilibrium reveal an important general property of bifurcations. A bifurcation is very often accompanied by *symmetry breaking*. Indeed, in the above quadratic equation case the most symmetric state (no solutions) changes into a state of two equal roots and again into an even less symmetric state of two different solutions. Similarly, Poincaré's pear-shaped figures are the least symmetric and Maclaurin spheroids the most symmetric in the sequence of rotating figures of equilibrium. Symmetry breaking is an example of a qualitative change. Appearance of new fixed points in a dynamical system also gives rise to a qualitative change in its phase-space structure. A bifurcation is always accompanied by a *qualitative* change in the dynamical system properties. It occurs when the system depends on one or several parameters. As the parameters are varied the system properties obviously change. Such continuous quantitative changes are not always essential, however. Bifurcations only occur at the parameter values when the system experiences an abrupt qualitative change. This is reminiscent of the phenomenon of a phase transition in condensed matter physics. Indeed, phase transitions and bifurcations of dynamical systems are mathematically very similar.

In this section we shall describe the basic types of bifurcations using simple ODE systems. In addition, a few general remarks will be made. The discussion will be rather heuristic, without the abstraction needed for mathematical theorems and proofs. For a more rigorous treatment the reader is referred to e.g., Drazin (1992), a very readable and useful mathematical introduction to nonlinear systems in general, and to the book by Iooss and Joseph (1980).

To set the stage let us consider an n-dimensional dynamical system dependent on l parameters, denoted by the vector $\mathbf{a}$, whose components are $(a_1, \ldots, a_l)$, thus

$$\frac{d\mathbf{x}}{dt} = \mathbf{F}(\mathbf{x}; \mathbf{a}) \qquad (3.64)$$

The equilibrium solutions (fixed points) of (3.64) are the solutions of the algebraic equations $\mathbf{F}(\mathbf{x}; \mathbf{a}) = \mathbf{0}$. The nature of these solutions generally depends on the value of the parameters. The investigation of qualitative changes of such special solutions of a dynamical system is the subject of bifurcation theory.

Let $X_0(a)$ be a fixed point of the system (3.64). For a particular value of the parameters, a_0, the equilibrium solution is $X_0(a_0)$. This solution is said to be *bifurcating* if the number of steady solutions in a small neighbourhood of X_0 changes when a is varied within a sufficiently small neighbourhood of a_0. The terms $X_0(a_0)$ and a_0 then give the *bifurcation point*. It is a point in a space composed of the system's phase space plus the parameter space. The parameters whose variations bring about bifurcations are usually called *control parameters*.

This definition has to be modified if we want to also extend it for non-steady (e.g., periodic) solutions of nonlinear dynamical systems, which change qualitatively, when the parameters vary. We shall consider in this section only bifurcating equilibrium solutions and thus we skip a detailed definition for non-steady solutions. Such cases will be discussed in due course.

Before considering in some detail the most basic bifurcations, we remark that the analysis used here will always be local. Global bifurcations are beyond the scope of this book. We shall see, however, that even locally defined bifurcations give rise to qualitative changes in phase space, which are almost always manifested in finite regions, i.e., globally.

Saddle-node bifurcation

Consider the system

$$\frac{dx}{dt} = a - x^2 \tag{3.65}$$

where a is a parameter. The fixed points of this system are $X_{1,2} = \pm\sqrt{a}$. Real solutions of this equation (fixed points of the ODE (3.65)) thus exist only for $a \geq 0$. It is not difficult to see that for $a > 0$ two fixed points exist, one of them is stable and the other one unstable. Indeed, writing $x = X_j + \delta x$ (for $j = 1, 2$) and linearising we get

$$\dot{\delta x} = -2X_j \delta x \tag{3.66}$$

with the solution $\delta x \propto \exp(-2X_j t)$. Thus the perturbation grows for $X_2 = -\sqrt{a}$ (instability) and decays for $X_1 = +\sqrt{a}$ (stability).

The value $a = 0$ requires special attention. Only one fixed point, $X = 0$, exists in this case and according to the above *linear* stability analysis it is neutral. However considering this case separately (i.e., $\dot{x} = -x^2$), it is easy to see that the solution of this equation is $x(t) = (x_0^{-1} + t)^{-1}$ and thus it blows up in finite time for any negative x_0, even starting from very close to $x = 0$.

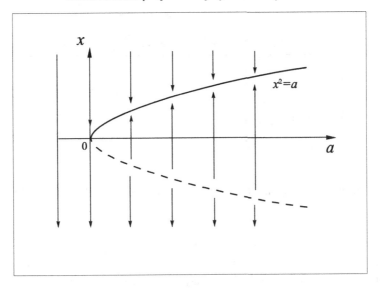

Fig. 3.10. Bifurcation diagram of a saddle-node bifurcation.

The situation can be summarised with the help of a *bifurcation diagram* (Figure 3.10). The horizontal axis is the parameter of the problem a and the vertical one is the solution value. As is customary, stable solutions are denoted by solid curves and unstable ones by dashed curves. The arrows indicate the direction of solution variation with time and so the stable solution is seen to attract, while the unstable one repels.

The value $a = 0$ and the corresponding solution constitute a bifurcation point and in this example it is a *turning point*. The bifurcation is said to be of the *saddle-node* type (sometimes also called *tangent* bifurcation). It is characterised by the appearance of two equilibrium solutions, one stable and one unstable, out of a situation with no solutions (as in this example) or with a unique solution.

Pitchfork bifurcations

This type of bifurcation occurs when a solution changes its stability at a point, the change being accompanied by the appearance or disappearance of two additional solutions, both with the same stability properties. The name *pitchfork bifurcation* describes the typical shape of the bifurcation diagram (see below). This is well illustrated by the system

$$\frac{\mathrm{d}x}{\mathrm{d}t} = ax - bx^3 \tag{3.67}$$

with $b \neq 0$ fixed and a being a parameter.

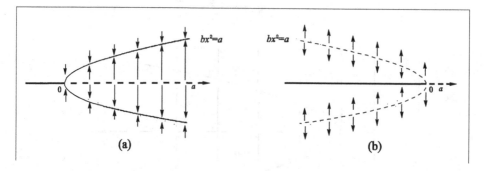

Fig. 3.11. Bifurcation diagram of a supercritical (a) subcritical (b) pitchfork bifurcation.

(i) First consider the case $b > 0$, or without loss of generality $b = 1$.

$X_0 = 0$ is a fixed point for all a, while for $a > 0$ there are two additional fixed points $X_{1,2} = \pm\sqrt{a}$. Linear stability analysis around a solution $x = X$ (X fixed) gives

$$\dot{\delta x} = a\delta x - 3bX^2\delta x \qquad (3.68)$$

Thus for $b = 1$ and $a < 0$, $X_0 = 0$, the only fixed point, is stable ($\dot{\delta x} \propto -\delta x$). If $a > 0$ this solution is unstable but the additional fixed points $X_{1,2} = \pm\sqrt{a}$ are stable since $\dot{\delta x} = -2a\delta x$ in this case. The null solution is marginally stable at $a = 0$ and it loses its stability when the value of a is *above* this critical value, concomitantly with the appearance of two additional stable solutions. This case, illustrated in the bifurcation diagram (Figure 3.11(a)), is called a *supercritical* pitchfork bifurcation.

(ii) If $b < 0$ (WLG $b = -1$) the situation is identical with regard to the null solution, as is apparent from 3.68 for $X = 0$. This time, however, the two additional solutions $X_{1,2} = \pm\sqrt{-a}$ exist only for $a < 0$ and are both unstable ($\dot{\delta x} = -2a\delta x$ with $a < 0$). The resulting bifurcation diagram reveals a *subcritical* pitchfork – Figure 3.11(b).

Pitchfork bifurcations occur often as a manifestation of symmetry breaking. Landau used an equation like (3.67) in his attempts at modelling fluid turbulence in 1944. This type of bifurcation occurs in most of the double-well-potential problems like the Duffing system (2.10).

Transcritical bifurcations

This type of bifurcation can be regarded as intermediate, in some sense, between the saddle-node and pitchfork bifurcations. We mention it briefly here for the sake of completeness. The characteristic feature of the transcritical bifurcation is the crossing of two equilibrium solutions (of opposite stability) at the critical value of the parameter, whereupon each of the solutions changes its stability. It can best be seen with the help of the equation

$$\frac{dx}{dt} = ax - bx^2 \tag{3.69}$$

where again $b \neq 0$ is fixed and a is a parameter. We encourage the reader to find the shape of a transcritical bifurcation diagram.

Hopf bifurcation

Until this point we have used one-dimensional dynamical systems to illustrate the different bifurcation types. The important case of a Hopf bifurcation, named after the mathematician who showed its generality, does not occur in one-dimensional systems. The simplest example will thus be two-dimensional.

Consider the system

$$\frac{dx}{dt} = -y + (a - x^2 - y^2)x$$
$$\frac{dy}{dt} = x + (a - x^2 - y^2)y \tag{3.70}$$

where a is a real parameter.

The only steady solution of this system is the origin, $x = y = 0$. Linearisation around this fixed point yields the following pair of linear equations

$$\dot{\delta x} = -\delta y + a\delta x$$
$$\dot{\delta y} = \delta x + a\delta y$$

which can also be written as $\dot{\delta \mathbf{x}} = \mathcal{J}\delta\mathbf{x}$, where the stability matrix $\mathcal{J}$ is

$$\mathcal{J} = \begin{pmatrix} a & -1 \\ 1 & a \end{pmatrix}$$

The eigenvalues of this matrix are $s_{1,2} = a \pm i$. The real part of both eigenvalues is a and thus the fixed point is stable for $a < 0$ and unstable for $a > 0$. No new fixed points appear at the critical point, $a = 0$, but as will soon become apparent a *periodic* solution (limit cycle) emerges when the critical point is crossed. The point $a = 0$ is therefore a bifurcation, when the definition of a bifurcation point given above is extended to also include periodic solutions. That a stable limit cycle is present for $a > 0$ is best seen when equations (3.70) are transformed into polar coordinates (r, θ) with $x = r\cos\theta$, $y = r\sin\theta$. Using complex notation $z \equiv x + iy$ we get, using (3.70),

$$\frac{dz}{dt} = \frac{dx}{dt} + i\frac{dy}{dt} = i(x + iy) + (a - x^2 - y^2)(x + iy) \tag{3.71}$$

Writing now $z = x + iy = re^{i\theta}$ and thus $\dot{z} = (\dot{r} + ir\dot{\theta})e^{i\theta}$ we get upon dividing by $e^{i\theta}$

$$\frac{dr}{dt} + ir\frac{d\theta}{dt} = (a - r^2)r + ir \tag{3.72}$$

The real and imaginary parts of this equation yield the equivalent system

$$\frac{dr}{dt} = r(a - r^2) \tag{3.73}$$

$$\frac{d\theta}{dt} = 1 \tag{3.74}$$

This system can be solved analytically and the exact solution, for the initial values (r_0, θ_0) at $t = 0$, if only $a \neq 0$ is

$$r^2(t) = \frac{ar_0^2}{r_0^2 + (a - r_0^2)e^{-2at}}; \quad \theta(t) = t + \theta_0 \tag{3.75}$$

Thus for $a < 0$, r decays exponentially to zero with θ growing linearly. The origin is thus a stable fixed point. This confirms the findings of the linear stability analysis performed above. For $a > 0$, we get $r \to \sqrt{a}$ exponentially as $t \to \infty$. A *stable limit cycle* is approached and has the form of a circle. For $a = 0$, the critical point, the exact solution is $r^2(t) = r_0^2(1 + 2r_0^2 t)^{-1}$ and thus the origin is stable. Phase portraits of the system for $a < 0$ and $a > 0$ are plotted schematically in Figure 3.12.

At $a = 0$ a bifurcation occurs, as a result of which a fixed point changes its stability and a limit cycle appears. Such a transition is called a *Hopf bifurcation*. It is possible to find it in any system in which self-excited oscillations occur. The bifurcation diagram for a Hopf bifurcation must be three-dimensional and it is customary to draw it in perspective as in Figure 3.13. The Hopf bifurcation depicted in this diagram is a supercritical one. Subcritical Hopf bifurcations are also possible, but we shall not discuss them here.

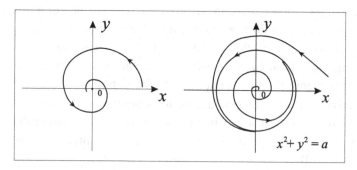

Fig. 3.12. Schematic phase portraits for the system (3.70). Some orbits are shown for $a < 0$ (left) and $a > 0$ (right).

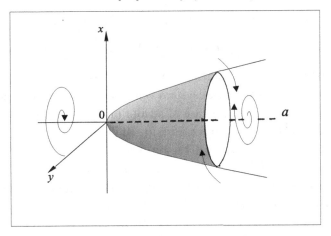

Fig. 3.13. Bifurcation diagram of the Hopf bifurcation for the system (3.70).

General results and remarks

In our discussion of bifurcations above we have exclusively used ODE systems (flows) in the examples. The bifurcation phenomenon occurs also, however, in maps, as we have already seen in the logistic equation, $x_{i+1} = 4rx_i(1 - x_i)$ (Chapter 2).

Before discussing the nature of the first two bifurcations of the logistic equation we define $a \equiv 4r$, for convenience. The map has two fixed points $x_1^* = 0$ and $x_2^* = 1 - a^{-1}$. As we have seen in Chapter 2, for $0 < a < 1$ the fixed point x_1^* is stable and the other one is unstable. For $1 < a < 3$ the situation is the opposite, with stability exchange of the two fixed points at the critical value of the parameter $a = 1$. Thus at $a = 1$ there is a transcritical bifurcation. At $a = 3$, the null fixed point continues to be unstable, x_2^* becomes unstable too and a stable 2-cycle appears. The appearance of a periodic solution in a one-dimensional map is not a Hopf bifurcation and is called, rather, a *flip* bifurcation. As the parameter is increased a series of period-doubling flip bifurcations occur, on the way to chaos. These first two bifurcations of the logistic map are not shown in the bifurcation diagram of the logistic map given in Figure 2.2, which concentrates on the period-doubling bifurcations. That figure starts only from $a = 3$ ($r = 0.75$), where the flip bifurcation occurs.

The essential property determining the bifurcation type in all the cases discussed above (in ODE and in maps) is connected to the nature of the stability matrix eigenvalues at the fixed points. These eigenvalues are in general complex and both their real and imaginary parts depend on the parameter. As the control parameter is varied the eigenvalues change and the stability of a particular mode may change.

In the case of flows we may write the eigenvalues in the usual way

$$s_k(a) = \sigma_k(a) + i\omega_k(a) \quad \text{for } k = 1, 2, \ldots$$

with the dependence on the parameter a explicitly marked. For the examples in the previous subsection the dynamical system was two-dimensional at most and thus $k \leq 2$. It is easy to see that in the case of a Hopf bifurcation σ_k changes its sign at the critical point, with $\omega_k \neq 0$. In contrast, in saddle-node, pitchfork and transcritical bifurcations, $\omega_k = 0$ always and the bifurcation occurs when σ_k changes sign (as in the Hopf case). The difference between the latter bifurcation types is in the *number* of steady solutions on both sides of the critical parameter value.

The behaviour and the local bifurcations of a nonlinear system in the vicinity of a fixed point are governed by the linearised system. Since many different nonlinear systems have very similar linearised systems, the dynamics near a fixed point falls into several generic categories, (e.g., of the type given above). This is the reason that archetypal simple equations have been defined for each type of bifurcation. These go under the name *normal forms*. Every system that undergoes a specific type of bifurcation behaves, in the vicinity of the bifurcation point, similarly to the normal form characteristic of the relevant bifurcation type. We list below the normal forms of the bifurcations described above. Often the normal form is identical with the equation we have used above as an example to demonstrate it. The normal form corresponding to the Hopf bifurcation is slightly more general than our example in the previous subsection. In the following list of normal forms we use a standard notation with the control parameter denoted by λ and all the other parameters being constant.

(i) *Saddle-node*:

$$\dot{x} = \lambda + \alpha x^2$$

(ii) *Transcritical*:

$$\dot{x} = \lambda x - \alpha x^2$$

(iii) *Pitchfork*:

$$\dot{x} = \lambda x + \alpha x^3$$

(iv) *Hopf*:
   (a) Cartesian form:

$$\dot{x} = \lambda x - \omega y + (\alpha x - \beta y)(x^2 + y^2)$$
$$\dot{y} = \omega x + \lambda y + (\beta x + \alpha y)(x^2 + y^2)$$

   (b) Polar form ($x \equiv r\cos\theta; \quad y \equiv r\sin\theta$):

$$\dot{r} = \lambda r + \alpha r^3$$
$$\dot{\theta} = \omega + \beta r^2$$

If the dynamical system is multidimensional there are more than just two linear modes and thus eigenvalues. It often happens, however, that at the critical value of the parameter the stability of only one or two modes changes. The bifurcation can thus be classified using the normal forms given above. To see this explicitly we assume that our system is n-dimensional and in some range of the parameter(s) all modes are (asymptotically) stable. Concentrating again on the case of a flow depending on just one parameter a, this means that the real part of all the eigenvalues is *negative*, that is, $\sigma_k(a) < 0$ for all $k = 1, \ldots, n$, in some range of a, say $a < a_c$. As the parameter is varied it can reach the value $a = a_c$, at which one or more modes become neutral and the rest are still stable. Thus the fixed point has stable and centre manifolds. There are theorems that under quite general conditions guarantee that, close enough to marginality, the solution of the dynamical system lives on the centre manifold (since the stable modes decay in time). These ideas will be mentioned again and explained in more detail in the next subsection, where they will be used in an effort to reduce the dimension of a multidimensional dynamical system and simplify it near marginality. The theorems mentioned above also have a very powerful role in singularity (or catastrophe) theory, which deals with bifurcations. Here we just illustrate the ideas mentioned above by a qualitative example of a multidimensional dynamical system, whose solution bifurcates as a parameter is varied. Consider a dynamical system describing a stellar model with k discrete mass shells. The state variables are the k shell positions, velocities and specific entropies, say. The system is thus $n = 3k$ dimensional. In practical applications a large number ($k \sim 100$) of mass shells is used in a typical numerical calculation. An equilibrium stellar model is a fixed point of the multidimensional dynamical system. Assume that this model is asymptotically stable. This means that all the eigenvalues of the stability matrix have negative (and nonzero) real parts.

If one considers a one-parameter family of such models, the parameter being the effective temperature, for example, it is possible that when the parameter approaches a critical value the real parts of some eigenvalues become marginal. We assume that the eigenvalues are discrete and nondegenerate (as is usually the case) and examine the situation at marginality. If one real eigenvalue is zero (and all the others have negative real parts) the system undergoes a bifurcation of either a pitchfork or a transcritical type. This type of instability is usually called secular or thermal instability in astrophysics. The type of bifurcation depends on the availability of other equilibrium solutions and their stability. For example, if the stellar model joins some other branch of stable models without any 'jumps' the bifurcation is transcritical. If no other neighbouring stable solutions exist, it may be a subcritical pitchfork bifurcation. If two stable solutions are possible after the parameter passes its critical value, the bifurcation is of a supercritical pitchfork type. If two complex conjugate eigenvalues become critical first (these go in pairs) the bifurcation is of the Hopf type. The star becomes linearly pulsationally unstable and a periodic solution appears. The solution close to the bifurcation point lives in a two-dimensional space (the centre manifold), in which the limit cycle resides.

If several eigenvalues become marginal together the situation is more complicated but still the centre manifold has a relatively low dimension, compared to the original system. The technique of reducing the original system close to marginality to a low-dimensional one is called *dimensional reduction* and will be discussed in the next section and used also in Chapter 7 for the derivation of amplitude equations. It will be applied to stellar pulsation in the second part of the book, where we shall see that it greatly simplifies the analysis of a multidimensional physical system.

We conclude this section by remarking that we have discussed here only the simplest types of bifurcations, occurring when one parameter is varied (*codimension one* bifurcations). More complicated cases may arise when one varies several parameters. The simplest types discussed above have however helped us in introducing several basic but important concepts of bifurcation theory. These types of bifurcations occur in a wide range of problems, described by multidimensional ODE (or even infinitely dimensional PDE). It turns out that essential properties of more complicated bifurcations are similar to the simple ones described here.

3.4 Dynamical systems near marginality

The concept of a marginal (with respect to linear stability) state of a dynamical system has already been mentioned several times in this chapter. We are now sufficiently acquainted with basic mathematical concepts and techniques of dynamical systems theory to be able to demonstrate the importance and usefulness of marginal states.

Low-dimensional dynamics

Assume that only a small number, k say, of modes (of the linearised system around a fixed point) become marginal at the critical parameter value and that the other modes are 'very' stable. The locations of the stability matrix eigenvalues in the complex plane in such a marginal situation are sketched in Figure 3.14.

At the critical value of the parameter the linear centre invariant subspace is thus k-dimensional and a centre manifold, for the nonlinear system, exists (at least in the neighourhood of the fixed point) and is k-dimensional as well. In this neighbourhood of the fixed point the situation is only weakly nonlinear and the centre manifolds of the original and linearised systems are close to each other.

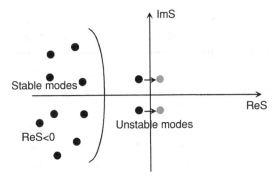

Fig. 3.14. Stability matrix eigenvalues as a pair of complex conjugate eigenvalues is close to marginality, while all the other eigenvalues are 'very' stable.

If we consider the dynamics close to the fixed point, only the neutral modes count, as the other ones are assumed to have sufficiently large negative real parts and thus decay fast. The fast decaying modes are called *slave modes* and as a result of their fast decay they can be eliminated. This means that the motion is attracted to and proceeds effectively on the centre manifold. Since we have assumed that k is small, the dynamics in the vicinity of the fixed point is *low-dimensional*.

If the centre manifold can be found (at least close to the fixed point) the system's dimension can easily be reduced (to the centre-manifold dimension) by projecting it onto the centre manifold. Moreover, it turns out that in many cases there is no need to construct the centre manifold explicitly, and it is still possible to reduce the dimension of the system by perturbation analysis. All such procedures are referred to as *dimensional reduction*.

To be explicit, we consider a somewhat simplified case where the n-dimensional dynamical system, given by

$$\dot{\mathbf{x}} = \mathbf{F}(\mathbf{x}; a) \tag{3.76}$$

has the following properties:

(i) it depends on only one parameter a,
(ii) $\mathbf{F}(\mathbf{0}; a) = \mathbf{0}$, that is, the origin is a fixed point for every a,
(iii) at marginality ($a = a_c$) one real eigenvalue or two complex conjugate ones of the stability matrix cross the imaginary axis (have zero real part).

Our goal is to significantly simplify the original multidimensional dynamical system for a close to a_c, by reducing its dimension and effectively replacing it by a small number of simple ODEs. Several methods for such dimensional reduction and simplification exist. General and direct methods, like the Liapunov–Schmidt method or centre-manifold reduction are usually quite technical and we shall not bring them here in their full generality.

Below we shall give only one specific example of centre-manifold reduction. The interested reader is referred to the book by Guckenheimer and Holmes (1983) for a full discussion. In the direct methods one actually constructs the centre manifold in the vicinity of the fixed point and the flow on this manifold is found. It is then generally possible to further simplify the system (by 'coordinate' transformations) and bring it to a normal form, of the type introduced in the section on bifurcation theory. It is not surprising that the problem is related to bifurcation theory since the existence of a centre manifold relies on some of the modes being neutral. This means that at this value of the parameter a the fixed point changes its stability properties. It is thus a bifurcation point with generic properties as we have explicitly seen in the previous section. If, for example, two complex conjugate eigenvalues have zero real part for some value a_c of the parameter a, and this real part is negative (positive) for $a < a_c$ ($>a_c$), a Hopf bifurcation occurs at $a = a_c$ and it is reasonable that close to the bifurcation point the system will be faithfully represented by the normal form corresponding to this bifurcation.

If it is too difficult to find the centre manifold or we are not interested in constructing it, it is still possible to perform a dimensional reduction using perturbative techniques, like the method of multiple scales, for example. This technique has already been used in Section 3.1 to calculate the limit cycle of the Van der Pol oscillator. As the original system in that case was two-dimensional the dimension of the system was not reduced. The method of multiple scales can be applied to multidimensional systems as well, with the hope of attaining dimensional reduction. It seems to be the simplest such method, as it allows us to reduce the dimension of the dynamical system without the need for finding the centre manifold itself. We shall demonstrate below the use of the method of multiple scales in reducing the dimension of a general dynamical system near a Hopf bifurcation. The Van der Pol limit cycle calculation of Section 3.1 will be seen to be a particular case of this procedure.

Centre-manifold reduction

We consider, as an example of the centre-manifold reduction technique, the famous Lorentz system (Lorentz, 1963). This system arose from the study of thermal convection in the atmosphere and has become one of the most widely studied dynamical systems. The Moore and Spiegel oscillator (one of our examples in Chapter 2), which was independently proposed several years later to model a similar physical phenomenon in an astrophysical context, did not enjoy similar attention but was later shown to belong (mathematically) to the same family of systems.

We shall derive the Lorentz model for a well-defined physical system and discuss it in detail in the second part of the book. Here we shall just use the Lorentz equations as a mathematical system and furthermore simplify them by fixing two parameters (the original model had three independent parameters) and consider them as a third-order dynamical system dependent on one control parameter r,

$$\dot{x} = y - x$$
$$\dot{y} = rx - y - xz$$
$$\dot{z} = -z + xy \qquad (3.77)$$

It is not difficult to see that the origin is a fixed point of this system and that it is asymptotically stable for $r < 1$ and unstable for $r > 1$. The stability matrix is

$$\mathcal{J} = \begin{pmatrix} -1 & 1 & 0 \\ r & -1 & 0 \\ 0 & 0 & -1 \end{pmatrix}$$

whose eigenvalues are $s_0 = -1 + \sqrt{r}$, $s_1 = -1 - \sqrt{r}$, $s_2 = -1$. We are interested in the behaviour of the system near marginality i.e., close to $r = 1$, so we do not consider negative values of r. It is convenient to redefine the control parameter as $\rho \equiv +\sqrt{r}$. At $\rho = 1$ ($r = 1$) the system undergoes a bifurcation since the fixed point becomes unstable. The eigenvalue s_0 is marginal and the other two are negative. The origin becomes a saddle point for $\rho > 1$. The eigenvectors corresponding to the eigenvalues are

$$\mathbf{e}_0 = (1, \rho, 0)^T, \quad \mathbf{e}_1 = (1, -\rho, 1)^T, \quad \mathbf{e}_2 = (0, 0, 1)^T$$

For $\rho < 1$ the linear stable subspace consists of all space. For $\rho > 1$ there exists a linear unstable subspace, in the direction of $\mathbf{e}_0$, the eigenvector of the positive eigenvalue. The stable subspace is spanned by $\mathbf{e}_1$ and $\mathbf{e}_2$.

At the critical value $\rho = 1$ there is a centre subspace, spanned by the eigenvector $\mathbf{e}_0 = (1, 1, 0)^T$, corresponding to the null eigenvalue $s_0 = 0$. The other two eigenvectors span the stable subspace. The centre manifold also exists (at least locally) and is tangent at the origin to the linear centre subspace, which consists of the line in the direction of $\mathbf{e}_0$ ($y = x$, $z = 0$).

In order to simplify the construction of the centre manifold near the origin, and the form of the reduced system, it is convenient to define new coordinates in the $x - y$ plane. The new orthogonal coordinates, (u, v) say, should be normal coordinates for the linearised system, that is, the linearised system of (3.77) should be diagonal when transformed to them. This is done using standard methods of linear algebra. The eigenvectors are used to construct the transformation matrix and for any ρ the transformation is

$$x = u + v; \quad y = \rho(u - v)$$

The system in these coordinates (z is a normal coordinate to start with) is

$$\dot{u} = s_0 u - \frac{1}{\rho}(u + v)z$$

$$\dot{v} = s_1 v + \frac{1}{\rho}(u + v)z$$

$$\dot{z} = -z + \rho(u^2 - v^2) \qquad (3.78)$$

where s_0 and s_1 are the above mentioned eigenvalues and the third eigenvalue has been explicitly substituted. The diagonality of the linearised system of (3.78) is apparent.

We know that the centre manifold exists near the origin and that it is tangent to the linear centre subspace at the origin itself. The centre subspace at criticality can be expressed in the transformed coordinates as $v = 0$, $z = 0$. It is thus simply the u-axis. A surface tangent at the origin to the u-axis can be faithfully represented, close to the origin, by polynomials in u, starting from the quadratic term. Thus we write

$$v = P_v(u) = q_v u^2 + c_v u^3 + \cdots$$

$$z = P_z(u) = q_z u^2 + c_z u^3 + \cdots \qquad (3.79)$$

where q and c are suitable constants. We keep terms up to third order which is equivalent to assuming that the nonlinearity is cubic at most.

Now we project the system (3.78) on to the centre manifold given by (3.79). This is done by taking the time derivative of equations (3.79) and substituting for $\dot{u}$ from the first equation in (3.78) (using 3.79 again to express v and z as functions of u). This gives

$$\dot{v} = (2q_v u + 3c_v u^2)\left\{ s_0 u - \frac{1}{\rho}[u + P_v(u)]P_z(u) \right\}$$

$$\dot{z} = (2q_z u + 3c_z u^2)\left\{ s_0 u - \frac{1}{\rho}[u + P_v(u)]P_z(u) \right\}$$

These expressions must be equal to $\dot{v}$ and $\dot{z}$ coming directly from the last two equations in (3.78) with v and z on the right-hand side expressed by (3.79). All the expressions are polynomials in u and we can thus compare both sides in the two equations term by term. This gives the constants

$$c_z = 0, \quad q_z = \frac{\rho}{2\rho - 1}, \quad c_v = -\frac{3\rho - 2}{2\rho - 1}$$

The reduced system near marginality ($\rho \approx 1$) and close to the origin ($|u|$ small) follows from the first equation of the system (3.78) upon using (3.79) up to third order in u

$$\dot{u} = \lambda u - \frac{1}{1 + 2\lambda} u^3 \tag{3.80}$$

where the control parameter has been redefined again $\lambda \equiv \rho - 1 = \sqrt{r} - 1$.

As we have seen in the previous section this is a normal form for a pitchfork bifurcation. Indeed, the Lorentz system undergoes a pitchfork bifurcation at the origin for $r = 1$. With the help of centre manifold reduction we were able to identify the essence of the two-dimensional dynamics near the fixed point when the system is close to marginality. The procedure applied above is an example of a very general technique, which can be applied for other cases as well.

Perturbative dimensional reduction

We turn now to an example of perturbative dimensional reduction of a rather general system near a Hopf bifurcation, using the method of multiple scales. We shall not construct any approximation to the centre manifold itself, but rather use a bona fide perturbative approach to find the solution near the bifurcation and close to a fixed point. We consider a general system defined in (3.76) with the assumptions listed after the equation, where we pick the case of a pair of complex eigenvalues near marginality.

We first identify the small parameters in the problem. We are interested in solutions that are close in phase space to the fixed point (the origin), where the centre manifold of the nonlinear system is close to the linear centre subspace. To express the fact that we are looking for solutions very close to the origin, which is a fixed point for any a, we write

$$\mathbf{x}(t) = \epsilon \mathbf{y}(t)$$

where ϵ is small and is actually used as a bookkeeping device, to remind us of the order of various terms. Expanding the system (3.76) around the origin we get

$$\dot{\mathbf{y}} = \mathcal{J}\mathbf{y} + \epsilon \mathcal{Q}(\mathbf{y}, \mathbf{y}) + \epsilon^2 \mathcal{C}(\mathbf{y}, \mathbf{y}, \mathbf{y}) + \cdots \tag{3.81}$$

where $\mathcal{Q}$ and $\mathcal{C}$ symbolise vector valued bilinear and trilinear forms generated by the appropriate partial derivatives in the Taylor expansion. They and the Jacobian matrix $\mathcal{J}$ are evaluated at the origin. The terms depend also on the parameter a, but for the sake of simplicity we do not write it explicitly. We keep only the quadratic and cubic nonlinearity and neglect higher-order terms. The above expansion is not a linearisation but rather can be perceived as a systematic 'nonlinearisation', that is, the identification of quadratic, cubic etc. nonlinearities term by term.

We now Taylor expand each term on the right-hand side of (3.81) with respect to the parameter around its critical value, a_c. To decide on the order of this expansion we reflect the closeness of a to the critical point by writing

$$a - a_c = \epsilon^2 \lambda$$

where λ is the shifted (and stretched) new parameter. This particular ordering (ϵ^2) is convenient, but we must remember that at this stage it is arbitrary. It means that we are, in some sense, closer to marginality than to the origin, and so the influence of the control parameter will be realised at the same order as the cubic (and not quadratic) nonlinearity. This gives

$$\dot{\mathbf{y}} = \mathcal{J}\mathbf{y} + \epsilon \mathcal{Q}(\mathbf{y}, \mathbf{y}) + \epsilon^2 \mathcal{C}(\mathbf{y}, \mathbf{y}, \mathbf{y}) + \epsilon^2 \lambda \mathcal{P}\mathbf{y} + \cdots$$

where $\mathcal{P} \equiv \partial_a \mathcal{J}$ is a constant matrix and all matrices and forms are evaluated at the bifurcation point $\mathbf{y} = \mathbf{0}, \lambda = 0$.

Turning now to the method of multiple scales we choose two time scales $t_0 = t$ and $t_2 = \epsilon^2 t$, which will be treated as independent variables. It will be apparent soon that this is consistent with the ordering of the control parameter (ϵ^2). In any case, perturbation methods are judged by their result and sometimes the choice of the correct ordering is more of an art than an exact science. Using the notation $D_k \equiv \partial/\partial t_k$ for the partial derivatives we write

$$\frac{\mathrm{d}}{\mathrm{d}t} = D_0 + \epsilon^2 D_2$$

and seek an expansion of the form

$$\mathbf{y}(t) = \mathbf{y}_0(t_0, t_2) + \epsilon \mathbf{y}_1(t_0, t_2) + \epsilon^2 \mathbf{y}_2(t_0, t_2) + \cdots \tag{3.82}$$

Substituting these expansions and collecting terms multiplying the same powers of ϵ we get

$$D_0 \mathbf{y}_0 - \mathcal{J}\mathbf{y}_0 = 0 \tag{3.83}$$

$$D_0 \mathbf{y}_1 - \mathcal{J}\mathbf{y}_1 = \mathcal{Q}(\mathbf{y}_0, \mathbf{y}_0), \tag{3.84}$$

$$D_0 \mathbf{y}_2 - \mathcal{J}\mathbf{y}_2 = -D_2 \mathbf{y}_0 + \lambda \mathcal{P}\mathbf{y}_0 + 2\mathcal{Q}(\mathbf{y}_0, \mathbf{y}_1) + \mathcal{C}(\mathbf{y}_0, \mathbf{y}_0, \mathbf{y}_0) \tag{3.85}$$

Up to this point the procedure has been quite general. We now specify that the system undergoes a Hopf bifurcation at the origin for $\lambda = 0$. This means that $\mathcal{J}$ has a pair of purely imaginary eigenvalues, which we denote by $s_{1,2} = \pm i\omega$. All the other modes are stable and their eigenvalues are well to the left of the imaginary axis. This means that it is possible to write the solution of the lowest order homogeneous equation (3.83), neglecting all the decaying modes, as

$$\mathbf{y}_0(t, \tau) = A(\tau)\mathbf{e}_r e^{i\omega t} + \text{cc} \tag{3.86}$$

where we use a more convenient notation for the two time scales, ($t = t_0, \tau = t_2$), $A(\tau)$ is a slowly varying (in general complex) amplitude and 'cc' means complex conjugate. As the matrix $\mathcal{J}$ is not self-adjoint in general, $\mathbf{e}_r$ is the *right* eigenvector corresponding to the eigenvalue $+i\omega$, that is

$$\mathcal{J}\mathbf{e}_r = i\omega\mathbf{e}_r$$

Equation (3.86) separates the slow evolution from the fast one and furnishes an approximate (to order ϵ) solution to the original problem, provided $A(\tau)$ is known. The amplitude can be found from the higher-order equations (3.84) and (3.85). We first substitute the solution (3.86) in the expression on the right-hand side of (3.84) and using the fact that $\mathcal{Q}$ is bilinear, this gives

$$D_0\mathbf{y}_1 - \mathcal{J}\mathbf{y}_1 = \mathcal{Q}(\mathbf{e}_r, \mathbf{e}_r^*)|A|^2 + \mathcal{Q}(\mathbf{e}_r, \mathbf{e}_r)A^2 e^{2i\omega t} + \text{cc}$$

The right-hand side of this inhomogeneous equation does not include any resonant driving terms ($e^{\pm i\omega t}$) and thus the equation is solvable in terms of bounded functions. The solution is

$$\mathbf{y}_1(t, \tau) = \mathbf{u}_0|A|^2 + \mathbf{u}_2 A^2 e^{2i\omega t} + \text{cc} \tag{3.87}$$

where the constant vectors $\mathbf{u}_0$ and $\mathbf{u}_2$ are the solutions of the algebraic matrix equations

$$\mathcal{J}\mathbf{u}_0 = -\mathcal{Q}(\mathbf{e}_r, \mathbf{e}_r^*)$$

$$(2i\omega - \mathcal{J})\mathbf{u}_2 = \mathcal{Q}(\mathbf{e}_r, \mathbf{e}_r)$$

In (3.87) only a particular solution of (3.84) is included. The general solution is of the same form as (3.86) and can be absorbed in that (lowest-order) solution.

So far there is no information regarding the amplitude of the lowest-order solution. This is, however, supplied by the equation of the next order (3.85). When we substitute into this the solutions for $\mathbf{y}_0$ and $\mathbf{y}_1$, using (3.86) and (3.87), the right-hand side will contain resonant driving terms. Solvability conditions have thus to be imposed to eliminate these terms, so that the solution for $\mathbf{y}_2$ does not contain secular terms, which break the validity of the original expansion (3.82). As equation (3.85) will only be used to eliminate the resonant terms we write only these and denote all the other terms by 'nst' (not producing secular terms). The result is

$$D_0\mathbf{y}_2 - \mathcal{J}\mathbf{y}_2 = -\mathbf{e}_r\frac{dA}{d\tau}e^{i\omega t} + \mathcal{P}\mathbf{e}_r\,\lambda e^{i\omega t} + [4\mathcal{Q}(\mathbf{e}_r, \mathbf{u}_0) + 2\mathcal{Q}(\mathbf{e}_r^*, \mathbf{u}_2)$$

$$+ 3\mathcal{C}(\mathbf{e}_r, \mathbf{e}_r, \mathbf{e}_r^*)]A|A|^2 \times e^{i\omega t} + \text{cc} + \text{nst}$$

The conditions for the elimination of secular terms, i.e., the solvability conditions, are in this case more involved than the ones employed in the simple case of the Van der Pol oscillator (Section 3.1). There we had scalar equations and thus the requirement for the vanishing of the terms producing secular behaviour.

Here we are dealing with vector functions satisfying linear differential equations and so the solvability conditions are more general and are based on the notion of the *Fredholm alternative* property of linear operators (see Friedman, 1990 for an excellent account on this topic as well as the theory of linear operators in general). This means that the elimination of the secular terms is guaranteed if the term multiplying $e^{i\omega t}$ is orthogonal to the null-space of the adjoint homogeneous problem.

Let $\mathbf{e}_l$ be the *left* eigenvector of $\mathcal{J}$ corresponding to the same eigenvalue $+i\omega$, that is

$$\mathbf{e}_l^T \mathcal{J} = i\omega \mathcal{J}$$

We assume that the corresponding left and right eigenvectors are normalised so that $\mathbf{e}_l^T \cdot \mathbf{e}_r = 1$. Multiplying (scalar product) from the left the term with $e^{i\omega t}$ by $\mathbf{e}_l^T$ and equating the result to zero, the following sufficient condition for the elimination of secular terms is found

$$\frac{dA}{d\tau} = \lambda p A + q A |A|^2 \tag{3.88}$$

where the complex constants p and q are given by

$$p = \mathbf{e}_l^T \mathcal{P} \mathbf{e}_r$$

and

$$q = \mathbf{e}_l^T [4\mathcal{Q}(\mathbf{e}_r, \mathbf{u}_0) + 2\mathcal{Q}(\mathbf{e}_r^*, \mathbf{u}_2) + 3\mathcal{C}(\mathbf{e}_r, \mathbf{e}_r, \mathbf{e}_r^*)]$$

Equation (3.88) provides the desired information on the slow amplitude variation and is an example of what we shall later call *amplitude equations*. After it is solved, $A(\tau)$ can be substituted in the lowest-order solution (3.86).

It is instructive to express (3.88) in its real form. Letting $A = re^{i\theta}$ and writing

$$p = p_R + i p_I, \qquad q = q_R + i q_I,$$

the real and imaginary parts of (3.88) give

$$\frac{dr}{d\tau} = \lambda p_R r + q_R r^3$$

$$\frac{d\theta}{d\tau} = \lambda p_I + q_I r^2 \tag{3.89}$$

It is not surprising that (3.89) is in the form of a Hopf bifurcation. Interestingly, though, we see that the perturbative approach gives this reduced (to just two out of n dimensions) system in the slow time variable. The method of multiple scales allows us to separate out the fast variation and captures the slow evolution near marginality and close to the bifurcating fixed point. It faithfully represents the slow dynamics on the centre manifold even if the structure of the manifold is not known.

3.5 Fractal sets and dimensions

> There is geometry in the humming of the strings.
> There is music in the spacings of the spheres.
> Pythagoras, quoted in Aristotle's *Methaphysics*.

Chaotic behaviour is intimately connected with fractals. As we have seen in examples in the first chapter, complexity of nonlinear oscillations is characterised by the attraction of the dynamical system flow in phase space to an object of a non-integral geometrical dimension, a *fractal*. Conservative dynamical systems may also yield fractal structures, appearing, for example, in their Poincaré sections. Fractals appear not only in flows, but in chaotic maps as well. It is thus important to consider the mathematical notion of a fractal set and to define precisely what is meant by its non-integral dimension.

Again, it is not our purpose here to be absolutely mathematically rigorous. We shall try, however, to be as clear and precise as possible. It is intuitively clear that a fractal set should be a geometrical object exhibiting some sort of self similarity. That is, the structure should look more or less the same on all scales, so that it is impossible to say what is the degree of zooming in, when looking at it.

A classical practical example in this context could be the western coastline of Britain. The coastline looks similar when viewed at different magnifications. Repetitive zooming in reveals finer and finer details and the coastline looks rugged on all scales. The zooming procedure in this example cannot be continued indefinitely, of course. It is thus only an approximation to a true mathematical fractal object.

The basic concept of topological *dimension*, d, of a geometrical object is usually approached in a rather intuitive way. In a three-dimensional Euclidean space R^3, for example, a finite set of points has $d = 0$. A curve in such a space is one-dimensional ($d = 1$), the interior of any closed curve in a plane has $d = 2$, and the interior of a closed surface has $d = 3$, like the space itself. The idea of topological dimension goes back to Euclid and Descartes and it was formalised in 1905 by Poincaré in the following way. Points are *defined* as having zero dimension. Now, any object that can be divided into two separate parts by cuts of dimension $d = n$, has the dimension $d = n + 1$. This definition uses mathematical induction and it can easily be applied to the examples above.

A non-integral dimension, the defining property of fractals, cannot be obtained from the above reasoning. Hausdorff, one of the founders of modern topology, defined in 1919 what is now called the *Hausdorff dimension*, D_H. The exact definition of D_H requires some mathematical notions, which are beyond the scope of this book. We shall thus define here the *box-counting* (also called *capacity*) dimension. This dimension, which we shall denote simply as D, is related to the Hausdorff dimension and is equal to it in all the cases we shall be interested in.

Consider a geometrical object – a well defined set of points, E, embedded in a 'normal', three-dimensional Euclidean space. We would like to completely cover the set E with a collection of cubes of edge length ϵ. Let $N(\epsilon)$ be the *minimal* number of such cubes needed for the above purpose. For example, if E is a line of length l, say, one needs at least $N(\epsilon) = l/\epsilon$ cubes. If E is an object of volume V, one can easily deduce that $N(\epsilon) \propto V/\epsilon^3$, with the constant of proportionality depending on the detailed shape of E. Clearly, for any set E as $\epsilon \to 0$, $N(\epsilon)$ scales as ϵ^{-D} for some fixed D. This D is defined as the box-counting dimension of the set E. This definition carries over to any dimension of the embedding space. Since from the scaling it follows that $\ln N(\epsilon) = \text{const} + D \ln(1/\epsilon)$, we can formally define

$$D = \lim_{\epsilon \to 0} \frac{\ln N(\epsilon)}{\ln(1/\epsilon)} \qquad (3.90)$$

It can be shown that for objects of a definite integral topological dimension d, the box-counting dimension D coincides with d. For example, if the set E contains just two distinct points, it is completely covered by just two cubes. As the value of the cubes' edge decreases $N(\epsilon) = 2$ remains fixed. Hence

$$D = \lim_{\epsilon \to 0} \frac{\ln 2}{\ln(1/\epsilon)} = 0$$

This result also carries over to the case of any finite number of points. Thus we have in this case, as well as for any integral dimension object, $D = d$.

For fractal sets, however, d cannot be figured out, while D is computable, at least in principle, and has a well defined meaning. Perhaps the most elementary example of a fractal set is the *Cantor middle-thirds set* (called also 'Cantor's dust'), proposed by the great mathematician Cantor, one of the founders of set theory, in 1883. Consider the unit interval $K_0 = [0, 1]$. Now remove the middle-third of this interval to obtain the two intervals $[0, 1/3]$ and $[2/3, 1]$. We shall name the set of points composed of the *union* of these intervals K_1. The next step will be to remove the middle thirds of the two intervals in K_1 to obtain four intervals of length $1/9$ each, the union of which is called K_2. The repetition of this procedure (illustrated in Figure 3.15 for the first few steps) indefinitely, yields as its limit the desired Cantor's middle-thirds set K_∞.

Clearly, K_∞ contains an infinite number of intervals, each of length zero. In an effort to compute D for this set we notice that the limiting procedure of equation

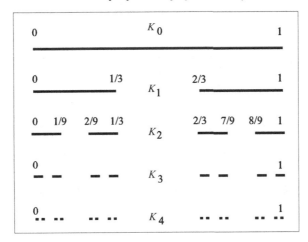

Fig. 3.15. Construction of Cantor's middle-thirds set.

(3.90) is naturally provided by the construction procedure outlined above. At the nth step we have K_n, which is composed of 2^n intervals each of size $1/3^n$. Thus, to cover K_n, we need $N(n) = 2^n$ cubes, each of size $\epsilon(n) = 1/3^n$. As $n \to \infty$, $\epsilon \to 0$ and the box-counting dimension of K_∞ is

$$D = \lim_{n \to \infty} \frac{\ln(2^n)}{\ln(3^n)} = \frac{\ln 2}{\ln 3} = 0.630\,93\ldots$$

Thus we see that Cantor's dust is something between a finite collection of points ($D = 0$) and a continuous line segment ($D = 1$). Its box-counting (and Hausdorff) dimension is non-integral $0 < D < 1$. The set K_∞ is a fractal set. This set lives, or can be *embedded* in a one-dimensional Euclidean space (a line) whose dimension is $d = 1$, but $D < d$ and is fractional. It is rather difficult to think of a physical example, to which Cantor's dust can be applied directly. It is described here because of its simplicity and its historical importance.

We shall now give one more example – a fractal which is qualitatively similar, at least in principle, to the above mentioned issue of the nature of the coastline of Britain (or maybe boundaries of clouds). This is the *Koch curve*, proposed by von Koch in 1904. The construction of this set begins, as in the previous example, with the unit segment being divided into three equal parts. The middle segment is removed again, but here it is replaced by *two* such equal segments arranged above the missing segment as two sides of an equilateral triangle (see Figure 3.16). The procedure is now repeated for each of the four segments resulting from the first step, yielding in the second step a curve composed of 16 segments of length $1/9$ each. Calling the successive curves S_n we find that S_n contains 4^n segments of length $1/3^n$. Koch's curve, also called Koch's snowflake, is S_∞. Its box-counting

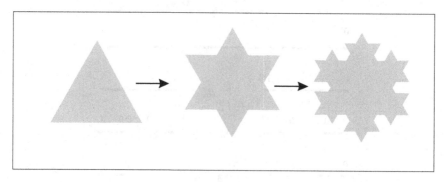

Fig. 3.16. Construction of the Koch snowflake, starting from three equal segments arranged in a triangle.

dimension is evaluated similarly to that of the Cantor dust thus

$$D = \lim_{n \to \infty} \frac{\ln(4^n)}{\ln(3^n)} = \frac{2\ln 2}{\ln 3} = 1.261\,86\ldots$$

The fractal dimension of Koch's curve (incidentally twice the dimension of Cantor's middle thirds set) is thus between 1 and 2. It can be regarded as something 'more' than a curve, but 'less' than an area filling set. Clearly, it looks the same on all scales. The curve is continuous but it can be said that it is 'trying', in some sense, to cover an area. In this way it is similar to the coastline example, but, in the latter case one cannot really go to the limit of $\epsilon \to 0$ or $n \to \infty$. At a certain stage the size of individual sand grains will be reached and the procedure will lose its meaning. 'Physical' fractals retain their self-similarity only within a finite range of orders of magnitude (albeit for at least several decades). Note that the length, L_∞, of Koch's curve is infinite, despite being restricted by its two boundary points, a unit length apart. Indeed

$$L_\infty = \lim_{n \to \infty} L_n = \lim_{n \to \infty} \left(4^n \frac{1}{3^n}\right) = \infty$$

The coast of Britain, in contrast, is obviously finite. This is also a result of the above-mentioned difference between a physical entity and its mathematical, true fractal, idealised model.

In many cases the computation of the box-counting dimension is not as easy as in the above two examples. In typical applications to dynamical systems, the relevant fractal set (a strange attractor, for example) is known only numerically. The dynamical system is integrated on a computer and the long time trajectory in phase space is sampled at discrete time intervals. Straightforward numerical algorithms devised to compute D may prove to be very inefficient. In most applications of fractals to nonlinear dynamical systems the fractal set is generated in some way by the dynamical system. The box-counting dimension concept does not utilise any information about the dynamical system generating the fractal. It is a purely geometrical concept and the attempts to count cubes covering the set may be rather wasteful. A strange attractor typically occupies some small region of phase space and most of the covering cubes will thus be empty.

We shall discuss shortly some alternative definitions of fractal dimension, devised to overcome the practical shortcomings of the calculation of the box-counting (capacity) dimension D in numerically studied dynamical systems. However, before addressing these practical issues we would like to give a general definition of (generalised) dimensions of order q (the box-counting dimension is the simplest member of this hierarchy, see below).

Returning to the discussion given just before the definition (3.90) we define the *generalised* dimension (of order q) of the set E to be

$$D^{(q)} = \frac{1}{q-1} \lim_{\epsilon \to 0} \frac{\ln \sum_{i=1}^{N(\epsilon)} [P_i(\epsilon)]^q}{\ln \epsilon} \tag{3.91}$$

where P_i is the probability that a point of the set falls in the ith covering cube, the total (minimal) number of which we have denoted as $N(\epsilon)$. It is not difficult to understand that $D^{(q)}$ defined in this way refers to the qth moment of the distribution of points in the set and for $q = 0$ we recover the box-counting dimension (3.90), that is, $D = D^{(0)}$. If the distribution of points (or more generally the *measure*, see below) is uniform, there is equal probability of finding a point in each of the $N(\epsilon)$ covering elements (cubes) and thus $P_i = 1/N(\epsilon)$. Substituting this result into (3.91) we get $D^{(q)} = D^{(0)} = D$ for any q. The higher-order dimensions reflect the nonuniformities in the distribution of points in the set and it can be shown that in general $D^{(q)} \geq D^{(q')}$ for $q \leq q'$.

In the definition (3.91) we have used a probability defined with the help of the actual density distribution of points in the set E. More generally, one can associate any invariant *measure* to each of the covering elements and define $D^{(q)}$ accordingly. For more complete discussions of this subject we refer the reader to the book based on the lectures of Ruelle (1989), and to Ott (1993).

We now address the problem of the practical computation of $D^{(q)}$ and limit ourselves to the first few moments only, since they are the most important in practice. In Chapter 5 we shall discuss the relevant methods used in the analysis of time series. The problem of finding the fractal dimension of attractors from discrete data sets will be illustrated there, but we would like to discuss one of its important aspects here at the outset.

We have seen that fractal dimension is defined as a rule through a limiting process, but it obviously must be actually computed using *finite* sets of points. The natural question that arises then is how many data points are required for a reliable determination of a fractal dimension? There is no definite answer to this question, but various estimates have been proposed in the literature. Most of these estimates can be incorporated in a statement that the number of points (we shall denote it as n_E) should not be smaller than Q^D, where D is the fractal dimension to be estimated and Q is a fixed number. There is no general agreement on what the exact value of Q should be and different authors estimate it to be between ~ 4 and ~ 40. On the other hand, there are estimates that there is no advantage in increasing n_E beyond a value for which the quantity $L/n_E^{1/D}$ (where L is some characteristic length in the set of points) is less than a characteristic length of the inherent inaccuracy ('noise') in the point positions. We shall discuss some practical aspects of this issue in Chapter 5 and in the second part of this book. In any case, a rough criterion of the type $n_E \sim 10^D$ probably provides a useful lower bound.

Assume that we know the positions of a sufficient number of points in the set, whose box counting dimension we wish to find. This may be the case when the set is created by a trajectory of a dynamical system and is sampled in a d-dimensional phase space yielding the (large) set E of n_E points. Let $\mathbf{x}_i$ be one of these points. We now surround this point with a sphere of radius ϵ and count the number of other points of E in this sphere. If the number of such points is $n(\epsilon, \mathbf{x}_i)$ the probability of finding any point of E in the sphere is $P(\epsilon, \mathbf{x}_i) = n(\epsilon, \mathbf{x}_i)/n_E$. The *pointwise dimension* of E is then defined as

$$D_P(\mathbf{x}_i) = \lim_{\epsilon \to 0} \frac{\ln P(\epsilon, \mathbf{x}_i)}{\ln \epsilon}$$

Since the above definition may depend on the particular point chosen, $\mathbf{x}_i$, this local pointwise dimension should be averaged over a large number of points, yielding the point-independent *average pointwise dimension*, D_P. In a practical calculation one cannot approach sufficiently closely the limit $\epsilon \to 0$ even if the set E is very large since, below a certain value of ϵ, there will be no points in the sphere. The standard trick (applied in other cases as well, see below) is to plot $\ln P(\epsilon)$ as a function of $\ln \epsilon$ for different embedding dimensions d. That is, taking only the first d elements of any point coordinates $\mathbf{x}_i$ for $d = 1, 2, 3 \ldots$. One needs then to find a scaling region in the plot where the slope of the different curves is independent of d. The slope of this portion is then the desired pointwise dimension D_P. Additional details of this technique will be described in Chapter 5. The properties of the pointwise dimension depend on the nature of the distribution of points in the set. It is possible to determine when $D_P(\mathbf{x}_i)$ is independent of the point (and thus is trivially equal to the average pointwise dimension) and also to state precisely how it is related to $D^{(0)}$ (see Ott's book).

The next member in the hierarchy, $D^{(1)}$, can be found in practice using the concept of the *information dimension*. Assume that the set E is covered by a number $(N(\epsilon)$, say) of spheres of radius ϵ. Let n_k be the number of points belonging to E, which are inside the kth sphere. The probability of finding a point in this sphere is thus $P_k(\epsilon) = n_k(\epsilon)/n_E$. Consider now the quantity $S(\epsilon)$ defined as

$$S(\epsilon) = -\sum_{k=1}^{N} P_k(\epsilon) \ln P_k(\epsilon)$$

This is the information entropy and its meaning is related to the amount of information contained in the specification of the system's state to accuracy ϵ. The information dimension, D_I is then defined as

$$D_I = \lim_{\epsilon \to 0} \frac{-S(\epsilon)}{\ln \epsilon} = \lim_{\epsilon \to 0} \frac{\sum_{k=1}^{N} P_k(\epsilon) \ln P_k(\epsilon)}{\ln \epsilon}$$

In practice the computation of D_I is performed in a similar manner to that explained above in the case of D_P. The quantity $-S(\epsilon)$ is plotted versus $\ln \epsilon$ for different embedding dimensions d and the slope of the d-independent region yields D_I. It can be shown that the information dimension is always smaller or equal to the capacity dimension, i.e., $D_I \leq D$, the equality being realised if the points chosen to compute the information dimension are uniformly distributed in the spheres. The fact that the information dimension is equal to $D^{(1)}$ can easily be seen by considering the limit $q \to 1$ of (3.91) and using L'Hospital's rule.

It is possible to show that the second member of the generalised dimensions hierarchy, $D^{(2)}$ is connected to the concept of correlation between pairs of points. The two-point correlation function, $\xi(r)$, is a well known concept in statistical physics. It is defined such that the joint probability of finding a set point in a volume element δV_1 *and* another point in a volume element δV_2, a distance r from the first is

$$\delta P = n_E^2 [1 + \xi(r)] \delta V_1 \delta V_2 \tag{3.92}$$

where n_E is the total number of the points in the set. It expresses the properties of the distribution of points in a set and is rather easily calculable.

A related quantity is the two-point correlation *integral*. Let $n_{\text{pair}}(\epsilon)$ be the number of pairs of points $(\mathbf{x}_i, \mathbf{x}_j)$ of E, whose distance is less than ϵ. Then the two-point correlation integral is defined as

$$C(\epsilon) = \lim_{n_E \to \infty} \frac{1}{n_E^2} n_{\text{pair}}(\epsilon) \tag{3.93}$$

where the number of relevant pairs can be written as

$$n_{\text{pair}}(\epsilon) = \sum_{i,j}^{n_E} \Theta(\epsilon - |\mathbf{x}_i - \mathbf{x}_j|) \tag{3.94}$$

with Θ being the unit step (Heavyside) function. The last expression merely defines the counting procedure.

It is not difficult to derive the relationship between this correlation integral and the two-point correlation function defined above, but we shall skip it here. Readers interested in this issue, as well as the discussion of higher-order correlation integrals (and functions), are referred to Murante *et al.* (1997) and references therein (see also Chapter 11). The above definition of the two-point correlation integral is strict only for infinite sets of points and is used in an approximate sense for n_E finite but large. A practical prescription for the calculation of $C(\epsilon)$ can be based on a probabilistic approach, similar to the one used in the definition of the information dimension above. The probability of finding a point within a distance ϵ from a given point $\mathbf{x}_m$ of the set is clearly

$$P_m(\epsilon) = \frac{n(\epsilon)}{n_E}$$

where $n(\epsilon)$ is the number of points in a sphere of radius ϵ around $\mathbf{x}_m$.

It can be shown that the two-point correlation integral is then given by

$$C(\epsilon) = \frac{1}{n_E} \sum_{m=1}^{n_E} P_m(\epsilon)$$

Again, for this relation to hold n_E must be very large and the above-mentioned criteria can be used to estimate its lower bound.

The *correlation dimension*, D_C, is defined by examining the behaviour of $C(\epsilon)$ as $\epsilon \to 0$. Thus

$$D_C = \lim_{\epsilon \to 0} \frac{\ln C(\epsilon)}{\ln \epsilon}$$

It can be shown that the correlation dimension too satisfies an inequality relating it to the capacity dimension and the information dimension, i.e., $D_C \leq D_I \leq D$ and that it is equal to $D^{(2)}$.

The ideas providing the basis for the above definitions of D_I and D_C and their relationship to $D^{(q)}$ were developed by Grassberger and Procaccia in early 1980s. This also enabled the design of efficient computational algorithms for calculating the fractal dimension of sets. The Grassberger–Procaccia method for finding the fractal dimension has since then become a standard tool. Readers interested in its details are referred to the original articles, e.g., Grassberger & Procaccia (1983) or some later books and reviews cited in this book.

We now turn to yet another definition of fractal dimension, which has been devised for finding the dimension of attracting sets in a dissipative dynamical system. It is related to the concept of Liapunov exponents. Let $\lambda_1 \geq \lambda_2 \geq \cdots \geq \lambda_n$ be the ordered set of the Liapunov exponents of some dynamical system in $\mathbf{R}^n$. If the system has an attractor, phase volume has to shrink and thus $\sum_{i=1}^{n} \lambda_i < 0$. If the system exhibits chaotic behaviour the largest Liapunov exponent is positive,

$\lambda_1 > 0$. By adding to it the next exponents in turn, the sum must eventually become negative. That is, there exists a natural number m for which

$$\sum_{i=1}^{m} \lambda_i \geq 0, \quad \text{but} \quad \sum_{i=1}^{m+1} \lambda_i < 0$$

Using this fact the *Liapunov dimension* for a dynamical system is defined as

$$D_L = m + \frac{\sum_{i=1}^{m} \lambda_i}{|\lambda_{m+1}|} \tag{3.95}$$

Clearly $m + 1$ is the smallest integral dimension in which the motion still contracts volume. An attracting set of the system must thus have a capacity dimension D, satisfying $m \leq D \leq m + 1$. Since all the negative Liapunov exponents, with indices above $m + 1$, are excluded from the definition we may reasonably assume that $D_L \geq D$. Kaplan and Yorke (1979) proposed that the capacity fractal dimension of the attracting set is actually given by $D = D_L$. In many cases it has been shown that the Kaplan–Yorke conjecture holds, although no general rigorous proof of this is known.

As we have said from the outset, fractal sets are objects exhibiting self-similarity. This fact is expressible in a relation known as *scaling* (an expression borrowed from statistical mechanics). Scaling implies power-law functional dependence of the relevant quantities, like the relation $N(\epsilon) \propto \epsilon^{-D}$ for the number of covering cubes (or $P(\epsilon) \propto \epsilon^{D}$ for the appropriate probability), which actually led to the definition of the box counting dimension (3.90).

We have seen that fractal sets, for which all the generalised dimensions are equal, are endowed with an exactly self-similar distribution of points. Such sets can be viewed as true fractals, as they have just one value of fractal dimension i.e., one scaling exponent. However, in sets arising from dynamical processes in nature it is unrealistic to expect this property to hold exactly. In many such sets $D^{(q)}$ decreases with q and this means that the higher moments are less and less 'space filling' with increasing q. It thus follows that we may expect a collection of different scaling exponents, each of which is applicable to the appropriate subset of the full set. Complicated sets, which are composed of what may be called many interwoven fractals and thus possess this property, are called *multifractals*. We may regard such sets as a superposition of different fractal sets, which can be parametrised by their scaling exponents (customarily denoted by α) and defined as

$$E(\alpha) = \{\mathbf{x}; \text{ such that } P(\mathbf{x}, \epsilon) \propto \epsilon^{\alpha}, \text{ for } \epsilon \to 0\}$$

To find the capacity dimension of the set $E(\alpha)$ it is necessary to determine $N(\epsilon, \alpha)$, the number of cubes of size ϵ needed to cover it. Assuming that scaling holds we must have

$$N(\alpha, \epsilon) \propto \epsilon^{-f(\alpha)}$$

where $f(\alpha)$ is the fractal dimension of the particular set and, in the context of the full multifractal set, can be considered as the distribution of fractal dimensions. The term $f(\alpha)$ is called the *multifractal spectrum* and characterises the multifractal set. We shall not go into any further details on this subject here, but refer the reader to, e.g., Ott (1993).

We conclude our discussion on fractals with a few general remarks. The idea of fractals and fractal geometry were introduced into physics by Benoit Mandelbrot. In his seminal works (see for example his book published in English in 1982) he demonstrated the ubiquity of fractals in nature. He also found particular fractal sets, referred to as Mandelbrot sets, in various mathematical applications. For readers wishing to gain a deeper knowledge of fractals we recommend the book by Barnsley (1988), which contains both the mathematical foundations of fractal geometry as well as stunningly beautiful visualisations of fractal sets. It is rather easy today to devise simple algorithms for home computers to produce visualisations of this kind.

The application of fractals to nonlinear dynamics is widespread. Fractals are found as strange attractors, in Poincaré maps of Hamiltonian and dissipative systems, in bifurcation diagrams – to mention just a few examples of concepts already introduced in this book. In this section we have defined their basic properties and they will appear quite frequently later in the book.

4

Properties of chaotic dynamics

Chaos is the score upon which reality is written.
Henry Miller, *Tropic of Cancer.*

In Chapter 2 we encountered examples of dynamical systems whose solutions may sometimes display very strange behaviour. This behaviour has been referred to as *irregular* or *chaotic*, without defining these terms exactly. Chapter 3 introduced the basic mathematical concepts used in the study of dynamical systems. Some of these concepts (like, e.g., Poincaré maps and Liapunov exponents) are directly applicable to qualitative and quantitative study of chaotic solutions and thus we are now sufficiently prepared for a detailed discussion of chaotic behaviour in dynamical systems.

One should keep in mind, however, that we have so far used rather informal language and will continue to do so for the remainder of this book. Readers who wish to see a mathematically more formal and rigorous exposition of the important concepts introduced in Chapter 3 and most of the material in this and the following chapter are referred to the mathematical literature, e.g., the book by Ruelle (1989), who is one of the founders of the ergodic theory of chaos.

We shall consider here flows

$$\frac{d}{dt}\mathbf{x} = \mathbf{F}(\mathbf{x}) \tag{4.1}$$

or discrete maps

$$\mathbf{x}_{i+1} = \mathbf{F}(\mathbf{x}_i) \tag{4.2}$$

as the two classes of dynamical systems whose chaotic dynamics we wish to explore. It is impossible to define chaotic solutions of flows or maps precisely and in a straightforward way, because such solutions cannot be represented analytically and their properties cannot be stated simply. From a practical point of view, however, chaos can be defined by stressing what it is not and by listing its special identifiable characteristics.

A chaotic solution must, first of all, be *bounded* and *aperiodic*. It therefore cannot be an equilibrium (fixed point) solution or a periodic solution or a quasiperiodic (multiperiodic with incommensurate frequencies, see below) solution. Nor can it be a divergent (i.e., growing above any finite bound) function or series.

If the dynamical system is *dissipative*, any volume of initial conditions in phase space contracts. We recall that this is guaranteed if in the flow (4.1)

$$D \equiv \nabla \cdot \mathbf{F}(\mathbf{x}) < 0$$

for all $\mathbf{x}$. In the case of a discrete map, like (4.2), the idea is the same, but the volume contraction proceeds in discrete steps. The condition is that the Jacobian determinant (see equation 3.7) satisfies

$$|J| \equiv |\det \mathcal{J}| < 1$$

Thus the motion in a dissipative system must approach a geometrical object of a dimension lower than that of phase space itself. This gives rise to, as we have already seen, attracting sets. In most cases the motion is quite regular and these sets have the form of stable fixed points, limit cycles or tori.

In some examples, however, we have encountered attracting sets with surprisingly complex geometrical properties. Such attracting sets will be seen to be intimately connected with chaos. We shall explore the general conditions for their occurrence in dynamical systems and try to characterise them precisely.

In spite of the absence of attracting sets in *conservative* systems, these may also display irregular behaviour. We have already seen one such example (the HH systems) and we will discuss in Chapter 6 those characteristics of chaotic dynamics that are applicable exclusively to Hamiltonian systems.

In this chapter we shall explore the geometry of attractors, quantify the property of sensitivity to initial conditions, discuss fractal structures in phase space, examine the transition to chaos and define the role of invariant manifolds intersections in chaotic dynamical systems. All these will serve as theoretical predictive criteria for chaotic motion. When the concepts or methods discussed are only applicable to dissipative systems we shall state this explicitly.

4.1 Strange attractors

> All strange and terrible events are welcome,
> But comforts we despise.
> W. Shakespeare, *Anthony and Cleopatra*.

We already know that chaotic behaviour of *dissipative* systems is related to the existence of very complicated attracting sets. To make this statement more precise we must define the notion of an *attractor* of a flow, or discrete map. The definition is based on the concepts of *attracting* and *invariant* sets of a dynamical system, which have already been defined in Chapter 3, but we rephrase those definitions here, using the concept of a dynamical system evolution operator (introduced in Chapter 1).

We consider a dissipative dynamical system of the type (4.1) or (4.2), denoted symbolically by

$$\mathbf{x}' = \mathcal{T}\,\mathbf{x}$$

where $\mathcal{T}$ is a continuous or discrete evolution operator, acting on states of the system and transforming them into other possible states. A set of states (points in phase space) $\mathbf{A}$ is an *attractor* of the dynamical system if:

(i) $\mathbf{A}$ is *invariant*, that is $\mathcal{T}\mathbf{A} \subseteq \mathbf{A}$,

(ii) $\mathbf{A}$ is *attracting*, that is, there exists $\mathbf{U} \supset \mathbf{A}$ (a neighbourhood of $\mathbf{A}$) so that $\mathcal{T}\mathbf{U} \subseteq \mathbf{U}$ and $\mathcal{T}\mathbf{U} \to \mathbf{A}$, as $t \to \infty$ (for flows) or $i \to \infty$ (for maps),

(iii) $\mathbf{A}$ is *recurrent*, that is, any trajectory initiated at a point inside $\mathbf{A}$ repeatedly comes arbitrarily close to the initial point,

(iv) $\mathbf{A}$ is *indecomposable*, that is, it cannot be split up into two nontrivial pieces.

Every attracting set has properties (i) and (ii), but only if properties (iii) (which rules out transient solutions) and (iv) (which implies a single basin of attraction, see below) hold, is the set an attractor.

This (rather formal) definition of attractors has to be supplemented by introducing the notion of an attractor's *basin* or *domain* of attraction. If $\mathbf{A}$ is an attractor of a dynamical system, the largest set $\mathbf{B}$, so that $\mathcal{T}\mathbf{B} \to \mathbf{A}$ as t or $i \to \infty$, is the basin of attraction of $\mathbf{A}$. Thus the basin of attraction is the set of all initial conditions in phase space for which the trajectory ultimately approaches a given attractor.

The above abstract definitions can readily be illustrated by using the now familiar logistic map (2.2) and the forced Duffing oscillator (2.15), introduced in Chapter 2 of this book. With their help the reader has already become acquainted with the intuitive meaning of attractors and transition to chaotic behaviour in dissipative dynamical systems. The above, more rigorous, definitions may not add as yet too much understanding, but they allow the following observations to be made.

The logistic map $x_{i+1} = 4r x_i(1 - x_i)$ was found in Chapter 2 to have two fixed points for $0 < r < 0.25$. In this parameter range the attractor of the mapping is the stable fixed point $x^* = 0$, its domain of attraction being the interval $[0, 1]$. The other fixed point is unstable and is thus not an attractor as it violates the property (ii), given in the definition. If $0.25 < r < 0.75$ the attractor of the map is the fixed point $x^* = 1 - 4/r$. For the range of r-values appropriate for a stable 2-cycle, this cycle (the two points) is the attractor. The fact that these are two distinct points does not violate property (iv), since each point by itself does not have its own basin of attraction and thus the attractor cannot be split. In fact, Figure 2.2 gives the attractor as a function of the control parameter r. For large r values, when the map is chaotic, the attractor looks rather complicated!

In the forced Duffing system (2.15), for $\epsilon = 0$ we have *two* attractors (the minima of the quartic double-well potential). The domain of attraction of each one can be found quite easily. The set consisting of *both* fixed points is not an attractor as it violates property (iv).

It can be shown by numerical integrations (see Chapter 2) that there exists a fixed value ϵ_1 such that if $0 < \epsilon < \epsilon_1$ the two attractors continue to exist. For sufficiently large values of ϵ in this range, transient and apparently irregular solutions appear, but the qualitative behaviour of an ultimate approach to one of the fixed points, remains. Other points on these, sometimes complicated, trajectories do not belong to any of the attractors, as they do not satisfy property (iii). Single periodic attractors (closed curves in the $x - y$ plane, encompassing the two potential minima) appear for still larger values of ϵ and they bifurcate to more complicated (but still periodic) attractors when ϵ is increased. Finally, for ϵ larger than some well defined value ϵ_∞ (which obviously depends on δ) an attractor with a very complicated structure appears.

Chaotic behaviour in dissipative systems results from the existence of an attractor that is not a simple geometrical object of integral dimension like a point, closed curve, or the surface of a torus. Chaotic attractors are complex objects with fractal dimension and they are not even smooth. They are called *strange attractors*. Such attractors can, as a rule, only be found numerically and their analytical properties are not easily defined. A considerable body of work has accumulated in recent years, regarding the mathematical properties of strange attractors. The theory is not yet complete and we shall omit most of its formal parts, as they are certainly beyond the scope of this introductory book. The presence of a fractal, strange attractor in a dissipative system will be for us a synonym for chaos in such a system.

The practical question is, of course, how can one determine that a given system has a strange attractor. It may not be a simple task to calculate the fractal dimension of a set of points (using the methods explained in Chapter 3), if they result from an experiment or a numerical calculation with a multitude of initial conditions. In the next chapter we shall discuss some explicit practical prescriptions for doing this. In what follows we shall discuss some important properties of strange attractors, and the ways in which they may emerge in the phase space of dynamical systems.

Qualitative features

We can examine the qualitative features of strange attractors by looking at their general appearance. Some basic properties are well demonstrated in the following two example systems. In Figures 4.1 and 4.2 typical strange attractors of a discrete

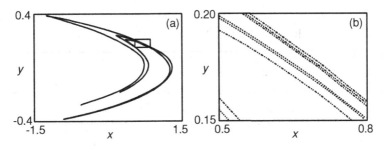

Fig. 4.1. The Hénon attractor.

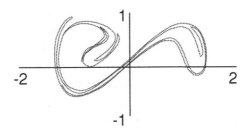

Fig. 4.2. The Duffing attractor. Poincaré map shown with the $x-y$ plane being the surface of section.

mapping, the Hénon map (3.28), and of a flow, the forced Duffing oscillator (2.15), are shown.

The Hénon map was introduced in Chapter 3 but we rewrite it here for convenience

$$x_{i+1} = 1 - \alpha x_i^2 + y_i$$
$$y_{i+1} = \beta x_i \qquad\qquad (4.3)$$

Clearly, the Jacobian determinant of this map (see 3.30) is $J = \beta$. Thus if $|\beta| < 1$ the map is dissipative. In the case studied in Section 3.1 ($\alpha = 0.1$, $\beta = 0.5$) we had two fixed points and one of them was stable (it is a 'simple' attractor). However, if we take, for example, $\alpha = 1.4$ and $\beta = 0.3$, the map is still dissipative, but the attractor is not a fixed point any more. Following the analysis done in Section 3.1 we get using (3.29) the two fixed points

$$X_1 = 0.661; \quad Y_1 = 0.198 \quad \text{and} \quad X_2 = -1.161; \quad Y_1 = -0.348$$

Examining the eigenvalues of the stability matrix for these fixed points (see Equation 3.31), we find that for the first point they are 0.150 and -2.001, and those for the second point are 3.341 and -0.090. Both fixed points are thus unstable (saddle points). It is not too difficult to iterate the Hénon map numerically. Starting from some initial condition (x_0, y_0) one can apply the map iteratively and get a sequence of points in the x–y plane. In Figure 4.1 the results of such iterations are shown. The full attractor is shown in the left panel. The right panel is the result of 'zooming in' on the small rectangular region marked on the left panel. The self-similar structure, characteristic of a fractal, can be demonstrated if one continues this process using the right panel. Subsequent 'zooming in' on a rectangle containing just a portion of the uppermost 'strip' will yield an almost identical picture, provided that a sufficient number of iterations is performed (ideal self-similarity is achieved only for an infinite number of iterations). The attractor of the Hénon map in this case is strange and the map therefore exhibits chaotic behaviour.

It is much more difficult to visualise the attractor of the forced Duffing system for parameter values for which the dynamics is chaotic. The system and its orbits are three-dimensional, and one way of trying to visualise the attractor is to plot a two-dimensional projection of chaotic orbits; but because of projection effects, spurious orbit intersections may appear and it is, of course, impossible to discover the fractal nature of the attractor. Some typical characteristics of chaotic attractors can however be observed if one marks the intersections of the orbit with some fixed plane. We have already discussed this technique (Poincaré map) in considerable detail in Chapter 3 and a plot of such a section, for this very system, was shown in Figure 3.8 Consider the forced Duffing system

$$\dot{x} = y$$
$$\dot{y} = x - x^3 - \delta y + \epsilon \cos z$$
$$\dot{z} = 1$$

i.e., with the choice $\omega = 1$ in (2.14). We have seen in Chapter 2 that for $\epsilon = 0$ (unforced oscillator) and for values of $0 < \delta < \sqrt{8}$, the two outer fixed points are stable spirals (see 3.27). If we take $\delta = 0.2$ and assume no forcing this would thus correspond to orbits spiralling (slowly) into one of the outer fixed points. However, a sufficiently large forcing does not allow the orbits to settle down. In Figure 4.2 a Poincaré map through the surface (x–y plane) of section is shown for $\epsilon = 0.3$ (this corresponds to the chaotic regime). Orbits started with different initial conditions are all attracted to an invariant complicated structure depicted in the figure (transient crossings are not marked). Note that this Poincaré section view of the Duffing attractor appears different from the one shown in Figure 3.8, which is a natural consequence of the fact that the attractor is more than two-dimensional and the planes of section in the two figures are different. Qualitatively, however, both sections reveal the typical 'leafy' folds of a strange and fractal attractor.

Although a definite diagnosis of chaotic behaviour must involve the calculation of Liapunov exponents (see below) and/or of the attractor's fractal dimension (see next chapter), the general qualitative appearance of strange attractors, supplemented perhaps by the typical shape of first return maps (as explained Chapter 3), is usually quite telling.

4.2 Sensitivity to initial conditions

> I am like that. I either forget right away
> or I never forget.
> Samuel Beckett, *Waiting for Godot.*

One of the primary characteristics of deterministic chaos is the *unpredictability* inherent in the solution. This may seem inconsistent, as the term 'deterministic' implies predictability, that is, given initial conditions uniquely specify the outcome. The meaning of unpredictability must thus be defined precisely in this context.

We may say that chaotic dynamical systems are predictable (since discrete maps or continuous flows are deterministic) only in an ideal sense. Any measurement, however accurate, has a finite error. Moreover, any numerical calculation suffers from the inherent inaccuracy of digital computers caused by the truncation error. If a dynamical system has the property of overwhelmingly amplifying tiny differences between two starting states of the system, it must be considered practically unpredictable. Chaotic behaviour of a dynamical system is thus characterised by *sensitivity to initial conditions* (SIC). That is, very small differences in the initial conditions are quickly amplified by the chaotic map or flow to create very significant differences in the orbit.

We have already encountered several dynamical systems having SIC. A typical example is the Bernoulli map (3.62). As we have seen, the repeated application of the Bernoulli map to a binary number between 0 and 1 brings to the 'front' of the decimal binary number more and more distant digits of the starting number. Thus two irrational binary numbers differing by no more than $2^{-100} \approx 10^{-28}\%$ may, after ≈ 100 iterations, differ by $\sim 100\%$! Here it is not a question of inaccurate instruments or computers. Our inability to even write or grasp an infinite sequence of numbers causes the unpredictability in this perfectly deterministic (and extremely simple) system.

Sensitivity to initial conditions is a general property of chaotic dynamics and occurs in all chaotic systems, not only those having strange attractors (dissipative). One of the first observations of SIC in a simple dissipative system was in the numerical calculations of Lorentz (1963), who tried to model atmospheric convection by three coupled nonlinear ODEs (see equation 3.77). He obtained *different* solutions when starting with the *'same'* initial conditions. The solutions progressively diverged from each other with time. The divergence of trajectories of Lorentz's

ODEs resulted from the finite accuracy with which he could specify the initial conditions (roundoff error of the computer), and not from any real errors he had suspected. As we have noted in Chapter 2, Moore and Spiegel (1966) independently discovered such behaviour in an alternative simple model of thermal convection, applied to an astrophysical system. In the context of weather prediction SIC has become popularly known as the *butterfly effect*. If the Earth weather system has the property of SIC, a perturbation caused by butterfly wings in China may cause a tornado in the US a few weeks later. Although formally deterministic, the weather system is in practice unpredictable for long times.

In Section 3.2 we defined the *Liapunov exponents* for discrete maps and continuous flows and have shown that the *largest* Liapunov exponent is the quantitative measure of local trajectory divergence, and hence of SIC. We repeat here the bottom-line conclusion given there, that a solution of a dynamical system is considered *chaotic* if the *largest* Liapunov exponent is *positive*. Except for very few simple cases (like the Bernoulli shift, for which it was shown in that section to be equal to ln 2) the largest Liapunov exponent must be numerically calculated, using the methods mentioned in Section 3.2. Before giving some actual results of such calculations we would like to discuss an important qualitative issue, using a typical example system.

The horseshoe map

Consider a dynamical system possessing the SIC property. Two points, however close to each other, must be carried by the system in such a way as to increase their separation exponentially (locally at least). Thus, in practice, the correlation between these points is lost after some finite time (or number of discrete iterations). We are interested, as a rule, in *bounded* motions, that is, solutions that do not cause 'blowup' of any phase-volume elements. In dissipative systems phase-volume elements actually shrink continuously. How can two neighbouring trajectories diverge along with simultaneous shrinking (or at least preservation, as in conservative systems) of phase volume? Clearly, a specified phase volume must be *stretched* (to increase the distance between any two points). The stretching (in some direction) cannot, however, simply continue since volume must shrink (or be conserved, or at least be bounded). It must thus be accompanied by some appropriate decrease of extension (in other directions). This can be achieved if the stretching is accompanied by *folding*. For example, if we consider a three-dimensional dissipative system, the seemingly conflicting demands of attraction and SIC can be reconciled if the transformation induced by the motion is topologically equivalent to some repetitive *stretch plus fold* operation. Attraction occurs in one dimension while divergence of trajectories occurs in another. This is an example of

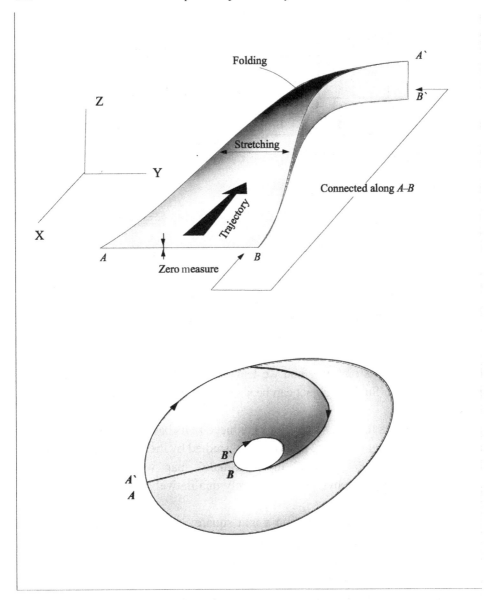

Fig. 4.3. Qualitative illustration of a strange attractor allowing divergence of trajectories together with volume shrinking.

hyperbolicity, a property we have already defined for fixed points (having a stable and an unstable invariant manifold, but no centre manifold). Figure 4.3 is a qualitative drawing of an object attracting neighbouring trajectories, but allowing them to diverge at the same time. We show the structure as if it was cut along AB. Actually, it has to be folded and joined so as to identify AB with $A'B'$. It is thus

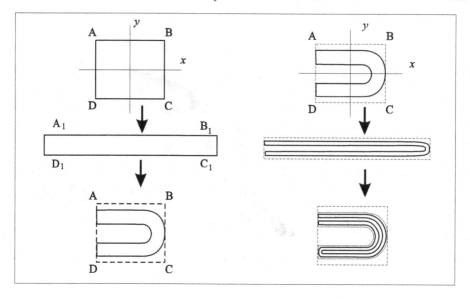

Fig. 4.4. First two iterations of the horseshoe map.

difficult to draw this on paper and we leave the resulting object to the imagination of the reader. Such an object can be a strange attractor of a chaotic dissipative flow.

The stretch plus fold transformation is reduced to its basics in a famous theoretical two-dimensional mapping, proposed and studied by the famous mathematician Stephen Smale in 1967. We shall discuss Smale's map, which is also called the *horseshoe map* (for an obvious reason), only qualitatively; this will be sufficient for our purposes here.

The horseshoe map is defined on a unit square. Its action in each iteration is composed of two stages. First, the square is *stretched* in one direction (horizontal, say) by some factor (2, say) and contracted in the other direction (vertical) by a larger factor (2α with $\alpha > 1$). This is shown schematically in the two upper parts of the left panel in Figure 4.4, where the square $ABCD$ is deformed into the rectangle $A_1B_1C_1D_1$. In the second stage the resulting rectangle is *folded* into the shape of a horseshoe and placed back onto the original square (the lowermost part of the left panel). The operation is then repeated on the square containing the horseshoe (see the right panel), and so on. The emergence of a layered and complex (but bounded) structure can already be seen after the first two iterations (the lowermost panel on the right side).

This simple-looking mapping captures the essence of chaotic dynamics. The horizontal stretching by a factor of 2, accompanied by the vertical shrinking by factor 2α give rise to the following two Liapunov exponents

$$\lambda_1 = \ln 2, \quad \lambda_2 = \ln \frac{1}{2\alpha} = -\lambda_1 - \ln \alpha$$

In the subsequent folding and fitting of the horseshoe in the square no surface area changes occur, and hence the above numbers are the Liapunov exponents of the mapping. We easily see that the map is dissipative – the area contracts by a factor $\alpha > 1$ at each stage. This is reflected by the sum of the Liapunov exponents being negative for $\alpha > 1$. The larger Liapunov exponent is positive and hence the map is chaotic. Its invariant attracting set has the above mentioned property of hyperbolicity.

It is not too difficult to actually find the strange attractor of the map (it exists only for $\alpha > 1$ when the map is dissipative). Imagine that a vertical cut is made through the image of the square, after many consecutive iterations. Such a cut of the repeatedly stretched and folded horseshoe clearly resembles a Cantor set. Using the definition (3.90) one can easily find the box counting dimension of this Cantor set (which we call H_∞^c, say)

$$D\left(H_\infty^c\right) = \lim_{\epsilon \to 0} \frac{\ln N(\epsilon)}{\ln(1/\epsilon)} = \lim_{n \to \infty} \frac{\ln 2^n}{\ln(2\alpha)^n} = \frac{\ln 2}{\ln 2\alpha}$$

(compare this with the calculation of D for Cantor's dust and the Koch curve in Section 3.5).

The Liapunov dimension of the full horseshoe attractor, H_∞ (H_∞^c was just a one-dimensional cut through it) can easily be found from the Liapunov exponents using the definition (3.95)

$$D_l(H_\infty) = m + \frac{\sum_{i=1}^{m} \lambda_i}{|\lambda_{m+1}|} = 1 + \frac{\lambda_1}{|\lambda_2|} = 1 + \frac{\ln 2}{\ln 2\alpha} \tag{4.4}$$

because in this case $m = 1$. It can be shown that the Kaplan–Yorke conjecture is valid in this case and therefore this is also the box-counting dimension of the horseshoe attractor. Thus we have $D(H_\infty) = 1 + D(H_\infty^c)$. For example, if $\alpha = 1.5$ the dimension of the horseshoe attractor is

$$D(H_\infty) = 1 + \frac{\ln 2}{\ln 3} = 1.630\,93\ldots$$

while the dimension of the cut is equal to that of Cantor's dust.

Note that as $\alpha \to 1$ the map is less and less dissipative (it is conservative for $\alpha = 1$) and $D(H_\infty) \to 2$, as it should. The conservative horseshoe map does not have an attractor since area (two-dimensional volume) is conserved. Still, it has SIC and is chaotic as the larger of the two Liapunov exponents is positive (their sum is zero). The stretching and folding operation gives rise to chaos in conservative systems as well as in dissipative ones. If we imagine that the upper half of the original square is white while the lower half is red (like the flag of Poland), for example, the number of red stripes will exceed that of the American flag (7) after only three iterations. After n iterations we will have 2^n narrow red stripes with white stripes between them. The conservative chaotic map is thus seen to be *mixing* (here the colours). This property, addressed here intuitively, will be further explained, and in a more precise manner, in Chapter 6, where we shall discuss chaotic behaviour in Hamiltonian systems and area-preserving maps.

The importance of the horseshoe map lies not only in its being such a neat example of chaotic behaviour (strange fractal attractor, fractal dimension of a section, SIC, a positive Liapunov exponent, chaotic mixing in a conservative system etc.). It turns out that Smale's horseshoe map is also a very useful theoretical tool in the investigation of complicated dynamical systems. As we shall see later in this chapter, the presence of horseshoe mappings in Poincaré sections of multidimensional flows is sufficient for chaotic behaviour. Horseshoes are thus generic in a large class of chaotic systems. A map related to Smale's horseshoe is the *baker's map* or transformation (deriving its name from the activity of stretching and folding dough). A good discussion of this mapping can be found in Ott (1993).

Estimating the divergence of trajectories

Horseshoe maps are very important, but looking for them is not always a straightforward task. The most direct procedure to find SIC, and thus chaos, is still the calculation of Liapunov exponents. This, as we have seen in Section 3.2, may be quite complicated. It is thus advantageous to attempt a simpler numerical calculation, which may indicate SIC. In this type of calculation one chooses two close initial conditions, separated by a distance d_0 say, in phase space. The dynamical system is then solved (integrated or iterated) numerically with the two initial conditions. The distance between the two trajectories is plotted as a function of time or iteration step. The exact form of the plot will obviously depend on the points chosen, and the graphs may look rather wiggly and complicated. However, the qualitative average shapes of the graphs for several pairs of close initial conditions are significant. The distance between two trajectories of a regular (non-chaotic) dissipative system should decrease on average. In contrast, if we find that this distance is increasing strongly, at least for some evolution period, we may suspect that the motion is chaotic. In Figure 4.5 we plot the results of the evolution of the logarithm of the amplification factor of the distance between two trajectories started at a small distance d_0, for a typical dynamical system in its chaotic regimes. The example here is the Rössler oscillator, which we take the opportunity to introduce

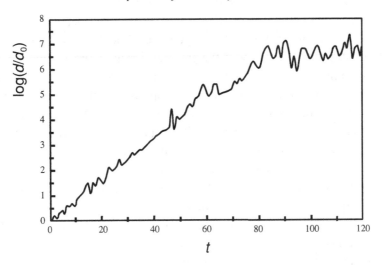

Fig. 4.5. The relative distance between two initially close phase points as a function of time for the Rössler system (with $a = 0.55$, $b = 2$ and $c = 4$). Chaotic behaviour is indicated by the divergence of nearby trajectories.

for the first time. It is defined by the following three equations:

$$\dot{x} = -(y + z)$$
$$\dot{y} = x + ay$$
$$\dot{z} = b + z(x - c) \qquad (4.5)$$

where a, b and c are constants.

Rössler, motivated by the study of chemical turbulence, found in 1976 chaotic behaviour in this simple three-dimensional system with a quadratic nonlinearity. The Rössler system has been extensively studied since then and has become one of the archetypes of dissipative chaos. The strange attractor of this system is very revealing and it consists of series of bands on a fractal funnel-like structure. The structure can be shown to result from the action of 'stretch and fold' dynamics (similar, in principle, to the horseshoe map). More details on the Rössler attractor, including some good visualisations, can be found in Abraham & Shaw (1992). We shall use this system in Chapter 10 to test a technique explained there.

The plot (in the chaotic regime) clearly shows the qualitative signature of SIC and thus chaos. For parameter values appropriate to regular behaviour the result (not shown here) is very different; the distance between the trajectories diminishes as the system is dissipative. We stress again that this method provides only a heuristic qualitative estimate if the system is chaotic, but a definitive answer, proving SIC, can be given only by an explicit calculation of the largest Liapunov exponent.

We conclude this section by quoting a few results from detailed Liapunov exponent calculations of this type. The Liapunov exponent for the logistic map has already been shown (as a function of the parameter r) in Figure 3.9. We remark here that in most computations Liapunov exponents are expressed in units of *bits/sec* or *bits/iteration*, i.e., binary (base 2) logarithms are used in the definitions (see Section 3.2) instead of the natural ones. This does not really matter, since the two differ by a multiplicative positive constant. Wolf *et al.* (1985) performed detailed calculations, using their algorithm, and obtained for the Hénon map (4.3) (for $\alpha = 1.4$ and $\beta = 0.3$) the exponents $\lambda_1 = 0.603$, $\lambda_2 = -2.34$ (in units of bits/iteration). For the Rössler system (with $a = 0.15$, $b = 0.20$ and $c = 10$) they obtained $\lambda_1 = 0.13$, $\lambda_2 = 0.0$ and $\lambda_3 = -14.1$ in units of bits/sec. In both cases the largest exponent is positive (chaotic behaviour) and the sum of all the exponents is negative (dissipative dynamics).

As explained in Section 3.2 the calculation of Liapunov exponents has to be divided into segments of duration Δt, usually referred to as 'restart time', and the trajectory divergence is computed numerically (by, e.g., a Runge–Kutta numerical integration) during each segment. Its logarithm is then averaged with all the previous data – see (3.63). These technical issues often pose nontrivial difficulties since the integration step, the restart time and the total integration time have to be chosen in an appropriate way so as to ensure good convergence of the results. We refer the interested reader to the book by Moon (1992), who explains in detail the procedure on the nonlinear oscillator system

$$\ddot{x} = -x^3 - \delta\dot{x} + \epsilon \cos \omega t \tag{4.6}$$

which resembles the driven Duffing oscillator (2.14) but differs from it due to the absence of the linear term.

In systems with several parameters a systematic exploration of the parameter space with a numerical algorithm for Liapunov exponent calculation may become prohibitive. Other tests discussed in this chapter and in the next one may then prove useful.

4.3 Fractal properties of chaotic motion

When a dynamical system exhibits chaotic behaviour, fractal sets may be found in the analysis of various different properties of the system. First and foremost, the existence of an attractor with fractal dimension has served us as the defining property of chaos in dissipative dynamical systems. A section through a fractal object is fractal itself (unless it is a point). We have seen this above in the case of the horseshoe attractor, whose dimension was $1 < D < 2$. It can thus be expected that fractal sets should be present in Poincaré maps of multidimensional chaotic systems.

In conservative systems attractors cannot exist, but aperiodic motion in these systems is usually associated with fractals as well. In Chapter 2 we have seen that the Hénon–Heiles system has aperiodic solutions above some value of the energy. This fact is reflected in the Poincaré sections of the system. It turns out that if one enlarges the region around one of the islands present in Figure 2.7 for $E = 1/8$ (when only a small fraction of the trajectories are aperiodic), a complicated sub-structure of new smaller islands, with irregular trajectories around them, can be seen. The presence of very intricate structures in the surfaces of section of con-servative systems exhibiting irregular behaviour is generic. It can be demonstrated well by examining a very simple area-preserving mapping, which is actually a non-linearly perturbed version of the linear rotation map (3.9). The map was originally proposed by Hénon and we shall call it the Hénon area-preserving map . It is writ-ten as

$$x_{i+1} = x_i \cos \alpha - \left(y_i - x_i^2\right) \sin \alpha$$
$$y_{i+1} = x_i \sin \alpha + \left(y_i - x_i^2\right) \cos \alpha \qquad (4.7)$$

where α is a parameter. Hénon introduced this map in his effort to study in detail the irregular type of behaviour he had found in his work with Heiles. The map is considerably more simple than the HH system and by using it Hénon was able to overcome the computational limits of the earlier calculations.

A typical phase portrait of the Hénon area-preserving map is shown in Figure 4.6. The qualitative similarity to the surface of section of the HH system (for $E = 1/8$) with its families of smooth curves, islands and aperiodic trajectories is quite striking. On this scale the intersections at the hyperbolic points look rather smooth, but when the resolution is increased a rich fine structure of islands inside a chaotic 'sea' appears. We shall discuss the origins of this complexity in Chapter 6 when the behaviour of Hamiltonian systems and area-preserving maps will be studied in detail. It turns out that as a parameter is varied the system may gradually become progressively chaotic, that is, a growing fraction of phase space becomes occupied by aperiodic trajectories. During such a process invariant surfaces (tori) corresponding to regular solutions are being destroyed. If one then examines the 'remnants' of the destroyed tori on the system's Poincaré map, a complicated, self-similar structure of new tori and heteroclinic tangles is apparent. We shall discuss this association of conservative chaotic dynamics with fractals in Chapter 6.

One additional (and here final) aspect of chaotic dynamics associated with frac-tals is that of a basin boundary. A dissipative dynamical system can have several attractors and each of them has its own basin of attraction. The boundary between

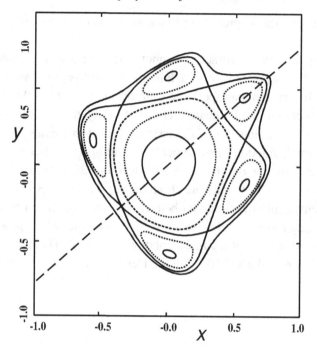

Fig. 4.6. The phase portrait of the Hénon area-preserving map (4.7) for $\alpha = 0.2144$. Five hyperbolic points are apparent in the figure. From Hénon (1969)

basins of attraction may be a simple curve, as is the case, for example, in the unforced Duffing system (2.10). In Figure 2.5 we have seen this system's phase portrait for $\alpha = \beta = 1$, i.e., for a symmetric double-well potential. It is not too difficult to delineate the boundary between the basins of attraction of the two fixed points. The basins boundary in this case is the stable manifold of the unstable (hyperbolic) fixed point. This was also the case in the oscillator (3.51) which has a triple-well potential.

In more general cases, however, the basin boundaries are more complicated and may become fractal. For example, consider the forced Duffing system (2.15), with the forcing amplitude ϵ in the region corresponding to the existence of two limit cycles around the two potential minima, i.e., for $\epsilon_1 < \epsilon < \epsilon_\infty$ (see Chapter 2). As long as ϵ is small enough, the boundary remains similar to the unforced case described above. When ϵ is increased, the basin boundary becomes fractal. The two sets of all initial conditions giving rise to orbits ultimately attracted to one or the other of the limit cycles become interwoven in a very complex way. The first calculations demonstrating the fractal nature of basin boundaries in the unforced Duffing system were made in 1985 by Moon and Li, using a grid of 400×400 initial conditions (1600 orbits were calculated). They also enlarged a small portion of phase space (by starting the calculation at 1600 points in the small region) and discovered an intricate structure with self-similar behaviour, typical of fractals. The reader is referred to the book by Moon (1992) for beautiful colour pictures of fractal basin boundaries in this and other systems. Note that the system in this case is not yet chaotic. We can say that basin boundaries in the forced dissipative double-well problem become fractal 'on the way' to chaotic behaviour. This brings us to the topic of the next section.

4.4 Transition to chaos

In the majority of the systems that we have seen so far, chaotic behaviour occurs only for some range of parameter values. The question of how a regularly behaving system changes when it becomes chaotic, is thus a natural and fundamental one.

We may hope at the outset that the transition to chaos is universal, namely independent of the particular system in question. Judging from our examples in Chapter 2 we can conclude that such an expectation is unfounded. In the logistic equation chaotic behaviour was preceded by a sequence of period-doubling flip bifurcations. In contrast, in the Hénon–Heiles system aperiodic orbits started appearing at some value of the parameter and gradually expanded to occupy eventually all of the phase space. A third type of behaviour was found to occur in the forced Duffing system when the period-doubling bifurcations and chaos were preceded by a regime in which the solutions only behaved chaotically for a finite time. The situation is however not as hopeless as it may seem. From what is known today not every nonlinear, potentially chaotic dynamical system has its own idiosyncratic route to chaos. On the contrary, dynamical systems seem to fall into just a few classes, as far as transition to chaotic behaviour is concerned. Similarly to the bifurcations, that we have classified in Chapter 3, we shall divide the transition to chaos in dynamical systems into a small number of classes having generic properties.

Since we shall have only three basic routes to chaos (with at most a few subclasses in some of them) the identification of a particular transition process in a given physical system should not be too difficult. The presence of such a process may then be considered as one of the indications of deterministic chaos. If the system can be studied experimentally in the laboratory, the typical procedure is to vary a control parameter and measure the outcome in terms of, e.g., a time-dependent signal. It is not easy (to say the least) to vary a control parameter of a typical astronomical system. This limitation may however be overcome in the usual way, i.e., by studying a multitude of similar systems in which the parameters are different.

Before describing the different types of transitions to chaos we remark that the subject is still far from being fully understood. Again using a zoological metaphor we may say that the different living creatures have been catalogued, but we do not know whether there exist other types of living creatures, nor which properties of the ones we know are necessary for life.

Period-doubling cascade

The *period-doubling* (or subharmonic) *cascade* is the best understood route to chaotic behaviour. It was discovered in the logistic mapping and subsequently in general quadratic maps and in many nonlinear flows. We shall discuss it using the example of the logistic equation (2.2). More general dynamical systems, like multidimensional dissipative flows, can be tested for the existence of the period-doubling cascade in their first return maps.

We already know (Section 3.1) that as the parameter of the logistic equation is varied, a sequence of period-doubling bifurcations occurs. When both fixed points of the map $x_{i+1} = 4rx_i(1 - x_i)$, i.e., $x_1^* = 0$ and $x_2^* = 1/4r$ become unstable, that is, for $r = r_1 = 3/4$ (see Chapters 2 and 3), a stable limit cycle appears. The limit cycle consists of just two points (it is a periodic orbit of a discrete mapping) and it is thus a 2-cycle. The two points of this limit cycle are fixed points of the appropriate second-generation mapping. The stability of the limit cycle of the logistic map is thus equivalent to the stability of the fixed points of the second-generation mapping. A straightforward calculation shows that both fixed points of the second-generation mapping lose their stability simultaneously at $r = r_2 = (1 + \sqrt{6})/4$. At this parameter value a stable 4-cycle of the logistic map appears, and all its four points are fixed points of the fourth-generation mapping. In Figure 4.7 we show plots of the function defining the fourth-generation mapping $H(x) = G(G(x))$ (in the language of Section 3.1), where $G(x) = F(F(x))$ is the function appropriate for the second-generation mapping of the original logistic equation $F(x) = 4rx(1 - x)$. The plot is for the parameter value $r = 0.875$, i.e., slightly above r_2 (the second period-doubling bifurcation point). We recall that the intersections of any function graph with the diagonal ($f(x) = x$) gives the fixed points of a map defined by $f(x)$, and that a given fixed point is stable if the slope of the function graph at the point is *smaller* that 1. It can be seen in the graph that the fourth-generation map has four stable fixed points. This means that we have here a stable 4-cycle of the original map and all fixed points of the lower-generation mappings can be shown to be unstable.

One can continue to increase r, and in this process period-doubling bifurcations will occur at an ever increasing density (on the r-axis). This is clearly seen in the bifurcation diagram shown in Figure 2.2. The values of the parameter r at each period-doubling (for the first few bifurcations) are given in Table 4.1. They must be calculated numerically, and we see that the series is rapidly convergent (e.g., the difference between r_4 and r_5 is in the fourth position after the decimal

Table 4.1. *The subharmonic sequence in the logistic equation*

Cycle appearing	Control parameter value
$2^1 = 2$	$r_1 = 0.750$
$2^2 = 4$	$r_2 = 0.862\,37\ldots$
$2^3 = 8$	$r_3 = 0.886\,02\ldots$
$2^4 = 16$	$r_4 = 0.892\,18\ldots$
$2^5 = 32$	$r_5 = 0.892\,472\,8\ldots$
$\vdots$	$\vdots$
aperiodic	$r = \lambda_\infty = 0.892\,486\,418\ldots$

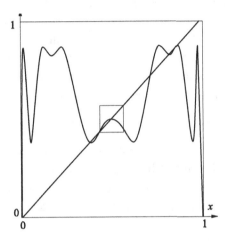

Fig. 4.7. Fixed points of the fourth-generation mapping of the logistic equation, given by the intersection of the function $F^4(x)$ with the diagonal. Note that in the inner rectangle a plot similar to that of the first-generation mapping (see Figure 2.1) appears.

point). The value of the ratio $\delta_i = (r_i - r_{i-1})/(r_{i+1} - r_i)$ measures the reduction in the difference between two consecutive bifurcations. In high precision but simple numerical calculation it was found to approach a *universal* constant.

Universal means, in this context, independent of the details of the mapping. For every mapping having a quadratic extremum the period-doubling bifurcations are such that

$$\delta \equiv \lim_{i \to \infty} \delta_i = \lim_{i \to \infty} \frac{r_i - r_{i-1}}{r_{i+1} - r_i}$$

is the same. This remarkable result, found by Feigenbaum in 1978, was a crucial breakthrough in the study of nonlinear dynamical systems. The universal scale reduction factor (also known as the *Feigenbaum number*) has been numerically computed to a very high accuracy. Its value is $\delta = 4.669\,201\,609\,102\,991\ldots$. The discovery of the universal properties of the period-doubling cascade gained even more significance when the cascade and the Feigenbaum number were found in experimental studies of diverse systems. Libchaber and Maurer (1980) were the first to find period-doubling bifurcations in a thermal convection experiment using liquid helium. As the Rayleigh number (the control parameter) was increased, subharmonic frequencies (period-doublings) appeared in the Fourier spectrum of the temperature–time signal. Moreover, the first few bifurcations that could be observed were consistent with the theoretical value of δ.

It is interesting to note that as the control parameter of the logistic map r is increased beyond λ_∞ chaotic behaviour is not uniform. On the basis of the bifurcation diagram in Figure 2.2 we can make the qualitative conjecture that for $r > \lambda_\infty$ there are 'windows' of regular behaviour. The results of numerical determination of the largest Liapunov exponent, given in Figure 3.9, support this conjecture. The largest Liapunov exponent is oscillating and becomes negative in a large number of segments of the r-axis. Interestingly, in the regular regions the map exhibits *odd*-cycle periodic behaviour. This property is also universal and is always present above the accumulation point of a period-doubling sequence. The importance of the period-doubling cascade with its universal properties cannot be overemphasised. A fluid at temperature 4 K, a population number of some species and many other seemingly unrelated systems have common basic characteristics, qualitatively and quantitatively identical to a simple quadratic function. This discovery is truly remarkable and it has brought the study of chaos in dynamical systems to the forefront of scientific activity since the 1980s.

Quasiperiodic route

As we stated at the beginning of this section the period-doubling cascade is not the only way to chaos. A separate class of theoretical, as well as experimental, dynamical systems exists, in which the transition to chaotic behaviour (as a parameter is varied) proceeds via a finite (and even small) number of Hopf bifurcations. In our discussion of dynamical systems in Chapter 3 we mentioned the generic form of a Hopf bifurcation at a fixed point. If we consider a multidimensional flow, a supercritical Hopf bifurcation at a fixed point (which becomes unstable) is accompanied by the birth of a stable limit cycle. The dynamics (at least close to the fixed point and near marginality) are thus essentially two-dimensional. In the flow's Poincaré section one then observes just one or several isolated invariant points. The stable periodic orbit in this case has a definite period, T say, and the corresponding angular frequency is $\omega = 2\pi/T$.

As the control parameter is varied further, the stable limit cycle may become unstable. We have not treated this case explicitly in Chapter 3 and only remarked that linear stability analysis of limit cycles can be performed using the Floquet theory. It is reasonable qualitatively that a limit cycle can also undergo a bifurcation of the Hopf type, losing its stability and being replaced by a stable doubly periodic trajectory. In general, an n-periodic function is defined as a function of n variables, $f(t_1, t_2, \ldots, t_n)$ that is *separately* periodic in its variables. This happens if for all natural j, satisfying $j \leq n$, a fixed number T_j exists so that

$$f(t_1, t_2, \ldots, t_j + T_j, \ldots, t_n) = f(t_1, t_2, \ldots, t_j, \ldots, t_n) \qquad (4.8)$$

One can thus define n angular frequencies $\omega_j \equiv 2\pi/T_j$ and consider f to be a function of just one variable t (replacing all t_j by $\omega_j t$). This possibility stems from the fact that a multiple periodic function can be written as a multiple Fourier series (see below in the next chapter).

Consider a one-parameter dependent flow, given by the state variable as a function of time $\mathbf{x}(t)$, having a stable limit cycle of frequency ω_0 for some values of the parameter. Let $\mathbf{X}_1(\omega_0 t)$ be the periodic orbit of the flow in this case. Now, if $\mathbf{X}_1$ loses its stability at a certain value of the parameter and is replaced by a doubly periodic stable solution, the orbit becomes $\mathbf{X}_2(\omega_1 t, \omega_2 t)$, where ω_1 and ω_2 are the two frequencies. We have already seen that $\mathbf{X}_2$ lies on the surface of a (possibly distorted) torus in phase space. If ω_1 and ω_2 are commensurate (their ratio is rational) $\mathbf{X}_2$ is actually periodic with one period. For example, if $n\omega_1 = m\omega_2$ the common single period is $T = 2\pi m/\omega_1 = 2\pi n/\omega_2$. In this case the orbit is a closed line on the torus. A more interesting case occurs when the two frequencies are incommensurate. In this case the function is irreducibly doubly periodic. Such functions are called *quasiperiodic* and the orbits they describe densely cover the torus. A Hopf bifurcation of a periodic orbit into a quasiperiodic one changes the flow from proceeding on a closed trajectory to one residing on and densely covering a torus surface. If the flow is dissipative, the closed curve before the bifurcation and the torus surface after it are the attractors of the flow. In conservative flows the flow actually lies on these invariant objects at all times. Quasiperiodic flows have closed curves in their Poincaré maps, resulting from a section of the invariant torus surface. It is possible to envisage, in analogy to the period-doubling cascade, an infinite series of Hopf bifurcations, at which the system acquires quasiperiodic orbits with a progressively growing number of frequencies. In such a scenario the invariant tori are replaced successively, at the bifurcation points, by new, higher dimensional tori. Landau proposed in 1944 that the transition of a fluid flow from laminarity to turbulence proceeds via an infinite sequence of such Hopf-type bifurcations. In this way, according to the *Landau scenario*, the fluid flow acquires an infinite number of frequencies associated with an infinite number of 'degrees of freedom'.

In 1971 Ruelle and Takens demonstrated that such an infinite sequence of bifurcations on the road to chaos is not always necessary. They found that chaotic motion can occur after only a small number of bifurcations. Ruelle and Takens proved that a chaotic attractor can appear, as a result of a perturbation on a dissipative quasiperiodic flow, proceeding on an n-dimensional torus, when $n \geq 4$. In several subsequent works chaotic behaviour was found to emerge, in some systems, directly from perturbed three- and even two-dimensional tori. In the context of fluid turbulence these findings form the basis of the *Ruelle–Takens–Newhouse scenario*. In this scenario a finite number of successive Hopf bifurcations can lead from a fixed point (an equilibrium solution) to a chaotic solution. Many experimentally observed transitions to turbulence have been interpreted using the Ruelle–Takens– Newhouse scenario. Turbulence in general and even just the issue of transition to a turbulent flow, remain to this day not fully understood. It is clear, however, that the Landau scenario is not universal. Some issues and questions relating to fluid turbulence will be discussed, in a more detailed way, in the second part of the book.

The quasiperiodic route to chaotic behaviour can be demonstrated using the example of the *circle map*. As is always the case, a particular feature of a dynamical system is best studied using a one- or two-dimensional discrete mapping. The reduction of a general flow to a mapping is almost always possible by using the surface of section technique. The Poincaré map of a quasiperiodic flow is characterised by the presence of invariant closed curves, which are gradually filled by the successive iterates. The circle map (also known as the *sine map*) is a rather general one-dimensional mapping, defined on the unit interval by

$$x_{i+1} = f(x_i) = x_i + \Omega + \frac{K}{2\pi} \sin(2\pi x_i) \quad (\text{mod } 1) \tag{4.9}$$

where Ω and K are constants and the result of the sum on the right-hand side is taken as modulo 1, so that all iterates remain in the unit interval. If we define an angular variable, $\hat{\theta}_i = 2\pi x_i$, the resulting mapping can be thought of as a transformation of a circle having some radius into itself, i.e.,

$$\hat{\theta}_{i+1} = 2\pi f(x_i) = \hat{\theta}_i + 2\pi \Omega + K \sin \hat{\theta}_i \quad (\text{mod } 2\pi) \tag{4.10}$$

To understand how the circle map is related to a doubly periodic flow on a torus say, we show schematically such a flow in Figure 4.8. Let ω_2 be the frequency of the winding motion and ω_1 (smaller than ω_2, say) be the frequency of the motion along the torus. Thus $\theta_1(t) = \theta_1(0) + \omega_1 t$ and $\theta_2(t) = \theta_2(0) + \omega_2 t$ are the angles along the torus and around it, respectively, as a function of time (see figure), expressed here

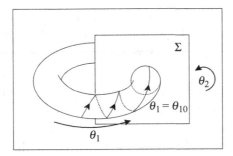

Fig. 4.8. A two-periodic orbit on a torus and its Poincaré section.

with some arbitrary initial values. These are obviously defined as modulo 2π. Note also the distinction between these continuous variables and the discrete angular variables of the circle map, which we have marked by a 'hat' symbol.

Choose now a plane Σ, defined at the initial θ_1 say (we call it in the figure, $\theta_{10} \equiv \theta_1(0)$) Clearly, Σ will be intersected at fixed time intervals $\delta t = 2\pi/\omega_1$. During this time the angular shift due to the winding motion will be

$$\delta\theta_2 = \omega_2 \delta t = 2\pi \frac{\omega_2}{\omega_1}$$

The Poincaré map (defined on the surface of section Σ) of the system can be written by using angular variables along the circle cut, and thus is equivalent to the value of θ_2 on discrete time intervals, as

$$\hat{\theta}_{i+1} = \hat{\theta}_i + 2\pi \frac{\omega_2}{\omega_1} \quad (\mathrm{mod}\, 2\pi)$$

This mapping is the linear (i.e., with $K = 0$) version of the circle map (4.9) with $\Omega = \omega_2/\omega_1$. If this ratio is rational, the map will consist of a finite number of points on a circle. If the frequencies are incommensurate, the quasiperiodic motion will result in a map filling an entire circle.

An important property of orbits of mappings like the circle map is the *winding number*, measuring the average rotation of successive iterates. It can be defined in general for a map given by the function f as

$$\rho = \lim_{n\to\infty} \frac{f^n(x) - x}{n}$$

where f^n denotes the nth iterate. The winding number is meaningful if it does not depend on the point of departure x, as it then becomes a property of the map itself. It is easy to find that the winding number of the linear circle map ((4.9) with $K = 0$) is $\rho = \Omega = \omega_2/\omega_1$. In the calculation of the winding number the modulo in the definition of f obviously has to be omitted.

The role of the circle map in our understanding of the quasiperiodic route to chaos is similar to that of the logistic map in the period-doubling cascade. The parameter K is considered as the control parameter and the circle map's behaviour is examined as K is varied. The parameter can be considered here, in contrast to the logistic map, as a nonlinear perturbation of a quasiperiodic solution. The relevant issue is the breakdown of a quasiperiodic solution (as, e.g., the one described above with $K = 0$) and its corresponding torus and the transition to chaos as K is increased. In the context of transition to turbulence K thus plays the role of the Reynolds number (see the second part of the book). We shall list below the most important results found in the study of the circle map. The behaviour of the circle map has been extensively studied by Jensen, Bak and Bohr. A good summary of the universal properties of the circle map and its comparison to the period-doubling cascade, together with the relevant references to original work, can be found in the book by Lichtenberg and Lieberman (1992). These properties and behaviour are considered to be generic in all systems, which become chaotic via the quasiperiodic route.

The circle map's behaviour is studied in its parameter space, i.e., the Ω–K plane. For $0 \leq K < 1$ the behaviour is similar, in principle, to that of the linear map. The map (4.9) is invertible and its behaviour depends on the winding number ρ, which is independent of the initial value. For $K = 0$ the winding number, as we have seen, is $\rho = \Omega$. If ρ is a rational number the map has periodic points. In contrast, when the winding number is irrational the iterates are dense on the circle and the map exhibits quasiperiodic behaviour. For $1 > K > 0$, ρ depends on both Ω and K and is rational in a *finite* range of Ω values. The regions in the Ω–K plane for which ρ is rational have the form of wedges, emanating from every rational number on the Ω-axis. This situation is shown schematically in Figure 4.9. A few representative wedge-shaped regions, called *Arnold tongues*, are shown. Within the Arnold tongues the orbits of the circle map are periodic and they are called mode (or frequency)-locked solutions. The boundaries of the Arnold tongues must, in general, be calculated numerically.

It is possible to show that the winding number in the circle map is rational, that is, $\rho = p/q$ with p and q integers, if and only if

$$f^q(x) = x + p, \quad \text{for some } 0 \leq x \leq 1$$

With the help of this condition the boundaries of the two Arnold tongues emanating from $\Omega = 0$ and $\Omega = 1$ can be calculated analytically. Indeed, for $q = 1$ the above relation implies that $\rho = p$ if and only if for some x

$$x + \Omega - \frac{K}{2\pi} \sin(2\pi x) = x + p$$

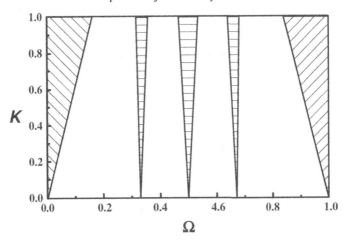

Fig. 4.9. Representative Arnold tongues in the Ω–K plane of the circle map, for K values up to 1.

i.e.,

$$\sin(2\pi x) = \frac{2\pi}{K}(\Omega - p) \qquad (4.11)$$

Now, Equation (4.11) has a solution in the unit interval, that is, is satisfied by some $0 \le x \le 1$ if $|2\pi(\Omega - p)/K| \le 1$. Consequently, the right boundary of the leftmost Arnold tongue ($p = 0$) is the line $K = 2\pi\Omega$, and the left boundary of the rightmost tongue ($p = 1$) is $K = 2\pi(1 - \Omega)$. This result is apparent in the figure.

The number of Arnold tongues, corresponding to rational numbers between 0 and 1, is a denumerable infinity. Their shapes are that of distorted triangles. As long as $K < 1$ the different tongues do not overlap. For small K, the probability (for a randomly picked Ω) that the winding number is rational is close to zero. In contrast, close to $K = 1$ this probability is close to one. At this point the set of all mode-locked motions is fractal and for $K > 1$ the tongues (regions of such motions) overlap and therefore different periodic motions coexist. Moreover, this also implies that the rotation number is no longer unique (it depends on the initial condition).

If we plot the winding number ρ as a function of Ω, for $K = 1$, a very interesting structure emerges (see Figure 4.10). The graph of $\rho(\Omega)$ has horizontal plateaux, corresponding to the rational values of ρ in the neighbourhood of rational values of Ω. In these regions the motion is mode-locked and is periodic. In between the plateaux, ρ increases with Ω, its value is irrational and the motion is quasiperiodic. The curious property of the graph is that if any of its sections is magnified, a similar structure at the new scale can be seen. This procedure can be continued

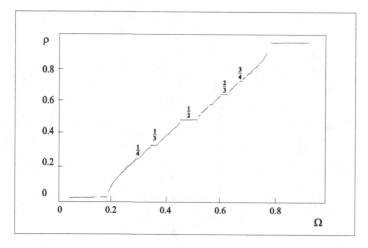

Fig. 4.10. The devil's staircase for the circle map, $K = 1$.

ad infinitum as there are infinitely many rational numbers and (nondenumerably) infinitely many irrational numbers between any two real numbers. The graph is fractal and its seemingly sinister appearance gave rise to its being called the *devil's staircase*. The fractal dimension of the quasiperiodic regions (the 'holes' in the devil's staircase at $K = 1$) has been found to be numerically equal to $D = 0.87$.

In spite of the complexity reflected by the devil's staircase, the dynamics remains regular (mode-locked periodic or quasiperiodic) as long as $0 \le K \le 1$. When $K > 1$ the Arnold tongues overlap, the circle map becomes noninvertible and the winding number is no longer independent of the iteration starting point. Since the winding number becomes ill defined, the orbits need not be periodic or quasiperiodic any more and chaos is possible. The closed curve in the circle map (or in the Poincaré map of a continuous flow) deforms and as K grows it develops wrinkles and breaks down. The transition to chaos is thus characterised by the non-invertibility of the Poincaré map, occurring at the critical value of the parameter, when the winding number becomes undefined. The quasiperiodic route to chaos has been studied theoretically in a number of systems and observed experimentally in a number of hydrodynamical settings, confirming the Ruelle–Takens–Newhouse scenario in these flows. The book by Bergé, Pomeau and Vidal (1986) contains a good summary of the quasiperiodic route.

Intermittency

The phenomenon of intermittency, characterised by a time signal having a regular appearance, interrupted by relatively short irregular bursts, has been observed in a large number of experiments. In particular, intermittency is a well known phenomenon in fluid dynamics. In a number of experimentally studied flows, laminar behaviour was interrupted by short outbreaks of turbulence as the Rayleigh or Reynolds number was increased. In these flows the duration of the turbulent bursts was found to increase with the control parameter on the way to a fully turbulent behaviour. This transition does not conform to the two universal routes to chaos described above and it was suggested in 1979 by Pomeau and Maneville as a third alternative. They investigated the Lorentz system (3.77) numerically and found intermittent behaviour on the way to chaos. In their pioneering work and in subsequent studies Pomeau, Maneville and collaborators proposed the intermittency route as a third universal mechanism for the transition to chaos.

The intermittency route is associated with a periodic attractor of a dynamical system bifurcating into a new, larger chaotic attractor, including the previous periodic orbit as its subset. The trajectory of the system can thus reside some time in the chaotic part of the attractor, but is ultimately attracted back to the periodic part. As the control parameter is increased, the relative importance of the chaotic part grows, ultimately taking over the whole attractor. This behaviour is, as usual, best studied with the help of discrete mappings. If we consider the Poincaré map of a flow, we may represent the flow's periodic solution by a fixed point on the map. The stability of this fixed point is governed by the size of the absolute value of the eigenvalues of the stability matrix (see Chapter 3). The loss of stability of a fixed point can occur, in the case of a discrete mapping, in three ways. Let q be a marginal eigenvalue of the stability matrix of a mapping. Then $|q|$ can become equal to 1 at the critical value of a control parameter if

(i) q is real and $q = 1$,
(ii) q is complex and both q and $q*$ are on the unit circle,
(iii) q is real and $q = -1$.

These three cases give rise to three different types of generic bifurcations, which were found to lead to intermittency. The intermittency mechanisms associated with these bifurcations have been classified as type I, type II and type III intermittency respectively. We shall discuss here only briefly a few important properties of the three types of intermittency.

Consider a mapping with a fixed point at the origin which becomes marginal as a parameter, p say, acquires the transition value p_T. Pomeau and Maneville found intermittency in the following generic maps, corresponding to the above behaviour of the eigenvalues. In their reduced form, valid when the iterates are close to the fixed point, the mappings are:

(i)

$$x_{j+1} = (1 + \lambda)x_j + x_j^2$$

(ii)

$$r_{j+1} = (1 + \lambda)r_j + r_j^3$$
$$\theta_{j+1} = \theta_j + \omega + r_j^2$$

(iii)

$$x_{j+1} = (1 + 2\lambda)x_j + x_j^3$$

where $\lambda \equiv p - p_T$ and ω is a constant. Note that (i) and (iii) are one-dimensional maps while the mapping (ii) is two-dimensional.

It is not difficult to identify these maps with the normal forms of bifurcations, given in Chapter 3. Indeed, for large j, the mappings can be approximated by differential equations with $x_{j+1} - x_j$ approximating the derivative with respect to a continuous independent variable. The bifurcations are thus classified as inverse tangent (i), subcritical Hopf (ii) and subcritical pitchfork (iii) respectively. In all the three cases, no stable solutions exist above the instability threshold. This is obviously necessary for intermittency to occur. The essence of intermittency is thus the destabilisation (or the disappearance) of a stable fixed point of the map, accompanied by the disappearance of one or two unstable fixed points or a limit cycle. Above marginality, but close enough to it, the attractor contains a 'ghost' of the stable solution which, before the bifurcation, was an attracting periodic solution of the flow. The repetitive return of the system to the neighbourhood of the 'ghost', called *reinjection*, is caused by a residual attracting structure of phase space. Since the limit cycle is no longer truly attracting, the trajectory ultimately leaves the *laminar* regime but is reinjected back into it after some time. As the control parameter is increased, the remnant of the attracting structure is totally destroyed and the system becomes fully chaotic. In Figure 4.11 we show schematically the bifurcation diagrams of the Poincaré maps of flows exhibiting the three types of intermittency. Each of these three types of intermittency transitions has its characteristic properties. In particular, it can be shown that the average time between irregular bursts scales as $\lambda^{-1/2}$ for type I intermittency and as λ^{-1} for the other two types. For a more comprehensive discussion of the various properties of aspects of the different intermittency types we refer the reader to the book by Bergé, Pomeau and Vidal (1986).

4.5 Bifurcations of homoclinic and heteroclinic orbits

In Chapter 3 we have seen that the existence of homoclinic and heteroclinic orbits may give rise to a transversal intersection of the stable and unstable manifolds of a hyperbolic point. While such a transversal intersection is impossible in two-dimensional flows, it may occur in two-dimensional maps. As was pointed out in

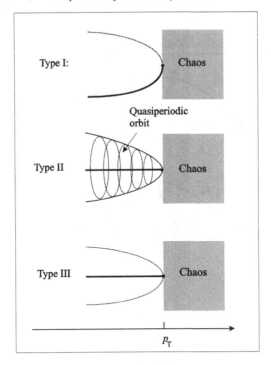

Fig. 4.11. Generic bifurcations of maps associated with the three types of intermittency, the point denoted by p_T on the parameter axis is the value at which the
transition occurs.

Chapter 3, transversal intersections of the stable and unstable manifolds result in
homoclinic and heteroclinic tangles. Such tangles are very complicated structures
with the stable and unstable manifolds intersecting an infinite number of times.
Since this complexity is possible in two-dimensional mappings it may occur in
the Poincaré maps of multidimensional flows. Homoclinic tangles in the Poincaré
maps generally reflect transversal intersections of invariant manifolds of unstable
periodic orbits in the associated flow. An unstable limit cycle may be considered
as a hyperbolic set of the flow, having its stable and unstable manifolds. If these
manifolds intersect transversally (the flow must be at least three-dimensional) it
is reflected in the surface of section by a homoclinic tangle. A pair of unstable
periodic orbits may similarly give rise to a heteroclinic tangle in the Poincaré map.
When the stable and unstable manifolds intersect, it is intuitively obvious that near
the fixed point (or points, in the case of a heteroclinic intersection) of the map,
the dynamics must be very complicated. In Figure 4.12 we show schematically the
formation of a homoclinic and heteroclinic tangle. Such tangles may appear in the
Poincaré map of a multidimensional flow when the control parameter is varied. In

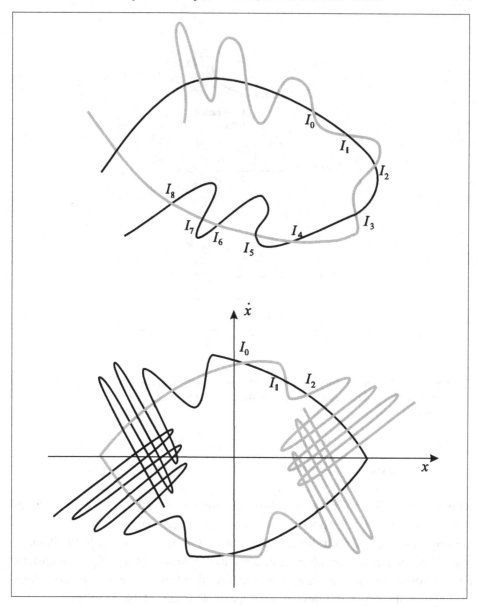

Fig. 4.12. Homoclinic (upper panel) and heteroclinic (lower panel) tangles in two-dimensional mappings (see also Figure 3.6). For details see text.

the upper panel the stable and unstable manifolds of a hyperbolic point intersect transversally at I_0, giving rise to an infinite number of intersections (on both sides of I_0) – a homoclinic tangle. In the lower panel two hyperbolic points are present and the transversal intersections of the stable manifold of one of them with the unstable one of the other gives rise to the heteroclinic tangle as shown. The axes drawn in the lower panel are only indicative, in the sense that the tangles may be seen in mappings resulting in the surface of section of a flow $x(t)$, arising from a higher than second-order ODE. Note that the shading of the orbits here is not significant.

Using the figure we consider a small element of phase volume, departing (i.e., taken as initial conditions) from the neighbourhood of a point near the transversal intersection I_0, along the unstable manifold of a fixed point. Clearly, when this volume element approaches the fixed point (homoclinic) or the other fixed point (heteroclinic) it will be distorted considerably, due to the repetitive intersections and stretching (and folding) associated with the relevant invariant sets. We may thus reasonably expect that the volume element will undergo an infinite sequence of stretch and fold transformations, not unlike the horseshoe map. This intuitive expectation is rigorously confirmed in the homoclinic case by a far reaching theorem, called the *Smale–Birkhoff homoclinic theorem*. The theorem implies that the dynamics close to a transverse homoclinic point (an intersection of the stable and unstable manifolds of a hyperbolic fixed point) is topologically similar to the horseshoe map. The corollary is that if a dynamical system exhibits a homoclinic tangle in its Poincaré map it has the property of SIC and hence is chaotic. The same conclusion is also valid in the case of heteroclinic tangles.

We have thus acquired a powerful diagnostic tool for finding chaotic behaviour. The existence of a transversal stable–unstable manifold intersection is sufficient for a confident prediction of chaos. Still, we face the nontrivial practical problem of ensuring the existence of a transversal manifold intersection. This problem may be approached numerically by computing the invariant manifolds of any particular system. Moreover, an analytical theory developed in 1963 by Melnikov may be effectively used. Melnikov's analytical technique provides a criterion which enables one to find the values of the different parameters for which *homoclinic* and *heteroclinic* bifurcations occur. At these bifurcations a homoclinic or heteroclinic orbit is created (or, if the approach is on the other side of the parameter value, destroyed) with all the accompanying dynamical complexity. The occurrence of horseshoe-like dynamics guarantees chaos, with its SIC property and the intermingling of stable and unstable (that is, bringing the system close to or far away from a fixed point) initial conditions.

Details of Melnikov's theory and the derivation of the so-called Melnikov function, necessary for the application of the criterion are beyond the scope of this book. The reader is referred to the book of Nayfeh and Balachandran (1995) for a lucid presentation of this topic. We shall conclude this discussion by quoting just one useful result, proved by Shilnikov in the late 1960s and generalised later on. He considered a general three-dimensional dynamical ODE system, dependent on one parameter λ and having a fixed point at the origin. We shall examine the case when the fixed point is a saddle spiral point (usually called saddle-focus). This means that one eigenvalue of the stability matrix is a positive real number and the other two are a complex conjugate pair with a negative real part. We may write these eigenvalues as $s_{1,2} = -\rho \pm i\omega$ and $s_3 = \sigma$ with ρ, σ and ω positive. Near the fixed point the system can be considered as the sum of the corresponding linearised system and the nonlinear part. It can be thus written as

$$\dot{x} = -\rho x - \omega y + f_x(x, y, z; \lambda)$$
$$\dot{y} = -\rho y + \omega x + f_y(x, y, z; \lambda)$$
$$\dot{z} = \sigma z + f_z(x, y, z; \lambda) \tag{4.12}$$

where the functions $f_\alpha(x, y, z; \lambda)$ for $\alpha = x, y, z$ and their first derivatives vanish at the origin for $\lambda = 0$. The additional assumption is that for $\lambda = 0$ the system (4.12) has a single orbit homoclinic to the origin. Shilnikov's work and later studies by Glendinning and Sparrow led to the following conclusion.

If $\sigma > \rho$, that is, the orbit's departure along the unstable direction is 'faster' than its spiralling in along the stable manifold, then for $\lambda = 0$ there are infinitely many unstable orbits near the homoclinic orbit, and it can in fact be said that there are infinitely many horseshoes. For $\lambda \neq 0$, the horseshoes still exist, but their number is finite and on one side of homoclinicity (with respect to the parameter λ) there are infinitely many more complicated homoclinic orbits.

In plain words this means that a situation, schematically depicted in Figure 4.13, implies chaos. The Shilnikov result can be generalised to systems of higher dimension as well, as long as the eigenvalues closest to the imaginary axis in the complex plane have the same properties as given above. This is so since the system

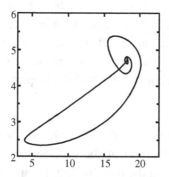

Fig. 4.13. A projection (onto a convenient plane whose axes are labelled in arbitrary units) of an orbit, homoclinic to a saddle-spiral fixed point, in an illustrative three-dimensional ODE system. The spiralling of the orbit into the fixed point is 'slower' here than its departure along the unstable direction, implying the existence of chaos. For details see text.

can then be reduced to a three-dimensional system like (4.12). The simple criterion, based on the Shilnikov theorem, is very useful if one wishes to get (or avoid!) chaotic behaviour in systems which contain some physical mechanism giving rise to a powerful excitation of oscillations, which slowly decay and are subsequently re-excited.

4.6 Theoretical predictive criteria of chaos

We conclude this chapter with a brief summary of the steps one may undertake in trying to diagnose a theoretical dynamical system (a given flow or discrete map) for chaos. Any similarity here to medical practice is not accidental and merely signifies that chaotic behaviour in dynamical systems is not yet fully and precisely understood.

A dynamical system exhibiting chaotic behaviour is, by definition, not soluble analytically. Its analysis must therefore be done using numerical computer calculations. The system can be numerically integrated or iterated, but since we are usually interested in the system's behaviour as some parameters are varied, a large number of orbits have to be computed. Moreover, it is not clear a priori what should be the initial conditions for the numerical calculations. This makes the number of orbits that have to be calculated even larger. In most cases chaotic behaviour occurs in just a portion of the parameter space and it may also be dependent on the initial conditions (this is the case in Hamiltonian systems, in particular). Thus the amount of numerical calculations needed to fully explore a dynamical system may be prohibitive. Brute force numerical calculation result in signals in the form of *numerical time series*. Such series are similar to the results of experimental measurements performed on natural systems. We thus defer the discussion on the methods for analysing these computer outputs to the next chapter, where we treat the analysis of time series in a general way.

The following list may be useful in the diagnosis of a dynamical system.

- Qualitative examination of representative numerical solutions.
  Stable equilibrium or periodic solutions can be easily identified. Intermittent behaviour should also be identifiable.
- Classification of the system and linear stability analysis.
  After the system is classified as dissipative or conservative, stable fixed points and/or limit cycles should be identified. It is important to find the critical value of the parameter(s), for which the fixed points and/or limit cycles lose their linear stability.
- Local bifurcation analysis.
  After the bifurcation points are found, they should be classified according to the generic normal forms. Dimensional reduction near the bifurcation point should be attempted. The analysis can then proceed on a low-dimensional system.

- Construction of Poincaré maps and the first return map.

  Poincaré maps of the system should be constructed using surfaces of section. The Poincaré maps can then be analysed instead of the original system. This is especially effective if the original system is multidimensional and cannot be easily reduced. In dissipative systems the Poincaré map can usually be further reduced to a one-dimensional first return map.

- Identification of universal transition behaviour.

  The Poincaré maps (or the original system) should be examined as a parameter is varied. If the typical features of a period-doubling cascade, or a quasiperiodic route, or intermittency are discovered, chaos is highly probable.

- Homoclinic and heteroclinic orbits and horseshoes.

  The system and its Poincaré maps may contain a transversal invariant manifold intersection for some range of parameters. This insures the presence of dynamics equivalent to the horseshoe map and thus chaos. A number of quantitative criteria due to Melnikov, Shilnikov and others may be useful in this context.

- Trajectory divergence.

  A heuristic (but easy) approach is based on calculating the distance between two prescribed neighbouring trajectories as a function of time. The resulting plots can be examined qualitatively. A quantitative (but usually difficult) approach gives the largest Liapunov exponent. This is a definitive test.

- Analysis of time series.

  The fractal dimension of the attractor can be computed directly or indirectly, using the methods explained in Chapter 5.

- Special methods for conservative systems.

  These will be discussed in Chapter 6, which is devoted in its entirety to such systems.

5

Analysis of time series

Time present and time past
Are both present in time future
And time future contained in time past.
. . . .
What might have been and what has been
Point to one end, which is always present.
T. S. Eliot, *Burnt Norton, I.*

Time-dependent signals are measured in many lab experiments and observations. This is certainly so if for the physical system studied it seems that theoretical modelling by a dynamical system is natural. A measured physical quantity (like, e.g., voltage, radiation intensity, temperature etc.) is usually recorded by a detector as a function of time. In most cases the signal is *digital*, that is, the numerical value of some physical variable is sampled at discrete time intervals. A digital signal is also obtained when a theoretical nonlinear dynamical system is studied numerically. Computer simulations are thus very similar, in this respect, to experiments or observations. There is, however, an important difference between the two. A theoretical dynamical system is, as a rule, well defined by a set of differential equations or maps. The dimension of the state space is known and a direct numerical simulation is just one of the possible approaches to study of the system. In contrast, when dealing with experimental or observational data, the dynamical system responsible for producing them is usually unknown, even though we may have some model in mind, and in most cases we do not even have any direct access to all the degrees of freedom of the system. Thus, even the dimension of the state space is generally not known.

Chapters 3 and 4 were devoted to descriptions of the different methods that can be applied in the study of well defined theoretical dynamical systems. To sum up in a few sentences a sometimes lengthy and complex investigation, we may (using some of the highlights of Chapter 3 and the list at the end of Chapter 4) mention here the main stages of such a procedure.

During the first steps in the characterisation of the dynamics one usually unfolds the properties of the system's phase space by analysing its fixed points, limit cycles and their stability, and by examining phase portraits. Of particular interest are bifurcations as a parameter is varied. Techniques of dimensional reduction can be applied close to marginality and these provide important information on the system's generic properties. If one wishes to discover chaos it is useful to construct surface-of-section Poincaré maps, first-return maps, and look for the possibility of homoclinic or heteroclinic tangles in them. Calculations of the fractal dimension of attractors in dissipative systems and of the largest Liapunov exponent in general systems are considered to be the ultimate tests for chaos.

Most of the above-mentioned powerful tools cannot be used directly, however, when the available information includes only series of measurements of physical observables. In many cases the experiments and observations are performed for the purpose of finding a suitable theoretical explanation of the data, usually by a model based on a dynamical system. If the signal is regular this task may be relatively easy, but it is obviously not so if the data are erratic. Is the irregularity of the signal reflecting an extremely complicated, highly dimensional system, that must be described by an enormous number of variables and couplings? Or maybe it is due to chaotic behaviour of a low-dimensional deterministic system? The answer to such questions is very important if we wish to apply dynamical system theory and chaos to natural phenomena.

These and related questions will be addressed in this chapter. We shall assume that a certain lab or computer experiment or observation provides a digital signal or *time series*, with a certain number of points at given values of the time variable. Methods to study such time series will be introduced, to provide quantitative criteria for deterministic chaos. The crucial problem in this context is the ability to extract a faithful representation of the data in some space, which can be perceived as the state space of the underlying dynamical system. Once this is done, the above mentioned techniques can be attempted. There are obvious limitations, stemming from the fact that given data sets may contain random experimental errors and they are given sometimes at too long a time interval. These difficulties can usually be overcome by accumulating a large amount of high quality data. In any case, the work of astronomers and experimental physicists is essential to the development of theory, and purely mathematical issues, however important, cannot replace physical theories.

Astronomers do not need, most probably, to be convinced of the importance of the data reduction methods that will be discussed here. The example of variable stars is classic in this context. If a signal from such a star is periodic or multiperiodic there is no problem. The frequencies are found by standard techniques. But is it wise to force such techniques onto irregularly variable signals? What is the use of finding 'time varying frequencies', especially when the time scale of such a variation is not too far from the period (of this frequency)? What can be deduced about the physical system from the fact that a signal produced by it is chaotic? We shall address these issues in astrophysical contexts (including variable stars) in the second part of this book, but first we introduce, in this chapter, the necessary concepts and methods.

5.1 Time histories and Fourier spectra

The examination of a long sequence of a time series, its *time history*, is the first step in an attempt to characterise the signal. The study of time series is a rather rich subject with a variety of applications in various branches of science and engineering. We shall only define here a number of basic useful concepts in this context. These definitions will be rather heuristic but they will be sufficient for the discussion in this chapter.

When examining a numerical or experimental signal, one is obviously concerned with numerical or measurement errors. Systematical errors must be avoided or eliminated and random errors (noise) have to be judiciously estimated. This is especially important for irregular signals (see below). The presence of noise limits the amount of information in a signal and new methods for noise reduction are now available (see the second part of the book). The signal may itself be *random* or *deterministic* (produced by a deterministic system). The methods for distinguishing between the two will be described below. In some laboratory situations one can subject the system to well defined deterministic excitations and check if the output is repeatable (within the bounds of experimental error). If it is not, the system is necessarily random. A signal from a deterministic or random system can be further classified as being *stationary* or *non-stationary*. In stationary signals the relevant properties such as Fourier spectra, averages, autocorrelation (see below) are time-independent. If these properties are time-dependent, but only during a finite period, and they gradually approach constant values, the signal is said to have *transient* behaviour.

The first and most basic tool in analysing a time history is the *Fourier* or *frequency spectrum* of the signal. Fourier spectra are useful for stationary signals. The mathematical basis for the Fourier transform was developed by the French mathematician Jean-Baptiste Fourier during his work on the heat equation. Although it is to be expected that most readers of this book have already been exposed to methods of Fourier analysis, we we shall give here, for convenience, some relevant results based on this mathematical theory.

The *Fourier transform* of a continuous signal $x(t)$ is defined as

$$\hat{x}(\omega) = \int_{-\infty}^{\infty} x(t)e^{-i\omega t}\,dt \tag{5.1}$$

where ω is the angular frequency and $\hat{x}$ is in general complex. The *frequency*, f, defined as $f \equiv \omega/2\pi$ is often used in (5.1) instead of ω.

In the above definition it is assumed that $x(t)$ is an integrable function. The inverse transform then exists and is given by the relation

$$x(t) = \frac{1}{2\pi} \int_{-\infty}^{\infty} \hat{x}(\omega)e^{i\omega t}\,d\omega \tag{5.2}$$

where the quantity in front of the integral is a normalisation constant and can be chosen in different (but consistent) ways in (5.1) and (5.2).

We further remark that if the function $x(t)$ is periodic with some period T, that is, if $x(t) = x(t + T)$ holds for all t, it can be written as an infinite Fourier series

$$x(t) = \sum_{j=-\infty}^{\infty} c_j \exp\left(i\frac{2\pi jt}{T}\right) \tag{5.3}$$

which converges if, in addition to being periodic, $x(t)$ satisfies some (not very stringent) requirements, called the Dirichlet conditions. The Fourier coefficients c_j are given by

$$c_j = \frac{1}{T} \int_0^T x(t) \exp\left(-i\frac{2\pi jt}{T}\right)$$

If one wishes to treat only real functions, separate sine and cosine Fourier series and transforms are defined. These are especially useful if the function is even or odd. Properties of the Fourier transform and series and the relevant mathematical theorems can be found in any book on mathematical methods (e.g., Boas, 1983). Here we shall discuss just the concepts relevant to the time-series analysis in our context.

First, we distinguish between the *amplitude* spectrum, $\hat{x}(f)$ (actually its real or imaginary part), and the *power* spectrum, defined as the function $P(f) \equiv |\hat{x}(f)|^2$. A single peak in the power spectrum at some f_1 indicates that the signal contains a periodic, purely sinusoidal, component with period $T = 1/f_1$. Next, we remark that a periodic signal whose period is T say, but is not a pure sinusoidal function, will contain in its power spectrum, in addition to the peak at $f_1 = 1/T$, peaks at integral multiples of f_1 also. These are the *harmonics* of the basic frequency and their occurrence is obviously the result of the fact that such a periodic function can be written as Fourier series, containing sinusoidal functions with these harmonic frequencies. The higher the harmonic frequency the smaller will be the height of the peak – a result of the fact that Fourier series of periodic functions converge.

Finally, we observe that in practice the signal is known for only a finite interval of time of duration t_{max}, say. Thus the 'theoretical' Fourier transform cannot be obtained and a *finite Fourier transform* is used

$$\hat{x}(f, t_{max}) = \int_0^{t_{max}} x(t)e^{-2i\pi f t}dt \qquad (5.4)$$

where $\hat{x}$ is now a (complex) function of the total time interval also.

Discrete Fourier transform

The signal $x(t)$ which we have used in the above definitions is in practice not only unknown for infinite time but also it is usually a *discrete* time series, that is, the value of the function $x(t)$ is known for only discrete time values. Assume that a given time series consists of n (usually quite a large number) data points, taken at fixed time intervals $\Delta t = t_{max}/n$, located in the finite time interval $0 < t \leq t_{max}$. We thus have only the n quantities $x_j \equiv x(j \Delta t)$, say, for $j = 1, 2, \ldots, n$. The *discrete Fourier transform* of the sequence x_j is defined as the series

$$\hat{x}_k = \sum_{j=1}^n x_j \exp\left(-i\frac{2\pi jk}{n}\right) \qquad (5.5)$$

for $k = 1, 2, \ldots, n$.

It can be seen that (5.5) is just the discrete approximation of (5.4) and tends towards it as $n \to \infty$ and $\Delta t \to 0$. The discrete transform consists of a series having n terms, each corresponding to the discrete frequency $f_k = k \Delta f$, with $\Delta f = 1/t_{max}$, and there is a maximal frequency in the series - $f_{max} = n \Delta f = 1/\Delta t$. The inverse transform is

$$x_j = \frac{1}{n}\sum_{k=1}^n \hat{x}_k \exp\left(i\frac{2\pi kj}{n}\right) \qquad (5.6)$$

Note that using the inverse transform, the values of the components x_j can be defined for any natural number j. Such a definition merely extends the domain of $x(t)$ by repeating it periodically. This can be easily seen from (5.6), which gives

$$x_{j+n} = x_j$$

for all j. Thus the finite discrete Fourier transform is just a finite Fourier series of an artificially extended periodic function.

The discrete Fourier transform has a number of characteristics that are worth stressing. In Figure 5.1 we plot, as an example, a certain discrete signal (time series), sampled on $n = 50$ points at equal intervals over time $1 \leq t \leq t_{max} = 50$, in some units, (i.e., $\Delta t = 1$) and its discrete power spectrum. The points on both

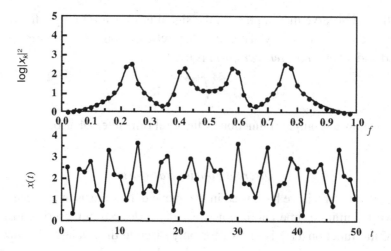

Fig. 5.1. Time series (lower panel) and its discrete power spectrum (upper panel).

graphs are joined by a continuous curve for the sake of clarity. As we have already noted, the spectral resolution of the power spectrum (the distance on the frequency axis between two adjacent points) is $\Delta f = 1/t_{max} = 1/(n\,\Delta t)$, and the spectral domain (the highest frequency of the spectrum) is $f_{max} = 1/\Delta t$. In the case shown in the figure $\Delta f = 0.02$ and $f_{max} = 1$ in the appropriate units.

To enlarge the frequency domain we must reduce the sampling interval Δt. Better spectral resolution will be obtained by increasing the product $n\Delta t$. Thus if Δt has a minimum determined by experimental constraints the spectral domain is fixed, but the resolution gets better with an increasing number of measurements n. An additional property apparent from Figure 5.1 is the symmetry of the power spectrum around its middle (at $f = 0.5\,f_{max}$). This is due to the fact that the time series in this example is real-valued. For real-valued functions $x(t)$, the Fourier transform has the property $\hat{x}(-\omega) = \hat{x}^*(\omega)$ and thus $|\hat{x}(\omega)|^2$, the power, is an even function of ω. The corresponding relationship for a finite interval discrete case is

$$\hat{x}_k = \hat{x}^*_{n-k} \longrightarrow |\hat{x}_k|^2 = |\hat{x}_{n-k}|^2$$

The actual computation of a discrete Fourier transform (with large n) of a signal may be very costly in computer time, if we apply formula (5.5) in a straightforward way. For $n = 10^4$, say (a reasonable number needed for a faithful representation of a signal) 10 000 sums, each with 10 000 terms, have to be calculated. The number of operations is thus of the order of n^2. Fortunately, however, when n is a power of 2, an algorithm, called *fast Fourier transform* (FFT), is available. It is an efficient computational scheme that takes advantage of certain symmetry properties of the circular functions. Fast Fourier transform requires only $n\log_2 n$ operations (instead of n^2) for the computation of a discrete Fourier transform of a signal sampled at n points. The gain in the number of operations is in fact a factor of $n/\log_2 n \approx 0.3n/\log_{10} n$. For $n = 10\,000$, for example, the factor is ~ 750 and it grows rapidly with n. This makes the computation of Fourier transforms feasible in practice and a number of standard software packages are available (e.g., in IMSL, MATLAB, Mathematica, etc.).

Before examining in some detail the properties of Fourier transforms of some special signal types, we define an additional useful notion. For a given signal $x(t)$, its *autocorrelation function* is

$$R_{xx}(\tau) = \lim_{T \to \infty} \frac{1}{T} \int_0^T x(t)x(t + \tau)\,dt$$

Clearly, this function measures an average of the product of the signal value at some time with its value at a time τ later. We can thus deduce from R_{xx} whether the signal at some time depends on (is correlated with) the same signal at another time. The discrete version of the autocorrelation function is

$$R_m = \sum_j x_j x_{j+m} \tag{5.7}$$

Since $x_{j+n} = x_j$ the discrete autocorrelation function is also periodic, i.e., $R_{m+n} = R_m$. It is not difficult to show that the power spectrum of a signal $|\hat{x}|^2$ is just the cosine Fourier transform of the autocorrelation function

$$|\hat{x}_k|^2 = \sum_{m=1}^n R_m \cos\left(\frac{2\pi mk}{n}\right) \tag{5.8}$$

This constitutes one of the forms of a general property of Fourier transforms, known as the Wiener–Khintchin relation (see e.g., Bergé, Pomeau & Vidal, 1986).

We conclude with a remark about notation. The discrete Fourier power spectrum which can be written as $P_k \equiv |\hat{x}_k|^2$ is defined for the natural numbers k. As we have seen before, each k corresponds to a frequency $f_k = k\Delta f = k/t_{\max}$. Thus we may well approximate the continuous power spectrum $P(f)$ by $P(f_k) = P_k$, provided the spectral resolution Δf is small enough.

Fourier power spectra of periodic signals

In order to identify the properties of the discrete Fourier transform of a periodic signal we consider the simplest signal of this sort, with period T, say, namely the purely sinusoidal complex function

$$x(t) = \exp\left(i\frac{2\pi t}{T}\right)$$

The complex exponential was chosen here for computational convenience and our qualitative findings, regarding the properties of the Fourier transform, will be valid also for real periodic signals.

Assume that the signal is sampled at n discrete times with equal intervals Δt, over the time $\Delta t \le t \le n\Delta t = t_{\max}$. Two obvious requirements are necessary for a meaningful discrete transform. They are that $t_{\max}$, the measurement time, is significantly larger than T, the signal period, and that the latter quantity is significantly larger than the time resolution Δt. Thus we assume

$$n \gg \frac{T}{\Delta t} \gg 1 \qquad (5.9)$$

The discrete signal can be written as

$$x_j = \exp\left(i\frac{2\pi j \Delta t}{T}\right) \qquad (5.10)$$

for $j = 1, \ldots, n$, and thus its discrete Fourier transform (see 5.5) is

$$\hat{x}_k = \sum_{j=1}^{n} \exp\left[i2\pi j \left(\frac{\Delta t}{T} - \frac{k}{n}\right)\right]$$

Define now

$$\psi_k \equiv 2\pi \left(\frac{\Delta t}{T} - \frac{k}{n}\right) \qquad (5.11)$$

and the Fourier transform is clearly the sum of a geometric series and can be readily evaluated analytically.

$$\hat{x}_k = \sum_{j=1}^{n} e^{ij\psi_k} = e^{i\psi_k}\frac{e^{in\psi_k} - 1}{e^{i\psi_k} - 1} = \exp\left[i\frac{(n+1)\psi_k}{2}\right]\frac{\sin(n\frac{\psi_k}{2})}{\sin(\frac{\psi_k}{2})} \qquad (5.12)$$

The power spectrum is thus

$$P_k = |\hat{x}_k|^2 = \frac{\sin^2\left(n\frac{\psi_k}{2}\right)}{\sin^2\left(\frac{\psi_k}{2}\right)}$$

For very large n it is easy to discover some of the important and instructive properties of this function. Let $t_{\max} = n\,\Delta t$ be fixed. The limit of $n \to \infty$ means then that $\Delta t \to 0$ and thus $\psi_k \to 0$ with $n\psi_k$ finite (see 5.11). In this limit the power spectrum is approximately

$$|\hat{x}_k|^2 \approx n^2\frac{\sin^2 z_k}{z_k^2} \qquad (5.13)$$

where

$$z_k \equiv n\frac{\psi_k}{2} = \pi\left(\frac{t_{\max}}{T} - k\right)$$

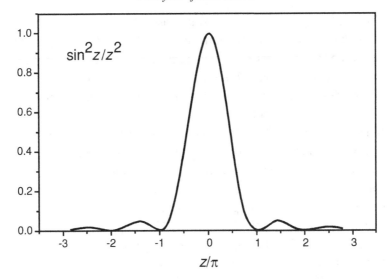

Fig. 5.2. The function $\sin^2 z/z^2$, which approximates well the discrete (normalised) Fourier transform of the discrete periodic signal (5.10) when the number of signal points is very large. See text for details.

If z_k were a continuous quantity (call it z) the power spectrum would peak at $z = 0$, i.e., for $k_0 = t_{max}/T$. In terms of the frequency this corresponds to

$$f_0 = k_0 \Delta f = \frac{k_0}{t_{max}} = \frac{1}{T}$$

This result is expected, as we obtain maximum power at the frequency of the original signal. Since z_k is discrete, the result depends on the spectral resolution. The larger the measurement time t_{max} the better the discrete transform represents the actual transform, but since t_{max} is finite we get, in addition to the peak at the signal frequency, a series of secondary maxima also, as can be seen in Figure 5.2, which is actually the plot of the continuous function $\sin^2 z/z^2$. These secondary maxima decay away from the peak and are called side-lobes. This process, where power at a certain frequency artificially (because of the finite time extent of the signal) 'leaks' to adjacent frequencies, is referred to as *leakage*. Leakage can be decreased by extending t_{max}, and it can also be eliminated altogether (see below).

The width of the central peak depends on the ratio t_{max}/T. It is largest when this ratio is half-integer. If the duration of the measurement t_{max} contains an integral number of periods, i.e., $n\Delta t/T = l$, with l an integer, $z_k = (l - k)\pi$, i.e., the numerator of the expression in the power spectrum (5.13) is always zero. The denominator is zero only for $k = l$ and thus the power spectrum contains just the central infinitely narrow (δ function) peak. Leakage is totally absent in this case.

However, this only happens in the (unlikely) case when an exact *resonance* condition, between the measurement duration and the period, is fulfilled. The side-lobes in the discrete Fourier transform are thus seen to be the result of a measurement over a time span that is not only finite, but also contains a non-integral number of periods of the original periodic signal. Practical techniques are available to reduce leakage when the duration of time series measurements is limited and the frequency is unknown a priori, but we shall not mention them here and refer the reader to any of the many books on the subject (e.g., Oppenheim & Schafer, 1975).

So far we have only considered here a discrete *sinusoidal* periodic signal. In the case of a discrete signal that is periodic with period T, but not purely sinusoidal, we can still obtain in the power spectrum peaks at harmonic frequencies (as in the continuous case explained above). This happens when a special resonance condition is satisfied, i.e., $n\Delta t = T$ (the measurement time is equal to the period) and the Fourier components $\hat{x}_k$ in the series (5.5) are located at frequencies $f_k = k\Delta f = k/T$. We thus have contributions at the fundamental frequency $f_1 = 1/T$ and its harmonics $2f, 3f, \ldots, nf$. If we have a measurement duration containing an integral number of periods, that is, $n\Delta t = lT$, with l being an integer, the frequencies of the peaks will be the same, and the spectral resolution will be better, but we will have $|\hat{x}_k|^2 = 0$ for all k that are not integer multiples of l. This result is obvious if one is familiar with the above mentioned concept of Fourier series expansion of a periodic function (see 5.3).

In summary, the discrete power spectrum of a periodic time signal with period T, sampled at a large enough number of points n in an interval of duration $t_{max} = n\,\Delta t$, contains a peak at the frequency $f = 1/T$, its side-lobes and possibly additional peaks (with their side-lobes) at harmonics of f. The central peak's width becomes very small and the side-lobes unimportant if one is close to a resonance condition (t_{max} contains an integral number of periods T). The power in the harmonics usually decreases very fast with their frequency. If the signal is purely sinusoidal, the harmonics are not present in the power spectrum.

Fourier power spectra of quasiperiodic and aperiodic signals

We shall now consider signals that are not simply periodic, and start by describing the qualitative features of the power spectrum of a *quasiperiodic* signal. We have defined quasiperiodicity in Chapter 4 and discussed its importance in transition to chaos. Here we would like to explore the characteristic features of quasiperiodicity in its Fourier transform. Consider a k-quasiperiodic signal, that is,

$$x(t) = x(\omega_1 t, \omega_2 t, \ldots, \omega_k t)$$

with $x(t)$ being separately periodic in each of its arguments with periods $T_j = 2\pi/\omega_j$ for $j = 1, 2, \ldots, k$, and where the frequencies ω_j are *incommensurate* (as defined in Chapter 4). A multiperiodic function can always be written as a k-tuple Fourier series (generalisation of 5.3).

$$x(t) = \sum_{j_1} \sum_{j_2} \cdots \sum_{j_k} c_{j_1 \, j_2 \ldots j_k} \exp\left[i\left(\frac{2\pi j_1 t}{T_1} + \frac{2\pi j_2 t}{T_2} + \cdots + \frac{2\pi j_k t}{T_k}\right)\right] \quad (5.14)$$

With $\omega_m = 2\pi/T_m$, this can be written in a more compact form as

$$x(t) = \sum_{\mathbf{j}} c_{\mathbf{j}} \exp\left[i\mathbf{j} \cdot \boldsymbol{\omega} t\right]$$

where the k-tuples $(j_1, j_2, \ldots, j_k)$ and $(\omega_1, \omega_2, \ldots, \omega_k)$ are denoted by $\mathbf{j}$ and $\boldsymbol{\omega}$ respectively. The Fourier transform of such a function may have a much more complex appearance than that of a periodic signal. It can rather easily be verified that the Fourier power spectrum of a continuous signal like (5.14) contains numerous peaks at all frequencies f, satisfying

$$2\pi f = |\mathbf{j} \cdot \boldsymbol{\omega}| = |j_1 \omega_1 + j_2 \omega_2 + \cdots + j_k \omega_k|$$

for any combination of the integers $j_1, j_2, \ldots, j_k$.

The complexity of such a spectrum can be appreciated even if we consider for simplicity just the biperiodic ($k = 2$) case. The frequencies are incommensurate and thus the peaks, which appear at all frequencies satisfying $f = |j_1 f_1 + j_2 f_2|$, with $f_m = \omega_m/(2\pi)$, will be a dense set. That is, between any two peaks there will be an additional peak. This does not mean that the power spectrum will look smooth. The different peaks may have very different amplitudes. It can be shown that in most cases the amplitudes of the peaks corresponding to high j_1 and j_2 values are very small indeed (exponentially small). Thus the power spectrum will be dominated by the peaks at frequencies composed of those combinations of f_1 and f_2 in which j_1 and j_2 are rather small. If the two frequencies are commensurate (f_1/f_2 is rational $= n_1/n_2$ with n_i integer), the signal is not quasiperiodic but actually periodic, its period being $T = n_1/f_1 = n_2/f_2$. All the peaks of the Fourier spectrum, in this case, will be harmonics of the lowest frequency $f_0 = f_1/n_1$. Thus the peaks will not be dense in this case.

These results obviously carry over to the case of $k > 2$ quasiperiodicity. If the signal is sampled at finite intervals, additional features like side-lobes (see above) further complicate the power spectrum of a quasiperiodic signal. Thus the power spectrum of a quasiperiodic signal may look very complicated indeed. In Figure 5.3 the power spectrum of a three-period quasiperiodic signal is shown.

If the signal is neither periodic nor quasiperiodic it is called *aperiodic*. The Fourier power spectrum of an aperiodic signal is continuous. Thus it may seem

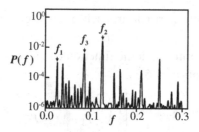

Fig. 5.3. Power spectrum (in arbitrary units) of a three-period quasiperiodic signal. The three incommensurate frequencies f_1, f_2, f_3 appear as the highest peaks.

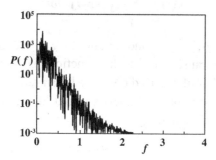

Fig. 5.4. Power spectrum (in arbitrary units) of an aperiodic signal (obtained from integration of the Rössler system in the chaotic regime).

that a simple Fourier transform can be used as the definitive diagnostic tool for determining whether a signal is aperiodic or not. The problem is that it is not easy to distinguish between a truly continuous spectrum and a dense quasiperiodic one. If the spectrum of an aperiodic signal of duration t_{max} is sampled discretely, we can find quite different amplitudes $|\hat{x}_k|^2$ for neighbouring values of k. A continuous-looking power spectrum will be obtained in this case if one averages over many spectra taken with different $\Delta t \ll t_{max}$ or alternatively locally, over a certain number of consecutive values of k. The spectrum of an aperiodic signal approaches a continuous function only in the mean. Still, if a quasiperiodic signal includes very many frequencies, and possibly also side-lobes, its power spectrum may easily be mistaken for an aperiodic signal spectrum. Figure 5.4 shows the power spectrum of a typical aperiodic signal. The specific power spectrum shown was obtained by numerical integration of the Rössler oscillator equations (4.5) in the chaotic regime. The time series were generated using several choices of the time averaging, and the power spectra for the different cases were then averaged (see above). In this way a power spectrum of 'broad-band' nature, which is characteristic of an aperiodic signal, can be clearly seen.

Even if it is certain that the signal has broad-band continuous characteristics and thus the original signal is aperiodic, the important question raised in the beginning of this section remains. How can one distinguish between an aperiodic signal resulting from deterministic chaotic dynamics and one that is truly random? A random signal, also called *noise*, results from a huge number of degrees of freedom, and the only useful approach aimed at analysing such a system must be probabilistic. In contrast, a low-dimensional deterministic system can be studied by the various methods described in this book. As we shall see shortly, the Fourier transform by itself cannot, unfortunately, provide a definite answer to the question posed above.

The most extreme case of a random system, 'white noise', is a signal devoid of any correlations. Let $x(t)$ be such a signal with zero mean and of a given variance $\sigma = \sqrt{\langle x^2 \rangle}$, where the meaning of averaging here is over different realisations of a discrete sampling (ensemble average). The absence of correlations in discrete time series $x_j = x(j\Delta t)$ can be expressed by the condition

$$\langle x_j x_{j+m} \rangle = \sigma^2 \delta_{m,0}$$

for all $j = 1, 2, \ldots, n$, where $\delta_{i,j} = 1$ for $i = j$, and it is zero otherwise. Thus the average of the autocorrelation function satisfies

$$\langle R_m \rangle = \sum_j \langle x_j x_{j+m} \rangle = n\sigma^2 \delta_{m,0}$$

Now using the inverse transform of (5.8) we get

$$\langle |\hat{x}_k|^2 \rangle = \frac{1}{n} \sum_m \langle R_m \rangle \cos\left(\frac{2\pi mk}{n}\right) = \sigma^2$$

So the average amplitude of the power spectrum is independent of k and thus of the frequency. The power spectrum of white noise does not have any structure – it is flat on average.

However, not all random systems produce white-noise signals. Power spectra of some of them have the appearance of a power law. Because of discrete sampling and experimental or numerical errors, it is often difficult to distinguish between the spectra of random and deterministic (aperiodic or even quasiperiodic with large number of frequencies) signals. If a signal has a continuous-looking power spectrum and it is impossible to determine definitely if is random or deterministic (quasiperiodic or aperiodic), we must have recourse to other methods of analysing the time series. We now turn to the description of such methods.

5.2 The pseudo-state space and embedding

When we consider a dynamical system defined by a set of differential equations, the natural mathematical space in which the motion is well defined is the system's *phase space* or *state space*. The notion of phase space is obvious in Hamiltonian systems and was also defined in this book for dissipative n-dimensional systems of the familiar form

$$\dot{\mathbf{x}}(t) = \mathbf{F}[\mathbf{x}(t)] \quad \text{or} \quad \dot{\mathbf{x}}_{i+1} = \mathbf{F}(\mathbf{x}_i) \tag{5.15}$$

as the Cartesian space spanned by the variables $\{x^k; \quad k = 1, \ldots, n\}$.

The notion of a *state space* is more general. It is sometimes advantageous to define the *state variables* (the set of variables needed to uniquely specify the state of the system) in some alternative way. The *state space* is then the space spanned by the state variables. For example, a one-degree-of-freedom oscillator described by

$$\ddot{x} = f(x, \dot{x})$$

where f is some suitable function, has a natural Cartesian phase space $\mathcal{R}^2$, spanned by the variables x and $y \equiv \dot{x}$. However, if we consider a forced oscillator, the system describing it is non-autonomous

$$\ddot{x} = f(x, \dot{x}) + \cos(\omega t)$$

say. As we have seen (e.g., the discussion of the Duffing system in Chapter 2), this system can be made autonomous by the definition $z \equiv \omega t$, and the additional ODE is

$$\dot{z} = \omega$$

The third state variable z, is clearly an angular variable, since $\cos(z) = \cos(z+2\pi)$. Thus, we may call it $\theta \equiv z \pmod{2\pi}$ for convenience, and we get a *cylindrical state space* $\mathcal{R}^2 \times \mathcal{S}^1$, where $\mathcal{S}^1$ denotes the one-dimensional sphere (a circle). The system evolution is represented by a trajectory on the surface of a cylinder. The invariant tori of Hamiltonian systems (see below in the next chapter) are also state spaces. These state spaces are manifolds, that is, local neighbourhoods of given points in such spaces always have the properties of a Euclidean space.

The geometrical features of flows in their state spaces are, thus, equivalent to the properties described in Chapter 3. Fixed points are points and limit cycles are closed curves and strange attractors have a definite fractal dimension. If the state space is two- or three-dimensional the dynamics of a system can be visualised directly. Together with the power spectrum a lot of information can be found about the system. In simple cases (periodic or low quasiperiodic motion) this may be sufficient. If the system has more than three dimensions we can only visualise projections of the dynamics and these may produce spurious features (like trajectory crossings). Still, such projections can be quite useful when they are complemented by the information extracted from the Fourier transform.

As we have seen, dissipative dynamical systems may have attractors (normal or strange) whose dimension is much lower than that of the full state space. It would thus be very useful to have a general method for the reconstruction of the important low-dimensional dynamics in a suitably chosen low-dimensional space. This task seems especially difficult if we do not explicitly know the system (5.15) and have only a single experimental or numerical signal resulting from it. Fortunately and quite surprisingly, such a method does exist and moreover it is guaranteed by exact mathematical theorems. We shall describe it below for the apparently most demanding case mentioned above (just a signal of an unknown system is available). The method can also be applied, of course, when the multidimensional dynamical system is explicitly known, since the signal can then be produced by a numerical experiment.

Assume that we can only measure one (scalar) variable arising from a system like (5.15), call it s say, and consider for the moment the case when this variable is known for all times, i.e., $s(t)$ is a continuous function. In practice s will usually be known only at discrete times and we shall discuss this case shortly. Let the original unknown (hypothetical) state-space variables vector be $\mathbf{x}(t)$.

The variable s is assumed to be a well defined smooth scalar function of the state vector, i.e.,

$$s(t) = S[\mathbf{x}(t)]$$

By choosing a suitable fixed time interval τ, called the *time delay*, one can construct the d-dimensional vector

$$\mathbf{y}(t) = \{s(t), s(t + \tau), s(t + 2\tau), \ldots, s[t + (d-1)\tau]\}^{\mathrm{T}} \qquad (5.16)$$

where the superscript T, meaning 'transpose', reflects as usual the fact that we are considering a column vector. In a deterministic system if the 'true' state vector $\mathbf{x}$ is given at some time, it uniquely determines the value of $\mathbf{x}$ at all other times. Hence $\mathbf{x}(t + m\tau)$ can be regarded as a unique (vector) function of $\mathbf{x}(t)$, i.e.,

$$\mathbf{x}(t + m\tau) = \mathbf{G}_m[\mathbf{x}(t)] \text{ for } m = 0, 1, \ldots, (d-1)$$

Thus all the components of $\mathbf{y}$ in (5.16) are functions of $\mathbf{x}$. For instance the kth component of $\mathbf{y}(t)$ can be written as

$$y^k(t) = s[t + (k-1)\tau] = S\{\mathbf{x}[t + (k-1)\tau]\} = S\{\mathbf{G}_{k-1}[\mathbf{x}(t)]\}$$

As a result, the d-dimensional *delay coordinate vector* can be expressed as

$$\mathbf{y}(t) = \mathbf{E}[\mathbf{x}(t)] \qquad (5.17)$$

where $\mathbf{E}$ is a smooth nonlinear vector (d-dimensional) function.

Now the **y** can be regarded as state variables, uniquely determined by the original state variables **x**. The space spanned by the state vectors **y** is called *pseudo-state space* and if certain requirements are met it can be considered as a faithful *reconstruction* of the original (maybe unknown) system state space. The pioneering idea that the pseudo-state space formed by the delay coordinate vectors can be used to reconstruct the original dynamics was introduced in 1980 by Ruelle and by Packard *et al.* Subsequent studies by Mañé and by Takens put this idea on a firmer footing and formulated the exact conditions under which the *essential* dynamics, that is, the invariant properties of the motion in the reconstructed state space, is *equivalent* to the original dynamics. In plain words these conditions are that time delay τ has to be chosen in a suitable way (it must be smaller than all relevant time scales), and d has to be, of course, sufficiently large. The equivalence of the dynamics means that there exists a one-to-one mapping (in the topological sense) between the relevant dynamical structures (like attractors) in the original n-dimensional state space and the corresponding structures in the reconstructed d-dimensional pseudo-state space. The above defined function **E** can serve as such a topological mapping, provided that the function is one-to-one, that is, $\mathbf{x} \neq \mathbf{x}'$ implies $\mathbf{E}(\mathbf{x}) \neq \mathbf{E}(\mathbf{x}')$. A mapping of this type is called *embedding*, and the topological issue involved in finding it is referred to as an *embedding problem*. In this context we remark here that general theorems in topology guarantee that *any* n-dimensional smooth manifold can be embedded in $\mathcal{R}^d$, where $d = 2n + 1$. A trivial example is a continuous curve (a one-dimensional manifold) that can obviously be embedded in a three-dimensional Euclidean space. As we shall see shortly, the above mentioned Takens theorem, and a theorem that generalised it, allow one also to apply conclusions in this spirit to *low-dimensional* dissipative dynamical systems, that is, to systems of type (5.15) whose *essential* dynamics takes place on invariant manifolds whose dimension is a rather small number, and may thus be significantly lower than n.

These considerations are very important and useful. By examining (experimentally or numerically) only a single variable, one can get the relevant information about the original dynamical system (even if it is unknown) by reconstructing the state space, using time delays. The problem of determining the embedding dimension d (we wish that d be as small as possible, but large enough to capture the dynamics) and the size of the time delay τ, will be addressed in the next section where we shall make use of the above mentioned theorems.

We now examine the modifications to the procedure explained above when the measured scalar quantity is known only at discrete times, as it happens in an actual experiment or observation (or if the original dynamical system is a discrete mapping to start with). Instead of the continuous signal $s(t)$ we just have the values of the time series at a large number, N, of discrete times (starting from $t = 0$ say)

$$\Delta t, \ 2\Delta t, \ 3\Delta t, \ldots, \ (N-1)\Delta t$$

Thus we have the time series

$$s_j = s(j\Delta t) \ \text{for} \ j = 1, 2, \ldots, N$$

As a result, the delayed coordinate vector (5.16) is known for discrete times only and instead of the continuous function of time $\mathbf{y}(t)$ we have the vector (d-dimensional) time series of N terms

$$\mathbf{y}_j = \mathbf{y}(j\Delta t) = \{s(j\Delta t), s(j\Delta t + \tau), \ldots, s[j\Delta t + (d-1)\tau]\}^\mathrm{T}$$

for $j = 1, 2, \ldots, N$, where obviously the time delay τ must be an integral multiple of the time series interval Δt, so that the components of $\mathbf{y}_j$ are members of the original, known time series. Again, as in the case of a continuous delay vector, the key issue is the existence of the one-to-one function $\mathbf{E}$ defined by (5.17) from the original state space to the reconstructed delay vector space.

The possibility of space state reconstruction thus amounts to the statement that an equivalent dynamics in the reconstructed space exists, that is, there exists a dynamical system

$$\dot{\mathbf{y}}(t) = \mathbf{F}^{(\mathrm{r})}[\mathbf{y}(t)] \quad \text{or} \quad \mathbf{y}_{i+1} = \mathbf{F}^{(\mathrm{r})}(\mathbf{y}_i) \tag{5.18}$$

where all vectors are d-dimensional, whose essential dynamics is equivalent to that of (5.15). The superscript (r) is written here to stress explicitly that the dynamics given by (5.18) is in the reconstructed pseudo-state space. Obviously this is possible because of the existence of the embedding $\mathbf{E}$ and formally the function $\mathbf{F}^{(\mathrm{r})}$ is nothing else but the functional product $\mathbf{E} \circ \mathbf{F}$. This possibility is crucial to the study of experimental or observational signals and it is also very useful for multidimensional theoretical dynamical systems. It provides a technique for reducing the dimension of the system to the minimal value necessary for capturing the essential dynamics.

A full account on the considerations described in this section can be found in the review paper of Abarbanel *et al.* (1993) and in the book by Weigend and Gershenfeld (1994).

5.3 The embedding dimension and the time delay

The general topological embedding theorem, when applied naively to the state space of the autonomous original system (5.15) merely guarantees that d can be no larger than $2n + 1$, and this doesn't seem to be very useful! However, if we assume that the *essential dynamics* of (5.15) takes place in only d_a (which is significantly smaller than n) dimensions, there is hope to find an embedding dimension d that is *smaller* than n. Indeed, in dissipative systems d_a may be significantly smaller than the formal dimension of the system. For example, if an 8-dimensional dissipative system has a strange attractor of fractal dimension $d_a = 2.5$ say, the essential dynamics should be faithfully described in d dimensions, where $d_a < d < 8$. When analysing the system by state-space reconstruction we must decide what is the minimal embedding dimension d necessary to capture the dynamics. If d is too small there is a danger that trajectories may cross because of projection effects and mathematically intolerable cusps may appear in them.

The work of Takens mentioned above (and its generalisation by Sauer, Yorke and Casdalgi in 1991) constitute what is known as the *fractal embedding theorem*. It guarantees that the embedding may be done if $d \geq 2d_a + 1$. For non-integer d_a, the minimal d is obtained by rounding off $2d_a + 1$ to its next integer value. Strictly speaking this is valid if d_a is the *box-counting* dimension D (see Chapter 3) of the attractor, that is, *if d is larger or equal to $2D + 1$ rounded to the next integer, then* the embedding is possible. Frequently, however, a significantly smaller value of d is *sufficient* for the embedding and often even as small as the smallest integer that is larger than d_a itself! The problem is that, in general, even if the dynamical system is known analytically, one does not know what d_a is. Only a rather large upper bound on d_a is known in dissipative systems, as it must be smaller than the dimension of the system. When analysing an experimental signal we do not even have an upper bound since the dynamical system itself is unknown.

In the remainder of this chapter we shall review some of the techniques that have been proposed for the purpose of overcoming this difficulty. Again, our discussion will be rather brief and the interested reader can consult Abarbanel *et al.* (1993) for details and references.

Saturation of the fractal dimension

The idea underlying a judicious choice of the embedding dimension is based on examining the saturation of system invariants, a concept which was already hinted at in Chapter 3. It relies on the observation that if the essential dynamics (like the attractor of a dissipative system or the fractal properties of the surfaces of section, for example) is completely unfolded by using the embedding dimension d, the invariant properties of the dynamics, like the Liapunov exponents or fractal dimension of the attractor, should not change as one increases d. Stated differently, this means that d is a good embedding dimension, if beyond it the invariant properties of the dynamics saturate.

To illustrate the idea we apply it to the calculation of the fractal dimension of an attractor. We assume that a certain dynamical system (explicitly known or unknown) gives rise to a discrete time series. By choosing an appropriate time delay (see below) we construct a sequence of d-dimensional vectors $\mathbf{y}_j$ in the reconstructed pseudo-state space (as explained in the previous section). We usually start with a fairly small $d = d_0$, e.g., $d_0 = 2$. The vectors $\mathbf{y}_j$ define a set of points $\mathbf{E}^d$ in the pseudo-state space. In this example we chose to compute the correlation dimension of the attractor, but the technique can also be applied to the other types of fractal dimensions defined in Chapter 3. To find the correlation dimension we have to perform a limiting process on the pair correlation function $C(\epsilon)$ calculated for the set of points $\mathbf{E}^d$ (see Chapter 3), that is,

$$D_C = \lim_{\epsilon \to 0} \frac{\ln C(\epsilon)}{\ln \epsilon} \tag{5.19}$$

In practice, the finite number of points in $\mathbf{E}^d$ and the presence of experimental or numerical errors give rise to a lower cutoff of ϵ. Thus one plots $\ln C$ as a function of $\ln \epsilon$ within reasonable limits (i.e., between the system size and several times the average distance between points). The slope of the plot is then an estimate for D_C. The procedure is subsequently repeated for higher embedding dimensions, i.e., for $d = d_0 + 1$, $d = d_0 + 2$ and so on, and the corresponding values for D_C are found. When we reach the stage at which D_C becomes independent (within some prescribed tolerance) of d, this value of D_C is considered to be the correlation dimension of the system's attractor. The procedure also yields the minimal embedding dimension, i.e., the dimension of the reconstructed state space needed to represent the dynamical system in a faithful way. In Figure 5.5

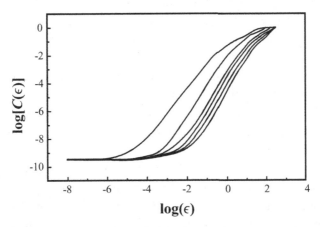

Fig. 5.5. Variation of the logarithm of $C(\epsilon)$ with the logarithm of ϵ, in arbitrary units, using different embedding dimensions for a chaotic solution of the forced Duffing system. The leftmost curve is for $d = d_0 = 2$, while the rightmost one is for $d = 7$. Saturation seems to appear for $d \geq 3$ (see next figure).

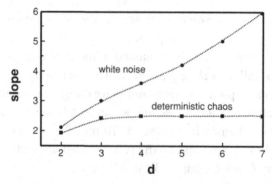

Fig. 5.6. Dimension calculated from the slope of $\log C(\epsilon)$ versus $\log(\epsilon)$ (see text) as a function of the embedding dimension for two systems. A chaotic signal from the forced Duffing system (see the previous figure) saturates at a dimension of approximately 2.5 (squares), while white noise (circles) does not show any saturation up to $d = 7$ (and in fact quite far beyond that as well).

six plots of $\ln C(\epsilon)$ versus $\ln \epsilon$ are shown for a typical dissipative chaotic system. The different plots are for the embedding dimensions, $d = 2$ through $d = 7$. The time series was generated by a numerical integration of the forced Duffing system in the chaotic regime (see Chapter 4). In this case the slope appears to saturate at the embedding dimension $d = 4$, and the correlation dimension of the attractor is $D_C \approx 2.5$.

Indeed, this expectation is fulfilled quantitatively if one plots the slope as a function of the embedding dimension. In Figure 5.6 the slope for the chaotic solution of the forced Duffing system (squares) is seen to saturate at the value of 2.5, which is thus a measure of this system's attractor fractal dimension. The data for a large embedding dimension are considered to be less accurate (the error is proportional to the length of the vector), but we have not depicted here the error bars as our discussion is largely illustrative in nature.

The second time series, for which the slope is shown in the figure (circles) as a function of the embedding dimension, came from a randomly generated signal (white noise). Here the dimension is seen not to saturate, at least not until $d = 7$. In fact, it can be shown without too much difficulty that this continues up to $d = 10$. For higher embedding dimensions the error becomes too large.

These examples demonstrate the difference between a low-dimensional dynamical system and a random system, and the technique described here is one of the ways to distinguish between signals, that both look erratic, but their origins are very different. One reflects deterministic low-dimensional dynamics, and the other is random (very highly dimensional).

False nearest neighbours

An alternative method for determining the minimal embedding dimension is based on the fact that in a space of too low a dimension the system trajectory has spurious intersections through projection effects. If the embedding dimension d is not large enough, points that are far apart in the full state space may fall close together in the reconstructed pseudo-state space. Points that are close to each other in an embedding dimension d say, but fall far apart when d is increased are referred to as *false nearest neighbours*.

The method of finding the correct embedding dimension is based on the elimination of false nearest neighbours. To do this one has to construct a numerical algorithm to search for false neighbours, that is, points that in higher embedding dimensions are not neighbouring anymore. When the embedding dimension is such that no nearest neighbours in the reconstructed state space are false, we may safely assume that the correct embedding dimension for the reconstruction of the system's dynamics has been reached.

Typically, an effective search routine of this sort should take of the order of $N \log N$ (and not N^2), where N is the number of points included in the numerical search algorithm, operations per search. The percentage of false nearest neighbours can be plotted as a function of the embedding dimension, and it decays to zero for the required correct minimal embedding dimension.

Autocorrelation of the data

The value of the time delay, τ, used in state space reconstruction is arbitrary in theory if a continuous $s(t)$ signal is known exactly, i.e., if an infinite amount of data with absolute precision are available. In practice this is, of course, impossible. The data contain errors (finite measurement precision, presence of noise) and their amount is finite (discrete time series). The time delay must be neither too small (the delayed coordinates are correlated) nor too large (complete lack of correlation). In addition if τ is accidentally close to a basic period of the system, the corresponding periodic behaviour cannot be represented in the reconstructed state space. Another basic problem is to determine how the time delay, τ, should compare to the sampling interval Δt. It is intuitively obvious that τ should be considerably larger than Δt, but not very much larger.

We briefly mention here just one technique for choosing the time delay, using the *autocorrelation function* of the signal. This notion has already been introduced for a discrete time series in equation (5.7). We consider a sampled data set (time series) $\{s_k\}$ with $k = 1, 2, \dots$, starting at some given time t_0. Assume that the sampling time interval is Δt and thus $s_j = s(t_0 + j\Delta t)$. We now look at N data points and examine the autocorrelation function of the variations from the average, that is, of

$x_j = s_j - \bar{s}$, where $\bar{s} = \sum_{j=1}^{N} s_j/N$. In practice it is convenient to normalise the autocorrelation function, so that it has a maximal value of unity. We thus have

$$C_m = \frac{R_m}{R_0} = \frac{1}{R_0} \sum_{j=1}^{N} x_j x_{j+m} = \frac{1}{R_0} \sum_{j=1}^{N} (s_j - \bar{s})(s_{j+m} - \bar{s})$$

where $R_0 = \sum_{j=1}^{N} (s_j - \bar{s})^2$.

For a given set of discrete data the autocorrelation function typically decays to zero, or becomes close to zero, for some fixed m. This determines the preferred value of the time delay. If we choose it to be $\tau \approx m \Delta t$ we may be confident that the time delay has been chosen in a reasonable way.

6

Regular and irregular motion in Hamiltonian systems

> There is no limit to how complicated things can get, on account
> of one thing always leading to another.
> E. B. White, *Quo Vadimus?*

Hamiltonian systems constitute a very important subclass of dynamical systems. The basic model systems of classical mechanics – point masses moving in external potentials (modelling, e.g., planets in the Solar System, oscillating bodies attached to springs etc.) – are among the simplest Hamiltonian systems. The Lagrangian and Hamiltonian formalism of Newtonian dynamics provides the mathematical framework for the study of such (literally) dynamical systems. As these were the first dynamical systems to be studied extensively, the subject owes its name to them.

The problems of classical mechanics are by no means the only examples of Hamiltonian systems. Hamiltonian maps and differential systems also arise in the study of such diverse physical systems as propagating waves, fluids and plasmas, electric circuits etc. In addition, the Hamiltonian formulation is central to such important branches of physics as statistical mechanics, field theory and, most notably, quantum mechanics. It is thus important to understand the fundamental properties of classical Hamiltonian systems and the possibility of chaotic behaviour in them.

A basic attribute of Hamiltonian systems is that they are conservative. As we have already defined, a dynamical system is *conservative* if it preserves phase volume. Volume preservation in Hamiltonian systems endows them with special properties. One such important feature is that they do not have attractors. As we have seen in Chapter 4, the existence of strange attractors in dissipative systems is a primary characteristic of chaos. The lack of attractors does not however rule out chaotic behaviour. Chaos is possible in Hamiltonian systems as well, as we have already seen in the example of the Hénon and Heiles system.

Hamiltonian chaos is a vast and extensively studied topic and the methods differ generally from the ones used in dissipative systems. In this chapter we give a brief summary of what is usually done in the study of nonlinear Hamiltonian systems. We shall avoid some of the advanced (and rather difficult) topics in this context and try to give just a concise overview of the subject. We refer the reader to the books by Lichtenberg and Lieberman (1992) and Gutzwiller (1990) for a much fuller description.

6.1 Properties of Hamiltonian systems

In this section we shall review briefly the formalism of Hamiltonian dynamics as it is used, for example, in classical mechanics. The review stresses the results that are needed for the understanding of the remainder of this chapter. It also establishes the notation we shall use and thus helps in making this chapter self-contained. Most statements and theorems will be given here without proof. For a full account of the subject the reader is referred to the classic book by Goldstein (1980).

Hamilton's equations of motion and their basic consequences

The state of a Hamiltonian dynamical system is specified by the values of its generalised coordinates, q_i, and momenta, p_i, with $i = 1, 2, \ldots, N$, where N is the number of the *degrees of freedom* of the system. We shall use the vector notation $\mathbf{q}$ and $\mathbf{p}$ to denote the N-tuples of coordinates and momenta respectively. The dynamics is completely specified by the Hamilton function or *Hamiltonian* of the system, $H(\mathbf{q}, \mathbf{p}, t)$, where t is the independent variable, of which the q and p terms are also functions. In a mechanical system t is, obviously, the time. The state of the system for any t is a point in the $2N$-dimensional *phase space* spanned by the generalised coordinates and momenta. The system follows a trajectory specified by the functions $\mathbf{q}(t)$ and $\mathbf{p}(t)$. Knowledge of these functions constitutes the solution of the dynamical problem.

The generalised coordinates and momenta satisfy the Hamilton equations of motion

$$\frac{d\mathbf{q}}{dt} = \frac{\partial H(\mathbf{q}, \mathbf{p}, t)}{\partial \mathbf{p}}$$

$$\frac{d\mathbf{p}}{dt} = -\frac{\partial H(\mathbf{q}, \mathbf{p}, t)}{\partial \mathbf{q}} \tag{6.1}$$

where a partial derivative with respect to a vector is understood as a vector – the appropriate gradient.

It can be shown that under a wide variety of circumstances, including all the cases that will interest us, Hamilton's equations of motion are equivalent to a variational principle. Namely, the path followed by the system between two points in the $\mathbf{q}$ space, corresponding to the positions of the system in two different times,

t_1 and t_2, is such that the integral, over this path, of the *Lagrangian* function, $L(\mathbf{q}, \dot{\mathbf{q}}, t)$, is extremal. The Lagrangian is a function of the generalised coordinates and *velocities* (time derivatives of the coordinates), and its relation to the Hamiltonian (function of coordinates and momenta) is given by the following (Legendre) transformation

$$H(\mathbf{q}, \mathbf{p}, t) = \mathbf{p} \cdot \dot{\mathbf{q}} - L(\mathbf{q}, \dot{\mathbf{q}}, t) \tag{6.2}$$

where the generalised momenta, *conjugate* to the generalised coordinates, are *defined* to be

$$\mathbf{p} \equiv \frac{\partial L}{\partial \dot{\mathbf{q}}}$$

The variational principle is thus written as

$$\delta \int_{t_1}^{t_2} L(\mathbf{q}, \dot{\mathbf{q}}, t) dt = 0 \tag{6.3}$$

where the integral is taken over a path connecting the two positions of the system at times t_1 and t_2, and the varied paths are only those that connect these two points; δ denotes a *variation*. The exact meaning of this variation is made clear in any text on variational calculus or analytical mechanics. Substitution from (6.2) gives rise to an alternative formulation of the variational principle, in terms of the Hamiltonian and its variables

$$\delta \int_{t_1}^{t_2} [\mathbf{p} \cdot \dot{\mathbf{q}} - H(\mathbf{q}, \mathbf{p}, t)] dt = \delta \int [\mathbf{p} \cdot d\mathbf{q} - H(\mathbf{q}, \mathbf{p}, t) dt] = 0 \tag{6.4}$$

The integrals in the last expression are again understood as path integrals, but now in phase space. The subtleties of what should be the variations of $\mathbf{p}$ at the end points ($\delta\mathbf{q}$ vanish there, similarly to 6.3) will not concern us here (see Goldstein (1980) for a detailed discussion).

A corollary of the variational principle (6.4) is that any function, differing from the Hamiltonian by a total time derivative of an arbitrary function of coordinates and momenta and time, can also serve as a Hamiltonian for the original system, yielding identical Hamiltonian equations of motion. The choice of the Hamiltonian is thus not unique.

From the Hamilton equations (6.1) the following is obvious.

(i) If the Hamiltonian is not an explicit function of time it is an *integral of motion*, that is, it remains fixed as $\mathbf{q}$ and $\mathbf{p}$ vary with time. This is so because if $H = H(\mathbf{q}, \mathbf{p})$ only, then by the chain rule

$$\frac{dH}{dt} = \frac{d\mathbf{q}}{dt} \cdot \frac{\partial H}{\partial \mathbf{q}} + \frac{d\mathbf{p}}{dt} \cdot \frac{\partial H}{\partial \mathbf{p}} = 0$$

by virtue of (6.1).

(ii) Phase volume is conserved since

$$D = \frac{\partial \dot{\mathbf{q}}}{\partial \mathbf{q}} + \frac{\partial \dot{\mathbf{p}}}{\partial \mathbf{p}} = 0$$

This result also follows from (6.1) and was already derived in Chapter 3.

These two observations are usually known in mechanics as the energy conservation (i) and Liouville (ii) theorems.

If we wish to give Hamilton's equations of motion (6.1) the look of a generic dynamical system, we may define the $2N$-dimensional vector $\mathbf{z}$ as composed of the two N-dimensional vectors $\mathbf{q}$ and $\mathbf{p}$, in this order

$$\mathbf{z}(t) \equiv \begin{pmatrix} \mathbf{q}(t) \\ \mathbf{p}(t) \end{pmatrix}$$

Then (6.1) can be written as

$$\frac{d\mathbf{z}}{dt} = \mathbf{F}(\mathbf{z}) \tag{6.5}$$

where the $2N$-dimensional vector function $\mathbf{F}(\mathbf{z})$ is

$$\mathbf{F}(\mathbf{z}) \equiv \mathcal{S} \cdot \frac{\partial H}{\partial \mathbf{z}} \tag{6.6}$$

with the matrix $\mathcal{S}$ defined by

$$\mathcal{S} \equiv \begin{pmatrix} \mathcal{O} & -\mathcal{I} \\ \mathcal{I} & \mathcal{O} \end{pmatrix} \tag{6.7}$$

$\mathcal{S}$ is called the $2N$-dimensional *symplectic matrix*, and is composed of the $N \times N$, unit ($\mathcal{I}$) and zero ($\mathcal{O}$) matrices.

This notation immediately reflects how special and restrictive Hamiltonian systems are. In a general dynamical system of the form (6.5) a $2N$-dimensional *vector function* $\mathbf{F}(\mathbf{z})$ has to be specified, whereas in a Hamiltonian system this function follows from a *single scalar function*, the Hamiltonian $H(\mathbf{z})$. All the components of $\mathbf{F}$ can be *derived* from H by (6.6). Hamiltonian systems satisfy (6.6) and it is usually said that they endow phase space with *symplectic structure*. In more advanced texts Hamiltonian systems are said to be defined on what is called *symplectic manifolds*.

Canonical transformations

Let $\mathbf{Z} = \mathbf{Z}(\mathbf{z})$ be a transformation of phase-space variables, $(\mathbf{q}, \mathbf{p}) \rightarrow (\mathbf{Q}, \mathbf{P})$. We single out those transformations that preserve the Hamiltonian form of the equations of motion (or the symplectic structure), i.e., such that the 'new' variables satisfy

$$\frac{d\mathbf{Q}}{dt} = \frac{\partial H'(\mathbf{Q}, \mathbf{P})}{\partial \mathbf{P}}$$

$$\frac{d\mathbf{P}}{dt} = -\frac{\partial H'(\mathbf{Q}, \mathbf{P})}{\partial \mathbf{Q}} \qquad (6.8)$$

where H' is the transformed Hamiltonian and we call them *canonical*. Obviously not all transformations of phase space have this property. Moreover, it is not clear a priori how to perform the transformation of the Hamiltonian.

There are a number of ways to verify that a given transformation is a canonical one. A fundamental property of canonical transformations is that they possess a *generating function*. If the transformation is canonical, the new variables must satisfy equations of motions in Hamiltonian form and thus a variational principle. For this, it is sufficient that the expressions under the integral sign of (6.4), formulated in the old and new variables, differ only by a *total differential* of some function, G say, of the coordinates, momenta and time. This is so since the integrals differ then by a constant (the difference between the value of G at the end points), contributing nothing to the variation. Thus the transformation is canonical if

$$[\mathbf{P} \cdot d\mathbf{Q} - H'(\mathbf{Q}, \mathbf{P})dt] - [\mathbf{p} \cdot d\mathbf{q} - H(\mathbf{q}, \mathbf{p})dt] = dG \qquad (6.9)$$

Let us choose $G(\mathbf{q}, \mathbf{P}, t)$ to be any sufficiently smooth function of the *old* coordinates, *new* momenta and possibly time. Observing now that

$$\mathbf{P} \cdot d\mathbf{Q} = d(\mathbf{Q} \cdot \mathbf{P}) - \mathbf{Q} \cdot d\mathbf{P}$$

and substituting it into (6.9) we get

$$\mathbf{Q} \cdot d\mathbf{P} + \mathbf{p} \cdot d\mathbf{q} + (H' - H)dt = d(\mathbf{P} \cdot \mathbf{Q} - G) \equiv dS \qquad (6.10)$$

where S can be made a function $\mathbf{q}, \mathbf{P}$ and t, if $\mathbf{Q}$ is expressed (by means of the transformation and its inverse) as a function of $\mathbf{q}$ and $\mathbf{P}$.

Writing the differential of $S(\mathbf{q}, \mathbf{P}, t)$ in the usual way (using the chain rule)

$$dS = \frac{\partial S}{\partial \mathbf{q}} \cdot d\mathbf{q} + \frac{\partial S}{\partial \mathbf{P}} \cdot d\mathbf{P} + \frac{\partial S}{\partial t}dt$$

and comparing with (6.10) we get

$$\mathbf{Q} = \frac{\partial S(\mathbf{q}, \mathbf{P}, t)}{\partial \mathbf{P}}$$

$$\mathbf{p} = \frac{\partial S(\mathbf{q}, \mathbf{P}, t)}{\partial \mathbf{q}} \qquad (6.11)$$

and

$$H' = H + \frac{\partial S(\mathbf{q}, \mathbf{P}, t)}{\partial t} \qquad (6.12)$$

Note that S is essentially arbitrary (because G is such) and the transformation *generated* by it (6.11) must be canonical. In addition, we get the prescription for properly transforming the Hamiltonian. We see that it is not sufficient in general to express the old variables in the old Hamiltonian by the new ones. To obtain the new Hamiltonian one must also add to it the partial time derivative of the generating function.

Obviously, the canonical transformation so defined is contained in (6.11) only implicitly. To get the transformation explicitly, that is, in the form $\mathbf{Q} = \mathbf{Q}(\mathbf{q}, \mathbf{p}, t)$ and $\mathbf{P} = \mathbf{P}(\mathbf{q}, \mathbf{p}, t)$, one needs to first solve the second equation of (6.11) for $\mathbf{P}$ in terms of $\mathbf{q}$ and $\mathbf{p}$ and t, and substitute this solution into the first equation of (6.11). This may be technically difficult.

The functional dependence of the generating function on its variables can also be different. A perfectly valid generating function can depend on $\mathbf{q}$, $\mathbf{Q}$ and t, for example, or other combinations of the old and new variables. The equations corresponding to (6.11) will then have a different form (the transformation of the Hamiltonian will be the same, however) and the transformation will be canonical. The remarkable property is that *any* (sufficiently smooth) function of the appropriate variables generates a canonical transformation. The upshot is that it not easy, in general, to solve the inverse problem to this procedure, i.e., to find the generating function of a specified canonical transformation.

A particular type of canonical transformation has far-reaching consequences in the theory of Hamiltonian systems. As a result of such transformations the Hamiltonian takes on a particularly convenient form, in which the new coordinates are all *cyclic* (do not appear explicitly in the Hamiltonian). The new momenta are then all constant and the solution of Hamilton's equations is trivial. The problem is, of course, how to find such a canonical transformation (or one of its generating functions). This task is not usually easier than solving the original Hamiltonian differential equations! Nevertheless the possibility of such a procedure has great formal power, as will be apparent shortly.

We finish our discussion on canonical transformations by remarking that the Hamiltonian flow (the solution of Hamilton's equations of motion) can be perceived as a continuum of canonical transformations. Indeed, let such a flow or solution (for a one degree of freedom system) be

$$q(t) = f_q[q(0), p(0), t], \quad p(t) = f_p[q(0), p(0), t] \tag{6.13}$$

where $q(0)$ and $p(0)$ are the initial conditions. We can regard these equations, for some fixed value of the time, t_c say, as a transformation of phase space onto itself, since the initial conditions can, in principle, take on all their possible values (span the full phase space). We can thus regard q_0 and p_0 as the old variables (q and p). The new variables are then $Q = q(t_c)$ and $P = p(t_c)$, and are both functions of the old variables by virtue of (6.13). It is not difficult to discover the remarkable property of the Hamiltonian in this context – it is the generating function of these transformations.

An important and general result, concerning any volume-preserving continuous one-to-one mapping that maps a bounded region of a Euclidean space onto itself, is due to Poincaré. It goes under the name of *Poincaré's recurrence theorem* and can obviously be applied to the above mentioned transformation, generated by the Hamiltonian flow in phase space. The recurrence theorem guarantees that for almost all initial conditions (i.e., except for a set of measure zero) the Hamiltonian flow will carry the system in finite time arbitrarily close to the starting point. This theorem has important consequences in the theory of Hamiltonian systems (see below and in Chapter 9).

Poisson brackets

In Hamiltonian systems the time dependence of dynamical quantities can be formulated very elegantly. A general dynamical quantity is usually defined as a function of the phase space variables and possibly the time. Consider such a quantity, functionally expressed as

$$\chi = \chi(\mathbf{q}, \mathbf{p}, t)$$

Using the chain rule we can write the total time derivative of χ as

$$\frac{d\chi}{dt} = \dot{\mathbf{q}} \cdot \frac{\partial \chi}{\partial \mathbf{q}} + \dot{\mathbf{p}} \cdot \frac{\partial \chi}{\partial \mathbf{p}} + \frac{\partial \chi}{\partial t}$$

Since the phase variables satisfy Hamilton's equations of motions we can substitute for the time derivatives of $\mathbf{q}$ and $\mathbf{p}$ from these equations to get

$$\frac{d\chi}{dt} = \frac{\partial \chi}{\partial t} + \frac{\partial H}{\partial \mathbf{p}} \cdot \frac{\partial \chi}{\partial \mathbf{q}} - \frac{\partial H}{\partial \mathbf{q}} \cdot \frac{\partial \chi}{\partial \mathbf{p}}$$

where the last two terms are understood to be scalar products, i.e., they are equal to

$$\sum_i \left(\frac{\partial H}{\partial p_i} \frac{\partial \chi}{\partial q_i} - \frac{\partial H}{\partial q_i} \frac{\partial \chi}{\partial p_i} \right).$$

The *Poisson brackets* serve as a shorthand notation for such expressions. For any two functions, χ and ξ, of the phase space variables, these brackets are defined to be

$$[\chi, \xi] \equiv \frac{\partial \chi}{\partial \mathbf{p}} \cdot \frac{\partial \xi}{\partial \mathbf{q}} - \frac{\partial \chi}{\partial \mathbf{q}} \cdot \frac{\partial \xi}{\partial \mathbf{p}} \tag{6.14}$$

Using this definition the *equation of motion* for any dynamical quantity χ is written as

$$\frac{d\chi}{dt} = \frac{\partial \chi}{\partial t} + [H, \chi] \qquad (6.15)$$

The vanishing of the total time derivative of a dynamical quantity means that this dynamical quantity is a *constant of motion*. Note that we distinguish in this book between constants and integrals of motion. The word *integral* is reserved in this context for only those *constants* of motion that do not depend *explicitly* on time (they are functions only of the phase-space variables). We have already seen that if the Hamiltonian does not depend explicitly on time, it is an integral of motion. This fact can be seen clearly from the equation of motion (6.15) for $\chi = H$ itself. We get

$$\frac{dH}{dt} = [H, H] = 0$$

as it is obvious from the definition of Poisson brackets that the bracket of any dynamical quantity with itself vanishes identically.

The identification of independent integrals of motion is very important on the way to reducing and ultimately solving a dynamical system, and Poisson brackets are a useful tool in such a procedure (see below). More generally, the Poisson brackets, viewed as a binary operation, possess some interesting basic algebraic properties, and there is also a close analogy between the Poisson brackets in classical mechanics and the commutation relation between operators in quantum mechanics. The *fundamental* Poisson brackets (between the phase variables themselves) satisfy

$$[q_i, q_j] = 0; \quad [p_i, p_j] = 0; \quad [p_i, q_j] = \delta_{ij} \qquad (6.16)$$

for any i and j, as can be verified easily from the definition.

It can be shown that Poisson brackets are a *canonical invariant*, that is, their value is preserved when phase variables are canonically transformed. Conversely, if Poisson brackets are preserved in a transformation, it must be canonical. For that it is sufficient to show that just the fundamental brackets (6.16) relation is satisfied.

The Poisson brackets are therefore a very fundamental construction. They are canonically invariant and help to identify constants and integrals of motion. Canonical transformations (including, of course, flows in phase space) of Hamiltonian dynamical systems also possess a whole hierarchy of *integral invariants* (first studied by Poincaré). We shall not discuss them in detail here and only remark that phase volume belongs to this class of invariants.

Hamilton–Jacobi equation and action–angle variables

We have mentioned before a special type of canonical transformation, which is, in a way, the optimal type. Such a transformation makes the transformed Hamiltonian independent of all the new generalised coordinates, a very useful situation, as the solution of Hamilton's equations is then trivial. We shall now describe a systematic way of deriving the generating function of such an optimal transformation for a system with a given Hamiltonian. For simplicity we consider time-independent Hamiltonians.

As the new Hamiltonian is assumed to be independent of the new coordinates, their conjugate momenta are constants. Thus we may write $\mathbf{P} = \mathbf{a}$ to remind us that this is a constant vector (with N components). Therefore, choosing the generating function to depend on the old coordinates and new momenta we may write it as $S = S(\mathbf{q}, \mathbf{a})$. Using equations (6.11) we obtain the new coordinates and old momenta

$$\mathbf{Q} = \frac{\partial S(\mathbf{q}, \mathbf{a})}{\partial \mathbf{a}}$$

$$\mathbf{p} = \frac{\partial S(\mathbf{q}, \mathbf{a})}{\partial \mathbf{q}} \tag{6.17}$$

According to (6.12) the new Hamiltonian, H', is equal to the old one; it just has to be expressed in the new variables. As it must depend (by construction) only on the new (constant) momenta we can write

$$H'(\mathbf{a}) = H(\mathbf{q}, \mathbf{p})$$

Noting that H' is also constant (dependent on the N constants $\mathbf{a}$) we call it α, say. Substituting formally now the form of $\mathbf{p}$ from (6.17) into the last equation gives

$$H\left(\mathbf{q}, \frac{\partial S(\mathbf{q}, \mathbf{a})}{\partial \mathbf{q}}\right) = \alpha(\mathbf{a}) \tag{6.18}$$

Equation (6.18) is a first-order partial differential equation in N independent variables for the function $S(\mathbf{q}, \mathbf{a})$. It is known as the *time-independent Hamilton–Jacobi equation*. Such equations require for their solution N independent constants of integration, and these can be taken to be the N constant new momenta $\mathbf{a}$. If one is able to solve this (in general difficult) nonlinear PDE, the function S furnishes the canonical transformation and its inverse, by manipulation of equations (6.17) as explained before. The new Hamiltonian can then be easily found and Hamilton's equations are trivially integrated. The equations are

$$\dot{\mathbf{P}} = -\frac{\partial H'}{\partial \mathbf{Q}} = 0$$

$$\dot{Q} = \frac{\partial H'}{\partial a} \equiv b(a)$$

where b contains N constants, determined by the constants a.

The solution is $P(t) = a$ and $Q(t) = bt + c$. This solution depends on $2N$ integration constants (the vectors a and c). These can be furnished by the initial conditions of the original problem, when the solution is transformed back to the old variables.

The Hamilton–Jacobi equation for more than one-degree-of-freedom systems is generally very difficult to solve, thus the practical value of the formal procedure outlined above may be doubtful. However, there is an important case, in which we may simplify the procedure significantly. This is when the system is separable and its motion is periodic. These two concepts are explained below.

(i) A Hamiltonian system is called *separable* if the appropriate generating function, S, can be written as a sum

$$S = \sum_i S_i(q_i, a) \tag{6.19}$$

where each term depends on only one coordinate. A special class of systems for which this is so is one in which the Hamiltonian itself is separable, i.e., it can be written as

$$H = \sum_i H_i \left(q_i, \frac{\partial S}{\partial q_i} \right)$$

In such a case, if one makes the Ansatz (6.19) for S, the Hamilton–Jacobi equation splits into N independent equations

$$H_i \left(q_i, \frac{\partial S_i}{\partial q_i} \right) = a_i$$

with $i = 1, 2, \ldots, N$, since the q_i are independent. Each degree of freedom is independent of the others, but the constants a_i must satisfy the following constraint $a_1 + a_2 + \cdots + a_N = \alpha$.

(ii) A motion of a Hamiltonian system is called *periodic*, if either both variables in each pair (q_i, p_i) are periodic functions of time, with the same period (libration or oscillation), or the p_i are periodic functions of q_i (rotation).

The separated form of the Hamiltonian in (i) above seems very restrictive. However one may still try to achieve such a form of the Hamiltonian by coordinate transformation. This is the case, for example, in systems of coupled linear oscillators. The coordinates in which the Hamiltonian becomes separated are then called *normal* orthogonal coordinates, and they may be found by methods of linear algebra (solving an eigenvalue problem, see e.g., Goldstein, 1980).

The choice of the new momenta as the constants a_i (the separation constants) is arbitrary. One could just as well choose as the new momenta any N independent functions of the a_i. If conditions (i) and (ii) are satisfied a special choice of this nature is very useful. Let the new momenta, denoted by $\mathbf{J}$, be defined as

$$J_i \equiv \frac{1}{2\pi} \oint p_i dq_i = \frac{1}{2\pi} \oint \frac{\partial S_i(q_i, \mathbf{a})}{\partial q_i} dq_i$$

where the integral is taken over a closed path reflecting the periodic nature of the system. The fact that the separation of the generating function is possible implies, by one of the generating function relations (6.17),

$$p_i = \frac{\partial}{\partial q_i} S_i(q_i, \mathbf{a})$$

that each p_i is a function of the corresponding q_i and thus the path integral is well defined.

In this way we get $\mathbf{J} = \mathbf{J}(\mathbf{a})$. Inverting, we can get the a_i in terms of the J_i. Thus the generating function can be transformed to these new variables

$$S(\mathbf{q}, \mathbf{J}) = S[\mathbf{q}, \mathbf{a}(\mathbf{J})]$$

The new coordinates are then given by the appropriate generating function relations. Denoting the new generalised coordinates, conjugate to $\mathbf{J}$, by $\boldsymbol{\theta}$, we now have a useful set of variables $(\mathbf{J}, \boldsymbol{\theta})$. These variables are called *action–angle* variables – $\mathbf{J}$ are actions and $\boldsymbol{\theta}$ are angles. This nomenclature will soon become self explanatory. To sum up, a separable and periodic system with a time-independent Hamiltonian can be systematically reduced to action–angle variables in the following way.

(i) N independent Hamilton–Jacobi equations are written as

$$H_i\left(q_i, \frac{\partial S_i(q_i, \mathbf{a})}{\partial q_i}\right) = H_i'(\mathbf{a}) \quad \text{for} \quad i = 1, 2, \ldots, N$$

where the generating function S_i for each degree of freedom is a function of only one coordinate.

(ii) For each degree of freedom the action variables are defined as

$$J_i = \frac{1}{2\pi} \oint \frac{\partial S_i}{\partial q_i} dq_i$$

and the substitution of $\partial S_i/\partial q_i$, derived from the corresponding Hamilton–Jacobi equations, into these definitions, yields N equations, after performing the cycle integrations. From these equations the a_i can be expressed in terms of the J_k.

(iii) The generating function relations (6.17) in the action–angle variables can then be obtained, and they read

$$\theta_i = \frac{\partial S(\mathbf{q}, \mathbf{J})}{\partial J_i} = \sum_{k=1}^{N} \frac{\partial}{\partial J_i} S_k(q_k, \mathbf{J})$$

$$p_i = \frac{\partial S(\mathbf{q}, \mathbf{J})}{\partial q_i} = \frac{\partial}{\partial q_i} S_i(q_i, \mathbf{J}) \tag{6.20}$$

The first of these relations gives the new coordinates $\boldsymbol{\theta}$.

(iv) Hamilton's equations with the transformed Hamiltonian, $H'(\mathbf{J}) = \sum_i H_i'(\mathbf{J})$, then give

$$\dot{J}_i = \frac{\partial H'(\mathbf{J})}{\partial \theta_i} = 0$$

$$\dot{\theta}_i = \frac{\partial H'(\mathbf{J})}{\partial J_i} = \text{const} \equiv \omega_i \tag{6.21}$$

The constants of the second set of equations, ω, are *not* N independent new constants. They depend on $\mathbf{J}$ so that $\omega = \omega(\mathbf{J})$. The solution is trivial

$$\boldsymbol{\theta}(t) = \omega(\mathbf{J})\,t + \boldsymbol{\delta}$$

with $2N$ integration constants $\mathbf{J}$ and $\boldsymbol{\delta}$.

(v) The meaning of the N constants ω is clarified by considering the first generating function relation (6.20)

$$d\theta_i = \sum_{k=1}^{N} \frac{\partial}{\partial q_k} \frac{\partial S(\mathbf{q}, \mathbf{J})}{\partial J_i} dq_k$$

Integration of the last equation over one cycle of the original system gives

$$\Delta\theta_i = \frac{\partial}{\partial J_i} \sum_{k=1}^{N} \oint \frac{\partial S_k(q_k, \mathbf{J})}{\partial q_k} dq_k = \frac{\partial}{\partial J_i} \sum_{k=1}^{N} \oint p_k dq_k = 2\pi$$

where the last equality follows from the definition of J_i. Now, from the solution $\boldsymbol{\theta}(t) = \omega t + \boldsymbol{\theta}_0$, it is obvious that the change of each θ_i over its cycle (of period T_i, say) is $\Delta\theta_i = \omega_i T_i$. Thus ω_i are the angular frequencies $(2\pi/T_i)$ of the periodic motions of each degree of freedom.

We conclude this section with an illustration of the above procedure in the case of a one-dimensional harmonic oscillator. The Hamiltonian of the problem is obviously

$$H(p, q) = \frac{1}{2}(p^2 + \omega^2 q^2) \tag{6.22}$$

where q is the displacement, p the momentum and ω the angular frequency.

The Hamilton–Jacobi equation is

$$\frac{1}{2}\left[\left(\frac{\partial S}{\partial q}\right)^2 + \omega^2 q^2\right] = H' \tag{6.23}$$

where the $H' \equiv E$ is a constant (equal to the energy).

The action variable can thus be defined as

$$J = \oint \left(\frac{\partial S}{\partial q}\right) dq = \oint \sqrt{2E - \omega^2 q^2}\, dq$$

where the integration has to be performed over a complete cycle, i.e., from $q = -\sqrt{(2E)/\omega}$ through the other turning point, $q = +\sqrt{(2E)/\omega}$, and back. The integral is elementary and we get $J = E/\omega$. Thus the new Hamiltonian is

$$H'(J) = E = \omega J \tag{6.24}$$

and the Hamilton equations are

$$\dot{J} = 0$$
$$\dot{\theta} = \omega \tag{6.25}$$

giving, as required, constant actions. The angle variable can also be trivially found. We thus get the solution

$$J = \text{const} = \frac{E}{\omega}$$
$$\theta(t) = \omega t + \delta \tag{6.26}$$

The two constants, J, δ (or E and δ), are determined, in a particular problem, from the initial values of the original coordinates and momenta.

The motion in the original variables can be found using the Hamilton–Jacobi equation (6.23), where the constant Hamiltonian is expressed in terms of the action variable (see 6.24). We get for the generating function

$$S(q, J) = \int_{q_0}^{q} \sqrt{2\omega J - \omega^2 q'^2} \, \mathrm{d}q'$$

The integral is elementary and after performing it the generating-function relations give

$$q(\theta, J) = \sqrt{\frac{2J}{\omega}} \sin\theta, \qquad p(\theta, J) = \sqrt{2J\omega} \cos\theta \tag{6.27}$$

where the (arbitrary) value of θ, corresponding to $q = q_0$, has been set to zero.

6.2 Integrable systems

We are now in a position to define when a given Hamiltonian is called integrable. An N degree of freedom Hamiltonian (or the corresponding Hamiltonian system) is said to be *integrable* (*completely integrable* is a term sometimes used) if there exist N *independent* integrals of motion. The integrals of motion are regarded as independent if they are *in involution*, i.e., their Poisson brackets with each other vanish. Such independent integrals of motion are also called *isolating integrals* of the motion. Let a_i, with $i = 1, 2, \ldots, N$, be such integrals. The involutive property makes it possible to regard these constants as a complete set of generalised momenta in canonically transformed phase space. Any complete set of N functions of the a_i can serve as an equivalent set of integrals. In particular, the action variables, J_i are such a set.

It is useful to understand the geometrical meaning of integrability. The existence of each of the N isolating integrals effectively reduces by one the dimension of the portion of phase space that is accessible to the system. This means, in more abstract language, that the existence of N integrals forces the trajectories of the system to be confined to an N-dimensional manifold, a subspace of the $2N$-dimensional phase space. As the integrals of motion are N independent functions of the coordinates and momenta, their being constant supplies N constraints and reduces the motion to the $2N - N = N$-dimensional manifold. As a trivial example consider a Hamiltonian system with one degree of freedom. Phase space is two-dimensional, but the fact that the Hamiltonian (if it is not an explicit function of time) is a constant reduces the motion to lie on a curve $H(p, q) = $ const.

The N-dimensional manifold, to which the motion of an integrable Hamiltonian system is confined, is topologically equivalent to an N-dimensional *torus*. The following arguments are very suggestive (it is not a rigorous proof) that this is so. Define the N vectors

$$\mathbf{u}_k \equiv S \cdot \nabla a_k, \quad \text{for} \quad k = 1, 2, \dots, N \tag{6.28}$$

where S is the symplectic matrix, defined in (6.7) and the gradient operator is with respect to the $2N$ variables $\mathbf{z} = (\mathbf{q}, \mathbf{p})$ (compare with 6.6), that is

$$\nabla = \left(\frac{\partial}{\partial q_1}, \dots, \frac{\partial}{\partial q_N}, \frac{\partial}{\partial p_1}, \dots, \frac{\partial}{\partial p_N} \right)$$

If the Hamiltonian is one of the integrals of motion, a_1, say, then for $k = 1$ (6.28) is equivalent to the Hamiltonian flow itself, as can be seen by comparing it with (6.5) and (6.6). The term $\mathbf{u}_1$ is then the phase space velocity vector $\dot{\mathbf{z}}$. This velocity field is clearly *tangent* to the manifold, on which the motion proceeds and on which the Hamiltonian is constant. All the other $\mathbf{u}_k$, corresponding to the other integrals of motion (which are constant on the manifold) are tangent to the manifold as well. Thus the manifold on which the trajectory must lie possesses N linearly independent tangent vector fields. The linear independence of the N 'velocity' vector fields defined by (6.28) follows from the involutive property of the integrals. A fundamental theorem in topology (known as the 'hairy ball theorem') states that a d-dimensional manifold, for which d-independent, nowhere vanishing, tangent vector fields can be constructed, is topologically equivalent to a torus. In two dimensions this theorem can easily be demonstrated by comparing a 'hairy' sphere surface to a 'hairy' torus. The hairs on a torus can be combed smooth to become flat (tangent to the surface) everywhere. In contrast, as every hairdresser knows, combing the hair on a human head (topologically equivalent to a sphere, in most cases) cannot give such smooth results. At least one singular point must exist.

A topological equivalence between two spaces means that there exists a smooth deformation of one space into the other. Clearly, a torus is not topologically equivalent to a sphere as there exists a basic topological difference between them. A sphere has no 'holes' while a torus has exactly one 'hole'. An N-dimensional manifold is topologically equivalent to an N-torus if and only if it can be smoothly deformed into an N-cube, whose opposite sides are 'glued' to each other.

Thus each trajectory of an integrable Hamiltonian dynamical system having N degrees of freedom lies on an N-dimensional manifold, topologically equivalent to a torus. The tori on which all trajectories must lie are *invariant*, as a trajectory starting on any particular one of them remains for ever on that torus. The existence of these tori is closely connected to the existence of a set of action–angle variables. On an N-torus one can find exactly N independent closed paths that cannot be deformed into each other or shrunk to zero by any continuous smooth transformation. The set of action variables can thus be defined as

$$J_i = \frac{1}{2\pi} \oint_{\Gamma_i} \mathbf{p} \cdot d\mathbf{q}$$

for the set of the N independent closed paths, Γ_i, on the torus. Hamilton's equations of motion can be written in the action–angle representation, i.e., have the form (6.21) and they can be trivially integrated.

If a system is integrable the whole phase space is filled with invariant tori. The set of initial conditions fixes a particular value for each of the integrals and determines the invariant torus (i.e., the set $\mathbf{J}$). The value of the angle variables $\theta_i(t)$ at a given time determines the position on the trajectory, fixed by the corresponding J_i, at that time.

To sum up, the following statements can be made on the basis of the above discussion. For any general Hamiltonian system with N degrees of freedom, the motion takes place in a $2N$-dimensional phase space. If $H = E$, i.e., the Hamiltonian is an integral of motion, the system remains on a $2N - 1$-dimensional *energy shell*. If, in addition, the system is integrable, the motion takes place on N-dimensional invariant tori. We shall return to the implications of these statements later.

A separable system is necessarily integrable. A systematic way of transforming it into an action–angle representation was described in the previous section and this is a sufficient condition for integrability. Non-separable systems may, however, also be integrable. We leave out the issue of how to decide if a non-separable Hamiltonian is integrable, which is beyond the scope of this book, and conclude this section by providing an example of a well-known integrable system, in which the Hamiltonian is not naturally written in a separated form. This does not mean that it cannot be separated in some other set of variables. Instead of trying to do this first we shall find the action variables at the outset (by taking advantage of known integrals of motion), and thus show that the system is integrable.

Consider the motion of a point mass m in a Keplerian potential (e.g., a planet revolving around a star). It is well known that this motion is planar. Let r and ϕ be the polar coordinates in the orbit plane with the origin of r being the one of the potential $V(r) = -K/r$ (K constant). The Hamiltonian of the system is

$$H(r, \phi, p_r, p_\phi) = \frac{1}{2m}\left(p_r^2 + \frac{p_\phi^2}{r^2}\right) - \frac{K}{r}$$

where p_r and p_ϕ are the momenta conjugate to the corresponding coordinates.

As the Hamiltonian is not an explicit function of time, it is an integral of motion (the total energy), that is, $H = E$, with E constant. We focus on the case of bounded motions, i.e., $E < 0$. An additional constant of motion is obviously the angular momentum $p_\phi = l$, since the Hamiltonian does not depend on ϕ (it is a cyclic coordinate).

The original Hamiltonian is not in a separated form, but we already know two integrals of motion and can see that the system is integrable by directly obtaining the action variables in terms of the two integrals l and E. Using these two integrals of motion we can express the two momenta as

$$p_\phi = l \quad \text{and} \quad p_r = \sqrt{2m\left(E + \frac{K}{r}\right) - \frac{l^2}{r^2}}$$

The action variables are thus

$$J_1 = \frac{1}{2\pi}\oint p_\phi d\phi = l$$

$$J_2 = \frac{1}{2\pi}\oint p_r dr = \frac{1}{2\pi}2\int_{r_{\min}}^{r_{\max}}\sqrt{2m\left(E + \frac{K}{r}\right) - \frac{l^2}{r^2}}\,dr$$

where $r_{\min}$ and $r_{\max}$ are the turning points, i.e., the solutions of

$$Er^2 + Kr - l^2/(2m) = 0$$

The result for the second integral is

$$J_2 = -l + K\sqrt{\frac{m}{2|E|}}$$

Thus the transformed Hamiltonian is

$$H'(J_1, J_2) = -\frac{mK^2}{2(J_1 + J_2)^2}$$

The system is obviously integrable and can be solved quite easily.

We summarise the discussion of integrable Hamiltonian systems by pointing out a few significant properties of the motion and the tori on which it takes place. In all cases the Hamiltonian is time-independent and is thus an integral of motion.

(i) If the system has only one degree of freedom, phase space is two-dimensional (plane). The energy shell is one-dimensional and this is also the one-dimensional torus (a closed curve), on which the motion proceeds. The full energy shell is explored by the system (such systems are called *ergodic*).

(ii) If the system possesses two degrees of freedom, phase space is four-dimensional. The energy shell is three-dimensional and the motion is on two-dimensional tori embedded in the energy shell. Each two-dimensional torus divides the energy shell into two separate parts (outside and inside the surface of the torus).

(iii) In systems with $N > 2$ degrees of freedom the energy shell is $2N - 1$-dimensional and each N-dimensional torus leaves out regions of the energy shell that are neither inside nor outside it. This is so because the dimension of the energy shell differs from that of a torus by more than 1.

(iv) Since the motion of an integrable Hamiltonian system proceeds on invariant tori, any dynamical variable is multi-periodic in time (it is quasiperiodic if the frequencies are incommensurate). Thus it can be expanded in a multiple Fourier series (see equation 5.14). For example, any generalised coordinate, which for convenience we write without the index i, can be expanded in terms of the angle variables $\boldsymbol{\theta} = (\theta_1, \theta_2, \ldots, \theta_N)$

$$q(t) = \sum_{k_1} \sum_{k_2} \cdots \sum_{k_N} a_{k_1 k_2 \ldots k_N} \exp\left[i\mathbf{k} \cdot \boldsymbol{\theta}\right] \tag{6.29}$$

where $\mathbf{k} \cdot \boldsymbol{\theta} = k_1 \theta_1 + k_2 \theta_2 + \cdots + k_N \theta_N$, and each $\theta_j = \omega_j t + \delta_i$ (the δ_i are constants). The Fourier coefficients of this expansion are functions of the action variables and are given by the N-dimensional integral

$$a_{\mathbf{k}}(\mathbf{J}) = \frac{1}{(2\pi)^N} \int_0^{2\pi} d\boldsymbol{\theta} \, q(\mathbf{J}, \boldsymbol{\theta}) \, \exp[-i\mathbf{k} \cdot \boldsymbol{\theta}] \tag{6.30}$$

where $q(\mathbf{J}, \boldsymbol{\theta})$ is the generalised coordinate expressed in terms of the action–angle variables (by means of a suitable canonical transformation) and the vector notation is obvious. If the motion is quasiperiodic, a trajectory densely covers the torus and the flow is then called *ergodic on the torus*. If the frequencies are commensurate, the motion will eventually repeat itself and the orbit on the torus is thus a closed curve.

(v) An integrable Hamiltonian system is called *nondegenerate* if the frequencies on different tori are different. Since the sets $\mathbf{J}$ label the tori, nondegeneracy is guaranteed if

$$\det\left(\frac{\partial \omega_i(\mathbf{J})}{\partial J_k}\right) = \det\left(\frac{\partial^2 H(\mathbf{J})}{\partial J_i \partial J_k}\right) \neq 0 \tag{6.31}$$

A system of linear oscillators, for example, is degenerate, as frequencies do not depend on the actions. It can be shown that in a nondegenerate system only a set of measure zero of the tori contain periodic orbits (since rational numbers are a measure-zero subset of real numbers). On the rest of the tori the motion is ergodic.

Many important physical Hamiltonian systems are not integrable. The three-body (or more generally the $n > 2$-body) problem is a convincing example. Such problems can be approached by numerical methods. However, as anyone who has simulated the gravitational n-body problem on a computer must know painfully well, they are very difficult even then. Any analytical knowledge on such systems is extremely valuable since the numerical analyst can use it to judiciously choose the appropriate method and test his results. Even the seemingly simple three-body problem is known to have irregular solutions, and its general case remained unsolved for centuries, in spite of the efforts of such analytical giants as Newton, d'Alembert, Laplace, Lagrange and Poisson. To appreciate the effort invested in this problem it is sufficient to consider the work of Delaunay in the 1860s. He published a book of approximately 2000 (!) pages in an effort to calculate the Moon's orbit (the other two bodies were the Earth and the Sun), Fourier expanding some of the relevant functions in hundreds of terms, and finding a special canonical transformation for each term (see Chapter 9).

A much less ambitious step is to ask the question what happens if a Hamiltonian system is not completely integrable, but it is 'very close' to being so? Consider, for example, an integrable Hamiltonian system perturbed by a small nonlinear perturbation, making the system non-integrable. What is the fate of the tori under such a nontrivial perturbation? Even this question has resisted a definitive answer despite being one of the main research topics of theoretical mechanics in the last two centuries. The breakthrough came in 1954, when Kolmogorov suggested a theorem, which was rigorously proved and generalised by Arnold and Moser in the 1960s. The KAM theorem, as it is now called, is basic to the understanding of chaos in Hamiltonian systems. This topic will be discussed in the remainder of this chapter.

6.3 Nearly integrable systems and the KAM theorem

The fundamental problems of mechanics, like the one-dimensional motion of a point mass in an external potential, the two-body problem, coupled harmonic oscillators and rigid-body rotations, are integrable. Such problems have been applied to various branches of physics and have helped to model, for example, the motion of the Moon around the Earth or that of a planet around the Sun, the hydrogen atom or the hydrogen molecule ion, the vibrations and rotations in various mechanical and quantum-mechanical systems, and so on.

When considering more complicated systems, it is natural to exploit our knowledge of the standard cases and try to base the research on known solutions. This seems possible if the problem considered differs only slightly from some solved case. As we have already remarked in Chapter 3 the relevant mathematical technique is then *perturbation theory*. The use of perturbation theory in many important physical problems is perfectly justified and it has been very successful in practical applications. Quantum mechanics is the primary example in this context, but many problems in classical physics have also been successfully approached, using the right perturbative technique.

The perturbation method that is suitable for problems in the Hamiltonian formulation is *canonical* perturbation theory (see Goldstein, 1980; Lichtenberg & Lieberman, 1992; for a detailed discussion of these techniques, which exploit the action–angle formulation of Hamiltonian integrable systems). The Hamiltonian of a nearly integrable system is composed, by definition, of an integrable part plus a small perturbation. The integrable part of the Hamiltonian can be written as a function of the appropriate action variables (as we have seen in the previous section). The perturbation can also be written as a function of the actions, but in general it will depend on the angles as well. Thus

$$H = H_0(\mathbf{J}) + \epsilon H_1(\mathbf{J}, \boldsymbol{\theta}) \tag{6.32}$$

where $\mathbf{J}$ and $\boldsymbol{\theta}$ are the action–angle variables of the unperturbed, integrable problem and ϵ is a small dimensionless parameter. If H_1 is independent of $\boldsymbol{\theta}$, the full problem (defined by H) is integrable, and we discard this case as trivial.

The general (i.e., for n-dimensional systems) machinery of canonical perturbation theory can be found in one of the standard references mentioned above. To illustrate its application to a nearly integrable system and the difficulties that may occur in the process, we shall work out explicitly, as an example, the problem of two harmonic oscillators with a (small) nonlinear coupling between them. The Hamiltonian of this problem, expressed by the generalised coordinates $\mathbf{q}$ and momenta $\mathbf{p}$ is

$$H(q_1, q_2, p_1, p_2) = \frac{1}{2}\left(p_1^2 + p_2^2 + \omega_1^2 q_1^2 + \omega_2^2 q_2^2\right) + \epsilon H_1(\mathbf{q}, \mathbf{p}) \tag{6.33}$$

where ϵH_1 is the small nonlinear coupling term. We recall that the problem of a point-mass motion in a cylindrically symmetric harmonic potential, perturbed by a small cubic nonlinearity, which was studied by Hénon and Heiles (see Chapter 2), has a Hamiltonian of this form. The HH potential (2.19) was written as

$$V(x, y) = \frac{1}{2}(x^2 + y^2) + \epsilon y\left(x^2 - \frac{1}{3}y^2\right) \tag{6.34}$$

and was studied for $\epsilon = 1$ for different values of the energy. The non-dimensional parameter in (6.34), ϵ (it was called η in Chapter 2), reflects the relative importance of the nonlinear term. Explicitly, this parameter is defined as $\epsilon = L/\Lambda$, the ratio of the chosen length scale to what we have called the nonlinearity scale in (2.18). The irregular behaviour in the Hénon–Heiles system was found, as can now be expected, when the nonlinearity was significant, i.e., $\epsilon = 1$. However, for $\epsilon \to 0$ in (6.34), the system becomes linear and the two oscillators are uncoupled. In fact, we have already reduced the problem of one harmonic oscillator to the action–angle representation in Section 6.1. The generalisation to a two-dimensional separated system is trivial.

In the HH problem the two harmonic frequencies were equal (and set to unity). We prefer to make the present example somewhat more general, by retaining the two separate frequencies ω_1 and ω_2 in (6.33). Apart from this, our example problem will be identical to the HH case (6.34). The problem at hand is thus formulated in the following way. We seek a perturbative solution to the weakly nonlinear system whose Hamiltonian is written as

$$H(\theta_1, \theta_2, J_1, J_2) = H_0(J_1, J_2) + \epsilon H_1(\theta, \mathbf{J})$$

in terms of the action–angle variables of the unperturbed problem. We clearly have

$$H_0 = \omega_{01} J_1 + \omega_{02} J_2 \quad \text{and} \quad H_1 = x^2 y - \frac{1}{3} y^3 \tag{6.35}$$

where the coordinates x and y in the expression for H_1 have to be expressed in terms of the action–angle variables through the appropriate canonical transformation (like Equation 6.27 in the one harmonic oscillator case). The additional zero index on the frequencies has been added to stress the fact that these refer to the unperturbed integrable system.

Generalising from the single harmonic oscillator example we can, according to (6.26), write the solution of the unperturbed problem in terms of the action–angle variables as

$$J_i = \text{const}$$
$$\theta_i(t) = \omega_{0i} t + \delta_i \tag{6.36}$$

for $i = 1, 2$. The motion is on the surface of a two-dimensional torus. Phase space is four-dimensional and the energy shell three-dimensional. The two constant actions are not independent since a linear combination of the two is equal to the constant energy (Hamiltonian). Thus two independent integrals of motion are, e.g., H_0 and J_1. The frequency ratio determines the winding number of the quasiperiodic motion on the the the torus.

Using (6.27) separately for the two uncoupled oscillators we get

$$x(\theta_1, J_1) = \sqrt{\frac{2J_1}{\omega_{01}}} \sin \theta_1, \qquad y(\theta_2, J_2) = \sqrt{\frac{2J_2}{\omega_{02}}} \sin \theta_2 \tag{6.37}$$

Thus the perturbation can be written in terms of the unperturbed action–angle variables as

$$H_1 = y\left(x^2 - \frac{1}{3}y^2\right) = A\sin^2\theta_1 \sin\theta_2 + B\sin^3\theta_2 \tag{6.38}$$

where A and B are defined as

$$A = 2\sqrt{2}\frac{J_1}{\omega_{01}}\left(\frac{J_2}{\omega_{02}}\right)^{1/2}$$

and

$$B = -\frac{2\sqrt{2}}{3}\left(\frac{J_2}{\omega_{02}}\right)^{3/2}$$

Using the usual prescription of canonical perturbation theory, we proceed to find the generating function of the appropriate canonical transformation, which must transform the action–angle variables of the unperturbed problem to the new action–angle variables $(\mathbf{I}, \boldsymbol{\varphi})$, appropriate for the full problem. To first order in ϵ we get

$$\omega_{01}\frac{\partial S_1}{\partial\theta_1} + \omega_{02}\frac{\partial S_1}{\partial\theta_2} = H_1(\theta_1, \theta_2) \tag{6.39}$$

It is obvious from (6.38) that the (doubly) averaged H_1 is zero, and this is the reason for omitting it here. Owing to the relatively simple form of H_1, the coefficients of its double Fourier series expansion can easily be found. There is no need to fill 2000 pages with calculations (as Delaunay did for the three-body problem), one page should be sufficient. We shall actually need much less even than that. Writing

$$H_1 = \sum_{k_1}\sum_{k_2} H_{1k_1,k_2}\exp(ik_1\theta_1 + ik_2\theta_2)$$

and using the identity $\sin\alpha = (e^{i\alpha} - e^{-i\alpha})/2i$ in (6.38) we easily see that this double Fourier series contains at most eight terms. The pairs of indices (k_1, k_2), for which the Fourier coefficients do not vanish, are $(0, \pm1)$, $(0, \pm3)$ and $(\pm2, \pm1)$. Thus the first-order generating function has a double Fourier expansion of the following form

$$S_1 = i\sum_{k_1}\sum_{k_2}\frac{H_{1k_1,k_2}}{k_1\omega_{01} + k_2\omega_{02}}\exp(ik_1\theta_1 + ik_2\theta_2)$$

which satisfies (6.39).

In this Fourier series there is a term with $2\omega_{01} - \omega_{02}$ in its denominator. If the unperturbed integrable problem is such that $\omega_{02} = 2\omega_{01}$, that is, its frequencies are *resonant*, canonical perturbation theory breaks down in the first order. Integrability is lost with even the slightest perturbation of the original tori!

From the simple example given here it may seem that the failure of canonical perturbation theory happens only in rather contrived circumstances. The truth is, however, that the so called *small divisors problem* is basic and generic. In a general integrable system the frequencies depend on the actions, that is, they are in general different for different tori. It is thus conceivable that a fair number of the tori are resonant. A torus of an integrable system is called *resonant*, if for the actions $\mathbf{I}$ defining it, the frequencies are commensurate, that is, a condition $\mathbf{k} \cdot \boldsymbol{\omega}(\mathbf{I}) = 0$ is satisfied for some set of integers $(k_1, k_2, \ldots)$. Even if a particular torus is nonresonant, small divisors are expected for a sufficiently 'large' $\mathbf{k}$, i.e., far enough into the Fourier expansion. The problem of small divisors (or denominators), identified as the *fundamental problem* by Poincaré, plagued the development of Hamiltonian dynamics for many years. Rigorous mathematical proofs were given for the divergence of the perturbation series arbitrarily close to any convergent case. If every integrable system becomes non-integrable by a small perturbation, that is, its tori are very easily *destroyed*, the consequences for the stability of the Solar System, for example, could be unpleasant. On the other hand, it is possible that the small divisors problem, at least for nonresonant tori, is nothing but an analytical difficulty.

Two extreme opposing views regarding the physical meaning of the fundamental problem of canonical perturbation theory, were held. The apparent stability of the Earth's orbit (very close to a Kepler ellipse) and other constituents of the Solar System supported the view that small perturbations are unable to destroy all the integrals of motion of an integrable system. On the other hand, the success of the basic postulate of statistical mechanics, i.e., that a dynamical system with a large number of particles is ergodic and explores all the energy shell (there are no other integrals of motion), was seen as an indication that the opposite is true. The issue was decided by Arnold and Moser, who succeeded in the 1960s in accomplishing (mathematically rigorously) a programme based on the ideas and conjectures of Kolmogorov. We shall not deal here with all the mathematical subtleties of the resulting KAM (Kolmogorov, Arnold, Moser) theorem which, in addition to its great formal value, has important consequences for practical physical problems. In what follows we shall formulate the basic findings of this theorem and its implications in informal language. A more detailed discussion can be found in the books of Tabor (1989), Lichtenberg & Lieberman (1992), or in the appendices of the book by Arnold (1978).

The KAM theorem implies that neither of the above two extreme views is justified. The truth lies somewhere in between. More explicitly, the theorem deals with a Hamiltonian system, whose Hamiltonian consists of an integrable part and a small Hamiltonian perturbation of a general nature. This is written as in equation (6.38), where it is additionally assumed that the unperturbed Hamiltonian is nondegenerate (see 6.31).

The theorem states that under these conditions (and some additional requirements regarding the analyticity of the Hamiltonian), *almost all tori survive the perturbation* for a sufficiently small ϵ. A number of clarifications should be made here. A torus is said to *survive* if it is merely continuously distorted by the perturbation. This also means that the frequencies on such a distorted torus, which can be written as $\omega(\epsilon)$, approach ω_0 (the frequencies on the unperturbed torus) as $\epsilon \to 0$. The term 'almost all' means that the tori that do not survive constitute a set of measure zero (like the rationals among the real numbers). Moreover, the theorem supplies a quantitative condition on the frequency ratios (see below). The theorem thus ensures that all tori of the unperturbed system, the frequencies on which are 'sufficiently' incommensurate, survive a small perturbation. The orbits on these tori continue to be integrable, in this sense, even if the torus is distorted by the small perturbation.

The tori whose survival cannot be guaranteed by the KAM theorem are precisely the resonant ones. The theorem, however, alleviates the apprehension, based on the divergence of perturbative series, that integrability is lost arbitrarily close to all tori. The proof of the KAM theorem is based on the construction of an alternative convergent procedure for the nonresonant tori, and this is obviously far from trivial. The process also defines precisely 'how incommensurate' the frequencies (or 'how irrational' their ratios) must be, in order to ensure the survival of the tori. The irrationality measure condition, necessary for the preservation of the tori, is that the ratio of any two frequencies ω_1/ω_2 cannot be approximated by a rational number l/m (with l and m integers) to better than

$$\left| \frac{\omega_1}{\omega_2} - \frac{l}{m} \right| > \frac{K(\epsilon)}{m^{5/2}} \quad \text{for all integers } l \text{ and } m \qquad (6.40)$$

where $K(\epsilon) > 0$ for all $\epsilon > 0$ and $K \to 0$ as $\epsilon \to 0$. This condition is related to a famous number-theoretic issue of how closely irrational numbers can be approximated by rational ones.

With the help of the KAM theorem it is possible to understand the transition to chaos in Hamiltonian systems. An integrable (non-chaotic) Hamiltonian system can change its behaviour under a continuous perturbation (or a parameter change) in the following way. When a resonant torus, or a torus that is not sufficiently nonresonant in the above sense, appears in the system, it is readily destroyed and the corresponding trajectory becomes chaotic or ergodic. This does not mean that the other tori are destroyed as well; all but a set of measure zero of initial conditions give rise to periodic or quasiperiodic orbits. Additional tori may gradually become unstable in this sense as the parameter ϵ is increased. The coexistence of regular and chaotic orbits usually remains for a finite range of the parameter values, and we have already seen this behaviour in the Hénon–Heiles system. Such a situation is called *soft chaos* (see below). Eventually, the fraction of phase space occupied by the surviving tori becomes very small and essentially all the system trajectories ergodically fill the energy shell.

6.4 Chaos in Hamiltonian systems

In this section we describe in more detail the characteristics of chaos in Hamiltonian systems. As usual, we shall use relatively simple systems as examples. Hamiltonian systems give rise to area-preserving Poincaré maps. It is much easier to study a two-dimensional map than a multidimensional dynamical system. Moreover, most of the characteristics of chaotic behaviour are reflected already in these area-preserved mappings. We shall thus use two-dimensional area-preserving-mappings to illustrate the concepts described in this section.

We shall start by considering soft chaos, that is, the situation (mentioned above in discussion of the KAM theorem) in which resonant tori break up, but the nonresonant ones survive. Consequently, regular and irregular orbits coexist in this case. In the second part of this section we will examine what happens when all KAM surfaces are destroyed and the system thus becomes irregular globally.

Soft chaos

The KAM theorem does not guarantee the preservation of resonant tori under a small perturbation. This means that the tori of the unperturbed system, for which the frequencies are commensurate, may be unstable. We would like to understand what happens to the unstable tori as they break up as a result of a small perturbation. This will be the first step in the description of Hamiltonian chaos. The simplest way to visualise what happens to the resonant tori is by examining the surface of section of the system. As we have already seen (Chapter 4) periodic or quasiperiodic motion on a torus gives rise to a circle map in its surface of section. In dissipative systems there is usually only one torus, and it is the attractor of the flow for all initial conditions. In Hamiltonian flows, when the system is integrable, all phase space is filled with invariant tori. The motion proceeds on a particular torus, which is chosen by the point of origin (initial conditions) of the trajectory. The initial conditions determine the set of action variables, which define the torus. The motion on each torus is reflected in the surface of section by a linear circle map, as in Equation (4.4), with the frequency ratio Ω depending on the particular torus. Thus the relevant map, appropriate for a surface of section through a continuum of nested

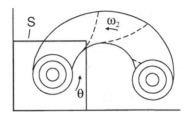

Fig. 6.1. Nested tori of an integrable Hamiltonian system and its Poincaré section.

tori (see Figure 6.1), is

$$r_{i+1} = r_i$$
$$\theta_{i+1} = \theta_i + 2\pi \Omega(r_i) \qquad (6.41)$$

where the variable r, characterising the particular torus, is its radius and θ is the angular variable taken here, as in (4.9), modulo 2π. The two-dimensional area-preserving map (6.41), which can be compactly written as $(r, \theta) \mapsto T(r, \theta)$, was introduced by Moser in 1973 and is called the *twist map*. The 'twist' is obviously produced by the dependence of Ω on r, giving rise to different angular shifts $(\theta_{i+1} - \theta_i)$ for different radii.

It can be quite easily seen that the frequency ratio on a torus, Ω, indeed depends on r, when we consider an integrable system with just two degrees of freedom. The tori in such a system are two-dimensional, and as the surface of section we may choose a surface S, say, cutting the nested tori as in Figure 6.1. The ratio of the frequencies is clearly a function of the two actions since

$$\Omega = \frac{\omega_1}{\omega_2} = \frac{\partial H(J_1, J_2)/\partial J_1}{\partial H(J_1, J_2)/\partial J_2} = \Omega(J_1, J_2)$$

Since the energy is an integral of the motion we have $H(J_1, J_2) = E$, so only one of the actions is an independent integral. Thus J_2, for example, is actually a function of J_1 and therefore $\Omega = \Omega(J_1)$. Now the action variable J_1 is actually

$$J_1 = \frac{1}{2\pi} \oint p_1 dq_1$$

and the value of the integral is just the area of the cut of the particular torus (whose radius is r) through the section S (see Figure 6.1), i.e., πr^2 in this case. Thus $\Omega = \Omega(r)$ because J_1, or alternatively r, label the tori.

The twist map is useful in studying the fate of the tori when the integrable system is perturbed. Consider a torus for which the unperturbed frequency ratio is rational, that is $\Omega = l/m$ say, with integer l and m. Since the orbits on this torus are periodic, each point on the corresponding circle on the surface of section is a fixed point of the Poincaré map, defined by m successive applications of the twist map, T^m.

Explicitly, for any point (r, θ) on the torus we get

$$(r, \theta) \mapsto T^m(r, \theta) = (r, \theta + m\, 2\pi l/m) = (r, \theta)$$

Consider now the perturbed twist map, T_ϵ, defined by

$$
\begin{aligned}
r_{i+1} &= r_i + \epsilon g(r_i, \theta_i) \\
\theta_{i+1} &= \theta_i + 2\pi \Omega(r_i) + \epsilon h(r_i, \theta_i)
\end{aligned}
\tag{6.42}
$$

where the functions g and h, determined by the perturbed Hamiltonian, do not violate the property of area preservation.

A rigorous theorem, named after Poincaré and Birkhoff, guarantees the existence of an even number of fixed points of the perturbed twist map T_ϵ, if in the corresponding unperturbed map T, Ω is rational. This means that not all fixed points lying on the circle $r = r_0$ say, are destroyed by the perturbation. It is not difficult to understand this result with the help of an appropriate graphical construction, but we shall not discuss it here and refer the interested reader to Tabor (1989). Moreover it is also possible to show (and this follows from the Poincaré–Birkhoff theorem as well) that the fixed points of the perturbed twist map are alternatingly elliptic and hyperbolic. The elliptic points are essentially centres of new nested tori, while the hyperbolic points give rise to heteroclinic tangles (discussed in Chapter 4 of this book). This situation occurs for every resonant torus and is shown schematically in Figure 6.2. The complicated structure depicted at the bottom of the figure reflects the coexistence of regular and chaotic orbits in the vicinity of a resonant torus, which has been broken up as a result of a perturbation or a parameter change. The discussion given above can be repeated for each new set of tori, and thus the complicated structure is in fact self-similar on all scales. This is reflected by the successive zooming depicted in the figure.

Summarising, we may say that the twist map reveals the remarkable structure of soft chaos in Hamiltonian systems. The regular and irregular orbits of the perturbed system are densely interwoven near the resonant tori of the unperturbed system. The nonresonant tori are deformed by the perturbation, but they survive as long as the perturbation is small. This description provides a qualitative understanding of the transition to chaos when an integrable Hamiltonian system is perturbed, as was the case, for example, in the Hénon–Heiles system for intermediate energies.

As the perturbation is increased, we may reasonably expect that the relative size of regions occupied by chaotic trajectories grows at the expense of the regions in which nonresonant KAM tori are present. The HH system accords with this expectation and, as we have seen in Chapter 2, the relative area occupied by KAM tori is found to decrease from 1 (for $E \approx 0.11$) to very close to zero (for $E \approx 0.16$). We stress again that increasing E in the HH model with ϵ fixed is equivalent to increasing ϵ (the nonlinearity) with E fixed.

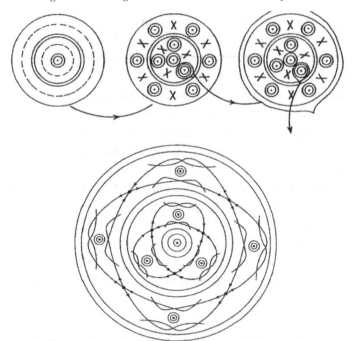

Fig. 6.2. Breakup of resonant tori (dashed circles on the leftmost panel) by a perturbation to an equal number of elliptic and hyperbolic fixed points of the perturbed map and the concomitant creation of new tori around the centres and heteroclinic tangles between the saddle-points, as reflected in the perturbed twist map. The structure is self-similar.

The chaotic regions, which are trapped between the KAM tori and reside near the separatices between their different families, are expected to 'break out' at some value of the parameter, and engulf the surviving KAM tori. This qualitative picture is confirmed by a detailed study of yet another area-preserving mapping, which is a particular realisation of the strongly perturbed twist map (with ϵ of the order of unity). The mapping, known as the *standard map*, was first studied by Chirikov. On the basis of the behaviour of this map Chirikov proposed in 1979 a criterion determining the transition to *global stochasticity* in Hamiltonian systems.

As we have seen the twist map of a system with two degrees of freedom is essentially a mapping of angles and actions (the actions are determined by the radii of KAM tori). The variables of the standard mapping are angles (θ) and actions (I) as well, and the map is defined by

$$
\begin{aligned}
I_{j+1} &= I_j + K \sin \theta_j \\
\theta_{j+1} &= \theta_j + I_{j+1}
\end{aligned}
\tag{6.43}
$$

where both θ and I are taken modulo 2π. Obviously, this map's behaviour depends on the value of K, the *stochasticity parameter*. We leave out several quantitative issues related to the fixed points of the standard map and simply quote the qualitative findings of Chirikov and his followers.

The *overlap criterion* for the transition to global stochasticity, that is, to a situation in which chaos 'breaks loose' in the above sense, is based on the notion of resonance overlap. For $K = 0$ the standard mapping is trivial (compare to the unperturbed twist map 6.41). When K is positive but very small, families of extended KAM tori are present with islands of smaller tori between them, and so on. Near the centre of each family of nested nonresonant tori there is a fixed point or a periodic point of the map. Such a point corresponds to a periodic orbit of the Hamiltonian flow. Actually the points inside the primary tori are fixed points (simple periodic orbits of the flow), and those inside the secondary tori are periodic points (higher-order periodic orbits of the flow). These points are referred to as resonances (primary and secondary, respectively). The criterion proposed by Chirikov is based on overlap of resonances in this sense, that is, when as a result of the perturbation such points come close enough to each other.

The situation is shown schematically in Figure 6.3. In case (a), that is, for small K, the shaded regions which are occupied by chaotic (stochastic) orbits, are detached from each other. When K is large, case (b), the situation is very different. The chaotic regions are interconnected and there is global stochasticity. A sufficient condition for global stochasticity is the overlap of primary resonances and this is realised when two adjacent stochastic regions just touch. The last KAM surface between the resonances is then destroyed and the remaining KAM tori are just islands in the sea of chaos.

The way in which a KAM invariant torus is being destroyed is quite interesting. After being distorted more and more as the stochasticity parameter is increased, a torus eventually breaks up. This originally invariant surface, which now becomes corrugated on all scales, has an infinite number of holes and is actually a fractal set. Thus the remnants of destroyed tori are called *cantori* and the trajectories can diffuse across them (see the discussion of destruction of the last KAM surface, given below).

The overlap criterion can be made more precise when it is formulated for secondary resonances. As can be seen qualitatively in the previous figure, a smaller K is sufficient for the overlap of a primary resonance with a secondary one, in comparison with the K necessary for the overlap of primary resonances. With the help of numerical calculations Chirikov estimated that the transition to global stochasticity is very close to $K = 1$ (hence the values of K in the figure). The details of the procedure require a rather lengthy discussion which we shall omit here (see, e.g., the book of Lichtenberg and Lieberman, 1992). We stress,

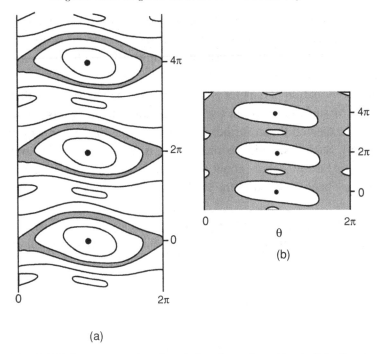

Fig. 6.3. KAM (nonresonant) tori and chaotic regions (shaded) in the standard mapping. Schematic drawing for (a) $K < 1$ (expanded, so as to show more detail); (b) $K > 1$.

however, that the overlap criterion has frequently been used as a quantitative measure of stochasticity in Hamiltonian dynamical systems (see Chapter 9 for astrophysical applications).

We conclude the discussion of what we have called soft chaos with several additional remarks. We have discussed above the perturbed twist map and the standard map as Poincaré maps of Hamiltonian systems with two degree of freedom (four dimensions). In such systems, if they are nearly integrable, each two-dimensional invariant torus cuts the three-dimensional energy surface in two. The chaotic orbits resulting from the breakup of resonant tori are thus trapped between the surviving nonresonant tori. For systems with more degrees of freedom, however, the gaps between the surviving tori form a single connected region (in the phase space of the full system). Chaotic trajectories may thus wander through the entire irregular region giving rise to what is called *Arnold diffusion*. Arnold diffusion is not a result of the stochastic layers breaking out through destroyed KAM surfaces (resonances overlapping), it is merely an effect of high dimensionality (at least three degrees of freedom, that is six dimensions) of phase space. In spite of the survival of KAM surfaces (small perturbations) the stochastic regions can be interconnected into one web. It is intuitively clear that as the Arnold diffusion must proceed only *along* surviving KAM surfaces and obviously not across them, it should be relatively slow. In addition, if the value of the stochasticity parameter is small, the measure of the chaotic regions is not large.

When the perturbation parameter is increased, resonances overlap, that is, more and more KAM surfaces between the corresponding fixed points are destroyed. It can then be expected that a further increasing of the parameter will ultimately destroy all tori. Such a situation is referred to as *strongly irregular* motion. It is instructive to add a few details about the destruction of the *last* KAM surface, a process that has been clarified in the studies of Greene in 1979 and Shenker and Kadanoff in 1980.

Any irrational frequency ratio $\Omega = \omega_1/\omega_2 < 1$ (and indeed any irrational number between 0 and 1) can be written in terms of a *continued fraction expansion*, i.e.,

$$\Omega = \cfrac{1}{a_1 + \cfrac{1}{a_2 + \cfrac{1}{a_3 + \cdots}}}$$

where all the a_i are positive integers. It is well known that this expansion is unique, and a straightforward iterative algorithm for obtaining numerically the a_i can easily be devised. The successive iterates of such a procedure converge to Ω, while alternating around it. If the a_i for $i > N$ (for some fixed N) rapidly increase with i, the convergence is very fast. If Ω is rational the a_i, from a certain N and on, are actually infinite. In this sense the irrational number between 0 and 1, which is *the most removed* from the rationals, is the one in which all the a_i, in its continued fraction expansion, are the smallest possible, that is, all $a_i = 1$. Clearly, this is the celebrated *golden mean*, the limit of the Fibonnaci series, $(\sqrt{5} - 1)/2$. Thus from all invariant tori of an integrable system, the last surviving one, as the perturbation parameter increases, is the torus whose unperturbed frequency ratio is the golden mean. Numerical work has verified this expectation and has shown how this torus develops corrugations, to make room for the stable periodic orbits corresponding to the successive rational truncations of the golden mean. At the critical value of the stochasticity parameter the torus has corrugations on all scales (it is fractal) and all the nearby period orbits become unstable. Above the critical value, an infinite number of holes appear in the distorted torus and it becomes a Cantor set (a cantorus). Remarkably, this exquisite abstract mathematical structure has relevance to real astrophysical systems, like stellar orbits in a galactic potential (see Chapter 9).

Strongly irregular motion

We now briefly sketch a few important results and definitions related to strongly irregular motion in Hamiltonian systems; that is, a situation in which there are no invariant tori in phase space. A detailed discussion of this issue can be found in Lichtenberg and Lieberman's book. As before, a lot of information can be extracted from area-preserving mappings. The standard map with large K can be exploited in this context, but as we are only interested in the basics we shall consider here a simpler and more amusing map – Arnold's cat map.

The cat map is well suited for the study of strongly irregular systems as all the fixed points of this map and its higher generations are hyperbolic (we saw above that elliptic fixed points give rise to regular motion and hyperbolic ones to chaos). Arnold's cat map, $(x, y) \mapsto A(x, y)$, is defined on a 2-torus by

$$x_{i+1} = x_i + y_i \quad (\text{mod } 1)$$
$$y_{i+1} = x_i + 2y_i \quad (\text{mod } 1) \tag{6.44}$$

so that x and y are angular coordinates on the torus, with the angles measured in units of 2π radians.

The map is area-preserving since the Jacobian of the transformation is unity. The eigenvalues of the Jacobian are $\sigma_{1,2} = (3 \pm \sqrt{5})/2$ and hence all the fixed points are hyperbolic (one eigenvalue is positive and the other one negative). The action of the map acting on the 'spread-out' torus (the unit square) is shown in Figure 6.4. The picture of a black cat face C is seen to be greatly distorted already after two iterations ($A^2 C$). The 2×2 square in the upper part of the figure is shown to demonstrate just the distortion caused by the cat map (a mapping of this operation is symbolically denoted by A_0), before the cutting and folding action of the modulo 1 operation. As an animal lover myself I apologise to the readers whose feelings may be hurt by Arnold's mapping. One may obviously replace the cat's face by a favourite enemy, or by an inanimate object. After a few iterations there is no way to tell anyway. The dissociation of the initial structure arises from the purely hyperbolic nature of the cat map, causing two initially close points to be pulled far apart.

The cat map is obviously chaotic and this can be seen from the positivity of one of its Liapunov exponents (it is $\lambda = \ln[(3 + \sqrt{5})/2]$). The fixed points of A^n are all hyperbolic and it can be shown that for $n > 1$ heteroclinic tangles occur on the torus, as is typical in chaotic maps. Since the cat map can be considered as representative of Hamiltonian systems (it is an area-preserving map) we have here a case with no regular orbits at all, and we proceed to investigate the properties of such strong irregular motion, in contrast to the soft chaos case. The cat map is also obviously *ergodic* on all its domain of definition. This means that an orbit started at

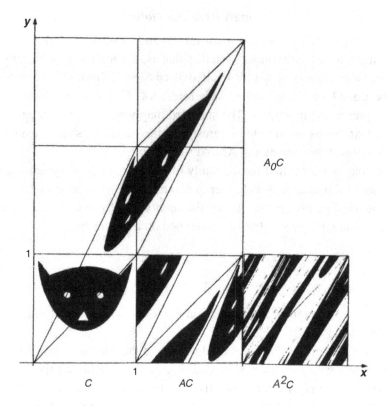

Fig. 6.4. Two iterations of the Arnold cat map (lower panels).

a point (x_0, y_0) with x_0/y_0 irrational will densely cover the torus (all phase space). Ergodicity is usually defined by the equivalence of long time averages to phase-space averages. In discrete maps a time average must be replaced by an average over iterations (as the iteration number goes to infinity) and thus the usual definition holds.

The cat map has, however, a 'stronger' property than ergodicity – *mixing*. Any given area of phase space is so greatly distorted by successive iterations as to be spread homogeneously over all phase space. One of the ways to define mixing precisely is by using the autocorrelation function (defined in Chapter 5). Mixing means that the autocorrelation between the initial point and the nth iterate decays to zero as $n \to \infty$. In mixing systems any dynamical variable approaches a constant value and this means that the system approaches thermodynamic equilibrium.

An additional property of the cat map is the exponential divergence of trajectories (positive Liapunov exponent, see above) for all initial conditions. Such systems are called C-systems. Not all ergodic systems are mixing and not all mixing systems are C-systems. The converse statements are however true and it is sufficient for a system to be a C-system to exhibit mixing and this, in turn, is sufficient for ergodicity. In fact, additional classes of systems may be defined in a hierarchical order of increasing randomness. We conclude by quoting such a hierarchy of randomness in Hamiltonian systems.

(i) Recurrent systems – the orbit returns infinitely many times arbitrarily close to its initial point (according to Poincaré's recurrence theorem any Hamiltonian system has this property for almost all orbits).

(ii) Ergodic systems – time average equals phase space average (e.g., an integrable system with incommensurate frequencies is ergodic on the torus, but not on the entire energy shell).

(iii) Mixing systems – any initial volume spreads over the whole phase space, i.e., the system approaches equilibrium. It is quite difficult to prove rigorously that a given system is mixing in this sense because the stochastic orbits must mix over the entire accessible phase space.

(iv) K-systems – trajectories in a connected neighbourhood diverge exponentially *on the average*. These systems are so called after Kolmogorov, who (together with Sinai) defined the Kolmogorov–Sinai (KS) entropy, a quantity measuring the average exponential divergence of trajectories. Positive KS entropy defines K-systems.

(v) C-systems – global instability, i.e., *every* trajectory has a positive Liapunov exponent (e.g., the cat map). Such systems were studied by Anosov and are sometimes named after him.

Additional material on the criteria for the different types of behaviour can be found in Tabor (1989) and in more detail in Lichtenberg & Lieberman (1992). As pointed out by Contopoulos (2002), the above classification actually pertains to a special limit of chaotic systems. In most systems, that have been numerically studied in detail so far, both chaotic and regular orbits have been found and therefore a broader classification is called for. He proposed that chaotic systems should be subdivided into two cases – *general* and *limiting*, and then the hierarchy given above is applicable only to the limiting cases, when no regular orbits exist in phase space.

For the more mathematically minded reader we should mention that the quantitative measure of the exponential divergence (or instability leading to chaos) of trajectories can be most rigorously treated by using the notion of curvature of the suitable configuration manifolds (see, e.g., Arnold, 1978).

7
Extended systems – instabilities and patterns

> Not to be born is the best for man
> The second best is a formal order
> The dance's pattern, dance while you can.
> Dance, dance for the figure is easy
> The tune is catchy and will not stop
> Dance till the stars come down with the rafters
> Dance, dance, dance till you drop.
> W. H. Auden, *Letter from Iceland*.

Spatially extended physical systems are naturally described by partial differential equations (PDE), which reflect the spatial structure of the system in addition to its time evolution. Classical examples in this context are fluid systems, described by the Euler or Navier–Stokes equations. Extended systems having several constituents that can react with each other and diffuse in space, are another well known example. These systems (like, e.g., mixtures of appropriate chemical species) are described by PDEs of the *reaction–diffusion* type. Many systems are only properly described when several types of process (e.g., fluid–dynamical *and* reactive–diffusive) are taken into account. A fluid system with heating and cooling processes as well as heat conduction operating within it is an example of such a system.

When we are dealing with only *temporal* evolution of a physical system, the dynamical system describing it consists of ODE or discrete maps. We have seen in Chapter 3 that near the threshold of an instability it is often possible to reduce the dimension of such systems and obtain a small set of *amplitude equations*, which faithfully describe the essential features of the system near marginality. A spatio-temporal PDE (the independent variables include coordinates and time) can be regarded as the infinite limit ($n \to \infty$), of a system of n-coupled ODEs. Indeed, a practical approach to an extended system may include spatial discretisation, yielding a 'coarse-grained' set of a finite (albeit very large) number of coupled temporal ODEs. Such sets arise naturally when one attempts to solve numerically a PDE dynamical system on a computer. Being a limit, in this sense, of ODE systems, a PDE dynamical system can first be examined using the ideas and techniques described in Chapter 3.

Such a treatment is effective when the spatial structures present in the system are 'frozen', i.e., they may develop in time but do not move, because of boundary effects (spatial confinement), for example. This usually arises when the lateral extension of the physical system, L, is not much larger than the characteristic spatial scale Λ of the instability involved. The system is then said to be *confined* and it contains only a few spatial structures, which are locked by boundary constraints and thus remain frozen. For example, when the cell in an experiment on convection has a sufficiently small aspect ratio (width/height), only two (say) convection rolls may be present, and they stay in fixed positions. The spectrum of the stability eigenvalues of such a confined system is, as a rule, discrete, that is, the linear eigenmodes are isolated. If only very few modes are excited it is possible to derive equations for their amplitudes, in a similar way to the ODE case (dimensional reduction).

In the opposite case $L \gg \Lambda$, one needs an explicitly spatially dependent treatment and it is reasonable to idealise the system as being infinite in space. Such systems are called *weakly confined* or *extended*. Extended systems can support propagating structures (travelling waves in the simplest case), and this must be reflected in an appropriate way in the reduction method to be employed. The basic idea is to extend the notion of mode amplitudes to a spatio-temporal amplitude, or *envelope*. By using suitable methods (such as singular perturbation techniques) it is usually possible to derive equations (PDE) for the envelope structure and evolution. While amplitude equations are the result of dimensional reduction of an infinite-dimensional PDE to a few ODEs, spatio-temporal envelope equations are themselves PDEs. These PDEs are, however, simpler and more basic than the underlying system and are thus usually more tractable.

When a spatially extended dynamical system has translational symmetry (this concept will be discussed in detail later on), discretisation of the PDE, in order to solve it numerically, generally yields difference equations (in the simplest case only one equation) defined on a *lattice* (a discrete spatial coordinate system). These equations essentially provide a rule (or set of rules) to advance in time the value of dynamical variables at a given lattice point, using their value in the neighbouring points. The resulting discrete dynamical system has the appearance of a spatio-temporal map, that is, a mapping with two types of indices, one defining the time level and the other (whose number depends on the spatial dimension) the spatial position. Such systems are called *coupled map lattices* (CML) and can be studied relatively easily. Coupled map lattices constitute a class of dynamical systems of their own and have been studied in various cases, especially when they include

known chaotic maps. These studies have provided useful insight mainly into correlations among different spatial parts of the system in generic chaotic systems with
spatial couplings.

Before starting a detailed description of the above mentioned ideas and methods
we should set the stage by introducing some basic concepts. We shall refer to PDEs
describing the detailed physics of a system as *microscopic equations*. Obviously,
this does not mean that these are true microscopic equations, reducing the problem
to its atomic constituents. We merely mean that these are the starting equations
including the elementary building blocks of the physical system they describe. The
Navier–Stokes equations, describing a fluid, are 'microscopic' in this sense and so
are reaction–diffusion equations in the context of reacting chemicals or heated and
cooled extended media, for example.

The microscopic equations are generally rather complicated nonlinear PDEs,
with only a very limited number of known analytical solutions. Such equations do
not yield to straightforward analytical treatments and they are often approached by
brute-force numerical integrations. Because of the enormous progress in computer
power, reliable solutions are now often available for a variety of cases. These can
serve as very valuable numerical experiments, but are rather limited in their ability
to provide physical insight and general theoretical understanding of the underlying
dynamical systems. It is therefore important to develop and use alternative techniques for the purpose of providing the knowledge mentioned above.

A useful approach starts from examining infinite, spatially uniform states of the
system, which are usually trivial solutions of the microscopic PDE. The next step
is to establish the *linear stability* properties of the uniform solutions. Interesting
systems have such instabilities for some range of *control parameters* (coefficients
in the PDE). Next, one proceeds by classifying the linear instabilities, similarly to
the classification given in Chapter 3 for ODEs and maps. The usefulness of such a
classification is based on the conjecture that PDE systems having the same linear
instability properties behave similarly in the (at least weakly) nonlinear regime, i.e.,
that the instability class (see below) indicates generic behaviour. The difference in
comparison with ODEs or maps is that here the small perturbation can have spatial
dependence, i.e., it must be written in the linear regime (compare with equation
3.16) as

$$\delta\mathbf{u}(\mathbf{r}, t) = \mathbf{D}\exp(i\mathbf{k}\cdot\mathbf{r} + st) \tag{7.1}$$

where the state variable is now generally a vector *function* $\mathbf{u}$, and $\mathbf{r}$ is the position
vector. The meaningful novelty here is the dependence of the exponential term
linearly on $\mathbf{r}$. The vector coefficient $\mathbf{k}$ is called the *wavevector*. Linear instabilities
are classified according to the values of the wavevector, $\mathbf{k}$, and frequency $\omega \equiv$
Im(s), at the *instability threshold*, namely when the control parameters pass their

critical value. At this point $\sigma \equiv \mathrm{Re}(s)$ of the marginal mode passes through zero. Such a classification will be made and discussed in the first section of this chapter.

Because of the conjecture that systems with similar linear instabilities also behave similarly in the nonlinear regime one can study, instead of the true microscopic equations, some simpler, generic *phenomenological* or *model equations.* Such equations should have the same linear instability properties as the original microscopic equations, but can be much more tractable analytically and/or numerically. We mention here just one famous example of a model equation, which has been extensively studied in many physical contexts, the Ginzburg–Landau equation. Be it a microscopic or a model equation, it is useful to examine its behaviour near the instability threshold. Some general properties of a pattern-forming system and the nature of the possible patterns can be deduced from this behaviour. We shall discuss this issue in Section 2.

As we have stated above, it is often possible to obtain (using perturbation methods or other reduction techniques) simple equations that either already have, or can be further reduced to, a universal form. The important property of these equations is their simplicity and their generic form, describing the behaviour of the original system near the instability threshold and reflecting the mechanisms of nonlinear saturation of the perturbation amplitudes. If solved, these equations yield the basic spatio-temporal structures or *patterns* present in the system near the linear instability threshold. We shall demonstrate the methods for the derivation of amplitude and envelope equations on example systems in the third section of this chapter.

Far from the linear instability, in the strongly nonlinear regime, simple equations can sometimes also be derived, again by perturbation techniques (but perturbing now around some ideal nonlinear structure). These go under the name of *phase equations* and are very useful in understanding patterns in this domain. We shall discuss this topic briefly in the fourth section.

It is important to remark here that sometimes the reduction procedure outlined above repeats itself. For example, while the reduced equations resulting from some microscopic or model equations are sometimes ODEs, they are usually (when the spatial pattern moves) PDEs (envelope equations). If they are nontrivial, they can be regarded again as model equations and be subject to further study, using the same methods again. The procedure aims, in general, to reduce the problem as much as possible. We wish to go from complicated PDEs to simpler ones and from them to low dimensional ODEs and finally to generic ODEs (normal forms) or even discrete maps, which are the most primitive (in the positive sense), best understood, dynamical systems.

In two futher sections of this chapter we shall discuss the very important aspect of patterns, namely their imperfections. In many systems departures from regularity of an ideal spatial pattern are observed. If these perturbing structures are *localised* and *persistent* they are called *defects* or *coherent structures*, and their study contributes a lot to the understanding of the system. We shall explore in some detail examples of one-dimensional defects. Multidimensional defects will be discussed only briefly. One-dimensional fronts and pulses as well as two-dimensional target and spiral patterns are observed in a variety of seemingly unrelated systems, and their theoretical study confirms the generic aspects of their nature.

We shall conclude this chapter with a brief discussion of CMLs. These systems can be considered as model equations and are introduced directly on the discrete mapping level. Rather than starting from a microscopic or a model PDE and reducing them in one of the ways explained above, CMLs are specifically devised to study systems that display temporal chaos locally and possess some spatial coupling. Although CMLs may sometimes be related to discretised PDEs, their (mostly numerical) study aims at the exploration of the qualitative features of spatio-temporal complexity.

In following this agenda we shall be illustrating the methods and concepts using particular examples. Some of the ideas developed here may be applicable to extended astrophysical systems. We shall mention a few such possibilities in the second part of the book and hope that this chapter provides the basics necessary for their understanding. Here, we shall only include a collection of rather well-established and general results of pattern-formation theory in different physical applications. For fuller and more systematic expositions we recommend the book by Manneville (1990) and the extensive review paper by Cross and Hohenberg (1993).

7.1 Linear instabilities

Consider a general set of nonlinear PDEs of the form

$$\frac{\partial \mathbf{u}}{\partial t} = \mathbf{F}\left[\mathbf{u}, \frac{\partial \mathbf{u}}{\partial x}, \ldots, a\right] \tag{7.2}$$

for the functions $\mathbf{u}(x, t) = [u_1(x, t), u_2(x, t), \ldots, u_n(x, t)]$. The vector functional $\mathbf{F}$ depends upon $\mathbf{u}$ and its spatial derivatives and the parameter a. We have explicitly expressed here the dependence of the functions on only one space variable x. A generalisation to several space variables (and several parameters) should be straightforward.

It is sometimes convenient to separate the linear and nonlinear parts of the right-hand side of equation (7.2) thus

$$\partial_t \mathbf{u} = \mathcal{L}(\mathbf{u}, a) + \mathcal{N}(\mathbf{u}, a) \tag{7.3}$$

Here we use the shorthand notation for partial derivatives – $\partial_t \equiv \frac{\partial}{\partial t}$, $\mathcal{L}$ is a linear (possibly differential) *operator* and $\mathcal{N}$ a strictly nonlinear one, and both may depend on the parameters.

Without limiting the generality of our discussion we may assume that the uniform constant state $\mathbf{u} = \mathbf{0}$ is a solution of the system for all values of the control parameter a. This solution is the analogue of a fixed point in ODE systems. The state space is here the infinite-dimensional function space, in which a given function $\mathbf{u}$ is a point. For the mathematical problem to be well posed we must specify definite *boundary conditions* applied on the boundaries of the functions' domain of definition and *initial conditions* at some time ($t = 0$, say).

Classification of instability types

For the purpose of performing linear-stability analysis of the null solution we look at a small perturbation of the form (7.1), depending on only one spatial-independent variable x. Substituting this into the original equation (7.2) and linearising (keeping terms up to first order in the perturbation) we get

$$s\delta\mathbf{u} = \mathcal{J}\delta\mathbf{u} \tag{7.4}$$

where the stability matrix $\mathcal{J}$ here consists of *functional derivatives*, that is,

$$\mathcal{J}_{ij} \equiv \frac{\delta F_i}{\delta u_j} \tag{7.5}$$

(compare with equation 3.7), evaluated for some a at $\mathbf{u} = \mathbf{0}$. The alternative form of (7.4) when the original PDEs are written as in (7.3) is very similar to this, but with the linear operator $\mathcal{L}$ replacing the stability matrix (7.5). We remind the reader that the functional derivative of a functional, $\mathcal{F}$ for example, with respect to a function $v(x)$ say (the generalisation to vector functions and several space variables is straightforward), is defined by

$$\frac{\delta\mathcal{F}}{\delta v} = \lim_{|\eta|\to 0} \frac{\mathcal{F}[v+\eta] - \mathcal{F}[v]}{\eta} \tag{7.6}$$

where $\eta(x) \equiv \delta v$ is an arbitrary variation of $v(x)$.

Linear stability is determined by the eigenvalues of the matrix $\mathcal{J}$ (or the linear operator $\mathcal{L}$), which depend (in view of the assumed form of the perturbation, see (7.1)) on the absolute value of the wavevector (i.e., the wavenumber k), in addition to their obvious dependence on the parameter. If the real part of every eigenvalue is negative for *all* k the equilibrium solution is linearly stable. For instability it is sufficient that the real part of some eigenvalue be positive for *some* k.

Let s_0 be the eigenvalue with the largest *real* part σ_0 (for any k). This eigenvalue corresponds to the most 'dangerous' mode, that is, the first one to become unstable as the parameter is varied. Suppose that the dependence of the system on the *parameter* is such that for $a = a_c$, $\sigma_0 = 0$ for some k_c, while for $a < a_c$, $\sigma_0(k) < 0$ for all k.

Clearly, a_c is the *critical* point, where the eigenvalue with the largest real part becomes marginal. This is therefore the linear-instability threshold. It is convenient to define the *reduced* control parameter $\lambda \equiv (a - a_c)/a_c$ (if $a_c = 0$ we set $\lambda = a$). This definition shifts the instability threshold to the zero value of the parameter λ. As in the ODE case, the behaviour of a nonlinear PDE system near its marginal state depends on the nature of the linear instability. The different linear instability types of a PDE system like (7.2) are classified according to the behaviour of the function $\sigma_0(k)$ as λ crosses zero (i.e., at its critical value). In what follows we shall drop the zero subscript and remember that we are considering the least stable (most dangerous) mode.

In Figure 7.1 the two different typical types of behaviour near the instability threshold are illustrated. In the first case, depicted in the left panel of the figure, the instability threshold is crossed for $k_c \neq 0$. This type of instability is called *cellular* (or spatially periodic), since near criticality there exists a characteristic length scale $\sim 1/k_c$ in the problem and for $\lambda > 0$ there is an unstable band of wavenumbers $k_1(\lambda) < k < k_2(\lambda)$. In the other case (depicted in the right panel) $k_c = 0$, and thus there is no characteristic length scale at the instability threshold ($\lambda = 0$). Consequently, this type of instability is called *homogeneous*.

Actually, the homogeneous instability comprises two subclasses, depending on the behaviour of the growth rate at $k = 0$ as the parameter λ is varied. It may happen that the growth rate at $k = 0$ remains zero for all values of λ (contrary to

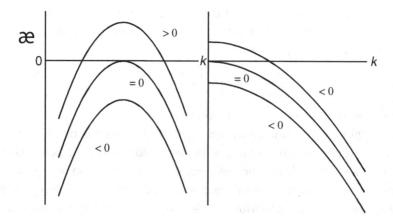

Fig. 7.1. The linear growth rate, σ, as a function of the wavenumber k for various values of the control parameter. The two cases – cellular instability (left) and homogeneous instability (right) are shown. See text for details.

the situation shown in the figure). This is usually caused by a physical constraint such as a conservation law. This case of homogeneous instability may be important in some applications but we shall not consider it further in this book.

The instabilities are further classified according to the imaginary part of the eigenvalue at instability threshold, ω_c. If $\omega_c = 0$ the instability is called *steady* and if $\omega_c \neq 0$ it is *oscillatory* (or temporally periodic).

The above classification of linear instability types for PDEs is more general than the classification done for ODEs. This is obviously because in PDEs the linear modes have spatial structure, reflected by the appearance of k, the wavenumber. Both the real and imaginary parts of the stability matrix eigenvalues are thus functions of k, and the linear stability condition is found from the *dispersion relation* $s = s(k)$. Still, the homogeneous types of instability are analogous to the ODE case because the mode which becomes critical has $k = 0$. The critical points of PDEs can thus be referred to as bifurcation points, similarly to the ODE case, and the homogeneous instability types correspond to bifurcations which we have already met in ODEs. For example, the homogeneous oscillatory instability is often referred to as a Hopf bifurcation (occurring simultaneously at all spatial positions). Before discussing further implications of linear instability classes and their symmetries on pattern formation in the nonlinear regime, we now give two examples of classification.

Examples

The first example is motivated by the PDE we used in Chapter 2 to describe a one-dimensional bistable medium with diffusion (2.33). We consider here a simpler model equation, whose properties near its linear instability (bifurcation) point resemble those of equation (2.33).

Let the function $u(x, t)$ satisfy the PDE

$$\partial_t u = \lambda u - u^3 + \partial_x^2 u \tag{7.7}$$

with λ being a parameter. This equation was proposed by Landau in the context of phase transitions and is known in the literature as the Landau equation. We wish to examine the existence and stability of the constant solutions of (7.7). In order to find them one has to solve the cubic equation obtained by the substitution $u = \text{const}$ in (7.7). The solutions are

$$u_0 = 0 \quad \text{and} \quad u_{1,2} = \pm\sqrt{\lambda}$$

Thus the null solution is present for all λ values, while the additional pair of solutions exist only for $\lambda > 0$.

Linear stability analysis of the uniform solution $u = 0$, performed by substituting a solution $\propto \exp(st + ikx)$, gives

$$s = \lambda - k^2$$

and thus if $\lambda < 0$ the null solution is stable. For $\lambda > 0$ there is instability for all wavenumbers satisfying $|k| < \sqrt{\lambda}$. As λ is increased from a negative value, the instability threshold appears at $\lambda = 0$ and $\sigma = 0$ at $k = 0$. At the critical value of the parameter we thus have $k_c = 0$, and consequently the instability is *homogeneous*. It is obviously also *steady*, as s is real ($s = \sigma, \omega = 0$) for all k. Such homogeneous steady instability cannot give rise to essential pattern formation in its vicinity, since no length or time scales exist in the problem.

We already know, however, that the original equation supports (for appropriate boundary conditions) stationary or moving fronts. These fronts are actually 'defects' connecting two homogeneous parts of the system. It should be remarked here that their significance is not limited to a state close to marginality. As we shall see later in this chapter and in the second part of the book, any ensemble of such one-dimensional fronts decays through slow annihilation of opposite fronts and a nontrivial pattern can persist only if external forcing (an additional term in the PDE) is present in the system.

As our second example we consider a PDE in one space dimension, which has been proposed in various contexts as a model equation (e.g., in the flame propagation problem) or a phase equation (e.g., in pattern-forming reacting chemicals). The equation is named after Kuramoto and Sivashinsky and in one of its variants it reads

$$\partial_t u + u \partial_x u = -au - \partial_x^2 u - \partial_x^4 u \tag{7.8}$$

where a is a parameter. Substitution of the perturbation (7.1) and linearisation gives

$$s = -a + k^2(1 - k^2)$$

As s is real for any k, the linear instability occurring for $s > 0$ is *steady*. The maximal growth rate occurs at the wavenumber maximising s, that is at $k = k_c = 1/\sqrt{2}$. The instability threshold ($s = 0$) thus occurs at $a = a_c = 1/4$ and we have instability for $a < a_c$ and stability for $a > a_c$. As $k_c \neq 0$ this instability is *cellular*.

7.2 General qualitative properties of pattern-forming systems

We start our discussion of the general properties of pattern-forming systems by mentioning their special important subclass – the *gradient* or *potential* systems. In a gradient system Equation (7.2) can be written in the form

$$\partial_t \mathbf{u} = -\frac{\delta \mathcal{F}}{\delta \mathbf{u}} \tag{7.9}$$

where $\mathcal{F}[\mathbf{u}]$ is a suitable scalar functional of $\mathbf{u}$ and possibly of its spatial derivatives.

Some important general properties of a gradient system can be predicted if the functional $\mathcal{F}$ is a Liapunov functional for the problem. The existence of a Liapunov functional has profound implications on *nonlinear* stability of solutions, and is a generalisation to PDEs of the same concept introduced for ODEs in Chapter 3. We shall now state without rigorous proof some mathematically exact results that are relevent to gradient systems of the form (7.9). A little reflection is enough, however, to make these results very suggestive. For simplicity we consider the case when $\mathbf{u}(\mathbf{r}, t)$ is a scalar function in one space dimension only, so that it can be written as $v(x, t)$ and the PDE in its gradient form is

$$\partial_t v = -\frac{\delta \mathcal{F}}{\delta v} \tag{7.10}$$

The mathematical statement is the following. If there exists a functional $\mathcal{F}[v]$ and it is *decreasing in time* and *bounded from below* for any $v(x, t)$ which is the solution of some particular PDE, then it is a Liapunov functional of the problem (analogous to free energy in statistical mechanics). Now, a Liapunov functional has the property that the function v, which minimises it (that is, makes its functional derivative to be equal to zero), is a *stable* and *steady* (time-independent) solution of the PDE.

For all gradient systems like (7.10) we have

$$\frac{\mathrm{d}\mathcal{F}}{\mathrm{d}t} = \int \mathrm{d}x \left(\frac{\delta \mathcal{F}}{\delta v}\right)\partial_t v = -\int \mathrm{d}x \left(\frac{\delta \mathcal{F}}{\delta v}\right)^2 \le 0$$

Thus if a PDE is a gradient system and can be written as (7.10), the functional $\mathcal{F}[v]$ does not increase in time for any function satisfying the PDE. It is thus sufficient to show that $\mathcal{F}$ is bounded from below (this depends on the particular form of the PDE) to verify the fact that $\mathcal{F}$ is a Liapunov functional for that particular problem. If this is so, the functional actually *decreases* for any solution, save the one that minimises it, i.e., the one for which $\delta \mathcal{F}/\delta v = 0$. The specific function $v_0(x)$, for which $\mathcal{F}$ is minimal, is a stable steady (equilibrium) solution of the PDE since the functional derivative of $\mathcal{F}$ at its minimum is zero, and it is positive for any other function. Consequently, the dynamics associated with a gradient system in such a case essentially consists of a relaxation towards the steady solution that minimises $\mathcal{F}$. Since this behaviour is reminiscent of the approach to thermodynamic equilibrium, $\mathcal{F}$ is sometimes referred to as the 'free energy' of the system. When a system develops so as to minimise its Liapunov functional, chaos and complexity are obviously impossible but the minimising function can still give rise to nontrivial patterns.

Out of the two examples given in the previous section, the first equation (7.7) is a gradient system with the functional $\mathcal{F}$ given by

$$\mathcal{F}[u] = \int dx \left[\frac{1}{2}(\partial_x u)^2 - \frac{\lambda}{2} u^2 \right]$$

In contrast, the Kuramoto–Sivashinsky equation (7.8) is not a gradient system and consequently complex behaviour may be expected.

Pattern-forming systems, modelled by gradient or more general PDEs, often possess various symmetries reflected by the fact that the PDEs are *invariant* under the relevant transformations. The most basic symmetry of an extended system is *translational symmetry*. This symmetry is associated with the invariance of the PDE under a coordinate shift, e.g., $x \mapsto x + l$, with l fixed. Both example equations considered in the previous section obviously have this symmetry as $\partial_x = \partial_{x'}$ if $x' = x + l$. This symmetry is *continuous*, since l can take on any value and the system remains invariant. If a given system is invariant for only some particular values of the displacement l, we talk of a *discrete* translational symmetry. Other types of continuous and discrete symmetries include, for example, rotational invariance, parity ($x \mapsto -x$), inversion ($u \mapsto -u$), Galilean invariance ($x \mapsto x + v_0 t$ and $u \mapsto u + v_0$, with v_0 fixed), and so on.

The spontaneous appearance of a pattern, when a highly symmetric state loses its stability as a result of a parameter change, is usually associated with *symmetry breaking* in a similar way to the bifurcations discussed in Chapter 3. For example, a stable uniform solution of (7.8) in an infinite domain is continuously translationally symmetric, as is the equation itself. As a decreases and reaches its critical value a_c the constant symmetric solution loses its stability and a definite (nondimensional) length scale appears in the problem, i.e., $\Lambda = 2\pi k_c^{-1} = 2\pi \sqrt{2}$. Thus, at marginality the continuous translational symmetry of the *solution* is broken and the equation admits solutions with a discrete translational symmetry $x \mapsto x + \Lambda$. In general, further bifurcations may result in less symmetrical patterns breaking even more symmetries of the system. Usually, however, we are interested in patterns that preserve some of the invariances of the original PDE, as was the case above. Such patterns are called *ideal* and they are usually fundamental to the understanding of less symmetric and more complicated patterns. The symmetry breaking described here is not only similar to the generic bifurcations, but is in fact connected with them. As we shall see shortly, the original PDE can be reduced in the vicinity of its critical points to equations which either have or are related to normal forms.

Assume now that a certain continuous symmetry is broken at some given parameter value λ_c, where we have a cellular ($k_c \neq 0$) instability. When $\lambda > \lambda_c$ the linearised system's modes will grow exponentially for all k such that $\sigma(k) > 0$. For example in the system (7.8) with $a < a_c$, there is a continuous band of unstable wavevectors satisfying $k^2 - k^4 > a$. If the system is confined, some particular modes may be selected by the boundary conditions. In extended systems all the above continuum of modes are in principle present, but this applies only to the linearised system. Nonlinearities usually give rise to saturation effects, which effectively select the excited modes and define the pattern. In addition, particular combinations of elementary patterns may be unstable, and one should consider only the regions in the parameter–wavenumber space where stable patterns exist. These regions are usually referred to as *stability balloons*. We shall explore several such possibilities later in this chapter.

We conclude by remarking again that boundary effects usually interfere with the formation of ideal patterns, by reducing the symmetry of the pattern from, e.g., continuously translational (or rotational) to discrete. Another effect that is observed frequently in laboratory pattern-forming systems is the appearance of *defects* or *localised structures* that can be defined as localised departures from an otherwise ideal pattern. We can consider the fronts between two possible ideal homogeneous solutions (like in the bistable medium) as elementary examples of defects. Pulses of higher amplitude in an underlying periodic wave pattern are also defects. Localised structures and their dynamics are important to the understanding of mechanisms of pattern selection and spatio-temporal chaos. This topic will be discussed in the last two sections of this chapter.

7.3 Amplitude and envelope equations

Our aim is to simplify microscopic or model PDEs by replacing them, at least close to marginality, with a low-dimensional ODE system (amplitude equations) or a generic simple PDE (envelope equation). The ideas and techniques are essentially the same as those used in the dimensional reduction of ODE systems, described in Chapter 3. We shall describe in this section two approaches to this problem. The first is based on the application of *projection* and *slave-mode elimination* techniques and will be demonstrated on a confined system. The second approach consists of a perturbative treatment and will be illustrated by a weakly confined (or extended) system. In both cases we shall use variants of the extensively studied Swift–Hohenberg model equation, which has been proposed in a variety of pattern-forming systems. The relevance of this model to convection is discussed in detail in Cross & Hohenberg (1993) (see also Manneville, 1990).

Dimensional reduction of a confined system

We consider a model PDE in one space dimension in the finite domain $0 \leq x \leq L$

$$\partial_t u + u \partial_x u = \lambda u - \left(\partial_x^2 + k_c^2\right)^2 u \tag{7.11}$$

with the boundary conditions $u = 0$, $\partial_x^2 u = 0$ at $x = 0$ and $x = L$. This equation is a particular variant of the Swift–Hohenberg equation mentioned above. It can be easily classified as having a cellular steady instability at $\lambda = 0$, with the critical wavenumber being k_c. The system is assumed to be confined by boundaries and consequently only normal modes having certain discrete values of the wavevector are possible. Linearisation around the null solution and substitution of a solution in one spatial dimension of the type (7.1),

$$u(x, t) = D \exp(ikx + st) \tag{7.12}$$

gives the dispersion relation for the linear modes

$$s(k) = \lambda - \left(k^2 - k_c^2\right)^2 \tag{7.13}$$

When $\lambda > 0$ all the modes with k in the band $k_c^2 + \sqrt{\lambda} > k^2 > k_c^2 - \sqrt{\lambda}$ are linearly unstable ($s > 0$).

The general solution of the linear problem is a superposition of the linear modes (7.12). Because of the spatial confinement, however, this solution is composed of only those modes whose spatial structure is consistent with the boundary conditions. Noting that s depends only on k^2 and thus the two modes with k and $-k$ correspond to the same s, we can write the general solution of the linearised problem as

$$u_{\text{lin}}(x, t) = \sum_k u_k(x, t)$$

with the individual eigenmodes given by

$$u_k(x, t) = (D_{k+}e^{ikx} + D_{k-}e^{-ikx})e^{st}$$

with $s(k)$ determined by the dispersion relation (7.13) and D_{k+} and D_{k-} constant. The boundary conditions require that for each eigenmode

$$u_k(0, t) = u_k(L, t) = 0$$

for all t. This gives

$$D_{k+} = -D_{k-} \equiv A_k$$

and

$$A_k(e^{ikL} - e^{-ikL}) = 2iA_k \sin(kL) = 0$$

that is, $k = n\pi/L \equiv k_n$ for $n = 1, 2, \ldots$. Inserting this into the dispersion relation gives the following growth rates for the discrete modes ($n = 1, 2, \ldots$)

$$\sigma_n = \lambda - \left(k_n^2 - k_c^2\right)^2 \tag{7.14}$$

The nth mode is marginal when $\sigma_n = 0$, that is for

$$\lambda = \lambda_n \equiv \left(k_n^2 - k_c^2\right)^2 = k_c^4 \left(\alpha^2 n^2 - 1\right)^2 \tag{7.15}$$

where the parameter $\alpha \equiv \pi/(Lk_c)$ is related to the relative size of the system. If the system is confined to a size comparable with the most unstable wavelength $2\pi/k_c$, we easily see that the discrete modes are well separated (in parameter space). For example, if the size of the system is $L = \pi/k_c$ say, the parameter values for which the first and second modes are marginal are

$$\lambda_1 = 0 \quad \text{and} \quad \lambda_2 = 9k_c^4$$

respectively. In contrast, if the system is not well confined (α small) the modes are dense and for $L \to \infty$ they become continuous.

When the most unstable mode is well isolated, in this sense, from the next discrete mode, it is possible to eliminate the stable (slave modes) in the following way. We seek a solution close to the relevant critical point, i.e., for small λ. To be specific, we assume henceforth that the size of the system is indeed $L = \pi/k_c$. The growth rates for the various modes follow from (7.14) and are given by

$$\sigma_n = \lambda - \left(k_n^2 - k_c^2\right)^2 = \lambda - k_c^4(n^2 - 1)^2 \tag{7.16}$$

This gives $\sigma_1 = \lambda$, $\sigma_2 = \lambda - 9k_c^4$, $\sigma_3 = \lambda - 64k_c^4$, for the first few modes, and so on. If λ is very small all the modes, save the first one, are 'very' stable.

We now attempt to express the complete solution of the *nonlinear* equation (7.11) near marginality as a superposition of all the *linear* spatial modes, but allow for a general form of time-variability of the amplitudes (absorbing in them the e^{st} term). Thus

$$u(x, t) = \sum_{j=1}^{\infty} A_j(t) \sin(k_j x) \tag{7.17}$$

where $k_j = j\pi/L = jk_c$ and the amplitudes are as yet unknown functions of time.

All the stable modes are decaying and the unstable ones (in our case only the first mode is unstable) become saturated as a result of the nonlinear term in (7.11). This can be seen by inserting (7.17) into the original system (7.11). Separating the coefficients of the different harmonics we get for each mode ($j = 1, 2, \dots$)

$$\frac{d}{dt} A_j = \sigma_j A_j + \frac{1}{2}\left(k_j \sum_{m=1}^{\infty} A_m A_{j+m} + \sum_{m=1}^{j-1} k_m A_m A_{j-m}\right) \tag{7.18}$$

where we have used (7.14) and elementary trigonometric identities.

The infinite system (7.18) of ODE, replacing the original PDE, is the desired amplitude equations set. The procedure described above (expanding the solution and deriving amplitude equations) is usually referred to as *projection*, since the dynamics of the original PDE system is projected onto the linear subspace spanned by the eigenvectors of the linear modes. Close to marginality of the first mode, that is, for small λ in (7.16), all the higher modes are stable and we may try to eliminate them and so reduce the order of system (7.18). We have already seen an example of dimensional reduction by the method of multiple scales in Chapter 3. Since the amplitude equations are an ODE system a procedure essentially identical to the one employed in Chapter 3 is also possible here.

We shall, however, exploit the present example for the demonstration of an alternative technique of dimensional reduction – the *elimination* of slave modes.

The first equation of the set (7.18) is

$$\frac{dA_1}{dt} = \lambda A_1 + \frac{k_c}{2}(A_1 A_2 + A_2 A_3 + \cdots) \tag{7.19}$$

while all the other equations (for the stable modes) have the form

$$\frac{dA_j}{dt} = \sigma_j A_j + f_j(A_1, A_2, \ldots) \tag{7.20}$$

with $\sigma_j < 0$ for $j = 2, 3, \ldots$ and where the f_j are known functions of the mode amplitudes. It is possible to show that if the modes are not resonant the solutions of equations (7.20) decay to zero after a rather short transient. This behaviour can be interpreted as the attraction of the dynamics onto the centre manifold, spanned by the marginal mode. Consequently, we may set $dA_j/dt = 0$ in (7.20), remembering that this is valid only for sufficiently long times, and also neglect λ in the expressions for σ_j ($j \geq 2$) in (7.16). If we also assume that the nonlinearity is weak, i.e., that $|u|$ is small, we can decide to keep only the lowest order (in the amplitudes) nonlinear terms in (7.19) and (7.20).

The first two algebraic equations resulting from (7.18) are then

$$A_2 = -\frac{1}{18k_c^3}A_1^2$$

$$A_3 = -\frac{3}{128k_c^3}A_1 A_2 \tag{7.21}$$

and the differential equation for A_1 (7.19) is

$$\frac{dA_1}{dt} = \lambda A_1 + \frac{k_c}{2}A_1 A_2$$

This procedure is consistent since the order of the amplitudes A_n for $n > 1$ is $A_n \sim A_1^n$.

Thus the dynamics on the centre manifold is given by the single amplitude equation

$$\frac{dA}{dt} = \lambda A - \frac{1}{36k_c^2} A^3 \tag{7.22}$$

where, for convenience, we have dropped the subscript from the amplitude of the only excited mode ($n = 1$).

As we have already seen several times in this book, the essential dynamics of a multidimensional system (here it is an infinite-dimensional PDE system) is governed, near marginality and close to an equilibrium solution, by a single generic ODE. Remarkably, the details of the particular dynamical system considered here are contained only in the coefficient of the normal form (7.22). The form itself is determined by the nature and symmetries of the bifurcation. The amplitude of the marginal mode undergoes, in this case, a supercritical pitchfork bifurcation and this is natural in view of the instability type (cellular steady). Slightly above the marginality of the first mode, and sufficiently close to the homogeneous solution $u = 0$, we may reasonably expect that the original confined system will have the solution

$$u(x, t) = A(t) \sin k_c x$$

with $A(t)$ given by the solution of (7.22). Thus $A \rightarrow A_\infty = 6\sqrt{\lambda} k_c$ as $t \rightarrow \infty$. The one-dimensional steady pattern will be sinusoidal with amplitude A_∞. This behaviour is typical for confined systems, whose linear instability is of the cellular steady type. It is possible in principle to provide a general prescription for dimensional reduction and derivation of amplitude equations for any confined system with a discrete mode spectrum. Since our aim here is to introduce the subject, we refrain from giving such a formal generalisation (see, e.g., Manneville's book). The case when several discrete modes become simultaneously unstable is of considerable importance as well. The interested reader is referred to the paper by Coullet and Spiegel (1983) for a general and complete derivation of amplitude equations in the case of two competing instabilities.

Envelope equations for an extended system

When we are dealing with an extended system, usually all modes in a continuous band are unstable above marginality. We have explicitly seen this in the example above, where the selection of only discrete modes was caused by confinement. We may approximate an extended system by an infinite one and try to examine the system's behaviour close to marginality of the most unstable mode. If k_c is the critical value of the wavevector, the pattern arising at marginality will obviously have a length scale associated with this wavenumber. As the parameter is increased, however, it is possible that the pattern becomes modulated (in space and/or time) by other solutions in the unstable band. As long as λ is small this modulation should be slow, that is, have large characteristic time and length scales.

As an explicit concrete example we consider again a variant of the Swift–Hohenberg model in two spatial dimensions, here, having a cubic nonlinearity instead of the advective term (compare with 7.11)

$$\partial_t u = \lambda u - u^3 - \left(\nabla^2 + k_c^2\right)^2 u \tag{7.23}$$

where $u = u(\mathbf{r}, t)$ with $\mathbf{r} = (x, y)$.

As the linear part of (7.23) is similar to that of (7.11), the linear instability arising at the threshold $\lambda = 0$ can be classified as being of the cellular steady type. The natural choice for the solution slightly above marginality is thus

$$u(\mathbf{r}, t) = A(x, y, t)e^{ik_c x} + \text{cc} \tag{7.24}$$

where the (generally complex) amplitude is considered here to be a slowly varying function of position (x, y) and time (t), because of the above mentioned effect of the other unstable wavenumbers in the band around k_c. Solution (7.24) is usually referred to as a set of *rolls*, and the variation of the amplitude A induces a slow modulation of these rolls.

We write now $\lambda = \epsilon$, to express the fact that we are close to marginality and note that the above solution should actually approximate to order ϵ (for $\epsilon > 0$) the true nonlinear solution of (7.23). The aim of the analysis is to obtain an evolution equation for A. Since the spatial dependence of A cannot be a priori separated from the temporal one, $A(x, t)$ is usually called the *envelope* and the desired equation is the *envelope equation*. The appropriate envelope equation can be derived in this case (and in other cases as well) using perturbation techniques like the methods of multiple scales (we discussed and used this in Chapter 3). We shall not give the details of the derivation here (it can be found in the appendix of Cross and Hohenberg's review article) but only remark that the analysis is done using stretched spatial coordinates *and* time. The scaling $X = \epsilon^{1/2}x$, $Y = \epsilon^{1/4}y$ and $\tau = \epsilon t$ works well in this case, and the envelope equation obtained from the appropriate solvability conditions, and after transforming back to the original variables, is

$$\tau_0 \partial_t A = \epsilon A - g_0 A |A|^2 + \xi_0^2 \left(\partial_x - \frac{i}{2k_c}\partial_y^2\right)^2 A \tag{7.25}$$

where the real constants are $\tau_0 = 1$, $\xi_0 = 2k_c$ and $g_0 = 3$. The fact that this equation is generic for all systems exhibiting a cellular steady instability at $\epsilon = 0$ should by now be no surprise. The details of the system are all contained in the values of the constants and (7.25) is a very general envelope equation. This equation is named after Newell, Whitehead and Segel who derived it in 1969 in their studies of weakly confined cellular convection, and it reflects the general symmetry of extended systems with a cellular steady linear instability type. The Newell–Whitehead–Segel (NWS) equation describes the modulation of the basic periodic pattern (7.24). In any particular case it is possible to set the three constants in the equation to 1, by properly rescaling $\mathbf{r}$, k_c, t and A. After doing this, for convenience, we examine first a very simple steady solution of the envelope equation for $\epsilon > 0$

$$A(x, y, t) = a_q e^{iqx} \tag{7.26}$$

The amplitude a_q must satisfy $a_q = (\epsilon - q^2)^{1/2}$ for (7.26) to be the solution of the nonlinear envelope equation (7.25 with $g_0 = \tau_0 = 1$). This elementary solution just induces a shift in the wavenumber of the original pattern (7.24) from k_c to $k = k_c + q$. It is also possible to consider solutions with a wavevector $\mathbf{q}$ having nonzero x and y components. In such a case, the envelope equation yields a rotation of the wavevector and a change in its magnitude so that instead of $\mathbf{k} = k_c\hat{\mathbf{x}}$ we have $\mathbf{k} = k_c\hat{\mathbf{x}} + \mathbf{q}$. Since there is a considerable freedom in the choice of $\mathbf{q}$ (the only constraint is $|\mathbf{q}| \leq \sqrt{\epsilon}$ so that a_q is real), it is important to check for which values of this wavevector shift, the solutions of the amplitude equations are stable. The linear stability analysis is not difficult and it yields an important and general result shown schematically in Figure 7.2, which depicts the q–ϵ plane.

The parabolic curve $\epsilon = q^2$, marked by N in the figure, shows the limits of validity of the nonlinear solution (7.26) of the envelope equation, which exists for $\epsilon > q^2$, that is, above the parabola. It turns out, however, that for $\epsilon > q^2 > \epsilon/3$ this solution is unstable to *longitudinal*, that is, of the form $\propto \exp(iKx)$ with K fixed, perturbations. This instability is called the *Eckhaus instability* and its growth rate is largest for the wavenumber

$$K_{\max} = \frac{3}{4q^2}(\epsilon + q^2)\left(q^2 - \frac{1}{3}\epsilon\right)$$

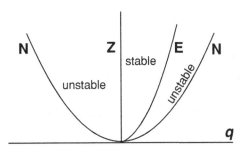

Fig. 7.2. Stability boundaries of spatially periodic solutions (7.26) of the Newell–Whitehead–Segel envelope equation (7.25) with $\tau_0 = \xi_0 = g_0 = 1$ in the q–ϵ plane. The limit of validity (N) as well as the limits of the Eckhaus and zig-zag instability are also shown. For details see text.

The curve $\epsilon = 3q^2$, marked in the figure by E delineates the region of this instability, which destroys the regularity of the pattern by modifying the scale of individual rolls. An additional instability is found for modes with $q < 0$. This instability corresponds to a growing *transverse* modulation of (7.26), that is, it occurs for perturbations $\propto \exp(iKy)$. Because of the form of the pattern it produces (parallel periodic pattern with wiggly shape) it is called the 'zig-zag' instability. The curve $q = 0$ (the ϵ axis), marked by Z in the figure delineates this instability. The entire diagram in the figure is called the *stability balloon*.

All of the above mentioned instabilities are generic and are found to occur in a variety of systems with a cellular steady instability. The various systems differ in this respect just in the parameter values. These instabilities have been found in experiments on pattern forming as well as in numerical solutions of a two-dimensional system, having a cellular steady instability. Examples of figures illustrating such destabilisation of rolls by the Eckhaus instability and the zig-zag instability can be found in the books by Manneville (1990) and Walgraef (1997).

There are obviously additional possibilities for the choice of the basic solution (7.24) of the original Swift–Hohenberg model. The plane-wave solution we had is just the simplest case – one set of *rolls*. One may examine the possibility of a superposition of several sets of rolls in different directions. The NWS equation is replaced then by a set of coupled envelope equations for the different families of rolls. The situation may become fairly complicated, but one particular possibility is worth noting. A superposition of three sets of rolls with equal-size wavevectors satisfying $\mathbf{k}_1 + \mathbf{k}_2 + \mathbf{k}_3 = 0$ gives rise to a pattern with hexagonal symmetry. This is especially valuable in the the study of the experimentally well known result of convection with hexagonal cells.

So far we have considered only the steady cellular instability. The ideas and methods described above can be applied to other types of instability as well. The simplest instability type is that of homogeneous steady, but this does not give rise to patterns near marginality. We shall briefly discuss now the form of the envelope equation for systems with a homogeneous oscillatory instability. Such instabilities arise, for example, in certain reaction–diffusion systems. We are interested in the generic envelope equation, valid near a critical point of such an instability, which, as we have said before, corresponds to a Hopf bifurcation.

The approximate solution of the original PDE near marginality, i.e., for $\lambda = \epsilon \ll 1$, and for a system in two space dimensions, can be written in this case as

$$u(\mathbf{r}, t) = A(x, y, t)e^{i\omega t} + \text{cc} \tag{7.27}$$

where $\omega = \omega_c - c_0 \epsilon$. Here ω_c is the frequency corresponding to the most unstable mode at marginality and c_0 is a constant, depending on the particular system we study. By setting $c_0 \neq 0$ we allow for a small frequency shift, in addition to the slow variation (in space and time) of the envelope A.

The resulting envelope equation is the celebrated (complex) Ginzburg–Landau equation (CGLE), which has been well studied in various applications. It is usually written as

$$\partial_t A = \epsilon A + \zeta_1 \nabla^2 A - \zeta_3 A |A|^2 \tag{7.28}$$

where the ζ_i are complex constants, which can be brought (by rescaling units) to the form $\zeta_1 = 1 + i c_1$ and $\zeta_3 = 1 - i c_3$, with c_i real. These constants are determined by the details of the original system. A lucid derivation of the CGLE can be found in the book by Kuramoto (1984), where it is obtained as a generic amplitude equation of general reaction–diffusion PDEs

$$\partial_t \mathbf{u} = (\mathcal{L} + \nabla^2)\mathbf{u} + \mathcal{N}(\mathbf{u}) \tag{7.29}$$

Here $\mathcal{L}$ and $\mathcal{N}$ denote, as before, a linear and a strictly nonlinear operator respectively, and they also depend on a control parameter. The CGLE is sometimes written in slightly different forms, resulting from various choices of the scaling of the independent variables and the amplitude. Kuramoto's book deals specifically with PDE systems of the above type (7.29), which for some critical value of a control parameter undergo a homogeneous oscillatory instability (Hopf bifurcation). The book contains a wealth of analytical derivations and examples of such systems, motivated in large part by experimental results on systems in which chemical species react and diffuse.

A simple but interesting class of solutions of the complex Ginzburg–Landau equation (7.28) are the travelling waves

$$A(\mathbf{r}, t) = a_q e^{i(\mathbf{q} \cdot \mathbf{r} - \Omega_q t)} \tag{7.30}$$

Such a travelling wave is a solution of the CGLE provided that $\Omega_q = -c_3 \epsilon + (c_1 + c_3)q^2$ and $a_q^2 = \epsilon - q^2$. It modulates the basic uniform oscillatory solution (7.27) and creates a pattern of long ($q \leq \sqrt{\epsilon}$) travelling waves.

One can examine next the stability of these solutions, as well as other possible solution types (e.g., standing waves), in a similar way to the one described above for systems having a cellular steady instability. An important type of instability found in this case, analogous to the Eckhaus instability in systems of the former type, is known as the *Benjamin–Feir instability*. The derivation of the conditions for the Benjamin–Feir instability is rather straightforward, but it is quite complicated algebraically and thus we shall not give it here. The physical cause for this instability is the resonant excitation of the sidebands, whose wavevectors and frequencies, $\mathbf{q}_i$ and ω_i with $i = 1, 2$, satisfy $q_1 + q_2 = 2q$ and $\omega_1 + \omega_2 = 2\omega$.

The envelope equations that can be derived for PDE systems near cellular oscillatory instabilities give rise to more complicated envelope equations. This is so because the equations have to account for the appearance (at marginality) of a basic pattern, having both a spatial variation and a time oscillation. The resulting equations are two coupled equations for the envelopes of a left-propagating wave and a right-propagating one. In one-dimensional systems the two equations are rather straightforward generalisations of the CGLE (7.28).

The analogue of the solution (7.27) in this case and for one spatial dimension is

$$u(x, t) = A_R(x, t)e^{ik_c x - i\omega t} + A_L(x, t)e^{-ik_c x - i\omega t} + \text{cc} \tag{7.31}$$

where k_c is the critical wavenumber and $\omega = \omega_c - c_0\epsilon$ as before. The equation for the envelope of the right-propagating travelling wave, A_R, is then

$$\partial_t A_R + v_0 \partial_x A_R = \epsilon A_R + \zeta_1 \partial_x^2 A_R - \zeta_3 |A_R|^2 A_R - \zeta_2 |A_L|^2 A_R \tag{7.32}$$

and a similar equation is valid for the envelope of the left-propagating travelling wave A_L as well. Both equations contain a coupling term (with a complex constant coefficient ζ_2) and an extra term of *advective* nature, since v_0 in the above equation actually denotes the linear group velocity $\partial\omega/\partial k|_{k=k_c}$.

A detailed discussion of these equations, the stability of some of their solutions and related issues can be found in Cross & Hohenberg (1993).

7.4 Phase equations and coherent structures

The envelope equations discussed in the previous section describe the evolution of a generally *complex* envelope, defining the pattern formed in the original microscopic or model system. These equations are derived for states close to the instability threshold and consequently are valid only slightly above marginality. The complex envelope can always be written as

$$A(\mathbf{r}, t) = a(\mathbf{r}, t)e^{i\phi(\mathbf{r},t)}$$

where $a = |A(\mathbf{r}, t)|$ is a real function. We wish now to focus on the phase ϕ and seek an equation for its evolution. For the sake of clarity we choose a specific example from the generic envelope equations given in the previous section: the NWS equation (7.25), which is valid near the threshold of a cellular steady instability. The simplest solution of the NWS equation that we have considered is the periodic solution (7.26), which we rewrite here for convenience

$$A(x) = a_q e^{iqx} \tag{7.33}$$

As we have seen in the previous section, this solution modulates the basic periodic pattern by shifting the wavevector from k_c to $k_c + q$ (in the one-dimensional case).

Now, instead of (7.33) we consider a slightly perturbed (in the phase) envelope function

$$A'(x) = a_q e^{i(qx+\delta\phi)} \tag{7.34}$$

with $\delta\phi(x, t)$ being a small phase perturbation, that is, varying on a large (as compared to q^{-1}) length scale. Substitution into the scaled version of the NWS equation ((7.25) with $\tau_0 = \xi_0 = g_0 = 1$) gives, by equating the imaginary parts of both sides,

$$\partial_t \delta\phi = \partial_x^2 \delta\phi \tag{7.35}$$

We note that, in the particular case when $\delta\phi$ is fixed in space (the phase shift is x-independent), equation (7.35) shows that $\delta\phi$ does not change in time. The real part of the NWS equation with the substitution of the solution (7.34) gives, in this case, the same conditions on q and a_q as in the unperturbed solution.

In the more general case, when $\delta\phi$ is a function of x, the above procedure of equating the real and imaginary parts of the NWS equation yields a PDE for the phase perturbation $\delta\phi$. In the example considered here the phase perturbation satisfies the diffusion equation (7.35). This equation is called a *phase equation* and we may also derive such an equation for the full phase, that is for $\phi(x, t) \equiv qx + \delta\phi(x, t)$.

The explicit form of the phase equation depends on the assumptions one makes about the dependence of $\delta\phi$ on spatial variables and on the order of approximations. We shall give a list of several different typical cases later on. The point we wish to make here is that a new and separate PDE can be obtained by such a procedure, and it contains additional information on the possible pattern shapes. The procedure outlined above relies on the envelope equation (here the NWS equation), which is only valid near the instability threshold. Consequently, the phase equation is only valid near this threshold as well.

We may, however, return to the original model equation and try to find a phase equation directly from it. The validity of such a phase equation should not be limited only to the neighbourhood of the instability threshold. We outline here the principles of such a derivation, which is somewhat subtle. Consider, for example, a model (or microscopic) PDE having an instability of the steady cellular type, and specifically again the Swift–Hohenberg model (7.23) treated above. We may consider a perfectly periodic stationary solution of this original equation, which will be a generalisation of the basic solution (7.24), used in the derivation of the NWS envelope equation.

Recall that in the basic solution (7.24) the choice $\mathbf{k} = k_c\hat{\mathbf{x}}$ was made, and so it was a one-dimensional periodic wave pattern having the critical wavenumber. This was natural in view of our desire to get an envelope equation for the pattern close to marginality. It may be possible, however, to consider a more general periodic reference solution whose phase is $\phi = \mathbf{k} \cdot \mathbf{r}$, with some general constant wavevector $\mathbf{k}$, not necessarily equal in magnitude to k_c. Such a solution can be assumed to serve as a reference, not only near marginality, but arbitrarily far from the threshold as well. This perfectly periodic reference solution can be written as

$$u_{\mathbf{k}}(\mathbf{r}, t) = u_{\infty}(\mathbf{k} \cdot \mathbf{r}) \tag{7.36}$$

where $u_{\infty}(\phi + 2\pi) = u_{\infty}(\phi)$, i.e., it is a periodic function.

The actual solution of the original microscopic or model PDE is then considered to be near the reference solution in the following sense. A small parameter, η, is assumed to exist in the problem and an approximate (up to order η) solution is sought, having the form

$$u(\mathbf{r}, t) = u_{\infty}[\phi(\mathbf{r}, t)] \tag{7.37}$$

where the phase satisfies

$$\nabla \phi(\mathbf{r}, t) = \mathbf{k}(\mathbf{r}, t) \tag{7.38}$$

with the wavevector $\mathbf{k}$ varying slowly, that is, on scale η^{-1}.

The small parameter η is not related to the distance from the instability threshold, and the solution (7.37) is not close to the reference solution in the sense discussed when dealing with situations near the instability threshold. This solution may describe a roll pattern with a direction varying through large angles, as long as it takes place slowly, that is, over many periods $(2\pi/k_c)$ of the basic pattern. A perturbative procedure is then possible and it yields the appropriate *phase equation* for the evolution of $\phi(\mathbf{r}, t)$, valid to order η arbitrarily far from the linear instability threshold.

We shall not discuss here the derivation of phase equations for the different types of system. These equations and their behaviour, referred to as *phase dynamics*, are discussed in Cross & Hohenberg (1993), Manneville (1990), and in Kuramoto's book, where a full account on the derivation of phase equations resulting from the CGLE (viewed as a model equation) and their classification for various cases is given. The derivation of phase equations can be carried out to higher orders in η as well, and the resulting equations have a generic form at least close to the instability threshold. Some of these phase equations are equivalent to famous, well studied nonlinear PDEs. We list below several examples of phase equations in one space dimension and mention some of their basic properties. This list is by no means systematic or complete and its purpose is just to illustrate what kind of phase equations can be expected.

(i) Cellular steady instability

(a) Lowest-order phase equation, valid arbitrarily far from the instability threshold:

$$\tau(k)\partial_t\phi = -\partial_x[kB(k)]$$
$$k = \partial_x\phi \tag{7.39}$$

where $\tau(k)$ and $B(k)$ are functions depending on the particular system.

(b) Close to marginality, the above lowest-order equation takes on the form of a diffusion equation:

$$\partial_t\phi = D(k)\partial_x^2\phi$$

(c) Higher-order phase equation, written in a form valid near marginality:

$$\partial_t\phi = \alpha\partial_x^2\phi - \gamma\partial_x^4\phi + \beta(\partial_x\phi)\partial_x^2\phi \tag{7.40}$$

(ii) Oscillatory instabilities

The reference solution, analogous to (7.36) is $u_\infty(\mathbf{k}\cdot\mathbf{r} - \omega t)$.

(a) Lowest-order phase equation, valid arbitrarily far from the instability threshold:

$$\partial_t\phi + \omega(k) = -\tau(k)^{-1}\partial_x[kB(k)]$$
$$k = \partial_x\phi \tag{7.41}$$

(b) For the homogeneous instability class and close to marginality the above equation becomes:

$$\partial_t\phi = \alpha\partial_x^2\phi - \beta(\partial_x\phi)^2 \tag{7.42}$$

(c) Higher-order phase equation for this instability class and close to marginality:

$$\partial_t\phi = \alpha\partial_x^2\phi - \gamma\partial_x^4\phi - \beta(\partial_x\phi)^2 \tag{7.43}$$

(d) Higher-order phase equation for the cellular instability class and close to marginality:

$$\partial_t\phi = \alpha\partial_x^2\phi - \gamma\partial_x^3\phi - \beta(\partial_x\phi)^2 \tag{7.44}$$

In all the above equations α, β and γ are suitable constants depending on the system studied. For special choices of the parameters and with the definition $v(x, t) \equiv \partial_x\phi$ we obtain a number of well known PDEs.

The *Burgers equation* is a particular case of (7.42) with $\beta = 1/2$ and $\alpha = \mu > 0$. It reads

$$\partial_t v + v\partial_x v = \mu\partial_x^2 v \tag{7.45}$$

A variant of the Kuramoto–Sivashinsky (KS) equation (see 7.8) follows from the phase equation (7.43) for $\alpha = -1$, $\beta = 1/2$ and $\gamma = \mu > 0$

$$\partial_t v + v\partial_x v = -\partial_x^2 v - \mu\partial_x^4 v \tag{7.46}$$

Finally, for $\alpha = 0$, $\beta = 1/2$ and $\gamma = \mu > 0$, a particular form of the famous *Korteweg–de Vries* (KdV) equation is obtained from (7.44)

$$\partial_t v + v \partial_x v = -\mu \partial_x^3 v \tag{7.47}$$

The Burgers, KS and KdV equations are endowed with Galilean invariance (see the discussion in the second section of this chapter). When they are perceived as phase equations, the 'velocity' variable v should be understood as the spatial gradient of the phase, reflecting the pattern local wavenumber.

The above three equations are famous and well studied because they arise in the modelling of different physical systems as well. The Burgers and KdV equations are model equations for nonlinear waves (in fluids, for example). Remarkably, both are *integrable*, that is, they have known exact solutions. These solutions have a finite extent and are essentially constant outside a finite spatial interval. The form of these localised solutions, which are called *solitons*, is typical. The Burgers equation supports moving fronts, while the KdV soliton has the shape of a pulse. The KS, and other equations of its family are not integrable. Nevertheless they also posses localised solutions, which are sometimes called *solitary waves*. Such solitary waves arise for example in the study of models of chemical and biological systems. They are dissipative and do not have the remarkable properties of solitons of integrable equations. Still, they may often be treated in the same way as solitons, at least approximately and during finite time intervals (when they persist).

It is instructive to understand the effect of the different *linear* terms in the above equations. If we consider the equations as describing nonlinear waves, it is natural to examine the stability of the null solution, $v(x, t) = 0$, to small spatially periodic perturbations

$$\delta v \propto e^{iQx+st}$$

We obtain in the three cases the following respective dispersion relations, resulting from the linear terms

$$\begin{aligned} \text{Burgers:} \quad & s(Q) = -\mu Q^2 \\ \text{KS:} \quad & s(Q) = Q^2 - \mu Q^4 \\ \text{KdV:} \quad & s(Q) = i\mu Q^3 \end{aligned}$$

So we see that in the Burgers equation the diffusion term is stabilising for any Q. Any homogeneous solution is linearly stable and fronts between two such homogeneous regions are widened by diffusion, and if they persist it must be due to steepening resulting from the nonlinear term. The KS equation has a negative diffusion coefficient causing instability, which can be controlled for sufficiently large Q (small scales) by the higher order (quartic) diffusion term. The KdV equation is neutral to linear perturbations, the eigenvalue s being purely imaginary for any Q.

The nonlinear term in all the above equations is of an advective nature and its physical meaning, at least in the case of fluids, is obvious. The persistence of localised structures is a nonlinear effect and is caused in all three cases by the interplay between the linear effects and the effects caused by the advective term.

Amplitude or envelope equations derived near the instability threshold are instrumental in understanding the formation of *ideal* patterns. As we have seen above, phase equations usually have a wider range of applicability and may be useful not only near marginality but also arbitrarily far from it. Localised solutions of phase equations, if they exist, correspond to an abrupt spatial transition from one well defined ideal pattern to another. These localised structures or *defects* may be stationary or moving. In many important cases these structures remain well defined as they move. The term *coherent structures* is then appropriate, emphasising the persistence of the local nature of the perturbation on an otherwise ideal background. The properties of single defects as well as their interactions and dynamics, when several defects are present, have paramount importance for the understanding of nonideal patterns. These depend, of course, on the properties of the relevant microscopic or model PDE. In most cases of interest the original PDE can only be fully solved by using brute-force numerical approaches. It is very important, however, to try to discover the essentials of the basic solutions without having recourse to full numerics, if possible. This can be done, as we have seen, with the help of envelope and phase equations. The structure and dynamics of defects provide additional important information on the departure of real patterns from their ideal state, and for the possible development of spatio-temporal complexity. Such analytical or semi-analytical approaches may provide a very useful tool for the understanding of numerical and experimental or observational results.

The dimension of defects must be smaller than the number of spatial dimensions of the PDEs. In PDEs with one spatial dimension, the dimension of defects must thus be zero – the defects are points. This actually refers to the defect centres, so that localised pulses or fronts on an otherwise uniform (or periodic) one-dimensional background are zero-dimensional defects. When the underlying PDE is spatially multidimensional the situation is more complicated. In three-dimensional systems, for example, two-dimensional front (the term *domain walls* is more appropriate) centres consist of two-dimensional surfaces. However in this case defects whose dimension is smaller than two are also possible (e.g., one-dimensional vortex lines). Also, d-dimensional defects living in an n-dimensional space (the space spanned by the spatial variables of the PDE) are said to have *codimension $n-d$*.

As stated above, the nature of the dynamics of localised structures provides important information on the properties of the system. It is thus advantageous to try to derive the properties and the dynamics of defects, and this is sometimes possible with the help of analytical (usually perturbative) methods. In the next two sections we address some rudimentary issues on this subject by treating example equations, some of which may actually arise from study of model astrophysical systems.

7.5 Defects and spatio-temporal complexity in 1D systems

It is only natural that the above mentioned analytical approaches to defect dynamics work best in one-dimensional (1D) problems, where the defects are always of codimension 1. To make our discussion here as simple as possible, we shall describe in some detail coherent structures in one spatial dimension, and the methods for studying their structure and dynamics. Multidimensional cases will be discussed only very briefly in the next section.

Structure and properties of defects

The following discussion of one-dimensional defects will be based on two particular examples, belonging to a rather general class of PDE, having the following form

$$\partial_t v = f(v) + \eta g(v, \partial_x v) \tag{7.48}$$

where $v = v(x, t)$ is a scalar field, and the function f depends only on v, while g may depend on v's spatial derivatives as well; η is a (constant) parameter. The Burgers, KdV and KS equations, introduced in the previous section, all have this form. Equation (2.33), introduced in Chapter 2 as a model of a one-dimensional thermally bistable medium, is also of this form.

In the limit $\eta \to 0$ Equation (7.48) describes the time evolution of the field $v(x, t)$ (which can be a physical variable in a microscopic equation, a variable in a model equation, or a phase variable) at position x, independent of the field values in the neighbourhood of x. In contrast, when $\eta \neq 0$ spatial coupling is introduced via the x-derivatives of v.

We shall consider only the cases when (7.48) has one or more *stable* constant solutions and then look for additional solutions, having the form of travelling waves, that is,

$$v(x, t) = D(\chi) \quad \text{with } \chi \equiv x - ct \tag{7.49}$$

D is a (as yet unknown) function of its variable and c is a constant. We wish that $D(\chi)$ should be significantly different from a constant solution of (7.48) in only a finite domain. In this sense D is a localised solution or a defect on an otherwise 'ideal' pattern (a constant solution in this case).

Substitution of (7.49) into the original PDE gives the following ODE

$$\eta g\left(D, \frac{dD}{d\chi}\right) + c\frac{dD}{d\chi} = -f(D) \tag{7.50}$$

We refer to this equation as the *associated* (to the original PDE) ODE and look upon it as a *boundary value* problem, defined in some fixed domain, that is, for $-L \leq \chi \leq L$, say. The boundary value problem is well posed only if suitable boundary conditions are given. The required number of boundary conditions depends on the order of the associated ODE, and we also have to remember that c is not determined a priori and is thus an eigenvalue of the problem.

In the context of localised travelling wave solutions of the PDE (7.48) it is useful to consider boundary conditions of the form

$$D(-L) = v_1 \quad \text{and} \quad D(L) = v_2$$

where v_1 and v_2 are the constant (not necessarily distinct) solutions of the PDE and L is very large (in a sense explained below). If any additional boundary conditions are required, they are chosen to consist of the vanishing of a suitable number of the derivatives of $D(\chi)$ at $\chi = \pm L$. Since v_i are constant solutions of the PDE, $D = v_i$ are fixed points of the associated ODE. It is not difficult to see that if v_i are stable solutions of the PDE, the corresponding fixed points of the associated ODE are unstable. It is thus clear that a solution of the relevant boundary value problem is an orbit connecting the two unstable fixed points (or one fixed point to itself) in the phase space of the associated ODE. It should be clear by now why we require L to be very large. This is because a truly heteroclinic or homoclinic orbit (see Chapter 3) of the sort we look for approaches a fixed point *asymptotically*, that is, for $\chi \to \pm\infty$.

To be explicit we illustrate these ideas using two examples, in which the above mentioned localised solutions can be found analytically.

Consider first the Korteweg–de Vries equation (7.47)

$$\partial_t v + v \partial_x v = -\mu \partial_x^3 v \tag{7.51}$$

We note from the outset that any constant value of v is a solution of this equation. This is a result of the above mentioned Galilean invariance of the KdV equation. We ignore this degeneracy for the moment and choose the null solution as the one in which we are interested. This obviously means that the boundary conditions for the associated ODE will be $v_1 = v_2 = 0$.

The associated ODE itself can be readily obtained by the substitution $v(x, t) = D(x - ct)$, but we can also use the fact that (7.51) is of the form (7.48) with $f(v) = 0$, $\eta = 1$ and the function g given by

$$g(v, \partial_x v) = -v \partial_x v - \mu \partial_x^3 v$$

Thus the associated ODE follows on from (7.50) and is

$$\mu \frac{d^3 D}{d\chi^3} + (D - c)\frac{dD}{d\chi} = 0$$

This ODE has an analytical solution satisfying null boundary conditions at $\chi = \pm\infty$ for *any c*

$$D(\chi) = \frac{c}{2} \operatorname{sech}^2(\gamma\,\chi) \qquad (7.52)$$

where

$$\gamma \equiv \left(\frac{c}{4\mu}\right)^{\frac{1}{2}}$$

Remembering that the hyperbolic secant is defined as

$$\operatorname{sech}(x) \equiv \frac{1}{\cosh(x)} = \frac{2}{e^x + e^{-x}}$$

we see that the solution (7.52) has the form of a symmetric pulse of maximal height $c/2$, localised around $\chi = 0$. The width of the pulse is $w \sim \gamma^{-1} = 2(\mu/c)^{1/2}$ and it decays exponentially to zero on both sides. Far from the *core* of the pulse (i.e., for $\chi = \pm L$, with $L \gg w$) the value of the function (and all its derivatives) is essentially zero. We can thus understand and now quantify the statement made above that L is large. The domain is very large compared to the defect core, $L \gg w$. The analytic pulse solution (a homoclinic orbit) is strictly only valid in the infinite domain; but in a sufficiently large finite domain it can still be considered as a very good approximation.

The corresponding solution of the PDE is

$$v(x, t) = D(x - ct)$$

and is thus a pulse of the above form, travelling in the positive x direction with speed c.

The KdV equation has both Galilean and translational invariance. Thus if $v = D(x - ct)$ is its solution,

$$v'(x, t) = D(x - ct - v_0 t + x_0) + v_0 \qquad (7.53)$$

where v_0 and x_0 are any constants, is also a solution. Boundary conditions obviously break the Galilean symmetry forcing the value of v_0 (we have chosen it at the outset to be zero).

The pulse solutions of the KdV equation are a paradigm of a remarkable class of coherent localised structures, known as solitons. As mentioned before, solitons retain their shape and do not dissipate even in extreme situations (like 'collisions' between themselves). They can thus be thought of as a type of particle, and their study has yielded many significant and beautiful results in the theory of nonlinear integrable PDEs. We refer the interested reader to the books by Whitham (1974) and Lamb (1980).

As our second example we consider a different particular case of equation (7.48) with $f(v) = \lambda v - v^3$ and $\eta g(v, \partial_x v) = \partial_x^2 v$. The resulting equation is identical to the example studied in the first section of this chapter, the Landau equation (7.7)

$$\partial_t v = \lambda v - v^3 + \partial_x^2 v \qquad (7.54)$$

At $\lambda = 0$ two stable constant solutions bifurcate from the null solution, which loses its stability. We have already seen that the instability is of the homogeneous steady type and thus no interesting ideal patterns are expected to arise close to marginality ($\lambda = 0$).

Let λ now have a fixed positive value. The constant stable solutions are then

$$v_1 = \sqrt{\lambda}; \qquad v_2 = -\sqrt{\lambda}$$

The question of which solution of these two (if any) is relevant to a particular case depends, of course, on the boundary conditions. Before dealing with this issue it is convenient to rescale the time variable t and the function itself v, by making the following replacements

$$v \mapsto \sqrt{\lambda} v; \qquad t \mapsto \frac{1}{\lambda} t$$

This gives

$$\partial_t v = v - v^3 + \frac{1}{\lambda} \partial_x^2 v \qquad (7.55)$$

From this form of the equation it is obvious that the significant length scale of the problem, set by the diffusion term, is of the order $w \sim \lambda^{-1/2}$. Having said this, we now assume that the system is weakly confined, i.e., it is subject to boundary conditions at 'infinity'; that is, at $x = \pm L$, with $L \gg w$. The constant solutions of the rescaled equation (7.55) are $v_{1,2} = \pm 1$ and they can be realised only if the boundary conditions are appropriate. For example, if $v(\infty) = v(-\infty) = 1$ the solution is simply $v(x, t) = 1$.

The constant solutions will not be realised for general boundary conditions and not much can be said about such cases at this point. A particularly instructive case exists, however, if $v(-\infty) = -1$ and $v(\infty) = 1$, for example. The fixed solution $v = -1$ should be valid near the left boundary and the other fixed solution $v = 1$ satisfies the right boundary condition. Somewhere in-between there should be a transition between the two solutions, and its width (the defect *core*) should be of the order of $\lambda^{-1/2}$. The boundary conditions in this case break the symmetry of the solution and force a *defect* in it. Our equation here is again simple enough to obtain the defect core structure analytically.

As in the previous example we first derive the associated ODE for $D(\chi)$, where $\chi \equiv x - ct$. In the present case it is

$$\frac{1}{\lambda}\frac{d^2 D}{d\chi^2} = D^3 - D - c\frac{dD}{d\chi} \tag{7.56}$$

This second-order ODE, together with the boundary conditions, constitute a well defined two-point boundary problem, and c is an eigenvalue. For the particular case given here there exists an analytical solution with $c = 0$

$$D(x) = K(x) \equiv \tanh(\gamma x) \quad \text{with} \quad \gamma \equiv \sqrt{\lambda/2}$$

This solution is clearly a heteroclinic orbit of the ODE between its two *unstable* saddle points. In the PDE it is a stationary $(c = 0)$ *front* (or *kink*) connecting the two *stable* uniform solutions. As expected, the width of the front is $w \sim \lambda^{-1/2}$. Thus the two uniform solutions, separated by a defect in the form of a front, is also a perfectly valid stationary solution of the PDE (7.55) in the infinite domain (with the boundary conditions given above).

Note that here a unique value of c is dictated by the boundary value problem. The position of the front is arbitrary, however, as the problem is translationally invariant. Thus

$$v(x) = K(x - x_0) = \tanh\left[\gamma (x - x_0)\right] \tag{7.57}$$

with any constant x_0 is also a stationary solution of the PDE, consisting of a front at $x = x_0$.

In a definite physical situation translational invariance must be broken by *initial conditions*. In addition, L cannot be truly infinite, and thus the hyperbolic tangent front can only serve as an approximation of the actual solution. It can be expected that for systems similar to (7.55), but with coefficients different from unity multiplying the terms on the right-hand side of that equation, we may get *moving* fronts (that is with the eigenvalue c of the associated ODE being different from zero). The Landau equation does not support true solitons; that is, although such moving localised structures remain coherent to a high degree, they do not survive collisions, for example. Nevertheless, it is still useful to consider these defects as a sort of 'particle', at least in some cases (see below).

In the above two examples we have obtained the defect structure analytically. In most cases, however, the associated ODE must be solved numerically. The advantage is that this endeavour is usually much simpler than a direct numerical simulation of the original PDE. Moreover, the analysis of defect solutions and their dynamics (see below) contributes a lot to the understanding of the basic properties of the PDE system.

A single defect in one spatial dimension (e.g., a front or a pulse) may be stationary or moving. As we saw in the last example of Chapter 2, its motion may sometimes be explored with the help of the associated ODE. The velocity of the defect is given by the particular value of the constant c (the eigenvalue), needed to obtain the desired heteroclinic (or homoclinic, in the case of a pulse) solution of the ODE. When the original PDE is a gradient system, it is often possible to understand the motion of a defect as the evolution of the system towards the minimisation of the appropriate Liapunov functional. To illustrate this idea in a more general way than in the case of the thermally bistable medium (Chapter 2), we consider an equation of the form (7.48) with $g(v) = \lambda^{-1}\partial_x^2 v$ and $\eta = 1$. The term $f(v)$ is left out as a general function and so equation (7.48) becomes

$$\partial_t v = f(v) + \frac{1}{\lambda}\partial_x^2 v \tag{7.58}$$

which is a generalisation of equation (7.55).

Assuming that f is a sufficiently well-behaved function, it is possible to define (up to an additive constant) a potential function $V(v)$ so that

$$f(v) = -\frac{dV}{dv} \tag{7.59}$$

We shall consider here only those functions $f(v)$ for which the potential V has a minimum.

Equation (7.58) can now be written as a gradient system

$$\partial_t v = -\frac{\delta \mathcal{F}}{\delta v} \tag{7.60}$$

with the functional $\mathcal{F}$ given by

$$\mathcal{F}[v] = \int dx \left[\frac{1}{2\lambda}(\partial_x v)^2 + V(v) \right] \tag{7.61}$$

where the integration is over the full domain of definition of the problem.

We have already shown (see Section 7.2) that in any system of the form (7.60), $\mathcal{F}[v]$ is *non-increasing*. The term $\mathcal{F}[v]$ is thus a Liapunov functional of our problem if it is bounded from below, and this obviously depends on the actual form of $\mathcal{F}$ given by (7.61). We may estimate a lower bound on $\mathcal{F}[v]$ as follows

$$\mathcal{F}[v] = \int dx \left[\frac{1}{2\lambda}(\partial_x v)^2 + V(v) \right] \geq \int dx \, V(v) \geq 2L V_{\min}$$

where $2L$ is the spatial extent of the system. From this we see that $\mathcal{F}$ in our system is bounded from below. We recall that we usually consider L to be very large with respect to the defect width (see the discussion above), but it is *finite*.

We proceed by examining a specific form of the function $f(v)$. For our purpose it will be sufficient to choose

$$f(v) = v - v^3 + 2\alpha v^2 \tag{7.62}$$

with $\alpha \geq 0$. This is a more general and richer case than the example given above (7.55) (which is recovered for $\alpha = 0$). The potential $V(v)$ can easily be found (we choose the integration constant to be zero)

$$V(v) = \frac{1}{4}v^4 - \frac{2\alpha}{3}v^3 - \frac{1}{2}v^2 \tag{7.63}$$

The constant solutions of (7.58) are the zeros of $f(v)$ and thus the extrema of $V(v)$. They are

$$v_{1,2} = \alpha \pm \sqrt{\alpha^2 + 1}; \quad v_3 = 0$$

Clearly, v_1 and v_2 are both local minima of the quartic (double-well) potential and the null solution is a local maximum. Similarly to the case with $\alpha = 0$ treated before, we note that v_1 and v_2 are stable constant solutions of the PDE (7.58). If $\alpha \neq 0$ we find that $V(v_1) \neq V(v_2)$ and the two minima are equal only if $\alpha = 0$, and this was the case in the Landau equation example. We recall that in that example we found that a front connecting the two stable solutions ($v_{1,2} = \pm 1$) was *stationary*, i.e., $c = 0$ was the eigenvalue of the boundary value problem given by the associated ODE.

Remembering our discussion on the thermally bistable medium model in Chapter 2, we can anticipate here the existence of *moving* fronts since for $\alpha \neq 0$ one of the solutions is 'less stable' (it is said to be *metastable*) than the other one. To check if this is so we first examine the associated ODE of (7.58) with $f(v)$ given by (7.62) for the case $\alpha \neq 0$. It is

$$\frac{1}{\lambda}\frac{d^2 D}{d\chi^2} = D^3 - 2\alpha D^2 - D - c\frac{dD}{d\chi} \tag{7.64}$$

The constant solutions of the PDE (7.58) are unstable fixed points of this ODE. Wishing to find front solution $D(\chi)$ similar to the ones found analytically above for $\alpha = 0$, we set the boundary conditions $D(-L) = v_1$ and $D(L) = v_2$, with $L \gg \lambda^{-1/2}$.

Similarly to the discussion in Chapter 2 we may regard (7.64) as a mechanical problem in which D is the 'displacement' and χ is the 'time'. The 'force' on the right-hand side of the equation includes a part derivable from a 'mechanical potential' U, defined by

$$U(D) = -V(D) = -\frac{1}{4}D^4 + \frac{2\alpha}{3}D^3 + \frac{1}{2}D^2$$

and a 'dissipative' part proportional to the 'velocity', with c being the dissipation constant. The fixed points $D = v_{1,2}$ are the *minima* of V and thus they are the *maxima* of U. The equivalent mechanical problem consists thus of the motion of a 'particle' with mass λ^{-1} ($\sim w^2$, where w is the width of the defect core), departing from one mechanical potential maximum (with negligible velocity and at a very large negative time) and arriving at the other maximum (with negligible velocity, at a very large positive time). As $\alpha \neq 0$, the mechanical potential maxima are unequal, and such motion is only possible if there is nonzero 'dissipation' (i.e., $c \neq 0$).

To simplify our calculation we consider the case $\alpha \ll 1$ and keep terms only up to the first order in α. Thus we have $v_1 = 1 + \alpha$ and $v_2 = -1 + \alpha$. The values of the mechanical potential at the fixed points are to this order

$$U_- \equiv U(v_1) = \frac{1}{4} + \frac{2}{3}\alpha$$

and

$$U_+ \equiv U(v_2) = \frac{1}{4} - \frac{2}{3}\alpha$$

Thus at the fixed point, which we have chosen as our left boundary condition (the initial condition of the mechanical problem), the value of the mechanical potential is higher than that at the other fixed point (the 'final condition' of the mechanical problem). Consequently, a *positive* dissipation constant, c, is needed to facilitate this motion. We are unable to write an analytical solution in a closed form for the front in this case, but it is obviously possible to calculate it numerically. The sign of c can be obtained, however, even if the solution is not known explicitly. Writing (7.64) as

$$\frac{1}{\lambda}\frac{d^2 D}{d\chi^2} = -\frac{dU}{dD} - c\frac{dD}{d\chi}$$

multiplying through by $dD/d\chi$ and integrating from $\chi = -\infty$ to $\chi = \infty$ we get

$$c = -\frac{1}{A}(U_+ - U_-)$$

where $A \equiv \int (dD/d\chi)^2 d\chi > 0$ (over the full domain) and the subscript $+$ ($-$) refers to the boundary value at the right (left) edge of the domain. This result follows from the fact that we require that the 'velocity' (and hence also the 'acceleration') is zero at $\chi = \pm\infty$. As $U_+ < U_-$ in our case, we have $c > 0$.

This means that the front in the PDE moves in the positive x direction ($\chi = x - ct$). We thus see that as a result of this motion of the front, the region in which $v(x,t) = v_1$ grows at the expense of the region where $v(x,t) = v_2$. This is indeed confirmed qualitatively by considering the Liapunov functional for this problem.

The potential $V(v)$ has an absolute minimum at v_1. Thus the Liapunov functional

$$\mathcal{F}[v] = \int dx \left[\frac{1}{2\lambda} (\partial_x v)^2 + V(v) \right]$$

attains its minimum for the stationary constant solution $v(x, t) = v_1$. As long as $v(x, t)$ is not this constant solution the Liapunov functional is not minimal. Consequently, a front solution cannot be stationary and the system must continue to evolve towards a state, which is as close as possible to $v(x, t) = v_1$. So a front connecting v_1 to v_2 has to move so as to decrease the domain of v_2.

The above rather simple one-dimensional examples have helped us to introduce some salient features of localised structures. Similar considerations can be applied to more general and complicated systems as well.

Interactions and dynamics of defects

The motion of solitons or solitary waves, discussed so far, resulted from the inherent properties of the PDE. We have considered only a single localised pulse or front, moving through an essentially uniform background. The next natural step is to consider cases in which several such defects are present. It is in this context that the so-called *effective-particle* methods are especially useful. In such approaches one considers a soliton or a sufficiently coherent solitary wave as a particle, and tries to derive the interactions between the defects and the resulting dynamics. To prevent confusion we remark that this has nothing to do with the mechanical analogies of the previous subsection. The idea is to examine special solutions of nonlinear PDEs, i.e., those composed of a number of localised structures, and look for a description of their interactions in ways analogous to many-body systems.

If a defect solution of a particular PDE with given boundary conditions is known (analytically or numerically) it is natural to consider, as the next step, a state composed of several such defects in a way consistent with the boundary conditions. For example, let us consider equation (7.55) in the infinite domain with the boundary conditions $v(\pm\infty, t) = -1$. A front (7.57) is a stationary solution of the equation for any x_0, but only for the boundary conditions $v(\pm\infty, t) = \pm1$. We also note that

$$v(x) = \bar{K}(x - x_0) \equiv -\tanh[\gamma(x - x_0)] \tag{7.65}$$

is a stationary solution corresponding to the boundary conditions $v(\pm\infty, t) = \mp1$. This solution is just the mirror image of the front (7.57) (remember that x_0 is arbitrary) obtained by the replacement $x \mapsto -x$, which leaves the equation itself invariant. In this sense it is referred to as an *antifront* or *antikink*.

Clearly a kink–antikink pair (with the kink on the left), or more generally, any number n of such pairs in succession, satisfies the boundary conditions we have chosen ($v = -1$ at $\pm\infty$). Consider thus such a configuration of defects and let them be centered on positions $x_1, x_2, \ldots, x_{2n}$, say. If the front–antifront separations $|x_{i+1} - x_i|$ are much larger than the front core width it could be naively expected that such a collection of fronts is also a stationary solution of the PDE. In the KdV case the freedom in constructing a multi-defect state is even larger. Owing to the translational *and* Galilean invariance of this PDE, the solitons (7.52) can be shifted and boosted and we may consider a collection of n solitons, centered on x_i and moving with velocities $c_i + u_i$ (see 7.53) with arbitrary c_i.

The above multi-defect states are completely described by specifying the values of what is called the *symmetry group parameters* for each defect. In the case of the translationally invariant Landau equation only one such parameter exists. It is x_i, the front (or antifront) position and the multi-front state we have considered is completely characterised by the $2n$ values of this parameter. The KdV equation has two symmetry groups (translations and Galilean boosts). Consequently, a multi-soliton state is defined by specifying the value of two parameters (position and velocity) for each of the pulses. In general, the number of independent group parameters is equal to the number of symmetries (invariance properties) of the equations.

If the defects are far enough from each other, i.e., the system of defects is *dilute*, the natural approach to the problem is perturbative. Singular perturbation methods, which are effective for such problems, have been applied in a variety of cases. A full description of these applications is too technical for this book and we refer the reader to the papers of Elphick, Meron and Spiegel (1988) and Elphick, Ierley, Regev and Spiegel (1991a) as examples. The former treats a PDE with translational invariance and the latter deals with a system with both translational and Galilean symmetry. References to other works on the subject can also be found in these papers.

Here we shall describe a more heuristic analysis of a simple multifront state in a simple gradient system, the Landau equation, done with the help of the system's Liapunov functional. This approach is sufficiently simple and transparent to demonstrate the essentials of the problem and can even provide some quantitative results.

We consider the Landau equation (7.55) with the boundary conditions $v(\pm\infty, t) = -1$. The simplest multi-front state of the type described above, and consistent with the current boundary conditions, is one consisting of just two fronts, actually a kink–antikink pair, centred on positions x_1 and x_2, say, with $x_2 > x_1$. We shall call this function $v_{k\bar{k}}(x, t)$.

If we assume that the system is dilute in the sense explained above, that is, the distance between the defects, $d \equiv x_2 - x_1$ satisfies $d \gg w$, where $w \sim \lambda^{-1/2}$ is the defect core size, it is obvious that up to order $\epsilon \equiv \exp(-d/w)$, the kink–antikink pair function can be approximated by

$$v_{k\bar{k}}(x, t) = K(x - x_1) + \bar{K}(x - x_2) - 1 \tag{7.66}$$

with $x_2 \gg x_1$ and where K and $\bar{K}$ are the front (7.57) and antifront (7.65) and are stationary if existing alone. Note that some time dependence of $v_{k\bar{k}}$ must be allowed because we are not sure a priori that the above pair solution is strictly steady. Such time dependence can be realised if the *positions* of the defects, x_1 and x_2, are weakly time-dependent. In the language of singular perturbation theory, the group parameters of the multidefect state are allowed to be slowly varying (i.e., to be functions of ϵt). Checking now if (7.66) is indeed a reasonable approximation for the kink–antikink state, we note that for $x \gg x_1$ we get $K(x - x_1) - 1 \approx 0$ and therefore the function is approximately equal to the antikink. Similarly, for $x \ll x_2$ we have $\bar{K}(x - x_2) - 1 \approx 0$ and thus the function is essentially identical to the kink.

We now examine the properties of the kink–antikink solution (7.66) using the Liapunov functional of the Landau equation, given in (7.61) with the potential

$$V(v) = \frac{1}{4}v^4 - \frac{1}{2}v^2 \tag{7.67}$$

Both single defect solutions, K and $\bar{K}$, minimise the Liapunov functional (they actually differ in the boundary conditions, but because $\mathcal{F}$ is invariant to $x \mapsto -x$, its value for both solutions is the same).

The functional $\mathcal{F}[v_{k\bar{k}}]$ may differ, of course, from the sum of such functionals for K and $\bar{K}$, and this difference reflects the presence of 'interaction energy', which we now define to be

$$\mathcal{U}_{\text{int}}(x_2 - x_1) \equiv \mathcal{F}[v_{k\bar{k}}] - (\mathcal{F}[K] + \mathcal{F}[\bar{K}]) \tag{7.68}$$

This is readily understood noting that our kink–antikink configuration (7.66) does not make $\mathcal{F}$ stationary and thus the fronts must move. The motion is driven by the interaction energy $\mathcal{U}_{\text{int}}$ and its nature is determined by the functional form of this energy.

In order to find $\mathcal{U}_{\text{int}}$ we now try to approximate the integrand in $\mathcal{F}[v_{k\bar{k}}]$, as defined in (7.61),

$$\mathcal{F}[v_{k\bar{k}}] = \int dx \left[\frac{1}{2\lambda}(\partial_x v_{k\bar{k}})^2 + V(v_{k\bar{k}}) \right] \tag{7.69}$$

in several different regions of the integration domain. To do this we recall that the single defects and their derivatives are known analytically in our case:

$$K(x - x_1) = \tanh[\gamma(x - x_1)], \quad \partial_x K(x - x_1) = \gamma \operatorname{sech}^2[\gamma(x - x_1)]$$

and

$$\bar{K}(x - x_2) = -K(x - x_2), \qquad \partial_x \bar{K}(x - x_2) = -\partial_x K(x - x_2)$$

where $\gamma \equiv (\lambda/2)^{\frac{1}{2}} \sim w^{-1}$.

Thus, far from the kink core, x_1, we have approximately

$$K(x - x_1) = \begin{cases} 1 - 2e^{-2\gamma(x-x_1)} & \text{if } x > x_1 \\ -1 + 2e^{-2\gamma(x_1-x)} & \text{if } x < x_1 \end{cases}$$

and

$$\partial_x K(x - x_1) = 4\gamma e^{-2\gamma|x-x_1|}$$

The corresponding expressions for the antikink and its derivative, approximately valid far from its core, x_2, are

$$\bar{K}(x - x_2) = \begin{cases} -1 + 2e^{-2\gamma(x-x_2)} & \text{if } x > x_2 \\ 1 - 2e^{-2\gamma(x_2-x)} & \text{if } x < x_2 \end{cases}$$

and

$$\partial_x \bar{K}(x - x_2) = -4\gamma e^{-2\gamma|x-x_2|}$$

Substituting these approximations in the Ansatz (7.66) we get close to the defect cores

$$v_{k\bar{k}} = \begin{cases} K(x - x_1) - 2e^{-2\gamma(x_2-x)} & \text{for } x \sim x_1 \\ \bar{K}(x - x_2) - 2e^{-2\gamma(x-x_1)} & \text{for } x \sim x_2 \end{cases}$$

and

$$\partial_x v_{k\bar{k}} = \begin{cases} \partial_x K(x - x_1) - 4\gamma e^{-2\gamma(x_2-x)} & \text{for } x \sim x_1 \\ \partial_x \bar{K}(x - x_2) + 4\gamma e^{-2\gamma(x-x_1)} & \text{for } x \sim x_2 \end{cases}$$

Far from both defect cores $v_{k\bar{k}}$ we have

$$v_{k\bar{k}} = \begin{cases} -1 + 2e^{-2\gamma(x_1-x)} & \text{if } x < x_1 \\ -1 + 2e^{-2\gamma(x-x_2)} & \text{if } x > x_2 \\ 1 - 2[e^{-2\gamma(x-x_1)} + e^{-2\gamma(x_2-x)}] & \text{in between} \end{cases}$$

and

$$\partial_x v_{k\bar{k}} = 4\gamma \left(e^{-2\gamma|x-x_1|} - e^{-2\gamma|x-x_2|} \right)$$

As the exponential terms in all the above expressions are small, we may use this fact and Taylor expand the terms $(\partial v_{k\bar{k}})^2$ and $V(v_{k\bar{k}})$ in the integrand of the Liapunov functional (7.69). Such a procedure gives

$$\mathcal{F}[v_{k\bar{k}}] = \mathcal{F}[K] + \mathcal{F}[\bar{K}] + \mathcal{F}_{1,2}$$

where $\mathcal{F}_{1,2}$, defined before as the interaction energy $\mathcal{U}_{\text{int}}$, is a small correction term, containing overlap integrals of the defect exponential tails. These overlap integrals can be calculated approximately and yield the result

$$\mathcal{U}_{\text{int}}(x_2 - x_1) = -g\exp[-2\gamma(x_2 - x_1)] \tag{7.70}$$

where $g > 0$ is a coupling constant, which can be expressed explicitly in terms of γ (see below). These considerations remain valid for any system in which the localised structures have exponentially behaving tails. The value of the coupling constant g depends on the detailed structure of the defect.

In ordinary particle dynamics the spatial derivative of the interaction energy gives the force acting on the particle and the acceleration is proportional to this force. Here we have a PDE that is *first order in time* and the only symmetry it has is translational invariance. Consequently, the defects (effective-particles) are described by only one group parameter (position), and their interaction energy gradient is proportional to the velocity (time-derivative of position). In systems endowed with Galilean as well as translational invariance, defects are described by two group parameters, position and velocity, and the force is proportional to acceleration, as in Newtonian dynamics. These observations are confirmed by detailed perturbative calculations (see, for example, the papers cited above).

The resulting effective-particle dynamics in our case is thus

$$\hat{\Gamma}\dot{x}_1 = -\partial_{x_1}\mathcal{U}_{\text{int}}(x_2 - x_1) = 2\gamma g\exp[-2\gamma(x_2 - x_1)]$$

and

$$\hat{\Gamma}\dot{x}_2 = -\partial_{x_2}\mathcal{U}_{\text{int}}(x_2 - x_1) = -2\gamma g\exp[-2\gamma(x_2 - x_1)]$$

where $\hat{\Gamma}$ is a constant reflecting the 'mass' of the effective particle (related to the defect width). This type of dynamics may be called 'Aristotelian', as we have the *velocity* being proportional to the force. The symmetry of the expressions is guaranteed by the fact that the localised structures in our example are symmetric fore and aft. Note that if the defects are not symmetric (as can well be the case in other systems), the force exerted by defect 1 on defect 2 need not be equal in magnitude to the force defect 2 exerts on defect 1. This can happen in systems with Galilean symmetry as well.

In our case we see that the two defects attract each other (the interaction is due to the overlap of their exponential tails), and the velocity at which the fronts approach is exponential in their separation. When the front and antifront come close to each other, the above approximate formulae are no longer valid, but it can be expected that eventually the front–antifront approaching motion continues until the defects annihilate each other and a uniform steady solution (minimum of the Liapunov functional) is achieved.

Before turning our attention to more complicated aspects of defects in multidimensional systems, it is instructive to examine the dynamics of a one-dimensional multifront system. Consider a collection of a large number of kink–antikink pairs in succession, arranged in such a way that the distances between any defect and its neighbours are large with respect to the defect core size, i.e., $\epsilon \ll 1$ (see above). In lowest order in ϵ, we may assume that each defect is attracted only by its nearest neighbours (on both sides), and thus its position x_i satisfies

$$\Gamma \frac{dx_i}{dt} = \exp[-2\gamma(x_{i+1} - x_i)] - \exp[-2\gamma(x_i - x_{i-1})] \tag{7.71}$$

where the new constant Γ is defined to be $\Gamma \equiv \hat{\Gamma}/(2\gamma g)$. A detailed perturbative calculation for our example of fronts in the Landau equation (7.55) furnishes the value of the constant as $\Gamma = (12\gamma)^{-1}$.

All the defects, save the leftmost and rightmost ones (which have a nearest neighbour on one side only), satisfy the above equation. To understand the evolution of an arbitrary pattern, composed of kink–antikink pairs, we focus on a region between two such defects and note from (7.71) that its size, $\Delta_i \equiv x_i - x_{i-1}$, changes according to

$$\Gamma \frac{d\Delta_i}{dt} = -2e^{-2\gamma\Delta_i} + e^{-2\gamma\Delta_{i+1}} + e^{-2\gamma\Delta_{i-1}} \tag{7.72}$$

If Δ_i is the size of a region where $v = 1$, say, then $\Delta_{i\pm1}$ are the adjacent regions at which $v = -1$. Calling, for the sake of convenience, the regions where $v = 1$ 'clouds', we immediately get an isolated single 'cloud' size, Δ, evolving as

$$\Gamma \frac{d\Delta}{dt} = -2e^{-2\gamma\Delta} \tag{7.73}$$

This follows from setting $\Delta_i = \Delta$ and $\Delta_{i\pm1} \to \infty$ in (7.72) and implies that an isolated 'cloud' of size $\Delta(0) = d$ disappears, due to the approach of its boundaries, in time

$$\tau = \frac{\Gamma}{4\gamma}\left(e^{2\gamma d} - 1\right) \tag{7.74}$$

We stress again that for our approximations to hold the cloud size d must be much larger than the front width $w \sim \gamma^{-1}$. Thus we get that τ is of the order of $\epsilon^{-1} \sim \exp(-d/w)$.

When the cloud is not isolated we return to (7.72) which includes, in addition to the 'self-interaction' term, the effects of the two neighbouring 'intercloud' (that is, zones where $v = -1$) regions. In particular, we observe that a strictly periodic pattern of 'clouds' and 'intercloud' regions, that is $\Delta_k = \Delta_0 = \text{const}$, is a steady solution of (7.72). It is very easy to see, however, that such a periodic solution is unstable. Thus it obviously does not minimise the Liapunov functional. If we start out with a distribution of many sizes of 'clouds' and their spacings, the smallest

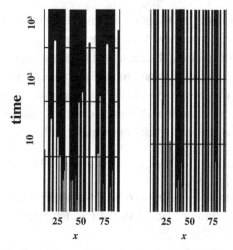

Fig. 7.3. Time evolution of a multifront system (initially 50 defects with random separations) without (left) and with (right) spatial forcing. 'Clouds', that is, regions for which $v = 1$, are marked black – for details see text. Reprinted, with permission, from Elphick, Regev & Spiegel (1991b), Blackwell Publishing.

objects will disappear first, as the system tends to become homogeneous (a solution minimising the Liapunov functional for the chosen boundary conditions, $v = -1$ at the boundaries). The time-scale of this transient behaviour becomes progressively longer during this process and the life-time of the whole chain will be of the order of $e^{2\gamma \Delta_m}$, where Δ_m is the system's largest 'cloud' size. Thus if we have initially N 'clouds', $\Delta_m \sim Nd$ and thus the system time is of the order ϵ^{-N}. The evolution can be described as an inverse cascade, in which small structures disappear and are replaced by progressively larger ones. Numerical experiments with equations (7.71) confirm these expectations and we refer the reader to Elphick, Regev & Spiegel (1991b) for a detailed description of the results in a similar system. Figure 7.3 (left panel) is an example of the results of such a calculation. The final large 'clouds' ultimately merge and disappear, but this process is not shown in the figure as it takes a very long time.

The rundown process described above does not give rise to a persistent steady pattern. An initial excitation (in the form of a sequence of front–antifront pairs) decays away and a uniform state is ultimately reached. If the boundary conditions at the two edges of the system are different, i.e., any state of the system necessarily has a number of fronts differing by one from the number if antifronts, the final state will be that of just one front. The process is analogous, in this respect, to what is known in condensed matter physics as spinodal decomposition.

Some aspects of the process may also be viewed as similar to the behaviour of turbulent fluids, in which the turbulent intensity always decays and a steady (in the statistical sense) situation can be maintained only if a source of persistent or recurrent excitation is present. In the case of three-dimensional turbulence the decay proceeds via a cascade from large scales to small ones, and consequently the system is usually driven on large scales. In our case the rundown process takes place via an inverse cascade (from small scales to large ones). It is thus logical to include forcing on small scales and to study the possibility of steady patterns. Physical considerations in any actual system should obviously provide the justification for the inclusion of such forcing (see the second part of the book). In this spirit we consider again the Landau equation (7.55), but now add to it a small spatial forcing term of a very simple form

$$\partial_t v = v - v^3 + \frac{1}{\lambda}\partial_x^2 v + A\sin(qx) \qquad (7.75)$$

where A is a constant of order ϵ and the spatial scale of the forcing, $\sim q^{-1}$, is assumed to be much larger than the front width, but obviously still much smaller than the full domain size. It turns out that a perturbative calculation of the front dynamics is not more difficult in this case than it is in the case in which $A = 0$ and the result is

$$\Gamma\frac{dx_i}{dt} = \exp[-2\gamma(x_{i+1} - x_i)] - \exp[-2\gamma(x_i - x_{i-1})] + a\sin(qx_i) \qquad (7.76)$$

where the constant a is proportional to A and is of order ϵ.

If we now look for steady solutions of (7.76), that is, the set of defect locations x_i for which $\dot{x}_i = 0$, we obtain that these locations are given by the following two-dimensional *mapping*

$$x_{i+i} = x_i - \frac{1}{2\gamma}\ln y_{i+1}, \qquad y_{i+1} = y_i - a\sin qx_i \qquad (7.77)$$

where we have defined $y_j \equiv \exp[-2\gamma(x_j - x_{j-1})]$.

This area-preserving mapping is the result of the reduction of the PDE (7.75) through an ODE (7.76) to a discrete map, describing a steady pattern. The map is thus called the *pattern map* of our problem. Interestingly, the mapping (7.77) has similar properties to the standard map associated with Hamiltonian systems (Chapter 6). It can be shown that this map has chaotic orbits (for appropriate values of a), that is, the pattern it describes is *aperiodic*. Numerical simulations of equations (7.76) (see, e.g., Elphick, Regev and Spiegel, 1991b) confirm this result and a spatially chaotic *steady* configuration is reached for any random initial condition, whose typical length-scale d_0 satisfies $\lambda^{-1/2} \ll d_0 \ll q^{-1}$.

The final steady state can be approximately described as a collection of fronts and their antifront counterparts locked by the external forcing, so that

$$x_i \approx x_{i-1} + \frac{2\pi n_i}{q}$$

where the n_i are arbitrary integers. This situation is shown in the right panel of Figure 7.3.

This possibility demonstrates that spatial complexity can arise even in a system that is very close to a gradient system, which does not form patterns close to marginality. This complexity is caused by external forcing (the forced system is not a gradient system) and is mediated by defects. One-dimensional chaotic wave trains (stationary or moving) and their stability have been studied in a variety of systems. For examples and descriptions of some general techniques we refer the reader to Pismen (1989) and to the review article by Meron (1992).

7.6 Defects in multidimensional systems

Defects arising in PDEs with one spatial dimension must have codimension 1. We have seen that such defects can take the form of fronts or pulses. In PDEs having $n > 1$ spatial dimensions we can also expect the existence of defects whose codimension is 1 (like, e.g., domain walls – the multidimensional analogues of fronts), but defects of higher codimension can exist as well. For $n = 2$, for example, codimension 2, point-like defects may be present.

In this section we shall discuss briefly defects in multidimensional PDEs. The behaviour of such defects is considerably richer than of their one-dimensional counterparts, and the analysis needed to describe their dynamics may be quite complicated. For the sake of simplicity and clarity we shall limit ourselves to only two examples of rather simple but non-trivial cases.

First we shall consider the generalisation to two spatial dimensions of the problem treated in detail in the previous section. The properties of one-dimensional fronts (line domain walls) in a two-dimensional system (defects of codimension 1) will be discussed. The second example will be of a defect type not treated so far - point defects in a two-dimensional system (defects of codimension 2).

Domain walls

Consider the multidimensional (in space) Landau equation

$$\partial_t v = v - v^3 + \frac{1}{\lambda} \nabla^2 v \qquad (7.78)$$

where $\nabla = \partial/\partial \mathbf{r}$. Similarly to the one-dimensional case (7.55), this equation has two stable uniform solutions $v(\mathbf{r}, t) = \pm 1$. The null solution is unstable. For the problem to be well-posed we have to specify the domain of definition of v and the boundary conditions. Similarly to the one-dimensional case we wish that the domain size (a square of length L, say) be large compared to the diffusion length (the front width), that is, $L \gg \lambda^{-1/2}$. The simplest boundary conditions are of

Dirichlet type, i.e., the function value is given on the domain boundary. We may choose, for example, $v = -1$ on the boundary and examine the simple situation in which a closed domain wall (front) encompasses a region of $v = +1$. Rubinstein, Sternberg and Keller (1989) proved by a rigorous asymptotic analysis that any such domain wall curve will have the tendency to shorten. This means that an initial condition of a single closed curve of the above type is not a steady solution of the equation. The domain encompassed by it, or a 'cloud' as we have been referring to such structures, will shrink on itself and disappear. The only possibility of a persistent front is when it is 'anchored' to the domain boundary by boundary conditions of a type different from the one assumed here (we shall not consider this case further).

We can demonstrate the essence of this result when remembering that equation (7.78) is a gradient system, that is

$$\partial_t v = -\frac{\delta \mathcal{F}}{\delta v} \tag{7.79}$$

with the functional $\mathcal{F}$ given by

$$\mathcal{F}[v] = \int d^n r \left[\frac{1}{2\lambda} (\nabla v)^2 + V(v) \right] \tag{7.80}$$

where n is the number of spatial dimensions and $V(v) = \frac{1}{4}v^4 - \frac{1}{2}v^2 + \frac{1}{4}$. Note that we have chosen the arbitrary constant in V to be such that $V = 0$ for $v = \pm 1$, and compare with the one-dimensional problem (7.9) and (7.61). Suppose now that there exists a steady solution $v = v_m(\mathbf{r})$ that minimises $\mathcal{F}$, that is we have for an arbitrary variation around this solution

$$\delta \mathcal{F}[v_m] = 0 \tag{7.81}$$

If we assume that the domain of integration is effectively infinite we may consider a uniform rescaling of all spatial coordinates by a parameter $\mu > 0$ say, and this should have no effect on the limits of the integrals. Thus we may write $\mathbf{r}' = \mu \mathbf{r}$ and

$$\mathcal{F}(\mu) \equiv \mathcal{F}[v_m(\mathbf{r})] = \int d^n r \left[\frac{1}{2\lambda} (\nabla v_m)^2 + V(v_m) \right] =$$

$$= \int d^n r' \, \mu^{-n} \left[\frac{\mu^2}{2\lambda} (\nabla' v_m)^2 + V(v_m) \right]$$

or

$$\mathcal{F}(\mu) = \mu^{2-n} \mathcal{T} + \mu^{-n} \mathcal{V} \tag{7.82}$$

where

$$\mathcal{T} \equiv \frac{1}{2\lambda} \int d^n r \, (\nabla v_m)^2 \quad \text{and} \quad \mathcal{V} \equiv \int d^n r \, V(v_m)$$

We now look at a special type of variation around $v_m(\mathbf{r})$, i.e.

$$\delta v = [v_m(\mu\mathbf{r}) - v_m(\mathbf{r})]$$

where we assume here that μ is very close to 1. Thus we have

$$\delta v = \left.\frac{dv_m}{d\mu}\right|_{\mu=1} d\mu$$

The corresponding variation of $\mathcal{F}$ is obviously zero and thus

$$0 = \delta\mathcal{F} = \left.\frac{d\mathcal{F}(\mu)}{d\mu}\right|_{\mu=1}$$

Taking now the derivative of $\mathcal{F}(\mu)$ (in 7.82) with respect to μ, evaluating it at $\mu = 1$ and setting the result to zero we have

$$(2 - n)\mathcal{T} - n\mathcal{V} = 0 \qquad (7.83)$$

Now both $\mathcal{T}$ and $\mathcal{V}$ are ≥ 0, since $\mathcal{V}$ is by our choice nowhere smaller than zero and so obviously is $(\nabla v_m)^2$. Thus the equality (7.83) for $n > 1$ can be satisfied only if $\mathcal{T} = \mathcal{V} = 0$. Consequently, the assumed solution v_m must be one of the stable uniform states. In one dimension ($n = 1$) (7.83) implies $\mathcal{T} = \mathcal{V}$, and this can be satisfied (as we have seen before) if v_m is one of the stable uniform states or a single kink (or antikink). Thus for a spatially multidimensional Landau equation the only steady stable state is one of the trivial uniform ones and the selection is made by the boundary conditions. In our case we have chosen $v = -1$ on the boundaries and thus any 'cloud' of $v = 1$ must shrink and disappear.

The shrinking of 'clouds', isolated by domain walls, in two (or more) spatial dimensional systems can also be understood as a curvature effect. As long as the local radius of curvature of a domain wall is finite, this curvature induces motion of the defect. This effect is similar to surface tension in fluids. A state without a boundary between states (domain wall) minimises the Liapunov functional (the analogue of free energy). If a curved boundary between two equally favourable states exists, the Liapunov functional is not minimal, due to the presence of a positive 'surface energy' in the defect. It turns out that it is possible to derive an equation for the local velocity of the front at each point. The asymptotic techniques for doing this are described, for example, in the book by Gringrod (1991). The *eikonal equation*, as it is called, for a general bistable PDE of the Landau type, i.e.,

$$\partial_t v = f(v) + \frac{1}{\lambda}\nabla^2 v$$

with $f(v)$ being derivable from a general double-well potential $V(v)$, gives the local normal velocity of the interface $u_\perp$ as

$$u_\perp = -\frac{1}{\lambda}K + c \tag{7.84}$$

where K is the local curvature of the interface and c is its velocity resulting from the asymmetry of the potential (see previous section). If the potential is symmetric, as it is in the Landau equation (7.55), $c = 0$ and the motion is purely curvature-driven. For example if we consider a circular 'cloud' of radius R we get

$$\frac{dR}{dt} = -\frac{1}{\lambda R}$$

and thus the radius of the cloud shrinks from its initial size R_0, say, as

$$R(t) = \left(R_0^2 - \frac{2}{\lambda}t\right)^{\frac{1}{2}}$$

Obviously, this formula remains valid only as long as $t \le \lambda R_0^2/2$, and in any case if R becomes too small the eikonal equation breaks down.

Shaviv and Regev (1994) examined some aspects of a two-dimensional problem of this sort, derived from an extended thermally bistable fluid system. We shall discuss this work in more detail in the second part of this book. At this point the reader can use it as an example of the application of the methods given in Gringrod (1991).

Defects of codimension 2

We have seen in the previous section that a phase variable, defined as the phase of some reference periodic solution of a microscopic, model, or envelope equation, itself satisfies a phase equation. Phase equations are valid if the wavevector is assumed to be slowly varying (in the sense explained in the previous section, see (7.38)). Consider now a multidimensional system and define the phase winding number W

$$W = \frac{1}{2\pi} \oint \nabla\phi \cdot d\mathbf{l}$$

that is, the line integral of the gradient of the phase (a vector) along a closed path. Clearly, if $\mathbf{k}$ in (7.38) is constant we have $W = 0$, since the pattern is then purely periodic with a constant wavelength. If $\mathbf{k}$ and thus $\nabla\phi$ is slowly varying, it is conceivable that over a large enough loop we may have $W = 1$, that is, a total phase change of 2π is accumulated as we return to the same point (W must be an integer, otherwise we have multivalued functions!). It is easy to see that such a situation cannot be consistent with the assumption of a slow variation *everywhere*. If the contrary were true, we could smoothly shrink the integration contour always through regions of slow variation and the winding number could not change. Eventually we would find an arbitrarily small loop around a point with a 2π phase winding. Such a point must clearly be a defect of the phase equation, where the slow variation and hence the whole phase description breaks down. In a two-dimensional system such a point defect has codimension 2 and is sometimes called a *topological defect*, reflecting the fact that it cannot be removed by any smooth deformation of the system.

Topological defects are known to exist also in equations describing phase transitions in condensed-matter theory and this fact has profound implications, but we shall not discuss this aspect of defects here. We shall concentrate on only two types of point defects in a particular phase equation, giving the interesting example of *target* and *spiral* patterns in the complex CGLE (7.28) equation. We refer the reader to Cross & Hohenberg (1993) for a systematic discussion of point defects in two-dimensional systems.

We have stated in Section 7.3 that, near a Hopf bifurcation, i.e., for systems with a homogeneous oscillatory instability, the generic amplitude equation (valid near marginality) is the complex Ginzburg–Landau equation. Assume that we are interested in the two-dimensional case, that is, the complex envelope $A(x, y, t)$ satisfies

$$\partial_t A = \epsilon A + \zeta_1 \nabla^2 A - \zeta_3 A|A| \tag{7.85}$$

with ζ_i being complex constants depending on the details of the original PDE.

In Section 7.4 we have given the phase equation, i.e., the equation for $\phi(x, t)$, resulting from (7.85) in one space dimension. The phase equation given there (7.42) can be derived by writing $A(x) = ae^{i\phi(x,t)}$ for the envelope in the CGLE, with a constant in space. It turns out that the form of the equation carries over to the two-dimensional case as well. If we add to it an external localised perturbation term (changing, for example, the frequency of oscillations at a given point) we may reasonably expect that a point defect will form. The relevant phase equation is

$$\partial_t \phi = \alpha \nabla^2 \phi - \beta (\nabla \phi)^2 - g(\mathbf{r}) \tag{7.86}$$

where the constants α and β depend, as usual, on the particular system studied (compare with equation 7.42) and $\alpha > 0$ and $\beta > 0$. The function $g(\mathbf{r})$ is assumed to be positive and centred at a point (the origin, say). This arrangement of a spatially uniform oscillating medium with a frequency perturbation located at a given point is sometimes referred to as an 'oscillating medium with a pacemaker' (see e.g., Kuramoto, 1984).

Note now that the transformation

$$\tilde{Q}(\mathbf{r}, t) \equiv \exp(-\beta \phi / \alpha) \tag{7.87}$$

linearises the phase equation (7.86) to give

$$\partial_t \tilde{Q} = \left[\alpha \nabla^2 + \frac{\beta}{\alpha} g(\mathbf{r})\right] \tilde{Q}$$

Setting next

$$\tilde{Q}(\mathbf{r}, t) = e^{-Et} Q(\mathbf{r}) \tag{7.88}$$

with E constant, we obtain a *time-independent Schrödinger equation* for the function $Q(\mathbf{r})$

$$-\alpha \nabla^2 Q + V(\mathbf{r}) Q = EQ \tag{7.89}$$

where the potential is given by

$$V(\mathbf{r}) = -\frac{\beta}{\alpha} g(\mathbf{r}) \tag{7.90}$$

Linearising transformations of the sort performed here (7.88) are very useful for a number of nonlinear PDEs, similar to the phase equations given in the previous sections, and are named after Hopf and Cole, who proposed them. For example, it was with the help of a Hopf–Cole transformation that the KdV equation soliton solutions were found.

Assume now that our pattern is such that the phase is circularly symmetric, that is, it is independent of the θ coordinate in planar polar coordinates (r, θ). Clearly, this is realised if g and thus V depend only on r. The time-independent Schrödinger equation in this spherically symmetric (two-dimensional) case is

$$-\frac{\alpha}{r} \frac{d}{dr}\left(r \frac{dQ}{dr}\right) + V(r) Q = EQ \tag{7.91}$$

and we may look, as in quantum mechanics, for the eigenvalues E and their corresponding eigenfunctions. Remember, however, that our procedure makes sense only if there exist real eigenfunctions that are *positive* everywhere, in view of (7.87). This is possible only if, using a quantum-mechanical language, there exist *bound states*, that is, solutions with $E < 0$.

As we have chosen $g(r) < 0$, and both $\alpha > 0$ and $\beta > 0$, we have a time-independent Schrödinger equation with an *attractive* potential and it is guaranteed (see e.g., any standard text on quantum mechanics) that there exists at least one bound state in this two-dimensional case. Assuming that the perturbation decays to zero away from $r = 0$, the asymptotic form of $Q(r)$ must be

$$Q(r) \to (\tilde{q}r)^{-\frac{1}{2}} \exp(-\tilde{q}r) \quad \text{as} \quad r \gg r_0$$

where $\tilde{q} \equiv \sqrt{-E/\alpha}$ and r_0 is the characteristic scale of $g(r)$.

The phase very far from the defect centre then follows from the inverse Hopf–Cole transformation, which for such a single eigenfunction solution is approximately

$$\phi(r, t) = -\tilde{\omega}t + qr \tag{7.92}$$

where $\tilde{\omega} \equiv \beta q^2$ and $q^2 \equiv -E\alpha/\beta^2 > 0$ (provided, of course, that $E < 0$). This means that the 'pacemaker' introduces an outgoing circular wave with a wavelength determined by E and a positive frequency (that is, a frequency larger than the one characteristic of the oscillatory medium, which in our scaling of the Ginzburg–Landau model was set to zero). A defect of this type located at $r = 0$ is called a *target* because of the characteristic shape of the wave pattern it induces (concentric circles around a point).

If the potential $V(r)$ does not admit bound states, positive eigenfunctions Q do not exist and the system does not have targets. The target defect in the case discussed above is induced by an external perturbation (the pacemaker) at the origin. It is possible in principle to induce target patterns intrinsically, that is, when $g(r) = 0$. In such a case we must allow for a spatial variation of the real amplitude a of the complex envelope A. The resulting phase equation in two dimensions is then (see Kuramoto, 1984).

$$\partial_t \phi = \alpha \nabla^2 \phi - \beta (\nabla \phi)^2 + 2\alpha a^{-1} \nabla a \cdot \nabla \phi \tag{7.93}$$

A suitable Hopf–Cole transformation that linearises (7.93) can also be found, and the issue of the existence of intrinsic target patterns depends on the asymptotic behaviour of $a(r)$ for $r \to \infty$ and $r \to 0$.

An obvious generalisation of a target pattern is a *spiral*. Indeed, targets and spiral waves are observed in experiments in a variety of systems (e.g., chemical reactions, biological excitable media, and so on). We could naively think that all that is needed is to repeat the above calculation done for a target, but allowing for an angle-dependence of the phase. The simplest choice of this is an $l\,\theta$-dependence of the phase with an integer l. The phase can then be written as

$$\phi(r, \theta, t) = -\tilde{\omega}t - l\theta + \tilde{\phi}(r)$$

(compare with 7.92) where $\tilde{\omega}$ and ϕ are to be determined. A nontrivial problem arises, however, because the Hopf–Cole transformation now gives an extra term in the effective potential of the radial problem (the centrifugal term) due to the fact that now

$$\nabla^2 Q = \frac{1}{r} \partial_r (r \partial_r Q) + \frac{1}{r^2} \partial_\theta^2 Q$$

Consequently, since

$$Q(r, \theta) = \exp\left(Et - \beta\phi/\alpha\right)$$

the radial time-independent Schrödinger equation is

$$-\frac{\alpha}{r}\frac{d}{dr}\left(r\frac{dQ}{dr}\right)-\left[V(r)-\frac{\beta^2}{\alpha}\frac{l^2}{r^2}\right]Q=EQ \qquad (7.94)$$

If $V(r)$ is bounded at $r=0$ (as can be expected for a physical system with a pacemaker) we have a singularity of the effective potential at the origin, i.e., it is infinitely attractive there, $V_{\text{eff}} \sim -r^{-2}$ as $r \to 0$. This results in an unbounded (from below) negative energy, a very sharply peaked eigenfunction, i.e., a complete breakdown of the phase equation (see Kuramoto, 1984). In the terms defined above we have here nothing else but a *topological* point defect.

Direct numerical simulations of the CGLE do however give rise to spiral patterns, and it is therefore logical to seek an analytical remedy for the above mentioned singularity. It is not too difficult to see that the situation becomes regular if we allow for a spatially varying amplitude $a(r)$. It can be shown that the potential in the Schrödinger equation, derived from (7.93) for a *spatially dependent* amplitude and no pacemaker ($g=0$), has the form

$$V(r;a)=\frac{\beta^2}{\alpha^2}a^{-1}\nabla^2 a$$

Thus, choosing $a(r) \sim r^l$ as $r \to 0$ gives $V(r) \sim r^{-2}$ near the origin and the divergent centrifugal term in the effective radial potential can be cancelled. If the outer asymptotic behaviour of the potential allows for the existence of a bound state in this case, a rotating l-arm spiral wave solution is obtained and it is regular everywhere in spite of the phase singularity at the topological defect located at its centre. The key to this construction is the assumption (see Cross & Hohenberg, 1993) that

$$a(r) \sim a_l r^l \qquad \text{for} \quad r \to 0$$

with $a_l > 0$ and

$$a(r) \sim a_\infty + a_{-1}r^{-1} \quad \text{for} \quad r \to \infty$$

with $a_\infty > 0$ and $a_{-1} < 0$.

Other approaches to the problem of spiral patterns exist as well; an especially fruitful one is based on the notion of curling of line defects. Simply put, this approach focuses on the development of a line defect (similar to the fronts in two dimensions discussed above), which can be followed with the help of an eikonal equation of a more general type than (7.84), i.e.

$$u_\perp = D_\perp K + c \qquad (7.95)$$

where in the case of line defects in the form of fronts in the Landau system we had $D_\perp = -\lambda^{-1} < 0$ and consequently the system had to evolve toward curve-shortening (i.e., extinction of 'clouds' embedded in the second stable phase). In systems having spiral patterns $D_\perp$ is positive and therefore any slight perturbation increasing the curvature locally will have a larger propagation velocity, resulting in a curling motion. This approach was pioneered by Meron and collaborators and we refer the reader to Meron (1992) for details.

When several point defects are present in a system (targets or spirals) it is important to investigate their dynamics. We have seen above that defect dynamics in one-dimensional systems can be investigated using the effective-particle approach. Such approaches are used in multidimensional cases as well and the analysis may naturally become rather involved. The subject of defect dynamics is growing rapidly and significant progress has already been made, but much remains to be done.

7.7 Coupled map lattices

A coupled map lattice (CML) is a discrete representation of a spatio-temporal dynamical system. Both the time and space variables are discretised and thus a CML is essentially a mapping having two types of indices, one of them representing the time evolution and the other reflecting the spatial couplings. Although one can think of a CML as a discrete approximation to a PDE, the discretisation is usually not a result of a systematic reduction procedure (of the kind discussed in this chapter so far). Rather, these systems are usually devised in a quite heuristic manner and the CML approach is largely a matter of convenience, as these systems are computationally very tractable.

We have seen that the essential properties of continuous temporal dynamical systems (ODE) can often be captured by appropriate mappings and their study may reveal low-dimensional chaotic behaviour. This was the case, for example, in the logistic or in the circle maps, which were studied in detail in Chapter 4. A CML is set to include, in addition to this dynamics (which may be viewed as *local*), *spatial* couplings as well. The spatial variables are discretised by introducing a *lattice* of sites, labelled by appropriate indices representing the (often very many) spatial degrees of freedom. In studying CMLs one hopes that at least some of the generic properties of physically relevant spatio-temporal dynamical systems can be found and qualitatively understood. These ideas will be clarified in what follows, when we shall discuss several typical examples of CMLs in one spatial dimension. Multidimensional lattices are possible as well, but the resulting CMLs are obviously more complex and we shall not discuss them here. Detailed discussions of CMLs can be found in the review of Crutchfield and Kaneko (1987) and in Kaneko (1989).

Consider the following CML dynamical system, written in a rather general form

$$u_k^{n+1} = f\left(u_k^n\right) + \epsilon_0 g\left(u_k^n\right) + \epsilon_R g\left(u_{k+1}^n\right) + \epsilon_L g\left(u_{k-1}^n\right) \tag{7.96}$$

for the spatio-temporal dynamical variable u, where the subscript k is the spatial lattice index, and n is the time level index. The function f determines the local dynamics, while g gives the coupling dynamics. The three constants $(\epsilon_0, \epsilon_R, \epsilon_L)$ constitute the *coupling kernel* and determine the nature of the spatial coupling of a given lattice site to its nearest neighbours (we shall consider only this case here).

As the CMLs were originally devised to study the implications of local chaotic dynamics, when a spatial coupling is present in the system, it is natural to select the local dynamics from the known classes of chaotic maps (like the ones mentioned above). The simplest coupling dynamics is *linear*, that is, $g(u) = u$. Among the more complicated possibilities we will mention the so-called *future coupling*, i.e., $g(u) = f(u)$ (see the references cited above). The coupling kernel can also be chosen in various ways, appropriate to the properties of the dynamical system. For example, the simplest choice $\epsilon_0 = 0$, $\epsilon_R = \epsilon_L \equiv \epsilon$ is referred to as *additive coupling*. The *Laplacian coupling*, i.e., $-\epsilon_0/2 = \epsilon_R = \epsilon_L \equiv \epsilon$ is frequently used as well and reflects diffusive spatial couplings (second spatial derivative).

CML with the local dynamics chosen to be the logistic map, that is, $f(u) = 4ru(1 - u)$, where the dynamical variable u and the parameter r both take their values in the interval $[0, 1]$ (see Chapters 2 and 4), and with different types of coupling dynamics and spatial coupling are referred to as *period-doubling lattices*. Such CMLs have been extensively studied (see the references cited above) and were found to exhibit spatially coherent domains and kinks separating them.

The simplest variant of a period-doubling lattice is the one with linear, additive coupling

$$u_k^{n+1} = f\left(u_k^n\right) + \epsilon\left(u_{k+1}^n + u_{k-1}^n\right) \tag{7.97}$$

where f is the logistic map. In order to iterate a CML like (7.97) one has to start with some initial condition and determine boundary conditions. Crutchfield and Kaneko summarised the results of such iterations for various cases of spatial couplings and we refer the interested reader to that article for details. Among the most important results found we should mention period doubling and merging of spatial domains and kinks, spatial-mode instability and spatio-temporal quasiperiodicity. These findings may provide qualitative insight into the behaviour of turbulent systems of various kinds.

Of particular interest is the CML in which the local dynamics is represented by the circle map (4.9)

$$f(u) = u + \Omega + \frac{K}{2\pi} \sin(2\pi u) \quad (\mathrm{mod}\ 1) \tag{7.98}$$

With the CML constructed using a Laplace future-coupled lattice one gets

$$u_i^{n+1} = (1 - \epsilon) f(u_i^n) + \frac{\epsilon}{2}[f(u_{i+1}^n) + f(u_{i-1}^n)] \tag{7.99}$$

When the nonlinearity parameter K is above a suitable threshold the circle map is chaotic and the iteration of CML (7.99) gives, depending on the parameter range, annihilation of kink–antikink pairs, and also sustained complexity of the localised structures dynamics. This type of behaviour in the *circle lattices*, as well as other possibilities (e.g., Pomeau–Maneville intermittency in the so-called *intermittency lattices*) were extensively studied, mainly by Kaneko (see the above cited references).

We conclude our discussion of CMLs with a simple example of spatio-temporal complexity arising in a system whose local dynamics is a version of the *linear* ($K = 0$) circle map. This model, called the *dripping handrail* by Crutchfield and Rössler, was later proposed in the modelling of irregularly accreting astrophysical sources (see Chapter 10).

The local dynamics of the dripping-handrail model

$$f(u) = su + \Omega \quad (\text{mod } 1) \tag{7.100}$$

has an additional parameter s, controlling the slope of $f(u)$. If $s \neq 1$ this map has a fixed point and it is very easy to show that the fixed point is stable for $s < 1$ and unstable for $s > 1$. In the case $s = 1$, the map, which is then the linear circle map, has no fixed points. It is possible to show (by calculating the Liapunov exponents, for example) that the map defined by (7.100) is chaotic for $s > 1$. Interestingly, however, a CML with local dynamics given by (7.100) and $s \leq 1$ is capable of producing spatial complexity, even though the isolated local dynamics has a negative Liapunov spectrum.

A CML whose local dynamics is given by (7.100), with $s \leq 1$, future-coupled dynamics ($g = f$) and the so-called 'totalistic' kernel $\epsilon_0 = -2/3, \epsilon_R = \epsilon_L = 1/3$, has been used to model the dripping handrail. The local dynamics models the increase of the water layer thickness (by Ω in each iteration) with sudden decrease (dripping) after the threshold has been reached. The spatial coupling mimics the replenishing flow of water *along* the handrail by a simple low-pass filtering of the pattern at each time step. Although the local dynamics for $s \leq 1$ is not chaotic, the system exhibits a rather long, very complex transient, before finally settling in some sort of ordered state. This result prompted Crutchfield and Kaneko to remark that complex behaviour in spatial systems may be dominated by very long transients, and thus the properties of a possible underlying attractor may be irrelevant to the observable behaviour. Consequently, this behaviour has been called *transient chaos* and it is important to keep this possibility in mind when considering some realistic systems.

Before concluding this chapter we would like to remark that pattern theory in general has begun to establish itself as an exciting branch of applied mathematics and interdisciplinary science. It turns out that pattern-forming systems in physics, chemistry, biology, material science etc. may be formulated in generic ways and have a lot in common. Investigations of the formation of regular spatio-temporal structures, their instabilities, the nucleation of defects in them and the resulting dynamics, are bound to go a long way towards the understanding of generic properties and the mechanisms shaping the macroscopic world around us with both its simplicity and complexity. This chapter was only a short introduction to this explosively growing research topic. We have mentioned several good references on the subject and the reader is encouraged to study them. In the second part of the book we shall return to pattern formation in some systems that may be applicable to astrophysics.

Part II

Astrophysical applications

8

Introduction to Part II

> There is nothing stable in the world; uproar's your only music.
> John Keats, *Letter to G. and T. Keats.*

The first part of this book contained a rather extensive survey of the mathematical background needed for the study of various types of dynamical systems. The following, second part will be devoted to applications of some of these (often quite abstract) mathematical ideas and techniques to a selection of dynamical systems derived from astrophysical problems. The connection between mathematics and astronomy is a well known fact in the history of science. Sir James Jeans, the great British astronomer and mathematician, illustrated this by writing in 1930 that '. . . the Great Architect of the Universe now begins to appear as a pure mathematician.'

The mathematical theory of non-integrable Hamiltonian systems indeed followed directly from the gravitational n-body problem – a model of the most ancient and basic astronomical dynamical system. Dissipative chaos was first explicitly demonstrated in numerical calculations of simplistic models of thermal convection, a fluid-dynamical phenomenon having an obvious relevance to astrophysics. It is thus only natural to look for additional applications of dynamical system theory in astrophysics. Astrophysics is a relatively young science and many of its achievements have been made possible by modern advances in technology. The space programmes of the competing superpowers gave rise to enormous progress in electronics, enabling the development of modern observational instruments and powerful digital computers. Astronomers are today able to collect a wealth of data in virtually all the bands of the electromagnetic spectrum and store and analyse them effectively. The development of fast electronic computers enables the theoretical astrophysicist to solve numerically complicated equations and thus simulate fluid and magneto-fluid flows, many-body dynamics, radiative processes, etc., which are operative in astrophysical objects.

Numerical analysis and computer simulations have revolutionised our science. The results of extensive calculations yield stunning graphic presentations of complex astrophysical objects and phenomena. However, it is difficult to avoid the feeling that the tongue-in-cheek directive 'Don't think! Compute!' occasionally starts to be taken too seriously. In a growing part of astrophysical literature the adjective 'nonlinear' is often a shorthand for 'that must be calculated numerically'. This is in my view a symptom of the trendy 'common knowledge' that a full scale, brute-force computer simulation is the ultimate goal of a successful theoretical astrophysicist.

There is little doubt that detailed numerical simulations, which include virtually all the relevant physical processes, are indispensable in applied research and engineering. Indeed, in the research accompanying the design of an aeroplane, for example, accurate simulations are a must. Astrophysics, however, still belongs to *pure* science. Consequently, our goal as astrophysicists is to *understand* the physics of astronomical objects, and not only simulate them numerically as accurately as possible. Indeed, if we just aim at reproducing all the observational details of some incredibly complex *particular* astronomical object (as important and exciting as it may be), by conducting mammoth numerical simulations, we may often not even know what are the relevant physical questions to ask. The physics of some simple-looking 'earthly' objects, like a pile of sand or a glass of beer, not to mention parts of our own body, still defies a complete understanding. I hope that nobody seriously thinks that by just 'reproducing' them numerically one can solve the mysteries of the intricate processes occurring in them. In astrophysical research numerical simulations are often very useful, but likewise, they alone cannot be the *key*. As unorthodox as this view may seem (these days), we should not forget some of the lessons of the past. For example, the basics of stellar structure were understood well before we could fully simulate stars numerically. Conversely, every serious stellar evolution code has shown that a red giant forms out of a star that has exhausted its nuclear fuel at the core. Should we give up the efforts to understand it physically because it can be accurately simulated?

I hope that the reader has been convinced by the first part of this book, that the theory of nonlinear dynamics has in its arsenal powerful analytical and semi-analytical methods, which can shed considerable light on complex problems before, during and after they are attacked by full-scale numerics. As was stated in the Preface, the main purpose of this book is to acquaint the reader with these methods, and this was the aim of the first part of the book. In this (less extensive) second part we turn to some representative astrophysical applications.

The modern ideas of Hamiltonian and dissipative chaos and pattern theory have now gradually begun to find their way into astrophysical problems. We shall review here a few problems of this kind. More importantly, we shall try to convince the reader that dynamical systems and pattern theory may be useful in resolving some of the outstanding problems of astrophysics. A prerequisite of such an approach would be the abandonment of some seemingly deep-rooted prejudices of at least some of today's astrophysicists. For example, a good 'toy model' should be considered as an essential tool for theoretical research and not just as a sort of game. Simplified models, capturing just a few of the essentials, have always been one of the keys to progress in theoretical physics. Such models certainly cannot reproduce all the features of a complicated physical or astrophysical system, but the understanding one can acquire from their study is extremely valuable. They can be used, for example, in our assessment of past numerical results and in planning and developing future computer simulations, on the way to a more complete physical understanding.

Essentially all the subjects of current theoretical astrophysical research contain nonlinear dynamical systems. Many exciting phenomena are related to instabilities of various kinds. In most cases, the linear regime of these instabilities has already been addressed in a multitude of works, and it would be fair to say that this kind of approach should now be supplemented and extended to beyond the linear. Nonlinear dynamics, initiated by Poincaré's seminal work, is the science of nonlinear instability. It is only natural to use it in the study of the World, in which nothing seems to be stable forever.

The choice of the subjects presented in this part of the book is obviously biased by my own preferences. Among the topics dicussed will be astrophysical realisations of the (gravitational) n-body problem, of dissipative systems exhibiting irregular temporal variability, and of extended pattern-forming systems. In the concluding chapter we shall discuss some fluid dynamical processes in the dynamical system perspective. We have naturally selected processes whose relevance to astrophysical systems is obvious.

To keep the text readable I have tried to avoid giving it a review article appearance. Consequently, not everything that has been done on chaos and complexity in astrophysics is mentioned and only a rather small number of selected works are discussed and referenced. I repeat here my apology (first given in the introduction of the book) to the authors of many relevant publications, whose work is not mentioned in this book. This is by no means a statement about the importance of these contributions, but rather a reflection of my conceptions or misconceptions about what can be didactically useful.

9

Planetary, stellar and galactic dynamics

She: Have you heard it's in the stars,
          Next July we collide with Mars?
He: Well, did you evah!
          What a swell party this is.
    Cole Porter, *Well, Did You Evah?*

Poincaré's important (and surprising) discovery that classical Hamiltonian systems, the paradigm of determinism, may be non-integrable and exhibit a seemingly erratic behaviour, had its roots in one of the most fundamental astronomical problems. In his book, entitled *New Methods in Celestial Mechanics*, Poincaré laid the basis for a geometrical approach to differential equations and made a number of significant mathematical discoveries. One of the consequences of these findings, phrased in a language that is relevant to us here, is that the gravitational n-body problem is generally non-integrable already for $n = 3$. We shall start this chapter with a short historical note on this subject.

Beyond their historical significance, Poincaré's findings have obvious relevance to astrophysics, as some of the most basic astronomical systems are naturally modelled by a number of mutually attracting point masses. For example, the Sun–Earth–Moon system can be viewed as a three-body problem, and its investigation was actually the initial motivation for Poincaré's work. The dynamics of the constituents of the Solar System (another obvious example of the n-body problem), and the question of the stability of this system as a whole, have always naturally enjoyed widespread attention. The masses of the objects in this system are very unequal and it is only natural to neglect the very minor planets, the planet moons and other small bodies if one is interested in the dynamics of the major planets. One is thus left with $3 < n < 6$, say. The motion of the minor bodies in the field of the massive planets is of interest as well, and it can be approached by considering them as test particles in a time-varying gravitational potential. The second section of this chapter is devoted to the description of the possible role of chaos in shaping the Solar System. In particular, we shall discuss the distribution of minor bodies (e.g., asteroids) and the stability of the planets' orbits. Some other specific problems, like that of the dynamics of a planet orbiting the Sun and one of its moons, including perhaps the complicating fact that the moon and/or the planet have finite size, will be discussed as well. In the latter case chaos can, as we shall see, appear in the rotational dynamics of the bodies.

On distance scales larger than the size of the Solar System the gravitational n-body problem is obviously applicable as well. Star clusters of varying richness abound in the Galaxy and in other galaxies. The simplest systems of this kind are binary stars and it is quite reasonable that some of them are not primordial but rather were formed by captures. The (completely solvable) gravitational two-body problem does not allow such a capture, that is, a change of an unbound orbit into a bound one is impossible without invoking external (e.g., due to the presence of additional stars in the vicinity) or internal (tidal) interactions. We shall discuss, in the third section of this chapter, one recent work on how chaos can play a role in the evolution of some types of binaries, which were formed by tidal captures.

Star clusters (in which n may vary from a few thousand to millions), an entire galaxy (with $n \sim 10^{11}$), or clusters of galaxies are all examples of the gravitational n-body problem with $n \gg 1$. Such systems are very complex and cannot be treated by perturbation methods, which are a natural tool in the case when n is quite small and many of the masses are very small relative to a few primary ones. Still, one may try an approximate method based on a 'mean field' approach, casting the attraction of all the point masses in the system into an average smooth effective potential, in which test-particle orbits can be calculated. It is, however, not clear a priori that the cluster can always be expected to relax into some sort of equilibrium configuration. Thus even the full n-body gravitational problem, with all its complexity, has obvious astrophysical interest as well. In the last section of this chapter we shall briefly comment on the role of chaos in galactic dynamics.

Before electronic computers became available the investigation of the above mentioned problems had largely been based on perturbation or mean field methods. The difficulties already encountered in finding perturbative solutions for $n \sim$ *a few* culminated in the celebrated KAM theorem, described in the first part of this book. The breakdown of integrability and the appearance of chaos in Hamiltonian systems was finally understood in an exact way. Some practical problems, like the important question, mentioned above, regarding the stability of the Solar System, and of orbits in mean potentials, were however not fully resolved by this new theoretical knowledge. In the last two decades or so such problems have been studied extensively by numerical calculations on the fastest computers available. The success of these calculations has relied heavily on the theory of Hamiltonian dynamical systems. The theory determines what are the relevant effects to look for and which quantities to monitor during the calculation and how to deduce global stability and other properties from the numerical results.

Numerical calculations of the dynamics of the Solar System and its constituents, integrations of stellar orbits in a galaxy and extensive simulations of the gravitational n-body problem (modelling star and galaxy clusters) have become a major part of mainstream geophysical and astrophysical research. As outlined above we shall review in this chapter several examples of meaningful applications of the theory of nonlinear Hamiltonian dynamical systems to such problems.

9.1 The *n*-body problem – a historical note

Most astrophysical applications described in this chapter are derived from the gravitational n-body problem. We therefore start with a brief survey of this problem's turbulent history. The discussion is divided according to the two basic questions which have driven the research in the n-body problem – that of *integrability* and of *stability*.

Integrability

It is well known (see Chapter 6) that the two-body problem is exactly reducible to an equivalent one-body problem (the Kepler problem, in the case of gravitational interaction) and is thus integrable, because the energy and the angular momentum are constants of the motion. The complete solution of the Kepler problem was first given by Johann Bernoulli in 1710. The natural quest for a similar general analytical reduction procedure for the n-body problem, with $n > 2$, was then undertaken by most of the great mathematicians of the 18th and 19th centuries. The side benefit of these studies was a significant progress in mathematics (mainly in the theory of differential equations), but definite answers were found in only some particular, rather limited cases.

We start with a description of the role played by the three-body problem in the discovery that a Hamiltonian system may be non-integrable. The simplest three-body problem is the so-called *planar restricted* one, in which all the three point bodies move in the same plane and one of them has a very small mass, in the sense that it does not influence at all the motion of the other two. This problem was the starting point of Poincaré, who applied to it his new geometrical approach to dynamical systems (see Chapter 3) and discovered that it is in general *non-integrable*.

Since the two large bodies move without any influence of the third one, their motion can be solved completely analytically. In particular, it is advantageous to consider the simple case in which the two primary bodies move in *circular* orbits around their centre of mass. It it then quite easy to describe the third body's motion in the rotating frame, in which the primary bodies are stationary. In this frame the small body is assumed to move in a plane and consequently it constitutes a well-defined two-degree-of-freedom dynamical system. Therefore its state can be completely specified by its trajectory in a four-dimensional phase space. Taking into account the fact that the energy of the two primary bodies is a constant of the motion, we can conclude that so also must be the energy of the small body. Thus we understand that its motion in phase space must be confined to a three-dimensional energy (hyper-) surface.

Using his surface-of-section concept, Poincaré reduced the problem even further, i.e., into a two-dimensional mapping. He found that an unstable equilibrium (saddle) point may sometimes exist in the relevant Poincaré map. Moreover, Poincaré was able to demonstrate in some cases the existence of a *transversal intersection* between the stable and unstable manifolds of the above hyperbolic fixed point. Poincaré did not have a computer, and thus could not calculate a detailed graph of the resulting homoclinic tangle in this case. He could however study in more detail a significantly simpler system – the forced pendulum, which is also similar in this aspect to the forced Duffing system (2.15) in the conservative case ($\delta = 0$). For the pendulum system, whose phase-space structure near a hyperbolic point is very similar to that of the restricted planar three-body problem, Poincaré described the tangle in words as 'a figure formed by infinitely many intersections, each corresponding to a homoclinic solution'. He added that 'neither of the two curves (the stable and unstable manifolds) can ever cross itself again and thus it must bend on itself in a very complex manner in order to cross the other one infinitely many times'.

We have seen in Chapters 4 and 6 of this book that homoclinic or heteroclinic tangles, occurring in the Poincaré sections of dynamical systems, are among the primary indicators of chaos in dissipative as well as Hamiltonian dynamical systems. The work of Birkhoff and Smale provided the rigorous formulation of this observation, utilising the generic properties of the horseshoe mapping; but it is obvious that Poincaré was the first to see the amazing geometrical structure of chaos, when he was considering one of the simplest three-body problems. Poincaré's discovery explained why the efforts to solve the general three-body problem by perturbation techniques could never succeed. The small-divisors problem (see Chapter 6), encountered sometimes in the expansions, was not just a mere technical difficulty. It was a symptom of a fundamental problem resulting from non-integrability.

We know today that chaos is inherent in the gravitational n-body problem as long as $n > 2$. The implications of this fact to some particular astrophysical systems will be discussed in the remaining sections of this chapter. However before this, we shall proceed with the historical survey and deal with some questions regarding the stability of n-body systems, as this issue has been very extensively studied over the years.

Stability – collisions and singular escapes

The issue of the gravitational n-body problem's stability has a longer history than the question of its integrability. It is not difficult to understand why the stability of a particular system of this kind, the Solar System, has been among the primary issues on the agenda of mathematicians, physicists and astronomers of the past three centuries. However, until Poincaré's work was completed, chaotic behaviour resulting from non-integrability had not been anticipated at all.

In practical terms, we shall define that the Solar System is considered *stable* in this context, if there are no collisions among the bodies and no planet ever escapes from the system. In Chapter 4 we have given some precise mathematical definitions of the *stability of solutions* (fixed points, limit cycles) of a dynamical system. The above practical definition of the *stability of the system* is obviously weaker than the statement that all solutions are stable, in the sense given in Chapter 4.

The meaning of the term *escape* also requires further elaboration. An escape of a planet from the Solar System is an instability in the above sense, but it is important to point out that two distinct classes are possible thereof. A *regular* type of escape is a situation in which the escaping planet's distance from the rest of the bodies gradually increases to infinity as $t \to \infty$. In a second type of escape, which signals a true mathematical *singularity*, the escape occurs in *finite* time. This can obviously only happen if the escaping body acquires an ever increasing speed. The latter case is the more interesting mathematical question and as such it has been intensively studied. In 1904 the Swedish astronomer and mathematician, H. von Zeipel, proved that any solution of the n-body problem can only ever become singular if a collision occurs, or if the motion becomes unbounded (infinite distances) in *finite* time. In other words, the possible mathematical singularities in the solutions only include collisions (the force becomes infinite) or singular escapes.

Three of the greatest French mathematicians, Laplace, Lagrange and Poisson, were the first to quantitatively address the escape question (whether regular or singular) using perturbation methods. The essence of their studies, conducted close to the turn of the 18th century, was to consider the elliptic (but nearly circular) Keplerian orbits of the planets revolving around the Sun in one plane and perturb them slightly, taking into account the influence of the planets on each other. They used as small parameters the eccentricity of an orbit, its inclination to the reference plane, and the ratio of a perturbing planet mass to that of the Sun. The idea was to look at the approximation series for the semi-major axes, say, and examine their time dependence, after all fast oscillatory (and thus bounded) behaviour has been *averaged* out. If *secular* (monotonically growing) terms are found in such an expansion, the planet is bound to leave the system and the latter can thus be considered unstable. Since then similar approaches have been extensively used in other problems as well, these are known collectively as the *method of averaging*, in perturbation theory (see Nayfeh, 1973; where a whole chapter is devoted to this topic). The perturbative expansions were carried out quite extensively but no secular terms appeared. Rigorously speaking, this was not proof of stability; however, since authorities of such calibre could not find instabilities resulting in the escape of planets from the Solar System, there was a general feeling of confidence. Poincaré's recurrence theorem (see Chapter 6), which had its starting point in Poisson's work, further strengthened the impression that there is no danger of planetary escapes, in spite of the fact that it too could not rigorously guarantee the lack of escapes, nor did it exclude the possibility of collisions.

The singularity issue was significantly advanced in 1895 by the French mathematician Painlevé, who first proved that the only singularities in the three-body problem are collisions. We remark here, in parentheses, that since Poincaré's treatment of the three-body problem (in its planar restricted variant) the problem has been extensively studied, in its different limits, by some very prominent dynamical astronomers and mathematicians. These studies have greatly contributed to dynamical astronomy and to the general understanding of Hamiltonian chaos, however we are unable to discuss this subject in any detail here. The fact that an entire book has quite recently appeared on just the restricted problem (Hénon, 1997) attests to the richness of the subject. A very good summary of the three-body problem, as well as of most other issues discussed in this chapter, can be found in the recent, very comprehensive book of Contopoulos (2002) and we refer the interested reader to that book, where a comprehensive list of references on the subject can also be found.

Returning now to Painlevé we add that he could not arrive at an exact result for $n > 3$ but he formulated a conjecture. It stated that for $n > 3$, solutions with singularities that are *not* collisions *do exist*. This conjecture was based on a scenario in which repeated close encounters (pseudo-collisions) cause ever increasing departures (with an increasing speed) of one body from the system, i.e., a singular escape. Painlevé's conjecture has now been rigorously proven for $n > 4$. The proof was completed in 1992 by J. Xia, ending a century of efforts. The case of $n = 4$ still remains undecided.

A particular three-body system, the so called Pythagorean or Burreau's problem, has played an important role in the development of the proof of Panlevé's conjecture. We shall discuss this briefly and explain how its solution led to Xia's proof.

Burreau's problem (and more general few-body problems) and their astrophysical implications were reviewed by Valtonen and Mikkola (1991). The Pythagorean problem consists of three gravitating point masses with mass ratios 3:4:5, whose initial positions are at the vertices of a Pythagorean triangle (side length ratios 3:4:5). Each body is placed at the vertex opposite to the corresponding side of the triangle and in the simplest case it is released from rest. Burreau stated the problem in 1913 and gave some useful insights regarding its solution, but a full solution was first found by Szebehely and Peters in 1967 with the help of numerical integration. They expected to find a periodic solution but found other rather surprising behaviour instead. The final result was an escape of the lightest body from the other two, which then formed a binary. This escape is obviously not a singular one (the only singularities in the three-body problem are collisions), but the way in which it occurs has relevance to the Painlevé conjecture.

The evolution of the system proceeds through an early formation of a binary, consisting of the two heavy bodies, with the third body repeatedly and frequently disturbing it. The lightest body affects the binary in a series of close encounters with the binary members, and occasionally the binary membership is exchanged (e.g., bodies 1 and 3 form a binary while the intermediate mass body recedes somewhat further away). These frequent encounters are then terminated when the lightest body is thrown out into a wide orbit, where it spends a rather long time (several orbital times of the binary) without disturbing the motion of the two heavier bodies. Such an event (a formation of an *almost* isolated binary) is sometimes called *ejection*, but the ejected body is bound to return. The close encounter resulting from this return finally gives rise to an *escape*, i.e., the lightest body acquires a large enough speed so as to never return. By examining in detail the system's configuration immediately prior to an ejection or escape event, it becomes quite clear that the light body acquires its high speed as a result of a close triple approach. This is sometimes referred to as the *slingshot effect*, and is thought to be an important mechanism in the formation of some astronomical binaries. Since the gravitational three-body problem is generally non-integrable and has chaotic solutions, it is endowed with the SIC property and thus it is not clear a priori if the escape found in Burreau's problem is an idiosyncratic or a generic result in any three-body problem. In a number of works performed in 1974 and 1975, the mathematicians R. McGehee and J. Waldvogel proved that in any *planar* three-body problem, if the three bodies come closely enough to each other, one of them generally escapes from the system with an *arbitrarily* large speed. This property provides an idea on how one may try to construct a non-collisional singularity (singular escape) in the gravitational n-body problem (for $n > 3$), in accord with Painlevé's conjecture. If one considers a system of $n = 5$ bodies, say, it may be possible to choose initial conditions in such a way that the light body, escaping from a binary (in a three-body subsystem), may be flung back towards it by one (or both) of the additional bodies. Then if such close triple approaches may be repeated again and again, with a significant slingshot effect (appreciable velocity increase) each time, a singular escape may in principle be possible. Indeed, Xia's proof was based on the construction of a pair of separate binaries and a fifth lighter body being continuously flung back and forth between the two. In this way the light body is accelerated at each encounter with the binaries, so that its oscillations between them become 'wild' and its speed increases without bound in *finite time*, i.e., a singular escape occurs.

Two remarks should be made here. First, as we are dealing here with a mathematical problem (point masses with Newtonian dynamics) one should not be alarmed by the occurrence of diverging speeds. Second, non-collisional singularities occur in the problem only for rather contrived boundary conditions. Xia showed that the removal of the undesired initial conditions (which lead to regular behaviour or actual collisions) from the manifold of all possible initial conditions leaves out a *Cantor set*. Thus only if the initial conditions are within this Cantor set does the system undergo a singular escape. The point here is that extremely close encounters are required to create a singular escape, but collisions must be avoided.

Physical bodies cannot accelerate without bound (relativistic effects become dominant) and they have finite size, making collisions likely during very close encounters. Thus singular escapes should not be expected in astronomical systems. Nevertheless, Xia's findings and subsequent work by J. Gerver and others, as well as parallel related lines of research, culminating in the KAM theorem for nearly integrable general Hamiltonian systems (see Chapter 6), have greatly enriched the mathematical theory of Hamiltonian dynamical systems. In what follows we shall describe some important consequences of this theory to astronomical systems. Although the most intriguing practical question, that of the stability of the Solar System, has not yet been fully settled, it seems very proper to conclude this section with the words of the mathematician J. Moser (the M of KAM): 'Is the Solar System stable? Properly speaking, the answer is still unknown, and yet this question has led to very deep results which probably are more important than the answer to the original question'.

9.2 Chaotic dynamics in the Solar System

We have today theoretical evidence, based largely on series of gradually improving numerical studies, that chaos permeates the Solar System and can be found essentially everywhere one looks. These calculations have recently been supplemented by analytical and semi-analytical studies.

The earliest definite (theoretical and observational) finding of chaotic motion among the permanent bodies of the Solar System was in the rotational dynamics of a planetary moon (Hyperion). Chaotic orbits of the bodies in the Asteroid and Kuiper belts and the Oort cloud have also been theoretically proposed, suggesting that long-range predictions of the motion of asteroids and comets are practically impossible. Moreover, numerical calculations have also indicated the presence of chaos in the orbits of the planets themselves. Pluto's orbit has been shown to be chaotic, and in more recent studies it became possible to extend the numerical simulations of the Solar System in such a way that the orbits of the inner planets could be examined as well. The results of these studies are very interesting but quite intriguing, because they indicate that the time for the loss of predicability due to the exponential divergence of trajectories is significantly shorter than the age of the Solar System.

We are thus left with the question of whether the sheer presence of chaos in the dynamics of the Solar System means that it is unstable, and if yes, in what sense? Some answers, based on tools from the theory of Hamiltonian chaos, have recently begun to emerge. We shall discuss several of these topics and questions in this chapter. A number of comprehensive reviews on the subject exist in the literature, e.g., Duncan & Quinn (1993), Laskar (1996), Lissauer (1999), Lecar *et al.* (2001), and readers interested in fuller and more systematic expositions, or in a complete literature survey are referred to these. It is clear that the achievement of all of the above-mentioned results would have been impossible without the use of modern analytical and numerical tools which were devised for the treatment of chaotic dynamics. The notions of Liapunov time (the inverse of the largest Liapunov exponent when it is positive), KAM tori, resonance overlap, Arnold diffusion etc., which we have introduced in the first part of this book, appear frequently in studies of this kind.

Sun–planet–moon system as a three-body problem

Planets and their moons (and obviously the Sun as well) are not point masses. This aspect of the problem is particularly important for the planet–moon pair as their sizes may not be negligibly small in comparison with their separation. Still, even if one ignores the finite size of the planet in question (for example Earth) and its satellite (e.g., the Moon) and their spin, assuming that the mass of these bodies as well as that of the Sun (the most massive body of the three) is concentrated in their centres of mass, one is left with a 'pure' gravitational three-body problem which, as discussed above, is still far from trivial. We shall be focusing in this subsection on the Moon–Earth–Sun (MES) problem, using it as an example in reviewing some interesting aspects of its development in the three-body problem perspective. Its history and present status have recently been comprehensively reviewed by Gutzwiller (1998).

The natural way of approaching the problem is to separate it into the motion of the Earth–Moon pair centre of mass around the Sun, described by the position vector $\mathbf{R}(t)$ say, and that of the Moon relative to the Earth, given by $\mathbf{r}(t)$. Since R is always much larger than r (typically by a factor of ~ 400) it is sensible to expand the appropriate terms in the relevant equations of motion in the small parameter r/R. This amounts to the usual expansion of the gravitational potential in multipoles. The lowest order (above the direct interaction term) multipole contribution in these equations is the quadrupole term (the dipole cancels out in an appropriate choice of coordinates). It is not difficult to estimate the relative size of the quadrupole term in the potential in the Moon's (relative to the Earth) and Earth–Moon centre of mass (relative to the Sun) equations of motion. The numerical values are $\approx 5 \times 10^{-3}$ in the former equation of motion and less than 10^{-7} in the latter.

If that very small quadrupole term is neglected in the equation of motion for $\mathbf{R}(t)$ this becomes a decoupled two-body Kepler problem. Indeed, in essentially all analytical studies of the MES three-body problem this approximation has been made at the outset. The objective of these studies has thus been *to find the motion of the Moon relative to the Earth when the centre of mass of the Earth–Moon system is assumed to move in a fixed Kepler ellipse around the Sun*. Together with the assumption that all three bodies are considered as point masses, this has been referred to as *the main problem of lunar theory*. In the equation of motion of the Moon relative to the Earth the multipole contributions are small as well, but neglecting them altogether would defeat the object of the study. The problem of lunar motion consists of finding their effect, but the natural tool for tackling the problem analytically is obviously perturbation theory. We have already mentioned in Chapter 6 the monumental work of Delaunay, which brought together more than a century of perturbative calculations by Euler, d'Alembert, Laplace and others. The novelty in Delaunay's work was the use of canonical transformations in perturbation theory with the purpose of reducing the problem to action–angle variables. The amount of canonical transformations required to effect this reduction was very large (505 in all!) because Delaunay, who was aiming at a significant improvement in accuracy, relative to his predecessors, had to include a substantial number of terms in his expansions.

Approximately a decade after the publication of Delaunay's work a new idea, devised to circumvent the problem of slow convergence which forced Delaunay to go to high order, was proposed by the American mathematician and astronomer G. W. Hill. He realised that the actual trajectory of the Moon should be close to a *periodic* orbit (but not a Kepler ellipse) whose period has the correct lunar value. He then proceeded to calculate this 'variational orbit', as it is now called. Hill showed that the displacement from the variational orbit satisfies a parametrically driven oscillator equation of the type that is now named after him. He discovered quite ingenious ways to solve his equation and was thus able to obtain the lunar orbit to a very high degree of accuracy. Hill's theory can be given a geometric interpretation, in which the possible orbits constitute a flow in a six-dimensional phase space. Since there exists one constant of motion (the Jacobian integral, a quantity related to the energy) the flow proceeds in a five-dimensional subspace. This description is quite complete as long as the actual orbit is close enough to the appropriate variational orbit. Ideas of this kind were used by Poincaré, more than 15 years after Hill, as a starting point for his new approach to mechanics and the theory of differential equations, but Poincaré was concerned with the problem of lunar motion per se as well. In 1908 he wrote a detailed paper on the small divisors problem in the theory of the Moon, a work in which he examined some important mathematical issues related to perturbative expansions of the Delaunay type. He concluded that expansions in the small parameter (see above) diverge if one pushes them too far. Terms with what he called *analytically very small divisors* (having *negative* powers of the small parameter) are bound to appear above certain order! Remarkably, Delaunay's results are still accurate; this is so precisely because he had to cut off his expansions rather early. This is reminiscent of asymptotic expansions (for a good introduction on this subject see Bender & Orszag, 1999 or Nayfeh, 1973), but the situation here is considerably worse as the coefficients of the divergent terms are *not* small.

Another American scientist, E. W. Brown, expanded Hill's work into a complete and accurate description of the motion of the Moon. In the resulting detailed lunar ephemiris, published in 1919, the main problem of lunar theory has essentially been completely solved for all practical purposes (long before the advent of electronic computers). There exists, however, quite a long list of influencing factors which are outside the strict 'main problem' or even the three-body problem. These include, to mention just the most obvious two, the Moon's secular acceleration due to the transfer of angular momentum (via tides) from the Earth's rotation to the Moon, and the perturbative action of the other planets on the Earth's orbit. Such effects are not easy to account for and are most efficiently addressed by combining phenomenological and numerical approaches. The astronauts of the Apollo missions left behind them on the surface of the Moon retro-reflectors that were used to achieve an astounding accuracy in distance measurements in the so-called Laser Lunar Ranging experiment. These data, together with new semi-analytical studies and full numerical integrations performed on advanced computers, have finally given rise to the modern ephemerides of the moon. These have replaced Brown's tables (and their improvements). The Moon–Earth–Sun problem is no longer on the agenda for theorists of Hamiltonian dynamical systems, but it is useful to remember the lesson we have learned from it.

It seems that the mass ratios and the present values of the relevant dynamical variables (which can be regarded as initial conditions) of the MES problem render a dramatic event, like the Moon's ejection or its collision with the Earth, rather unlikely. This is despite the fact that initial conditions leading to such an event are probably dense everywhere in phase space and constitute a fractal set (albeit of low dimension). Even when restricting ourselves to the pure main problem of lunar theory we note that the Moon–Earth 'planet' lives, as we have explained, in a five-dimensional slice of phase space. Allowing now for lunar motion that is not necessarily restricted to the ecliptic leaves out three dimensions for the Moon's invariant tori. Thus a possible chaotic trajectory is not trapped between the tori and Arnold diffusion can take place. This always takes a very long time but irregularities in the Moon's orbit, even if they do not involve collisions or escapes, cannot be ruled out. The finite size of both the Moon and the Earth and their rotation further complicate the problem and we shall now address this issue, in the general context of the rotational dynamics of planets and their satellites (i.e., moons). We have decided to discuss this problem in considerable detail because it yielded the first theoretical prediction of chaotic dynamics in the Solar System that was actually observationally verified.

Chaotic variations in the spin and obliquity of planets and moons

The planets of the Solar System and their moons are obviously not point masses and this fact is important if one is interested in their rotational dynamics. We shall refer to the rotation around the axis as *spin* and to the angle between the rotating object's equatorial and orbital planes as *obliquity*. From elementary rigid body dynamics it is known that when such a rotating body is placed in a uniform gravitational field (a 'heavy top'), its axis of rotation does not remain fixed in space and is generally subject to periodic motions – *precession* and *nutation*. The situation becomes significantly more complex if the gravitational field, in which the rotating body resides, is not constant. The fact that a planet or a moon is not perfectly rigid (some are even partly or entirely composed of fluid) and that some moons have quite irregular shapes, further complicates their rotational dynamics.

The conventional wisdom, regarding the evolution of the orbit, spin and obliquity of a satellite during its orbital motion around a central body, has been based on Sir George Darwin's work of over a hundred years ago. The basic physical mechanism invoked in Darwin's theory (which is an equilibrium theory) involves the tidal force, which distorts the bodies (notably on their surface). A pair of tidal bulges on the central body, resulting from this distortion, should point towards the satellite, but if the central body's rotational period is different from the satellite's orbital one (as is usually the case), the central body 'rotates through the tidal bulges'. As a result, energy dissipation due to internal friction in the central object causes a lag in the tidal bulges, and their symmetry axis is not aligned with the direction to the satellite's centre. If the satellite's orbital angular velocity is *smaller* than the spin of the central object, the lagging tide is in fact *ahead* of the satellite. In any case a nonzero lag angle allows for a nonzero tidal torque between the bodies. The main effect of this torque is angular momentum transfer from the central object's spin to the satellite's orbital motion (a spin–orbit coupling). The satellite itself may also be distorted by the tidal force, and frictional processes within it can give rise to changes in the satellite's spin angular momentum as well. These processes, which take place in both the central object and its satellite, are usually referred to as *tidal friction* or *dissipation* and the changes they induce in the rotational dynamics of the system are called *tidal evolution*.

A detailed description of the developments in the theory on this subject (up to the late 1960s) can be found in the review of Goldreich and Peale (1968). We shall omit most of the details of this theory and summarise here only some of the basic findings that will be relevant to our discussion. For the sake of simplicity we shall often refer to the satellite as a *moon* and to the central body as a *planet*, although the theory is also applicable to the case of a planet orbiting a star. The case of close binary stars, in which tidal evolution is also very important, is quite different from our case here, since the two stars usually have comparable masses and sometimes comparable sizes as well.

An important factor in the prediction of the consequences of tidal evolution is clearly its timescale. It depends on several factors and the strength of the tidal force (decreasing with the separation r as $1/r^3$) is one of them, but the dissipational processes, whose details are rather complicated, are important as well. Thus the timescale of tidal evolution is usually determined phenomenologically. In general, Darwin's theory predicts that, for a given fixed moon spin value, the obliquity should change in time and tend to an equilibrium value, between 0 and $\pi/2$ radians. The equilibrium obliquity value is monotonic with spin and consequently, if the spin is slowly decreasing by tidal friction, the obliquity should also decrease. For example, for a spin value of twice the mean orbital period (if the orbit is assumed to be fixed) the equilibrium obliquity approaches zero, and it remains so for smaller spin values as well. It is intuitively clear that tidal friction should cause the spin of the moon to decline steadily (a process called *despinning*) until its period becomes equal to the orbital one, that is, until one face of the moon is always pointing towards the planet (as is the case with our moon). This situation is called *synchronisation* or *synchronous lock* and the obliquity in it is zero. Clearly, the synchronisation time-scale for the Earth–Moon system is shorter than this system's lifetime.

A large majority of planetary moons in the Solar System fall into one of two well-defined categories. The first one includes moons whose spin has significantly evolved during their lifetimes and they are essentially in a synchronous lock. The spins of the moons that belong to the other category essentially have their primordial values. When examining the major planets it is apparent that none of them is in a synchronous lock and thus they seem to be satellites of the second category. Mercury however, the closest planet and the one whose synchronisation time is the shortest, is special. It appears to be locked in a 3/2 spin–orbit resonance, that is, it spins three times in every two orbital periods. Thus, the expectation that the de-spinning of a satellite necessarily continues until a state of a synchronous lock and zero obliquity is achieved, appears not to be realistic.

The case of Mercury prompted Goldreich and Peale (1966) to extend Darwin's work into a more complete spin–orbit coupling theory. They showed that if the figure of a moon is not too far from being spherical, and if its orbit has nonzero eccentricity, resonant lock states, having a spin rate of an integer multiple of half the mean orbital period, are dynamically stable. In this study they assumed that the orbit is fixed, which is reasonable because the spin changes of the *satellite* should have only negligible effects on the orbit. Goldreich and Peale also showed that in many cases such resonant states are stable against any further tidal evolution, and moreover they provided a way to estimate the probability of capture of a de-spinning satellite into a resonant state. In subsequent work Peale showed that the capture probabilities are very small for all natural satellites in the Solar System whose despinning time-scale is shorter than the age of the system. Indeed, all the *tidally evolved* satellites, whose rotation state was known at the time, were in a synchronous lock. Mercury, which inspired the whole theory, was the only exception.

So far we have not discussed any aspect of the problem that is explicitly related to our topic here, but if one examines the central differential equation of Goldreich and Peale (1966), hereafter referred to as GP, it is easy to recognise that it has the form of a driven nonlinear oscillator. Armed with the knowledge acquired in the first part of this book, where several nonlinear driven oscillators were discussed, we may at least suspect that chaotic solutions of this equation are possible for some parameter values. We thus turn to the discussion of this equation and the possibility of its having chaotic solutions.

Let I_A, I_B and I_C be the principal moments of inertia of a moon, with the last one being about the spin axis. The tacit assumption that the spin axis coincides with one of the principal axes can be justified by a symmetry argument. Assume that the moon orbits a planet in a *fixed* elliptical trajectory, specified by its standard parameters (a, the semi-major axis and e, the eccentricity). The position on the orbit is determined by the true anomaly ϕ, that is, the angle between the line joining the centres of the two bodies (this distance is denoted by r) and a fixed direction in an inertial frame (e.g., that of the periapsis). Obviously, both ϕ and r are *known* periodic functions of time.

To simplify the problem even more, assume now that the moon's *spin axis is perpendicular to the orbital plane*, that is, the obliquity is fixed at the value of zero. In the standard picture of tidal evolution (see above) the obliquity is driven to zero, therefore it is only natural to consider this reduced problem. Now let the orientation (with respect to a fixed direction in an inertial frame) of the principal axis having the smallest moment of inertia (A, say) be denoted by θ (a function of time). Using Euler's equations of the theory of rigid body motion one can then find the evolution equation for $\theta(t)$ to be

$$\frac{1}{n^2}\frac{d^2\theta}{dt^2} + \frac{\eta^2}{2}\left(\frac{a}{r}\right)^3 \sin[2(\theta - \phi)] = 0 \qquad (9.1)$$

$n \equiv \sqrt{GM_p/a^3}$ is the orbital *mean motion*, where M_p is the central object's (here a planet) mass and $\eta \equiv \sqrt{3(I_B - I_A)/I_C}$. The assumptions given above, together with the ordering $I_A < I_B < I_C$, indeed simplify the geometry of the problem so that the tidal torque can be written in a rather simple way and equation (9.1) results (see any standard text on celestial mechanics, e.g., Danby, 1988).

In the language of GP equation (9.1) is 'insoluble'. It describes a nonlinear oscillator subject to a periodic parametric forcing through the known functions $r(t)$ and $\phi(t)$, and indeed such problems are in general non-integrable. Since GP were interested in the above mentioned case of Mercury, they decided to focus only on the case in which the spin angular velocity $\dot{\theta}$ is close to being resonant with the mean orbital motion, that is, $\dot{\theta} \approx pn$, where p is a half-integer. Rewriting the equation of motion in terms of a resonance variable (sometimes called the *detuning parameter*), defined as $\gamma \equiv \theta - nt$, it is possible to simplify significantly the equations by eliminating the nonresonant high frequency contributions through the perturbative technique of averaging over the orbit (see also Nayfeh, 1973; where a detuning parameter was used for the same purpose but within the method of multiple scales). In this way GP were able to reach the conclusions mentioned above about the existence and stability of resonant lock states.

The averaging procedure is feasible since, by definition, $\dot{\gamma} \ll n$, but it should also be borne in mind that the 'strength' of each resonance depends on e, the orbital eccentricity and on the value of η (the extent of the departure from sphericity of the satellite's figure). In particular, if $(I_B - I_A)/I_C \ll 1$, the different resonances are well separated in phase space and there is no danger of resonance overlap. We have seen in Chapter 6 that resonance overlap causes chaos in nearly integrable systems (Chirikov's overlap criterion). Thus for a moon, which is in an eccentric orbit (see below) and has a sufficiently aspherical figure, Equation (9.1) may have chaotic solutions. If this is the case, averaging is not able to provide a good approximation of the solution.

Until 1981 all the available observational data on the spin of natural satellites in the Solar System seemed to indicate regular behaviour. The surprise came when Voyager 2 closed in on Saturn and photographed, among other important things, the inconspicuous small Saturnian moon Hyperion. Hyperion's orbit, residing well outside the Saturnian rings, is quite eccentric and its orbital period is approximately 21 days. The Voyager images indicated that Hyperion is unusually flat and its spin behaviour is quite strange. The spin period was established to be around 13 days and from examining Hyperion's figure it was found that $(I_B - I_A)/I_C \approx 0.26$. Moreover, it appeared that Hyperion was spinning neither around its longest axis nor around the shortest one, as could be expected (an elementary exercise in rigid body mechanics shows that rotation around any other direction is unstable).

Wisdom, Peale and Mignard (1984) analysed theoretically the strange case of Hyperion in a remarkable paper which is an exemplary application of nonlinear dynamics and chaos theory to astronomy. They examined the same equation (9.1) as GP but, instead of employing analytical approximation methods (like averaging), they approached the problem numerically. Equation (9.1) is a second-order *nonautonomous* ODE and consequently a meaningful and simple tool to study its properties is the Poincaré surface of section. Wisdom *et al.* (1984), hereafter WPM, chose to construct the section by marking the value of $\dot{\theta}$ and θ, obtained from a numerical integration of (9.1), once per every orbital period at periapse passage (this choice is arbitrary and any other definite orbital phase is equally as good). Such a stroboscopic plot should create a faithful surface of section because of the periodicity of $r(t)$ and $\phi(t)$.

For the sake of clarity, one can rescale the time variable by the period of the mean orbital motion, and this is trivially done by the formal replacement $t \mapsto nt$. The orbital period is then simply 2π. It is then easy to see that equation (9.1) is a Hamiltonian dynamical system, whose time-dependent Hamiltonian (in these units) is

$$H(x, p, t) = \frac{1}{2}p^2 + \frac{1}{4}\eta^2 \left[\frac{a}{r(t)}\right]^3 \cos 2[x - \phi(t)] \qquad (9.2)$$

where the conjugate variables are $x \equiv \theta$ and $p \equiv \dot{\theta}$. It is also useful to note that, because of the symmetry of the inertia ellipsoid, θ and $\theta + \pi$ correspond to dynamically equivalent configurations and it is therefore sufficient to examine the stroboscopic surface of section plot only in the strip $0 < \theta < \pi$. Wisdom and co-workers fixed the value of the eccentricity at $e = 0.1$ (the known value for Hyperion's orbit) and performed numerical integrations of (9.1) for different values of the asphericity parameter η. In each case they examined many solutions

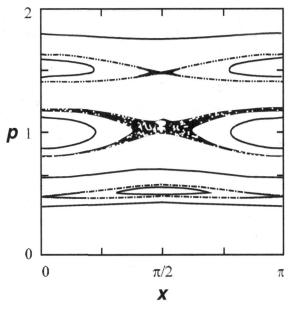

Fig. 9.1. Surface of section generated from numerical solutions of the Hamiltonian system resulting from (9.2) for $e = 0.1$ (an appropriate value for Hyperion's
orbit) and $\eta = 0.2$ (significantly smaller than Hyperion's asphericity). Quasiperiodic rotation and libration are seen here together with chaotic layers, which are
well separated. Reprinted, with permission, from Wisdom *et al.* (1984), Elsevier.

(trajectories), resulting from different initial conditions. The result for $\eta = 0.2$
(well below the appropriate value for Hyperion's out-of-round figure) is shown in
Figure 9.1.

It can be seen that orbits corresponding to quasiperiodic rotation and closed
librational orbits surround not just one point, but they appear around each of the
resonances ($p = 1$, $p = 1/2$ and $p = 3/2$ are shown here). More importantly,
even for quite a modest asphericity ($\eta = 0.2$) chaotic layers are present around the
librational trajectories. These layers are well separated by the regular trajectories
corresponding to quasiperiodic rotation.

As η is increased the appearance of the surface of section becomes more complex. The chaotic zones surrounding adjacent resonances overlap, creating a large
domain of irregularity. When WPM took $e = 0.1$ and $\eta = 0.89$ (parameter values
which are appropriate for Hyperion) they obtained the surface of section displayed
in Figure 9.2. As can be seen from this section the phase space near synchronous
rotation ($p = \dot{\theta} = 1$) is dominated by an extended chaotic zone, resulting from
the overlap of resonances $p = 1/2$ through $p = 2$. Several islands (remnants of
the regular solutions) can be identified in the 'chaotic sea'. These characteristics of
the surface of section are reminiscent, of course, of the transition to chaos in the

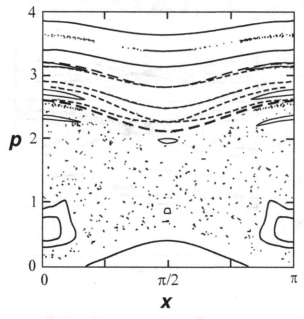

Fig. 9.2. Surface of section generated from numerical solutions of the Hamiltonian system resulting from (9.2) for $e = 0.1$ and $\eta = 0.89$ (both appropriate for the case of Hyperion). A dominant large chaotic zone is seen in the central region. Reprinted, with permission, from Wisdom *et al.* (1984), Elsevier.

generic case discussed in Section 6.4, where we introduced the Chirikov resonance overlap criterion (see Figure 6.3).

Note that for $e = 0$ (zero eccentricity) the problem is integrable (and therefore chaos is ruled out) irrespective of the value of η. This can easily be seen by writing $r = a$ and $\phi = t$ (circular orbit) and formally setting $n = 1$ (see above) in Equation (9.1). The equation can then be cast into the form

$$\ddot{\vartheta} + \eta^2 \sin \vartheta = 0 \tag{9.3}$$

where we have performed a change of variables $\vartheta \equiv 2(\theta - t)$. This equation is identical with that of the unforced pendulum problem (defined in Chapter 2), and the solutions consist of closed trajectories surrounding the origin in the $\dot{\vartheta} - \vartheta$ plane (libration) and open trajectories (rotation). There is no irregular motion as the equation has the obvious integral $I \equiv \frac{1}{2}\dot{\vartheta}^2 - \eta^2 \cos \vartheta$. The surface of section in the original $\dot{\theta}-\theta$ plane can thus be found easily.

These numerical results of WPM support their earlier analytical estimates, which were based on the Chirikov resonance overlap criterion, and the claim that the averaging technique, used by GP for Mercury, is not adequate for Hyperion (η is too large). Moreover, WPM showed that the state in which Hyperion's spin axis is perpendicular to the orbital plane (a crucial assumption in the derivation of the dynamical system (9.1), is unstable. This *attitude* instability was shown to be present in the chaotic zone and in the synchronous spin–orbit resonance, and in other resonances as well. Only the $p = 2$ and $p = 9/4$ states were found to be attitude stable. To find these instabilities it is necessary to consider the fully three-dimensional problem of a triaxial ellipsoid whose centre of mass moves in a fixed elliptical orbit, that is to use all of the three relevant Euler's equations. We shall skip here the details of this analysis and refer the interested reader to the original paper of Wisdom *et al.* (1984).

To obtain quantitative support for the results indicated by the surface of section, Wisdom and his collaborators calculated the Liapunov exponents (see Chapter 3), and showed that the motion within the region in which chaos is indicated by the surface of section plot is indeed fully chaotic (positive Liapunov exponents). The chaotic variations in the obliquity are rather large, with the spin-axis orientation changing by more than $\pi/2$ radians from its original, perpendicular to the orbital plane, direction. Remarkably, the timescale for this chaotic *tumbling* was found to be of the order of only several orbital periods! These results definitely indicate that during the synchronisation process, driven by tidal dissipation, Hyperion must enter the chaotic zone and tumble in an irregular manner. The probability of capture into one of the above mentioned attitude stable resonant states is extremely small, but it is possible that the tumbling moon may occasionally come close enough to these states and linger there for some time, with its spin axis having an obliquity close to zero. Ultimately (but most probably on a timescale longer than the age of the Solar System) Hyperion should be captured into one of these stable resonances.

Judging from their numerical experiments WPM concluded that Hyperion is expected to be found in a chaotically tumbling state, and this expectation was not only consistent with the early Voyager 2 data, but was also actually confirmed by later, more detailed and careful, observational studies. This discovery prompted Wisdom to examine the rotational dynamics of other irregularly shaped natural satellites as well. The width (in phase space) of the libration zones, corresponding to the resonances, was found in the original study to be proportional to the asphericity parameter η and to some power of the eccentricity. Thus extended chaotic zones, resulting from resonance overlap, could be expected only for moons having significantly out-of-round figures and eccentric orbits.

Wisdom (1987) thus considered all the other natural satellites in the Solar System that are known to be significantly out-of-round (including the notorious Martian moons Phobos and Deimos). The orbital eccentricity of all these moons is rather low and in all the cases in which their rotation is known, they are found in a state of attitude-stable synchronous rotation. Because of the small orbital eccentricity the chaotic layers in their surfaces of section in the $\dot{\theta}$–θ plane were found to be rather limited and (for a fixed zero obliquity case) there is no widespread chaotic zone as in the case of Hyperion. However, if one considers the rotational history of these moons, it is reasonable that during their tidal evolution, as they despin towards a synchronous lock, they are bound to enter the narrow chaotic zone surrounding the spin–orbit lock resonance. As in Hyperion's case, this chaotic zone is attitude unstable and the satellites residing in it should be subject to a period of chaotic tumbling. Moons whose orbit eccentricity is small do not suffer from very large deviations of their obliquity during the chaotic tumbling phase, however, complete rotations about the long axis may give rise to enhanced energy dissipation and thus affect the orbital evolution.

The next natural step in the research on the possibility of chaotic rotational motion of planetary moons is to consider the rotational dynamics of *planets* orbiting the Sun. It is quite obvious that the simplistic formulation (9.1) based on the Hamiltonian (9.2), while being sufficient for finding chaotic solutions in the rotational dynamics of irregularly shaped planetary satellites, cannot give rise to such behaviour in the planet–Sun case. That formulation was based on the approximation that the satellite orbits the central body in a fixed Kepler ellipse and this gives rise to a *periodic* forcing term in the relevant equations of motion for the rotational dynamics of the satellite. We have seen that if the asphericity parameter η is not large enough, chaotic behaviour of the obliquity is very improbable. In a typical planet $I_A \approx I_B$ and therefore $\eta \approx 0$ and thus chaotic changes in the obliquity, based on the same mechanism as in the irregularly shaped moons, cannot be expected. However, complications may arise if the orbits themselves are sufficiently complex.

Since the late 1980s a growing body of evidence on chaotic behaviour in the orbits of the outer planets has begun to accumulate, based on numerical experiments performed on rapidly improving computers. In the early 1990s studies of this general kind were also extended to the inner planets of the Solar System (and the Earth among them), and have indicated that chaotic behaviour in their *orbital* dynamics can be expected as well. We shall discuss these findings in more detail later on in this section. This result has however immediate implications for the rotational dynamics of the planets, which is the problem we are now considering. The appropriate Hamiltonian system for this problem is considerably more complicated than the one used by WPM for a moon's rotation, since the driving term, resulting from the orbital motion, is not periodic and not even *quasiperiodic*. Thus it is the complicated orbit (not just a fixed Kepler ellipse) and the resulting time dependence of the Hamiltonian (the driving term) that can be the source of complexity in this case. Laskar and Robutel (1993) considered the problem in detail and derived an appropriate model Hamiltonian dynamical system. They assumed that the principal moments of inertia of a planet orbiting the Sun satisfy $I_A = I_B < I_C$, with C being the spin axis, and proceeded to formulate the rotational dynamics of such a planet in terms of action–angle variables. Since the derivation of the full equations of precessional motion or the Hamiltonian of this problem is somewhat lengthy and technical, we refer the reader to the original paper for details.

To see the essence of the dynamics it is however not necessary to consider all the terms that finally went into the full numerical calculations. We shall thus concentrate here on only those physical ingredients that are essential for the understanding of how chaotic behaviour may appear in this system. As a first step we have to note that the orientation in space of a planet's spin axis can be fully determined by the instantaneous value of two variables, its *obliquity* (see above) and the *precession* angle. The knowledge of these two variables is equivalent to the determination of the motion of the planet's equatorial and orbital (ecliptic) poles. To be specific we shall formulate here the problem for the Earth with the tidal influence of the Moon included, in addition to that of the Sun. The perturbations due to the other planets are taken into account also, via their influence on the Earth's orbit; this can obviously be applied to other planets as well. Let $\epsilon(t)$ be the obliquity of the Earth and $\psi(t)$ its precession angle. Neglecting a number of small terms, including the ones arising from the eccentricity of the Earth and Moon orbits and of the inclination of the Moon, the Hamiltonian of the precessional motion reduces to

$$H(\psi, X, t) = \frac{1}{2}\alpha X^2 + \sqrt{1 - X^2}[A(t)\sin\psi + B(t)\cos\psi] \qquad (9.4)$$

where $X \equiv \cos\epsilon$ is the action variable associated with the angle ψ and the *precession constant*, α, is defined by

$$\alpha = \frac{3}{2}\frac{I_C - I_A}{I_C}\frac{1}{\omega_s}(n_s^2 M_s + n_m^2 M_m)$$

where ω_s is the Earth's spin angular velocity, n_s and n_m are the Sun's and Moon's mean motion *around the Earth* and M_s, M_m their masses, respectively. The functions $A(t)$ and $B(t)$ contain the information about the exact orbit of the Earth and render the dynamical system non-autonomous. If one is interested in the rotational dynamics of a moonless planet, the above Hamiltonian can also be used, but without the moon term in α.

The functions $A(t)$ and $B(t)$ have to be provided from numerical calculations of the Solar System, like the ones performed by Laskar himself (see below), and as we shall see they exhibit chaotic behaviour over sufficiently long timescales. It turns out, however, that a quasiperiodic approximation for these functions is already sufficient to allow for chaotic behaviour of the obliquity. Although the physical process causing chaos in the dynamical system governed by the Hamiltonian (9.4) is quite different from the case of the irregularly shaped moons, described by the Hamiltonian (9.2), the underlying mathematical structure of the emergence of chaotic behaviour is similar. This is not very surprising since (as we have seen in Chapter 6) one can expect that the onset of chaos in almost integrable Hamiltonian systems is always connected to the overlap of resonances.

With a quasiperiodic approximation for the forcing term we may write

$$A(t) + iB(t) \approx \sum_{k=1}^{N} \alpha_k e^{i(v_k t + \phi_k)} \tag{9.5}$$

where α_k, v_k and ϕ_k are constant coefficients, and thus the Hamiltonian (9.4) becomes

$$H(\psi, X, t) = \frac{1}{2}\alpha X^2 + \sqrt{1 - X^2} \sum_{k=1}^{N} \alpha_k \sin(\psi + v_k t + \phi_k) \tag{9.6}$$

It is possible to show that for $N = 1$ (periodic forcing) the above Hamiltonian is integrable, but for $N > 1$ integrability is lost and the resonance overlap criterion predicts the existence of chaotic zones in phase space.

Consider now the Hamiltonian (9.6) and assume for the moment and for the sake of simplicity that $|\alpha_k| \ll 1$ for all k. In this case, the second term of the Hamiltonian can be considered as a small perturbation and the first unperturbed part, $H_0 = \alpha X_0^2/2$, corresponds to a harmonic oscillator in the action–angle variables X_0 and ψ_0 (see Chapter 6) . The unperturbed problem is

$$\dot{X}_0 = -\frac{\partial H_0}{\partial \psi_0} = 0$$

$$\dot{\psi}_0 = -\frac{\partial H_0}{\partial X_0} = \alpha X_0$$

with the solution

$$X_0 = \text{const}; \qquad \psi_0 = \alpha X_0 t + \text{const}$$

giving this unperturbed oscillator's frequency as αX_0. Now, if the perturbation of the Hamiltonian contains in its quasiperiodic series a term whose frequency, v_j say, satisfies $v_j \approx -\alpha X_0$, a resonance can be expected since this causes a nearly stationary argument in one of the sines in (9.6). The existence of a resonance is a necessary condition for resonance overlap which in turn gives rise to the onset of chaos. These arguments and estimates are rather heuristic but this way of reasoning can provide some clues as to what can be expected from the full numerical calculations.

The functions $A(t)$ and $B(t)$ follow from the full orbital solution and their physical meaning can be understood by examining the approximate expression they satisfy (see Laskar & Robutel, 1993)

$$A(t) + iB(t) \approx 2\frac{d\zeta}{dt}$$

with the variable $\zeta(t)$ defined by $\zeta \equiv \sin(i^*/2)\exp(i\Omega)$, where i^* and Ω are two standard orbital parameters, the inclination and the longitude of the node, respectively.

The mean value of $\alpha X_0 = \alpha\cos\epsilon_0$ for the Earth has been estimated to be $50.47''/\text{yr}$ and one should therefore look for frequencies close to this in the appropriate quasi-periodic approximation (9.5). Using the frequency analysis method (essentially a Fourier decomposition over a finite time) Laskar found such a frequency, $\nu_{13} = -50.302''/\text{yr}$ (that of the 13th term in an order determined by the size of the amplitudes). This term arises from a combination of Jupiter's and Saturn's perturbation on the Earth's orbit. The Earth is however rather far from the main relevant planetary resonances, given by the leading terms in the quasiperiodic series.

Laskar and Robutel (1993), hereafter LR, numerically integrated the full equations of precessional motion for the inner planets. In an accompanying study (with Joutel) they investigated the influence of the Moon on their results for the Earth. The orbital solutions and analysis of Laskar (1989, 1990), which will be described in the last part of this section, were used as input. The integrations were done over $\approx 18\text{Myr}$ and were repeated for a variety of combinations of the two relevant parameters – the initial obliquity ϵ_0 and the precessional constant α.

In what follows we shall briefly summarise the results for Mercury, Venus, Mars and Earth. The results for Earth will then be further elaborated, examining the influence of the Moon's presence on them. However, before doing this we would like to explain the essence of the technique employed by LR to analyse their results. LR used the method of *frequency analysis* to look for chaotic behaviour in their numerical results. This method was originally proposed by Laskar in 1990 as a diagnostic tool for identifying chaos in the numerical calculations of planetary orbits (see in the last part of this section). It is well suited for Hamiltonian dynamical systems having many degrees of freedom, a case in which the usual Poincaré surface of section method cannot work well. Laskar's frequency analysis is however only applicable to systems which are close to integrable ones, i.e., if the motion is close to being quasi-periodic and the non-integrability or chaos is reflected by slow diffusion of trajectories away from the invariant tori (see Chapter 6). In such systems the Liapunov time is very long compared to the typical periods (the inverse of the frequencies) in the problem.

The essence of the method is the computation of the dominant n Fourier frequencies of the time-series in question over a *finite* time interval. These frequencies are then regarded as variables of a state space, the full time span of a computation is split into a sequence of many finite (partially overlapping) time intervals, and the frequencies are computed over each interval. If the original system is integrable, the corresponding frequencies should retain the same values for all intervals and thus reflect the true fixed frequencies of the motion on the invariant tori. If, however, the frequencies are found to slowly change in time (i.e., from interval to interval), the system is diagnosed as being non-integrable. A number of specific problems, such as computational accuracy, variability timescales and so on, must be carefully examined, though, to ensure the reliability of the frequency analysis method (see Laskar, 1993)

The precessional motion problem is close to being integrable. One can be convinced of this by examining the Hamiltonian (9.4) and noting, as we have mentioned above, that this Hamiltonian is strictly integrable if $A(t) + i B(t)$ is periodic. It turns out (see below in the last part of this section) that $A(t) + i B(t)$ for the inner planets is very close to being periodic and the deviations thereof, i.e., the terms with $k > 1$ in the series approximation (9.5), are all smaller by at least a factor of a million than the $k = 1$ term. As we have noted above a resonance in the system described by the Hamiltonian (9.6) is feasible, in the case of the Earth, if at least 13 terms in the quasiperiodic approximation of the forcing term (9.5) are retained, and the situation is similar for the other inner planets as well. Thus we have here a highly multi-dimensional, almost integrable, Hamiltonian dynamical system and therefore the frequency analysis method is naturally suited to this problem.

LR found the values of initial obliquity (ϵ_0) and precession constant (α) for which chaotic behaviour was indicated by frequency analysis of the numerical integrations, and delineated the zones (in the ϵ_0–α plane) in which 'large-scale' chaotic behaviour of the obliquity, that is, irregular variations over a substantial interval of ϵ, could be expected. We refer the reader to the original article of Laskar and Robutel (1993) where accurate, computer-generated graphs can be found, but will attempt also to describe here some important findings of this work. The most striking result of these is the presence of significant chaotic zones in the ϵ_0–α plane for all the four inner planets. In these zones the obliquity of the planet's rotational axis is bound to vary irregularly, by chaotic diffusion, within a sizeable interval. This variability was monitored by LR over more than 30 Myr and the extent of the chaotic zones should probably be even larger for longer times. The timescale for chaotic obliquity variations is typically a few million years and thus it is much shorter, in most cases, than the tidal despinning time. Consequently, irregular variations of the obliquity in the chaotic zones can be expected with α remaining essentially constant.

We turn now to a summary of the LR results and their possible consequences for the individual inner planets. It is worth stressing that the effects of possible impacts of significantly sized bodies are not taken into account here.

- *Mercury*

  This planet is today apparently trapped in a 3/2 spin–orbit resonance and thus, since the orbital period of Mercury is ~90 days, it turns out that its spin falls well below the edge of the chaotic zone found by LR. However, while the Goldreich and Peale theory explains the stability of this state, it does not exclude the possibility that Mercury's spin underwent significant changes during its history by means of tidal friction. If the primordial rotation rate of Mercury was significantly higher (as some estimates indicate) then during its tidal despinning it must have passed through a large chaotic region, encompassing essentially all obliquity values for spin periods between 100 and 300 hours according to the LR results. It is thus reasonable to say that Mercury suffered large scale chaotic variations of its obliquity, between 0° and more than 90°, on a timescale of a few million years (in its present state Mercury's pole points towards the Sun).

- *Venus*

  The present rotational state of Venus (178° obliquity, i.e., a *retrograde* spin) is rather puzzling, and it is usually attributed to dissipational processes within the planet (like core–mantle interactions) which however cannot explain changes in the obliquity from less than 90° to the present value. The primordial obliquity of Venus is unknown and if it was over 90° (with the obvious implications for theories of planet formation) the planet may have avoided the chaotic zone altogether. However, if Venus was born with a small value of obliquity and a spin period of below 20 h (according to Goldreich and Soter the primordial spin of Venus is 13 h), it must have passed through the chaotic zone while tidally despinning. Thus its obliquity may have changed, within a few Myr, up to a value ~90° and the planet may have left the chaotic zone with this high obliquity value, evolving subsequently to its present state by means of the above mentioned internal processes.

- *Mars*

  The case of Mars seems to be the most clear-cut. Being rather distant from the Sun and having no large moons, its spin is generally considered as primordial. Taking $\alpha = 8.26''/\text{yr}$, the appropriate current value for Mars, LR results indicate that the chaotic zone encompasses all initial obliquity values between $\approx 0°$ and ~60°. It appears thus that the motion may be regular for some very small obliquity values. LR examined this case more closely by extending the frequency analysis over a larger time interval (45 Myr) and repeating the calculations with orbital data obtained from somewhat different initial conditions. The result was that the left boundary of the chaotic zone shifted all the way to 0° obliquity. It can thus be rather safely concluded that the obliquity of Mars is currently varying chaotically between 0° and ~60° on a timescale of a few million years, and it has probably always been varying in this manner. This should obviously have significant impact on the climate and distribution of water on Mars, and possibly its internal 'geology' as well.

• *Earth*

The results of the LR analysis for the Earth are generally not very different from the ones for Venus. In particular, for the present values of spin period ($\approx$24 h) and obliquity ($\approx 23°$) it appears that the Earth's obliquity should currently be in the process of chaotic variation between the values of nearly 0° and up to approximately 80°, on a timescale of a few million years! This should obviously have a significant influence on the history of the Earth's climate and therefore this result must be examined in more detail. As indicated above, these results for the Earth are based on calculations done *without including the Moon*. Laskar, Joutel and Robutel (1993) therefore performed additional detailed numerical calculations, including this time the effect of the Moon. With α fixed at its best known present value, they found that the presence of the Moon's torque has a significant effect in diminishing the size of the chaotic zone. For all initial positive obliquities up to $\approx$60° frequency analysis indicated a regular behaviour of the obliquity. Chaos thus appears to be confined to the initial obliquity segment $60° < \epsilon_0 < 90°$. These results are thus quite comforting. It appears that the planet we live on is presently in a stable rotational state with the obliquity varying in a regular manner and by no more that $\pm 1.3°$ around the mean value of 23.3°. The confinement of chaotic behaviour to only rather high obliquity values is clearly due to the presence of the Moon. It is thus reasonable to conclude that the presence of the Moon is largely responsible for regulating the Earth's climate. Without its presence the obliquity of the Earth would be chaotic, with very large variations over a timescale of several Myr. It turns out, however, that the presence of the Moon does not confine the chaotic region to high ϵ_0 for all values of α. Extensive numerical calculations indicate that it merely shifts the chaotic region in the α–ϵ_0 plane and so our planet will probably ultimately enter the chaotic zone, when its spin becomes significantly smaller (longer days) due to tidal dissipation. Fortunately these processes are very slow.

Before concluding our discussion of the present topic we briefly mention that, according to LR, the rotational dynamics of the outer giant planets is quite different from that of the inner ones. The planetary forcing term $A + iB$ for these planets has only rather few narrow and well separated peaks in its Fourier spectrum, as their orbital motion is very close to being regular (see below). Thus resonances with the precessional motion, governed by the Hamiltonian (9.6), are quite unlikely and their overlap even more so.

Minor bodies in the Solar System

We have already made some general statements about the possibility of chaos in planetary orbits, a subject which will be discussed in more detail in the last part of this section. The reason for delaying this discussion is that the dynamics of the full planetary system is the most difficult among the problems addressed in this section and the results remain, at the time of the writing of this book, still somewhat unclear. It is mainly so because, in this case, one must account for the gravitational interactions of a nontrivial number of planets with the Sun and between themselves.

A simpler case is that of the dynamics of a *test particle* that is, of a body of a very small mass (so that its own gravitational influence is ignored), moving in the Solar System, whose dynamics is assumed known and serves as an input to the test-particle computation. If only one planet (in addition to the Sun) is assumed to dominate the test-particle dynamics (and Jupiter is naturally picked to be this planet), we recognise it as the *restricted* three-body problem. We know that even this problem is non-trivial because it is in general non-integrable.

It has been known for centuries that small rocky bodies (asteroids) swarm the Solar System, the vast majority of them in what is known as the *asteroid belt*, lying between Mars and Jupiter. Their masses are too small to affect the dynamics of the Solar System's major constituents (therefore they may indeed be viewed as test particles). The orbits of over 50 000 of them are known and have been documented by the Minor Planet Center. As has been known since the pioneering studies of Kirkwood (in 1867) the distribution of the asteroids in the Solar System is not uniform. The histogram of the number of asteroids, plotted as a function of the semi-major axis of their elliptical orbits, shows a number of conspicuous gaps in the distribution (see Figure 9.3). Already Kirkwood himself noticed that the gaps in the asteroid distribution coincide with commensurability between the periods of the asteroid orbits and Jupiter's mean motion, but neither he nor his followers could give a full and satisfactory explanation as to why resonances of this kind give rise to the gaps.

Prompted by the work of Chirikov on non-integrability in Hamiltonian systems, and the transition to global stochasticity by means of resonance overlap (see Chapter 6), Wisdom began in 1982 to apply these ideas to the problem of asteroid motion. In what follows we shall describe in considerable detail Wisdom's early and inspiring work on the 3/1 gap, which is clearly the most dramatic one (see Figure 9.3). More recent developments will then be discussed only briefly.

The analysis of asteroid motion can be performed within the framework of a restricted three-body problem, because the approximation that the asteroid can be viewed as a test particle moving in the field of just two major bodies (Jupiter and Sun) seems to be very reasonable. This three-body problem is however not in its simplest variant as the main bodies must be considered moving in elliptical orbits. The analytical knowledge on this problem is very limited, and so the studies aimed at understanding the Kirkwood gaps by means of the restricted planar-elliptical three-body model have largely been based on numerical studies. In an effort to make such calculations tractable, given the limited power of early computers, averaging techniques were generally used. This kind of a perturbative approach dates

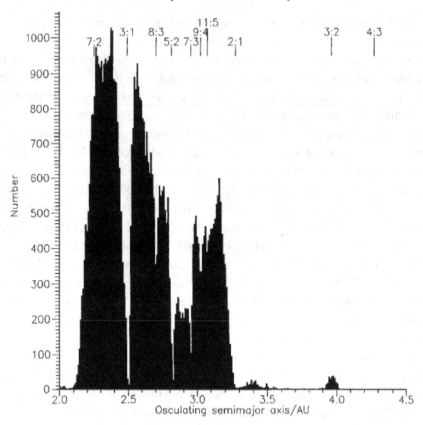

Fig. 9.3. The number of asteroids as a function of their semi-major axis (in units AU). The location of the Kirkwood gaps, marked by the corresponding resonances, is clearly apparent. Reprinted, with permission, from Lecar *et al.* (2001), ©*Annual Reviews*.

back to Lagrange and Laplace (see the first section of this chapter) and implies that the essentials of the dynamics are determined by the averaged equations of motion, and the periodic parts of the perturbation merely introduce periodic variations around the underlying evolution. Wisdom (1982) noted that the failure of these studies to account for the Kirkwood gaps may be due to their inability to allow for chaotic (Arnold) diffusion of the trajectories. As the averaged problem gives rise to an autonomous two degrees of freedom system (four-dimensional phase space), Arnold diffusion is impossible (see Chapter 6) and to allow for it, at least in principle, one has to treat the original unaveraged system or perhaps employ a *three-dimensional* elliptic restricted three-body model. Wisdom considered the unaveraged problem and, with the purpose of making it tractable for integration over a sufficiently long time, he proposed an equivalent discrete *mapping*, whose structure near one particular commensurability (the 3/1 resonance) is similar to that of the flow induced by the Hamiltonian of the restricted planar-elliptical three-body problem. Mappings are obviously simpler to study but there is no general way to derive them from the original Hamiltonian flow. In the derivation of Wisdom's mapping he first analytically averaged the Hamiltonian, removing from it the high-frequency contributions. As the next step he added to the averaged Hamiltonian a periodic series of Dirac delta functions (periodic 'kicks'), and thus transformed the original time-dependent Hamiltonian into another one. In this way the structure of the latter (new) Hamiltonian became naturally suited for the derivation of an equivalent mapping.

Wisdom's derivation of the mapping for the planar-elliptic restricted three-body problem is rather lengthy (several pages of algebra yielding a seven-step mapping), but the principle can be readily understood if we follow instead the derivation of Chirikov's standard mapping from a suitable time-dependent Hamiltonian (Chirikov was originally considering the dynamics of a charged particle in a confining magnetic field), which we shall give below. The full derivation of Wisdom's mapping can be found in Wisdom (1982) and in his subsequent papers.

We start from the following time-dependent Hamiltonian, written in the action–angle variables, I and θ

$$H(\theta, I, t) = \frac{I^2}{4\pi} + \frac{K_0}{2\pi} \cos\theta + \sum_{k \neq 0} K_k(I) \cos(\theta - kt) \tag{9.7}$$

If one assumes that that the K_k are all small, the sum on the right-hand side of (9.7) can be considered as a perturbation. We note, however, that if the argument of one of the cosines is almost stationary, that is, if the 'unperturbed frequency' $\dot{\theta} \approx I/2\pi$ is approximately equal to an integer, a resonance occurs. This is very similar to the case of the Hamiltonian used in the problem of the precessional motion, equation (9.6).

The system described by the Hamiltonian (9.7) can be approximated by using the *averaged* Hamiltonian

$$H_{av}(\theta, I) = \frac{I^2}{4\pi} + \frac{K_0}{2\pi} \cos\theta \tag{9.8}$$

The rapidly oscillating part of the original Hamiltonian is thus essentially ignored, because it can contribute only periodic oscillations around the average motion. The averaged Hamiltonian (9.8) is clearly equivalent to the Hamiltonian of a pendulum and is thus seen to be integrable.

Consider now another time-dependent Hamiltonian

$$H'(\theta, I, t) = \frac{I^2}{4\pi} + \frac{K_0}{2\pi} \cos\theta + \frac{K_0}{2\pi} \sum_{k \neq 0} \cos\theta \, \cos kt \tag{9.9}$$

It is obvious that the time-averaged version of (9.9) is identical to that of (9.8), and thus the motion induced by it is approximately equal to the motion resulting from (9.7). In other words, if the high-frequency terms can be ignored, suitably modifying them should be an equally good approximation. The advantage of H' over H is that a discrete mapping corresponding to it can be readily found. Using the well known Fourier representation of the periodic Dirac delta function (peaks at $t = 2\pi n$, where n is an integer)

$$\sum_n \delta(t - 2\pi n) = \frac{1}{2\pi} \sum_k \cos kt$$

we can write

$$H'(\theta, I, t) = \frac{I^2}{4\pi} + \frac{K_0}{2\pi} \delta_{2\pi}(t) \cos \theta \equiv H_\delta$$

with the shorthand notation $\delta_{2\pi} \equiv \sum_n \delta(t - 2\pi n)$.

Such a Hamiltonian is usually referred to as one of a *periodically kicked* system, and it is obviously integrable between the kicks with the trivial equations of motion

$$\dot{I} = 0 \quad \text{and} \quad \dot{\theta} = \frac{I}{2\pi}$$

The equations of motion across a kick are

$$\dot{I} = -\frac{\partial H_\delta}{\partial \theta} = K_0 \delta(t) \sin \theta$$

$$\dot{\theta} = \frac{\partial H_\delta}{\partial I} = \frac{I}{2\pi}$$

Since the duration of a kick is zero and I remains finite, the angle θ does not change during such a kick. Between the kicks (during a time interval $\delta t = 2\pi$) the change in θ is $I \delta t / 2\pi = I$. In contrast to the angle, the action I remains fixed between the kicks and changes by $K_0 \sin \theta$ across a kick. Thus the state of the system just after the $(j + 1)$th kick, say, can be explicitly written as a function of the state immediately after the jth kick

$$\begin{aligned} I_{j+1} &= I_j + K_0 \sin \theta_j \\ \theta_{j+1} &= \theta_j + I_{j+1} \end{aligned} \tag{9.10}$$

which is exactly the standard map (6.43). Wisdom's mapping is considerably more complicated, but it bears the same relation to the original time-dependent Hamiltonian of the planar-elliptic restricted three-body system as does the standard map to the Hamiltonian (9.7). By numerically studying his mapping (instead of the time-dependent Hamiltonian system) near the 3/1 commensurability, Wisdom gained a factor of $\sim 10^3$ in computer time and was able to iterate the mapping for periods corresponding to over a million years. This was a significant improvement over the best numerical integrations of even the *averaged* system available at that time. The results were surprising and encouraging. Test particles placed near the 3/1 commensurability in a low eccentricity orbit ($e < 0.1$) were found to hover around this eccentricity for almost a million years, and then experienced a sudden large increase of eccentricity (to a value above 0.3). Such an eccentric orbit would carry the asteroid past Mars and thus enable a close encounter or even maybe a physical collision with that planet. These findings thus provided an attractive mechanism for the explanation of the Kirkwood gaps: asteroids that originally reside near commensurabilities acquire chaotic orbits occasionally suffering from a sudden significant eccentricity increase. They are thus removed from their original position due to the interaction with Mars.

In the years following this work, Wisdom definitely proved the adequacy of his mapping by performing full numerical integrations of the unaveraged time-dependent system. The eccentricity of a test particle orbit for a particular initial condition, plotted as a function of time, looked very similar to the corresponding behaviour obtained by the mapping. In addition, Wisdom computed the maximal Liapunov exponent, using the results of his numerical integration and compared it to the corresponding Liapunov exponent computed from the mapping. He showed that the behaviour of the corresponding Liapunov exponents as a function of time is essentially identical, and they even seem to be approaching the same numerical value. These calculations not only proved that deterministic chaos can cause sudden increases of eccentricity in the orbits of asteroids near the 3/1 resonance, but also provided a way to determine the extent of the chaotic zones and the relevant timescales. The important result in the latter respect is that, although the Liapunov time is in this case only a few thousand years, a chaotic orbit may remain in a low eccentricity state for a duration of $\sim 10^5$ years (almost 100 Liapunov times) before suddenly jumping to a highly eccentric orbit. This type of behaviour, sometimes referred to as *stable chaos*, is found to occur quite often in simulations of weakly chaotic Hamiltonian dynamical systems (see below in the discussion of the full planetary problem).

To determine the extent of the chaotic zones for the purpose of quantitatively reproducing the Kirkwood 3/1 gap, Wisdom devised a more sophisticated three-dimensional mapping, taking into account additional terms, arising from the perturbation caused by Jupiter. Utilising this mapping he found that some test particles starting near the 3/1 commensurability reach eccentricities as high as $e > 0.6$, i.e., large enough to become Earth crossers, and they spend a significant fraction of time above the Mars crossing limit. It thus became possible that such asteroids suffer from successive interactions with Mars and are thus ultimately shifted to an orbit dominated by encounters with the Earth. Computations based on the iterations of the three-dimensional elliptic mapping, combined with the above considerations and several full numerical calculations, which included several planets, quantitatively confirmed Wisdom's scenario on the formation of the 3/1 Kirkwood gap. In addition, Wisdom presented a semi-analytical study in an effort to understand the mechanism leading to chaos in asteroid orbits. A detailed review of most of Wisdom's work on the 3/1 Kirkwood gap, as well as his subsequent attempts to use his mapping techniques for other commensurabilities can be found in his Urey Prize lecture (Wisdom, 1987).

Deterministic chaos has thus been shown to be of primary importance in the shaping of the asteroid distribution, but efforts to understand the mechanism behind its appearance, and the details of its applicability to all the gaps, have continued to prompt extensive numerical and analytical studies. A full description of these works is clearly outside the scope of this book: they are numerous, quite lengthy and rather technical. We shall make here only a few remarks on what seem to be the most important conclusions of these studies and refer the reader interested in greater detail to some recent reviews. Among the various findings, we choose to mention here the role of *secular* resonances, i.e., those which occur between the precession frequencies of the orbiting bodies. The resonances we have dealt with so far are the ordinary or *mean motion* ones, in which the orbital periods of the bodies are commensurate. Thus, when examining the possibility of resonance overlap the two types of resonances have to be taken into account. For details see the recent reviews by Moons (1997) and Ferraz-Mello (1999). For a recent work on the role of Saturn (i.e., three-body resonances) in causing chaos in the asteroid belt see Murray, Holman and Potter (1998).

So far we have discussed only those minor bodies (or test particles) that reside in the inner part of the Solar System, that is, inside Jupiter's orbit. It is well known, however, that small mass objects must abound in the region between and outside the giant planets as well. The sheer existence of comets is generally accepted as evidence of this fact. The pioneering studies on the origin of comets were carried out in the early 1950s by Oort, who proposed that comets originate in a vast external (to the Solar System) reservoir that now bears Oort's name. The *Oort cloud* is today thought to comprise of a large collection of comets, which formed as icy planetisimals at the outskirts of the primordial Solar nebula, and were repeatedly scattered by the growing outer planets to high values of orbital semi-major axis. After reaching a region in which planetary perturbations on them became no longer significant, their space distribution had been shaped by the gravitational effect of passing stars and Galactic tidal fields. Stellar perturbations, whose effects randomise the distribution of comets, are most probably responsible for the spherical shape of the Oort cloud. Its extension is estimated to be between $\approx$3000 AU (inner edge) and up to

$\approx$50 000 AU, with the density falling roughly as $r^{-3.5}$. Comets are thought to be injected from the Oort cloud into the Solar System by occasional close encounters with stars, and this scenario can account for the observations of all comets other than the short-period ones (period values below $\sim$200 years).

These short-period comets are postulated to originate in the *Kuiper belt*, a smaller, flat ensemble of low inclination comets, situated just outside the orbit of Neptune (between $\approx$35 AU and $\approx$50 AU). Such a belt could be formed from those planetesimals in the outer primitive Solar nebula, only slightly removed by Uranus and Neptune and which have reached a stable region just outside the Solar System, as was originally suggested by Kuiper in 1951. This topic is still quite controversial, but new observations seem to confirm the existence of objects just beyond Neptune. In parallel, extensive numerical calculations have recently been performed to better constrain the Oort cloud and Kuiper belt models. In the remainder of this discussion we shall only mention (very briefly) those aspects of cometary dynamics in which chaotic behaviour has been found to play a significant role. Broad reviews on the dynamics of comets can be found in Duncan & Quinn (1993) and Fernandez (1994). Nowadays, in general, most long-period comets coming from the Oort cloud are thought to have chaotic orbits. The origin of this chaotic behaviour is attributed simply to repeated close encounters with the planets. For example, it has been shown (by Sagdeev and Zaslavsky in 1987 and by others) that the possibility of comet–Jupiter encounters when the comet has an almost parabolic orbit can account for a large chaotic zone, extending up the Oort cloud. More specifically, attempts to numerically calculate the orbit of the most famous object of this kind, the Halley comet, close to the time of its latest visit in 1985, failed to agree with each other and with observations. This problem probably arises from the fact that Halley's orbit is in fact chaotic and this conclusion was supported by Chrikov and Vecheslavov, whose analysis in 1989 indicated that the timescale for practical unpredictability in this case could be as short as $\sim$30 years.

Regarding the short-period comets, whose origin must be in the periphery of the Solar System, numerical studies conducted in the early 1990s showed that there are practically no stable orbits among the outer planets. Stable orbits were found, however, above about 40 AU and it is thus reasonable that planetisimals could survive there for a very long time. Close to these stable regions there exist instability regions, from which comets could be injected into a more internal part of the Solar System by means of chaotic diffusion, and ultimately (but probably only temporarily) be captured, by interaction with the planets, into resonant orbits and appear as short-period comets.

More detailed discussions on minor bodies dynamics in the outer parts of the Solar System and relevant references can be found in Duncan & Quinn (1993) and in the recent extensive review by Lecar *et al.* (2001).

Stability of planetary orbits

We turn finally to the problem of the global stability of the planetary orbits themselves. The historical importance of this problem and the role it has played in understanding the mathematical aspects of non-integrability in Hamiltonian systems has already been mentioned in this book, and we shall not repeat these aspects here. We shall, however, start with a few introductory remarks, which have to be made for the purposes of setting the stage for our subsequent discussion.

The effects of the gravitational influence of the planets on each other and their back reaction on the Sun are obviously very small in comparison with the effect of the direct gravitational attraction of the Sun. The question of the long-term (with respect to the orbital period) evolution of planetary orbits has thus customarily been approached in the classical perturbative way (i.e., using canonical perturbation theory, see, e.g., Goldstein, 1980), by splitting the Hamiltonian into an unperturbed part, consisting of the planets' independent motions around the Sun, and a perturbation, called the *disturbing function*, containing all the rest of the dynamics. The disturbing function itself can be expanded in terms of a number of small parameters of the problem (like the planet to Sun mass ratios, planetary orbital eccentricities and inclinations), and one may seek approximate solutions in the form of perturbation series. Straightforward expansions of this kind may, however, include (often already in first order) terms which grow in time without bound – the so-called *secular terms*. The secular terms thus render a straightforward perturbative approach to be of little value if one is interested in the *long-time* behaviour of the orbits, and more sophisticated perturbative techniques have to be used. We are unable to pursue this subject in detail here (the equations are rather long and quite complicated) and refer the interested reader to classical texts on celestial mechanics, like Brouwer & Clemence (1961).

The standard technique, usually referred to (in the astronomical literature) as *secular perturbation theory*, consists of ignoring all the periodic terms in the disturbing function or equivalently averaging it over the mean motion, and is thus essentially based on the method of averaging in perturbation theory (see Nayfeh, 1973). If only the first-order terms in the averaged disturbing function (the secular part of it) are retained, the equations of motion for the orbital elements of the planets (the *secular system*) are linear and can be solved, at least in principle, analytically. Already Laplace found that the linear secular system admits only quasiperiodic solutions (and no secular behaviour) of the orbital elements of all the known planets, that is, these quantities undergo only small bounded variations around their mean value. The frequencies involved in these motions (actually the eigenfrequencies of the linear differential secular system) were found to correspond to periods ranging from tens of thousands to a few million years.

Laplace's famous assertion that one can predict absolutely the future of all planetary orbits in the Solar System by using only Newton's laws, and thus the world is essentially a deterministic and stable clockwork machine, was largely based on the above findings. The problem of small divisors, which appeared when attempts to carry the calculations to higher order were made, and Poincaré's subsequent fundamental discoveries (see above in the historical introduction), cast some shadow of doubt on Laplace's statement. Still, the modern developments in Hamiltonian system theory and in particular the KAM theorem (Chapter 6) seemed to guarantee that chaotic solutions described by Poincaré are, in a well-defined sense, very rare. The phase space of nearly integrable Hamiltonian systems is dominated by invariant tori, created by quasiperiodic trajectories and indeed the Solar System is undoubtedly nearly integrable.

It turns out, however, that the mathematical structure of the Hamiltonian appropriate for the Solar System does violate one of the conditions of the KAM theorem in its classical form. The unperturbed Hamiltonian of the Solar System planetary dynamics does not depend on all the independent action variables of the problem (it depends only on the actions related to the semi-major axes) and thus it is *degenerate* (see Chapter 6). Arnold himself was the first to try to extend the findings of the KAM theorem taking into account this degeneracy. His and his followers' work established the existence of isolated KAM tori in systems involving the Sun and *two* planets. Since the Solar System has more than just two degrees of freedom, these KAM tori could not exclude the possibility of Arnold diffusion, that is, the exploration of large portions of phase space by a chaotic orbit (see Chapter 6). In subsequent studies it has been shown that this diffusion is very slow, at least for trajectories starting very close to the invariant tori, having a timescale longer perhaps than the age of the Universe. Since the relevant mathematical theorems have included, as is usually the case, conditions on the perturbations to be 'sufficiently small', the issue of the stability of planetary orbits has relied mainly on the belief (or disbelief) that the findings of mathematical theorems are valid well beyond truly infinitesimal epsilons.

This was the state of affairs in the late 1980s, when computer technology reached the point at which direct numerical attacks on the problem became feasible. The simplest (at least in principle) approach is to solve numerically, using some standard multistep integrators, the equations of motion of the Sun and all its planets. The immediate difficulty encountered in such brute-force attempts stems from the fact that the orbital periods of the innermost (Mercury) and outermost (Pluto) planets differ by approximately three orders of magnitude. Thus the time-step required for solving for Mercury's orbit at reasonable accuracy is extremely small, as compared to Pluto's orbital period when about 10^5 steps are necessary for reproducing just one orbital revolution of the latter. As explained in the previous paragraph the expectation is that chaotic behaviour, resulting from Arnold diffusion, becomes significant only after a very long time relative to a typical natural period in the system (i.e., after an extremely large number of orbital revolutions). The CPU power available in the late 1980s was still insufficient to follow the integration of the full planetary system for long enough. Thus the numerical studies have often included only the outer massive planets and auxiliary analytical ideas and methods have been devised and incorporated into them. These techniques have remained valid until today since they enable longer and longer integration times irrespective of computer power.

The first numerical projects were conducted by Applegate *et al.* (1986) and by Milani *et al.* (1986) (the LONGSTOP project) which indicated that the expectation of finding multi-periodic motion (that is, one proceeding on tori in phase space) seems to be justified, although the longest integrations showed some departures from regularity in Pluto's orbit. For the purpose of examining these irregularities further, the LONGSTOP project calculations were extended to longer integration times and the secular (resulting from averaging over mean orbital periods) effects of the inner planets were included as well as general relativistic corrections. Still, the results of calculations spanning $\sim$100 Myr of evolution, which were published in 1989, could not definitely determine if the planetary orbits were chaotic, but they suggested that this might be the case.

Independently of the LONGSTOP project and at about the same time, calculations using a special-purpose parallel computer (with a separate CPU per planet), which had been developed at MIT, provided more conclusive results. The 'Digital Orrery', as this computer was called, enabled Sussman and Wisdom (1988) to carry their calculations up to almost 900 Myr and to discover chaotic behaviour (with a Liapunov time of 20 Myr) in Pluto's orbital motion. Despite some doubts, raised by a re-examination of the LONGSTOP results which seemed to indicate that the value of the Liapunov time may depend on initial conditions, subsequent and more general calculations by Sussman and Wisdom, as well as by others (see below), largely confirmed the early result.

With computer power rapidly increasing, the direct integrations could handle a truly enormous number of time-steps, but this immediately gave rise to a new worry – the numerical roundoff error. A realistic estimate indicated that this error (specifically, in the longitude) grows with the number of time-steps, n, as fast as n^2. It thus became clear that the hope for significantly longer integration times and/or the inclusion of the inner planets in the calculations (the LONGSTOP results suggested that their effects should not be completely ignored) relied on the development of sophisticated schemes instead of the direct brute-force ones. We list below three different approaches to addressing this problem, and summarise some of the most important findings from calculations based on schemes which incorporated these ideas.

(i) Quinn, Duncan and Tremaine applied corrections to the integration algorithms and employed a high-order symmetric scheme, maintaining a linear (in n) longitude error growth. In spite of this significant improvement the technique remained essentially as a direct numerical integration scheme, and as such was extremely time consuming. Still, the results of such direct calculations are the most accurate and they are extremely valuable, if only for comparison with other calculations over their common range (see below). For more details on these methods and the relevant references see the review by Duncan and Quinn (1993). Among the most recent of such calculations we should mention the one by Duncan and Lissauer (1998), which followed the Solar System (excluding Mercury and Pluto) for 1 Gyr., indicating that the orbits remain stable over very long times, that is, no escapes, collisions or other dramatic changes occur.

(ii) Sussman and Wisdom combined a scheme specifically adapted to the problem, a mixed variable *symplectic integrator*, with special purpose hardware, and succeeded in their effort to follow the full Solar System evolution for very long periods of time. Symplectic integrators are designed to maintain the Hamiltonian structure of the system at all times. These schemes are however quite complex and consequently of low order, and thus their accuracy tends to be poor in long-time simulations. Improved accuracy can however be achieved by employing a mapping method, similar in principle to the one used by Wisdom in the Kirkwood gaps problem (see above). The principle is to split the Hamiltonian into a Keplerian part and a perturbation, and to effectively replace the perturbation by a periodic sequence of kicks, in a similar way to the replacement of the Hamiltonian (9.7) by (9.9) in the derivation of the standard map (see above).

Results from calculations of this sort are given in Sussman & Wisdom (1992), confirming that the motion of Pluto is robustly chaotic with Liapunov time of 10–20 Myr, essentially independent of whether the rest of the planets behave chaotically or not. Moreover, positive Liapunov exponents (that is, chaos) were found for all planetary orbits with Liapunov times of $\sim$4 Myr for the inner planets, and varying between 3 and 30 Myr for the outer ones. In recent work Murray and Holman (1999) compiled the results from a large number of such extensive long-term numerical calculations of the outer Solar System, confirming the existence of chaos among the Jovian planets with a Liapunov time of $\sim$5 Myr. In this work Murray and Holman also proposed an analytic theory for the origin of this behaviour.

(iii) Laskar's approach to the problem of planetary dynamics has been quite different. In the spirit of the great analytical studies that had been conducted before him in his institution, the Bureau of Longitudes in Paris (Laplace and Lagrange themselves were among the founders of this institution), Laskar performed extensive traditional analysis before using fully-fledged numerical calculations. In a series of works conducted during the second half of the 1980s he embarked on an extensive averaging procedure for the planetary equations of motion. Keeping all terms up to second order in the planet-to-Sun mass ratios and up to fifth order in the orbital eccentricities and inclinations, he averaged the perturbing function over the rapidly varying orbital angles. Computer-aided symbolic manipulations facilitated the derivation of the secular system, comprising approximately 150 000 terms.

As the rapid orbital motions are absent from the secular system, the full system can be numerically integrated with relatively large step-size (e.g., ∼500 years). The first complete numerical calculations were done in 1989 on a supercomputer of that time and represented 200 Myr of Solar System evolution. The results revealed chaotic behaviour in all the inner planets with Liapunov times of ∼5 Myr. The analysis of the results was far from trivial, and Laskar devised his frequency analysis method (discussed above in the section on rotational dynamics, see Laskar, 1993 for details) for purposes of diagnosing chaos in this highly multi-dimensional problem. According to the most recent calculations (Laskar, 1994; 1996) the effects of chaos on the orbits of Mars, Venus and Earth are small for at least 5 Gyr, but the orbit of Mercury may change its eccentricity drastically in about 3.5 Gyr and thus this planet's fate may be quite catastrophic.

The above three types of approach, although quite different, produced results that were generally in rather remarkable agreement. When Laskar compared his code with that of Quinn and Tremaine over a period of their common validity, the values of eccentricity and inclination for the Earth and Mars, for example, differed by no more than a fraction of a percent of the mean values, despite rather wild chaotic variations! The Sussman and Wisdom results were in even better agreement with the calculations of Quinn, Tremaine and Duncan (see Table 1 in Sussman & Wisdom, 1992). The evidence for chaos in the long-term numerical integrations described above is thus quite convincing. Moreover, Sussman and Wisdom, and independently Laskar, have performed many tests to rule out the possibility that chaotic behaviour is produced by some numerical artifacts.

However, the fact that nothing truly dramatic happens in the long-term integrations and this is most probably true in the real Solar System as well is a bit unsettling (intellectually!). How can we reconcile the relatively short Liapunov times with the long-term stability of the orbits found? For example, Pluto's orbit was found to be undoubtedly chaotic with Liapunov time much shorter than The Solar System lifetime, yet it is still with us. In addition, some specific numerical integrations have followed its orbit for 5.5 Gyr and confirmed that it does not escape during such a long long period of time. The term *stable chaos*, coined by Milani and his collaborators in 1990 is appropriate for such behaviour. But a detailed understanding of the physical mechanism causing chaotic behaviour in the Solar System is clearly needed, an understanding that would also account for the stable chaos it manifests, which accounts for the system's long-term behaviour.

Significant theoretical progress in the investigation of this problem has recently begun to emerge. Specifically, it is now widely accepted that chaotic orbits result from overlapping resonances – three-body resonances in the outer Solar System (Murray & Holman, 1999) and secular resonances in the inner planets (Laskar, 1991). The question of the disparity between the Liapunov time, t_λ, of an orbit and the corresponding macroscopic instability time of this orbit, t_m, (e.g., ejection time) is also being addressed. Typically $t_m \gg t_\lambda$ and this relation is not only true for planetary orbits. Several asteroids are known to have very short Liapunov times ($< 10^5$ years) but they are still in the asteroid belt. The understanding of this phenomenon is still incomplete, but some useful relationships between the two timescales, as well as a possible physical basis for the phenomenon have been proposed (see Lecar *et al.* 2001 and Contopoulos 2002 and references therein).

In addition to its being a primary intellectual challenge, the understanding of chaotic behaviour of the Solar System's planets may provide constraints to theories of formation and evolution of this system, as well as of other, extra-solar, planetary systems. Lissauer (1995) proposed that chaos is a major factor in planet growth when they are forming from the large numbers of planetisimals travelling on intersecting, highly chaotic orbits (see Lissauer 1999 for a review on this topic and for references). In recent years many planet-like bodies have been discovered in the vicinity of scores of nearby stars. Intensive efforts are currently being made to observe external planets and to understand the processes responsible for the formation and evolution of such systems. For example, chaotic variability of the eccentricity has been suggested by Holman, Touma and Tremaine (1997) with the purpose of explaining the observations of one particular system of this type. More recently Murray *et al.* (1998) proposed a model in which, as a result of resonant interactions between a massive planet and planetisimals, the planet migrates to small orbital radii. Our understanding of the Solar System's complicated dynamics can be very useful in trying to model these extra-solar planetary systems. Conversely, we may also learn from these exo-systems something about the history of our Solar System.

Laskar's simulations indicate that the inner part of the Solar System is largely *full*, that is, chaotic diffusion causes the orbits of the inner planets to sweep over practically all of the available physical space in the system's lifetime. The test-particle simulations, mentioned at the end of the previous subsection, indicate that the outer Solar System is full in this sense as well. Do these findings have some bearing on the elucidation of the almost mystical Titius and Bode 'law'? In what sense is our planetary system generic? In what type of a planetary system can a planet with livable conditions retain a stable enough orbit and rotational obliquity for a sufficiently long time? These important (and many other) questions still await definite answers. There is no doubt in my mind that chaotic dynamics will play a primary role in solving these outstanding problems.

Before moving on to our brief account of the many applications of chaos in stellar and galactic dynamics, we shall now discuss in considerable detail one particular problem in binary-star dynamics, where chaos theory has recently been applied in an attractive and convincing way.

9.3 Chaos and the evolution of tidal-capture binaries

The simplest processes by which a gravitationally bound binary stellar system can form from two unbound stars is the so-called tidal capture. The probability of such a process occuring depends on the density of the stellar candidates for capture, as well as on their initial (when they are far apart) dynamical parameters and their radii. We shall consider here neutron star–low-mass main sequence binaries, since such objects are candidates for X-ray sources, which appear to be overabundant in dense stellar environments like globular clusters.

The tidal-capture process appears to be viable near the centres of rich globular clusters, but a binary so produced typically has quite a high eccentricity. The question of the subsequent dynamical evolution of such binaries has therefore been addressed in a number of studies. The subject is quite complex since the energy transfer during a periastron passage of a compact (modelled by a point mass) star and a main-sequence (modelled by a polytrope of finite radius) star depends, for example, on whether or not the extended star oscillates. Such oscillations are tidally driven by the previous periastron passage, and the question of how long they persist depends on the dissipation timescale as compared with the orbital time. We cannot discuss here all the details of this problem but will concentrate on the work performed by Mardling, published in 1995, who found that chaos can play a significant role in the evolution of such binaries. A detailed review of the history of the problem and its various aspects, as well as relevant references, can be found in Mardling (1995).

The problem we wish to discuss is formulated by considering a point mass M_2 (the compact star) and a polytrope of mass M_1 (the main-sequence star). Both stars move under their mutual gravitational attraction, and we also allow for distortions of the extended star because of tides raised on it by the compact star. If one assumes that the fluid composing the extended star is ideal, that is, there is no dissipation in the flow induced within it, it is possible to use a Hamiltonian formulation of the problem. This relies indeed on the fact that a Hamiltonian formulation of fluid dynamics is available, because an ideal fluid satisfies a variational principle. We shall touch upon the rudiments of this topic in Chapter 12, where the reader will also find the appropriate references for more complete study. As explained there (see Section 12.5), the Lagrangian appropriate for the fluid is composed of its total kinetic energy minus its gravitational potential energy and minus the total *internal* energy, and the natural way of writing this Lagrangian, by analogy with particle dynamics, is to consider fluid particles and follow for their motion (utilising the so-called Lagrangian picture, or notation, of fluid dynamics). It is, however, not very easy to transform the above mentioned Lagrangian to Eulerian notation, that is, the one appropriate for the formulation of fluid dynamics in which fluid properties are fields of space coordinates and time.

For the case considered here (a polytrope, which is a particular case of a barotrope, see Chapter 12), the internal energy depends on the density alone. In addition, the Kelvin circulation theorem (valid for ideal barotropic fluids, see Chapter 12) guarantees that if the flow is initially irrotational (as it obviously is here, if we assume that the polytrope starts from an equilibrium), the velocity is derived from a potential, that is $\mathbf{v} = \nabla\phi$ at all times. In this case, the 'fluid Lagrangian' acquires a reasonable form also in Eulerian notation (see Equation 12.100) and reads

$$L_{\mathrm{f}} = -\int \left(\partial_t\phi + \frac{1}{2}\nabla\phi \cdot \nabla\phi + \Phi + U\right)\rho \mathrm{d}^3\mathbf{r} \qquad (9.11)$$

where $\phi(\mathbf{r}, t)$ and $\rho(\mathbf{r}, t)$ are the Eulerian velocity-potential and density fields respectively, U is the specific internal energy (depending only on ρ for a polytrope), the gravitational energy is

$$\Phi(\mathbf{r}) = -\frac{G}{2}\rho(\mathbf{r})\int \frac{\rho(\mathbf{r}')}{|\mathbf{r} - \mathbf{r}'|}\mathrm{d}^3\mathbf{r}'$$

and the integration is over the whole volume of the fluid. Note that we suppress here, for the sake of conciseness, the possible explicit time-dependence of the functions.

Since we are considering here not only the fluid composing the polytrope, but also the point mass and consequently the motion of these two bodies and their gravitational interaction, we also have to include the 'interaction Lagrangian', which when written in coordinate frame whose origin is at the *centre of mass of the polytrope* is

$$L_{\mathrm{int}} = \frac{1}{2}\mu\dot{\mathbf{R}} \cdot \dot{\mathbf{R}} + GM_2\int \frac{\rho(\mathbf{r})}{|\mathbf{R} - \mathbf{r}|}\mathrm{d}^3\mathbf{r}$$

Here $\mathbf{R}$ is the radius vector from centre of mass of the polytrope to the point mass, $\mu = M_1M_2/(M_1 + M_2)$ is the reduced mass and the integration is over the volume of the polytrope. The second term is clearly minus the total gravitational interaction energy *between* the two objects. The interaction Lagrangian is thus essentially the usual Lagrangian of the two-body problem, with the gravitational energy properly modified (due to the finite extent of one of the bodies).

When writing the total Lagrangian of the system one has to remember that L_{int} is written in a non-inertial frame and thus additional terms may be needed, but it can be shown that these terms are proportional to the *linear* momentum of the fluid of the polytrope relative to its centre of mass, and thus drop out. Thus the Lagrangian of the entire system can be written as

$$
L = L_{int} + L_f = \frac{1}{2}\mu\dot{\mathbf{R}}\cdot\dot{\mathbf{R}} + \frac{GM_1M_2}{R} + GM_2 \int \rho(\mathbf{r})\left(\frac{1}{|\mathbf{R}-\mathbf{r}|} - \frac{1}{R}\right)d^3\mathbf{r}
$$
$$
- \int_V \rho(\mathbf{r})\left[\partial_t\phi + \frac{1}{2}\nabla\phi\cdot\nabla\phi + U - \frac{G}{2}\int_{V'}\frac{\rho(\mathbf{r}')}{|\mathbf{r}-\mathbf{r}'|}d^3\mathbf{r}'\right]d^3\mathbf{r}
$$
$$
(9.12)
$$

where $R \equiv |\mathbf{R}|$ and the modifications needed for obtaining this formula after substituting the above L_f and L_{int} are trivial.

As pointed out above we shall discuss in detail the Hamiltonian formulation of fluid dynamics later in this book. A detailed derivation, showing that the total Lagrangian of this problem is indeed as given here by (9.12), can be found in Gingold & Monaghan (1980).

If R remains large enough (with respect to the radius of the polytrope) throughout the motion, the fluid motion induced by the point mass can be treated in a perturbative way. The relevant perturbations are the deviation of the density from its value in an equilibrium spherical polytrope ($\eta \equiv \rho - \rho_0$) and the velocity (and therefore the velocity potential ϕ). These perturbations are quite general and one can expand them in spherical harmonics in the usual way (as in the problem of nonradial stellar oscillations), that is

$$
\eta(\mathbf{r}, t) = \sum_q b_q(t)\eta_{kl}(r)Y_{lm}(\theta, \varphi)
$$

$$
\phi(\mathbf{r}, t) = \sum_q a_q(t)\phi_{kl}(r)Y_{lm}(\theta, \varphi) \qquad (9.13)
$$

where the index q is a shorthand notation for the triple k, l, m. The generally complex amplitudes $b_q(t)$ and $a_q(t)$ satisfy appropriate relations (see Mardling 1995).

Following Gingold and Monaghan, Mardling performed an expansion of the Lagrangian (9.12) to *second order* in the perturbations and substituted the mode expansions (9.13). The gravitational potential can also be expanded in a standard way and after rather lengthy algebra, which we shall skip here, the following equations of motion for the mode amplitudes and the orbital radius vector are found:

$$\frac{d^2 b_{\mathbf{q}}}{dt^2} + \omega_{kl}^2 b_{\mathbf{q}} = \frac{M_2}{M_1} C_{\mathbf{q}} \frac{e^{-im\psi}}{R^{l+1}}$$

$$\mu \frac{d^2 \mathbf{R}}{dt^2} + \frac{GM_1 M_2}{R^3} \mathbf{R} = \frac{M_2}{M_1} \sum_{\mathbf{q}} D_{\mathbf{q}} b_{\mathbf{q}} \nabla_{\mathbf{R}} \left(\frac{e^{im\psi}}{R^{l+1}} \right) \qquad (9.14)$$

Note that the mode frequencies are written as ω_{kl} because they are independent of the index m (the unperturbed polytrope is spherically symmetric). The angle ψ is the azimuthal angle of the vector $\mathbf{R}$ and thus is essentially the true anomaly of the *orbit*, and the constants $C_{\mathbf{q}}$ and $D_{\mathbf{q}}$ depend on the mode and on the internal structure of the extended polytropic star.

The equations of motion (9.14) can be readily understood by noting that their homogeneous parts describe essentially unforced harmonic oscillators (for the different modes) and the Kepler problem, respectively. The inhomogeneous terms (the right-hand side) reflect the coupling between the orbital motion and the internal oscillatory modes of the extended star. The differential system (9.14) is thus composed of a system of *forced* oscillators and a perturbed orbital motion. It is a Hamiltonian dynamical system and as such it describes a physical system in which the total energy is conserved, because the Lagrangian does not depend explicitly on time. This is consistent with the fact that tidal (or any other form of) *dissipation* was ignored in the derivation of the Lagrangian. We have already seen in this chapter similar systems modelling chaotic behaviour of bodies in the Solar System and it is therefore reasonable that (9.14) admits chaotic solutions as well. Mardling indeed found such solutions and has shown that their existence can play an important role in the circularisation of the eccentric orbits, which are characteristic of binaries formed by tidal captures.

Before briefly describing Mardling's numerical findings and discussing them, we note that the system (9.14) admits two obvious integrals of motion – the total energy E and angular momentum J, and they both consist of their orbital and oscillatory parts and in the case of the energy an interaction term as well

$$E = E_{\text{orb}} + E_{\text{osc}} + E_{\text{int}}; \qquad \text{and} \qquad J = J_{\text{orb}} + J_{\text{osc}}$$

The form of the orbital parts is self-evident

$$E_{\text{orb}} = \frac{1}{2} \mu \dot{\mathbf{R}}^2 - \frac{GM_1 M_2}{R}; \qquad \text{and} \qquad J_{\text{orb}} = \mu R^2 \dot{\psi}$$

and the oscillatory and interaction terms can be written as

$$E_{\text{osc}} = \frac{1}{2} \sum_{\mathbf{q}} F_{\mathbf{q}} (\dot{b}_{\mathbf{q}} \dot{b}_{\mathbf{q}}^* + \omega_{kl}^2 b_{\mathbf{q}} b_{\mathbf{q}}^*); \quad E_{\text{int}} = -\frac{M_2}{M_1} \sum_{\mathbf{q}} D_{\mathbf{q}} b_{\mathbf{q}} \frac{e^{im\psi}}{R^{l+1}}$$

and

$$J_{\mathrm{osc}} = \sum_{\mathbf{q}} im F_{\mathbf{q}} \dot{b}_{\mathbf{q}} b_{\mathbf{q}}^*$$

where $F_{\mathbf{q}} \equiv C_{\mathbf{q}}/D_{\mathbf{q}}$. Because only the total E and J are conserved, energy and angular momentum can be exchanged between the orbit and the internal oscillations. Note that the angular momentum exchange here is not by tidal *dissipation* (which we have discussed earlier in this chapter in the context of planets and their moons). The process operating here has nothing to do with the tidal lag brought about by tidal friction; it is rather a dynamical effect arising from the asymmetric nature of the tides and the resulting oscillations. When integrating the equations of motion (9.14) Mardling found essentially three different types of behaviour, depending on the parameters of the problem and the initial conditions. Fixing the mass ratio at a given value and choosing a polytropic index for the extended star ($M_2/M_1 = 1$ and $n = 1.5$, for example) the integrations were started with different initial conditions for the orbit, which were parametrised by the pair (R_p, e), that is, the value of the initial periastron separation and the orbit eccentricity. The initial tidal energy was set to zero (by starting with an undisturbed polytrope).

The regions of the different possible outcomes are displayed in Figure 9.4. Solutions falling in the region marked 'quasiperiodicity' in the figure were the most

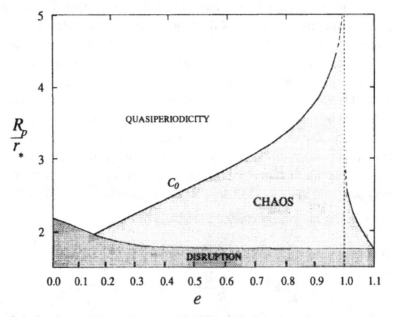

Fig. 9.4. Regions of the different types of behaviour in the R_p (initial periastron separation in units of the polytropic star radius R_*) $-e$ (orbital eccentricity) plane. These results correspond to the case $M_2/M_1 = 1$ and a polytropic index of 1.5. For details see text. Reprinted, by permission, from Mardling (1995).

regular. The orbital eccentricity was found to undergo oscillations of only small amplitude around the initial value and nothing dramatic happened. In contrast, if the initial periastron separation was small enough, the polytropic star was found to be tidally disrupted, irrespective of the initial eccentricity. This was the second category of solutions, falling in the lighter shaded region on the figure. The most interesting, at least as far as the subject of this book is concerned, regime is the one marked 'chaos', which is separated from the quasiperiodic region by the curve marked C_0. Solutions of this type had the SIC property and the eccentricity was found to vary in an irregular manner by a large factor.

The position of C_0 (the 'chaos boundary') was found to depend on the mass ratio and the polytropic index. As the coupling terms between the oscillations of the polytrope due to the tidal disturbances and the orbit are proportional to the mass ratio, C_0 was found to move up with the value of M_2/M_1. Conversely, increasing the polytropic index, and thus the central condensation of the polytrope, drove the chaos boundary downward. For additional details we refer the reader to the original papers, and we turn now to the possible effect of chaos in the problem of tidal capture.

For the purposes of investigating this issue Mardling considered a system of two $0.7 M_\odot$ stars initially on an almost bound, that is, parabolic orbit ($e = 1$) with a sufficiently small periastron separation as to place it well in the chaotic region ($R_p = 3R_*$, say). She was able to follow the evolution of the system for approximately 3000 orbits (overcoming some nontrivial computational problems) and found that the eccentricity varied chaotically between the initial value and $e \sim 0.5$. In such an evolution, however, the approximations leading to the differential system (9.14), and in particular neglecting mode–mode interactions and dissipation, clearly break down. When $e = 0.5$ in this case the tidal distortion is very large (of the order of the stellar radius) and it is not clear how the stars will respond to such violent tides. Still it is quite plausible that the tidal energy is deposited deep enough in the star, and so the binary may survive the chaotic variations of the eccentricity. If indeed it is so and tidal dissipation is operative (but not totally destructive), the eccentricity changes chaotically, the binary will leave the chaotic evolution regime with a given eccentricity $e < 1$. The duration of this phase depends on the nonlinear dissipation timescale and can be estimated to be of the order of a few hundred years in this case. After that the binary continues to tidally circularise in the usual regular way and slowly (over millions of years) it finally becomes an almost circular binary of a period of the order of a day or so.

What is the fraction of tidally captured binaries that can survive such a circularisation process (through the chaotic and regular phases) in a given astrophysical setting (e.g., a globular-cluster core)? The answer to this question has not yet been fully settled. It is however reasonable that future research on this topic should take into account the possibility that chaotic behaviour plays an important role in the process of tidal capture, and indeed it has already been incorporated in full n-body simulations of stellar clusters (Mardling & Aarseth, 2001). Moreover, as recently noted by Mardling, a formal correspondence between the binary-tides problem and the three-body problem exists, so that stability against escape in a triple may be predicted by a formula almost identical to that which predicts chaotic tidal interactions in a close binary. This development is very promising as it may shed some light on the enigmatic three-body problem, discussed at the beginning of this chapter.

9.4 The role of chaos in stellar and galactic dynamics

The mathematical problem resulting from modelling a galaxy or a stellar cluster by a very large number of point particles, moving under their own mutual gravitational attraction and possibly also in some external gravitational potential (e.g., that of dark matter), is obviously extremely complex. This problem is central to stellar and galactic dynamics, an important branch of dynamical astronomy and astrophysics in general. The n-body problem, as we already know, is in general intractable analytically for n as low as 3; in stellar clusters, however, we have to deal with perhaps $n \sim 10^5$ and in galaxies much more than that. It is therefore only natural that before computers became powerful enough to handle numerical calculations of such magnitude, the approach to the problem had largely been based on statistical methods. These methods were inspired by statistical mechanics, whose success when applied to systems of very many microscopic particles has been quite impressive. It has however been realised (even as early as the 1940s) that typical relaxation times t_R, relevant to dynamical astronomy, are very long (for detailed estimates see, e.g., Binney & Tremaine, 1987) and therefore stars could not have enough time to settle into the invoked (and also observed) statistical equilibria. The idea of *violent relaxation*, introduced by Lynden-Bell (1967), seemed to provide a way out of this problem, but any effort to understand the underlying mechanism (irregular fluctuations of the gravitational potential) called for a better examination of the detailed dynamics.

The attention began thus to shift towards the study of individual orbits of stars, i.e., *orbital dynamics*, the simplest version of which is essentially a mean-field approach, relying on the computation of test-particle orbits in a *known*, time-independent gravitational potential. Orbital dynamics became particularly appealing after digital computers had developed to a point at which numerical calculations of orbits were feasible. It is within this kind of approach that chaos made its first explicit appearance in galactic dynamics. Calculations of orbits by Hénon and Heiles (1964) (see Chapter 2) and later on by many others, who used more complicated and realistic potentials, clearly indicated the presence of chaotic orbits. The significance of such irregular orbits in the overall global dynamical properties of a stellar cluster or galaxy had, however, become the subject of some debate and the issue remained open.

While orbital dynamics has obvious advantages, the choice of the potential function used in this approach raises the question of *self-consistency* of the model. A reasonable possibility in this context is to 'derive' realistic potential functions from observations and/or employ iterative procedures to make the models based on orbital dynamics more self-consistent. Still, if chaotic orbits are sufficiently numerous they may be able to affect, on short enough timescales, the underlying mean configuration. This indeed had been indicated by various numerical studies and so the self-consistency problem became a major worry for galactic and cluster modelling based on orbital dynamics. We shall attempt to give, in the first part of this section, the main findings of the orbital dynamics approach in this context, including also ideas on how to improve self-consistency that have been proposed over the years.

The possibility that, even in realistic potentials used in orbital dynamics calculations in which special methods for self-consistency are used, the latter may not be guaranteed, has naturally led to a more thorough examination of this problem in itself. A truly self-consistent way to study the dynamics of a stellar cluster or galaxy involves an appropriate calculation to solve the full gravitational n-body (with a *very* large n) problem. At present, computer power is sufficient to allow for computations of not much more than $\sim 10^6$ bodies. In addition to the direct integration approach, however, several improvements (including judicious implementations of approximations, as well as sophisticated software and special purpose hardware) have been employed. The results of such numerical studies are directed towards the resolution of some highly non-trivial issues like the nature of equilibria, stability and evolution of collisionless many-body systems. This difficult topic and some of the recent contributions to the progress in its understanding will be briefly addressed in the second part of this section.

Our discussion here will necessarily be rather sketchy and brief. We refer the reader to the comprehensive books by Binney and Tremaine (1987) and Contopoulos (2002). In addition to being excellent sources for the study of the basics of galactic dynamics and chaos in dynamical astronomy, they include reviews of what work has been done on the topics discussed in this section and they also contain extensive lists of references. Moreover, these books obviously reflect the (often contradictory) views of their authors, leading contributors to the field, on the role of chaos in stellar and galactic dynamics.

Orbital dynamics

As explained above, the 'orbital dynamics' approach to the study of cluster and galactic dynamics is based on the *assumption* that the force field in which a particular star is moving can be cast into a smooth function, that is, the small scale spatial and/or temporal variations due to individual stars are averaged out in some sense, and only the large-scale forces resulting from the overall mass distribution of stars and dark matter are considered. The investigation of the stellar component of clusters and galaxies consists thus of a systematic exploration of stellar orbits in smooth potentials.

We have already mentioned the self-consistency problem and shall address this question in the next subsection in the context of *n*-body simulations. Here we shall review the main findings relevant to our interest (the role that chaos may play) of the orbital dynamics approach with galactic potential models that have reached an equilibrium configuration, despite the fact that two-body relaxation times are usually larger than the Hubble time. This can be achieved, as suggested above, by invoking a violent relaxation process and we shall elaborate on this in the next subsection. We shall include some results of calculations in which measures were taken to improve on the self-consistency of the models, and in this way hopefully increase the reliability of the orbital dynamics approach.

Various models of galactic potentials, from rather simple two-dimensional static cases all the way to fully three-dimensional (triaxial) ones, including possibly time dependence as well, have been used over the years in the study of clusters and galaxies by means of orbital dynamics. Our aim here is, in accordance with the subject of this book, to discuss only those aspects of orbital dynamics in which chaos could possibly be relevant. In some cases this will also require a brief review of the necessary background. The classification of orbits into various kinds has been one of the major tools in attempts to provide coherent and realistic galactic models. The natural tendency in these efforts was to look for regular orbits that could fit the classification schemes. Irregular ones, whose existence could not be ruled out in many cases, were considered hopefully to be rare and insignificant.

One of the simplest potentials, considered adequate for the investigation of planar orbits in very flattened systems, is the logarithmic potential

$$\Phi_L(x, y) = \frac{1}{2}v_0^2 \ln[R_c^2 + x^2 + q^{-2}y^2] \qquad (9.15)$$

where x and y are cartesian coordinates in the plane and v_0, R_c and q are constants. Clearly, the equipotentials have constant axial ratio, q. Thus the degree of non-axisymmetry (due to a bar, for example) is the same at all radii. For $r \equiv \sqrt{x^2 + y^2} \ll R_c$, Φ_L is approximately the potential of a two-dimensional harmonic oscillator (a potential generated by homogenous ellipsoids). In the opposite limit ($r \gg R_c$) and for $q = 1$ this potential gives rise to flat rotation curves. In planar polar coordinates (r, ϕ), Φ_L can be written as

$$\Phi_L(r, \phi) = \frac{1}{2}v_0^2 \ln[R_c^2 + ar^2 - br^2 \cos 2\phi] \tag{9.16}$$

where the positive constants a and b are expressible in terms of q in the following way: $a \equiv (1 + 1/q^2)/2$ and $b \equiv -(1 - 1/q^2)/2$. It can be shown (see, e.g., Binney and Tremaine's book) that all orbits in this potential, which are confined to a region not too far from the centre (i.e., within a few R_c), are *regular* (that is, confined to KAM tori). Some of them are *box orbits* (trace the well known Lissajous curves) and the others are *loop orbits* in which the particle rotates in a given sense around the centre while oscillating in radius.

However if one considers a slightly modified form for the potential, that is

$$\Phi_N(r, \phi) = \frac{1}{2}v_0^2 \ln\left[R_c^2 + ar^2 - br^2 \cos 2\phi - \left(\frac{r}{R_e}\right)r^2 \cos 2\phi\right] \tag{9.17}$$

a radically different behaviour of the orbits may appear as was found by Binney (1982). The new parameter R_e, determines the value of the radius from which escape of orbits to infinity is possible. When analysing in detail the behaviour of orbits arising in Φ_N, Binney found that the presence of the term involving R_e gives rise to radial variation of the axial ratios of the equipotential contours, and thus modifies the fundamental frequencies associated with the orbits. There is little doubt that the transition to non-integrability in this case is in accordance with the KAM theory. As Binney himself noticed, for $r \ll R_c$ and $q = 1$, Φ_N is formally quite similar to the Hénon and Heiles potential (compare with 6.34) and it is thus reasonable that the process of transition to non-integrability in a system governed by it is similar to that described in Chapters 2 and 6.

We have already mentioned in the second chapter of this book the notion of a third isolating integral in the context of galactic potentials. We have also seen in Chapter 6 the importance of the existence of isolating integrals in Hamiltonian systems. The existence of the third integral (even in some approximate way) thus has profound implications on the integrability (or near integrability), and therefore the regularity (that is periodicity or quasi-periodicity), of all orbits of some relevant Hamiltonian system. This issue, which has played a central role in orbital dynamics, is reviewed in detail in Contopoulos (2002), both in its general context as well as for particular types of galactic potential. With the first numerical calculations the question of integrability and of the existence of the third integral could be attacked directly and the pioneering study of Hénon and Heiles (HH) and a number of studies performed in the same period by Contopoulos and Woltjer and by Barbanis demonstrated the existence of irregular (alternatively referred to as 'chaotic' or 'stochastic') orbits. As discussed in Chapter 2, the HH problem has provided a clear-cut counterexample to the postulated omniexistence of a third isolating integral of motion in axisymmetric potentials. Moreover, this simple-looking problem has been very instrumental in the development of the general mathematical theory of the loss of integrability and the emergence of chaos in Hamiltonian systems – the KAM theory.

Although it is difficult to see how a 'realistic' galactic potential can be of the HH form, that work was clearly a breakthrough, not just of theoretical value. If chaotic orbits exist in such a simple axisymmetric potential, it is very reasonable to expect that such orbits will exist in more realistic applications (in which the potential is usually more complicated) as well. The above findings of Binney confirm this expectation. The transition from integrability to stochasticity (as a suitable perturbation parameter is varied) was first approached, in the astronomical context, by means of a qualitative theory of the third isolating integral.

The KAM theorem and subsequent mathematical developments (see Chapter 6), which have provided the theoretical basis for the understanding of the transition from integrability through near-integrability to global stochasticity in general Hamiltonian systems, have begun to find their way into galactic orbit dynamics, mainly through studies by Contopoulos and his co-workers. The review by Contopoulos (1987) and his recent book (2002) contain summaries of these studies and a complete list of references. Still, many researchers considered the direct relevance of this type of study to actual galactic models not straightforward (see the book by Binney & Tremaine, 1987).

It is rather clear that very close to integrability chaotic orbits are rare (the region of phase space occupied by them is small). Two-dimensional, time-independent potentials give rise to a two-degree-of-freedom Hamiltonian system, in which the orbits are confined to a three-dimensional energy shell in phase space, and the KAM tori are two-dimensional. Thus possible chaotic orbits are confined by the tori and transition to global stochasticity can occur only when the last KAM torus is destroyed (see the description of this process in Chapter 6). In a class of models of this kind it is possible to set the 'nonlinearity' or perturbation parameter to be actually the energy of the orbit (see the HH example), and thus only the high-energy orbits are globally stochastic. Even if we consider orbits whose energy is above the threshold for the existence of the last KAM torus, the timescale for the penetration of the orbit through its destroyed surface (the cantorus) may be very long (e.g., more than the Hubble time). It is thus important to calculate, within specific two-dimensional galactic models, what are the critical values of the energy for the appearance of cantori, and for what energies the penetration rates through the cantori are appreciable. These considerations are necessary if one wishes to find if and when stars can be expected to escape from a galaxy. Numerical studies (see Contopoulos, 1987) using two-dimensional potentials indicated that a very large number of orbital periods (and thus timescales significantly longer than Hubble time) are required for a stellar escape, even if the energy is both above the critical value for the destruction of the last KAM surface and above the escape energy. Thus although chaotic orbits were found and investigated in detail for many two-dimensional potentials (including the bar-like planar potential 9.17), their role in actual galactic models based on a two-dimensional potential seemed quite insignificant.

Even if one assumes that the galactic potential is triaxial, that is, three-dimensional, irregular orbits seemed at first to be quite insignificant to galactic dynamics, at least in the context of orbital dynamics. On the most elementary level, it is clear that if the galactic potential is three-dimensional (triaxial), KAM tori are three-dimensional and the dimension of the energy shell in phase space is five. Thus chaotic orbits are not confined by the tori and may wander in phase space by Arnold diffusion, even if KAM tori still exist. However, as we have already indicated several times in this book, this process is extremely slow and indeed specific studies have indicated that its extent in galactic models is insignificant in Hubble time. However, triaxial potentials admit a significantly larger variety of irregular orbits than axisymmetric potentials do. Indeed, in a pioneering study Goodman and Schwarzschild (1981) found that a significant fraction of the orbits in relatively simple triaxial potentials (those generated by a density law dating back to Hubble and having a smooth core) are stochastic. In this study it was, however, found that the stochastic orbits behave quite regularly, that is, do not significantly mix in phase space for as long as ∼100 orbital oscillations (i.e., more than Hubble time). These 'semi-stochastic' orbits, as they were called, did not fill all the domain visited by all unstable orbits of the same energy. As these orbits had the possibility of rapid spreading (by resonance overlap) and filling of all the accessible phase space, they could do so only very slowly, by means of Arnold diffusion. This study utilised a method, which had been proposed by Schwarzschild (1979), to improve on the self-consistency of the models based on orbital dynamics. The idea behind Schwarzschild's method was to calculate many orbits in a given density distribution, and then to populate the orbits in such a way as to reproduce the postulated distribution.

Experimenting with more complicated scale-free (that is, without a finite core radius like in Φ_L above) triaxial potentials, thought to be appropriate for modelling elliptical galaxies, Schwarzschild found in 1993 that in these models *most* box-like orbits are stochastic, the only exception being those orbits that lie near stable periodic orbits which always remain far enough from the centre. If the relative number of irregular orbits is large, the question of self-consistency of the orbital dynamics approach clearly emerges. Is it safe to assume that the underlying potentials persist over the required time? In the cases investigated in his study Schwarzschild gave an affirmative answer, based on the fact that the Hubble time is of the order of approximately only 50 orbits in his halo models. However, he restricted his conclusion to the consideration of self-consistent equilibrium of a galactic halo by itself, when possible instabilities and the process by which halos are formed are excluded. A number of careful simulations of single-particle motions, using the orbital-dynamics approach, performed subsequently seemed to indicate however that the collisionless approximation cannot guarantee that the density distribution would not evolve on a Hubble time. In these calculations ensembles of orbits were evolved with the orbits being carefully tested for stochasticity by using several criteria, like Liapunov exponents and Laskar's frequency-mapping technique (see above in this chapter). The potentials used in these calculations were chosen from among those which may have relevance to real galaxies. Hasan, Pfenniger and Norman (1993) found significant time evolution of the density distribution of orbits in a potential appropriate for barred galaxies, and Merritt and Fridman (1996) showed that this might be the case in elliptical galaxies as well.

In the mid 1990s it became observationally clear (using the Hubble Space Telescope) that central cusps in the light distribution are the rule rather than an exception in elliptical galaxies. In many cases, including spiral galaxies, indications of pointlike central mass concentrations (possibly very massive black holes) were found. These findings were clearly in contrast to the assumption that galaxies have a constant-density core and thus a harmonic potential near the centre (as in some widely used classes of integrable potentials allowing only regular orbits). Motivated by these findings Merritt and Fridman (1996) found that in potentials appropriate to a strongly cusped triaxial mass distribution the stochastic orbits seem to diffuse over the *entire* energy surface in less than a Hubble time (in contrast to above-mentioned result of Goodman and Schwarzschild). Moreover, in this study Merritt and Fridman could not find a self-consistent fully stationary solution as the timescale for chaotic diffusion, even though it is quite long compared to a crossing time, it may be short (near the centre) compared to Hubble time. The nature of orbits in triaxial models with high central condensations was carefully examined in subsequent studies by Merritt and Valluri (1996), and it was found that such orbits do indeed tend to be strongly stochastic and diffuse throughout all their available phase space on a timescale of the order of a few tens of dynamical times.

The effects of spatio-temporal irregularities in the underlying smooth potential, used in studies based on orbital dynamics, also became of interest in this context following the suggestion of Pfenniger (1986) that noise may significantly affect inherently unstable chaotic orbits in times shorter than t_R. Merritt and Valluri (1996) found that indeed random time-dependent perturbations in the potential are likely to enhance the rate of the diffusion of orbits they calculated. Habib, Kandrup and Mahon (1997) investigated this type of effect in more detail by explicitly including weak friction and noise in the calculation of orbits in non-integrable two-dimensional potentials. Habib and co-workers found that even in two-dimensional, but strongly chaotic, potentials, the statistical properties of orbit ensembles (and indeed the structure of phase space itself) were altered dramatically (on a timescale much shorter than t_R) by the presence of friction and noise.

All of these studies have essentially been modifications of the orbital dynamics approach, but it is reasonable to conclude from them, and some similar calculations, that the full gravitational n-body problem can be expected to contain a significant number of irregularities and to exhibit evolution on a short (compared to t_R) timescale. Thus, quite a strong evolution of the phase-space distribution is expected to occur already in time-independent (but non-integrable) potentials. The mechanism driving this evolution was identified as being associated with chaotic diffusion or *mixing*. The numerical experiments mentioned above suggest that mixing behaviour does indeed occur in some realistic potentials and moreover, it is significantly enhanced by cusps and irregularities (which have been modelled by noise and friction) in the potential.

What is the implication of these findings to galactic dynamics? It is quite clear that this question cannot be answered within the realm of orbital dynamics. If the stochastic orbits are causing a slow (but not too slow) evolution of the halo, chaos probably plays a role in galactic evolution. Moreover, it is possible that the underlying potential is not only evolving in time, but also cannot be assumed to be smooth. We sketch below some recent ideas on this subject drawing on n-body simulations.

Chaos and the evolution of n-body systems

As pointed out in the opening remarks of this section, direct n-body calculations are by definition truly self-consistent. With t_R appropriate for galaxies being much larger than Hubble time, these systems are essentially collisionless and thus the problem of relaxation, so central to the orbital dynamics approach, is naturally eliminated when one simulates the whole n-body cluster or galaxy.

The full gravitational n-body problem is, however, formidable and while it is clear that such a system must be chaotic (starting from such low values of n as three!) it is far from clear into which category of randomness in Hamiltonian systems (see Chapter 6) it should be classified. Do gravitational n-body systems belong to the *mixing systems* category or are they more (or less) random, and under what conditions? The answer to this is still largely unknown. Full numerical simulations of the gravitational n-body problem can obviously be very valuable in trying to find out what can be said about these aspects of the problem, but it is still not clear what is the effect of the potential softening (see below) and the relatively small number of particles employed in such calculations.

Even though gravitational n-body systems obey the collisionless approximation, the existence and stability of steady states is not guaranteed when a significant amount of chaos is present. If we want to invoke the lack of such equilibrium we need mixing in *phase space*, that is the spread of localised volumes over large portions of the full phase space of the problem. Moreover, this process must operate on sufficiently short timescales, and for this to occur, the system must be sufficiently chaotic (e.g., to belong to the mixing systems or even more random category, see Chapter 6).

It seems, therefore, that a serious analytical effort in the direction of examining the stability of the gravitational n-body system per se is necessary. Combined with judicious numerical simulations, it can perhaps lead us to a better physical understanding of the collisionless relaxation process, and provide an estimate for its characteristic timescale. We also have to identify what is the degree of randomness of the gravitational n-body system and under what conditions. Such studies are necessary to determine what is the role of chaos in galactic dynamics, but they are highly non-trivial, and this is perhaps the reason that at the time of writing this book, there are still no conclusive results.

Quite some time before detailed studies of chaos in physical systems became commonplace, Miller (1964) found that in a gravitational n-body system local exponential divergence of nearby trajectories occurs. Later on this result was shown to be an extremely robust phenomenon with an exponential timescale that is largely independent of initial conditions, perturbations, and the value of n. This instability has been attributed to close encounters and as such should not depend on the integrability, or lack thereof, of the mean gravitational potential.

From the theory of chaos we know that chaotic orbits deviate exponentially at a rate defined by the largest Liapunov exponent and this defines also a timescale, the Liapunov time t_λ, which should give an estimate for the time in which such chaotic orbits mix. It is therefore reasonable to think of the Liapunov time as the timescale relevant for collisionless relaxation in the gravitational n-body problem.

We have already mentioned the collisional relaxation time t_R of an n-body system and now we add, for reference, the rather trivial notion of a dynamical, or crossing timescale t_c, whose meaning is self-explanatory. In stellar clusters and galaxies obviously $t_R \gg t_c$ (e.g., for a typical galaxy composed of solar mass stars $t_R \sim 10^{14}$ years, while $t_c \sim 10^8$ years). It is also simple to show (see, e.g., Binney & Tremaine, 1987) that in general, for three-dimensional n-body systems, $t_R/t_c \propto n/\ln n$. If the Liapunov time is indeed the relevant measure for collisionless relaxation the main question is what is its value in comparison with t_c? This problem can be addressed numerically, by performing n-body simulation, or theoretically (see below) but, as suggested above in general, such questions are best addressed by combining both approaches. Numerical simulations of n-body systems today constitute an important branch of theoretical astrophysics. Several comprehensive reviews, written by the leading experts, on the various methods and results of such simulations, exist in the literature (e.g., Sellwood, 1987; Aarseth, 1998). We shall mention now some results that shed some light on the questions raised above pertaining to collisionless relaxation.

In a rather large number of studies conducted mainly by Kandrup and his collaborators (e.g., Kandrup *et al.* 1994 and references therein) it was generally found that t_λ is of the order of t_c, the ratio between the two decreasing weakly with n (tending to $t_\lambda / t_c \approx 0.68$ for $n > 200$). Goodman *et al.* (1993) found a similar result but they differed in the identification of the instability source. They emphasised the role of close encounters in driving the instability while Kandrup proposed a global instability due to the mean field, reinforced by close encounters. It may be that this difference is due to the different models they considered (see the discussion in Contopoulos, 2002 on this subject). In any case, it would be fair to say that numerical studies largely confirm that gravitational n-body systems are in general unstable.

On the theoretical side, most studies have employed a method pioneered originally by Krylov and introduced to the gravitational n-body problem by Gurzadayan and Savvidy (1986). It is a geometrical approach involving, in the case of the n-body problem in three spatial dimensions, the examination of the $3n$-dimensional Riemannian configuration manifold. When one introduces a metric on this configuration manifold, local quantification of the divergence of trajectories can be found by calculating the Riemann curvature at the desired point. For details see Gurzadayan & Savvidy (1986) or the summary in Contopolous (2002). A particularly comprehensive and lucid presentation of the rudiments of the mathematical material, together with an interesting application to the problem at hand can be found in El-Zant (1997). The *negativeness* of the Riemann curvature is known to be a measure of local trajectory divergences, since negative curvature implies what is called exponential instability. As hinted in Chapter 6, the classification of chaotic Hamiltonian systems can be performed using different measures of the curvature. For example, in the characterisation of the highly random C-systems (see Chapter 6) Anosov used the two-dimensional Riemann curvatures. Obviously, less random systems (like K-systems and mixing systems) are defined by properly defined average curvature of the configuration manifold. A concise introduction to the mathematics involved can be found, for example, in Appendix 1 of the book by Arnold (1978) or in Abraham & Marsden (1978).

Gurzadyan and Savvidy (1986) have found the relevant scalar curvature of the system and found a Liapunov time of $t_\lambda = <F>/<v>$, i.e., equal to the ratio of the average force to the average velocity. For the average force they have used a particular prescription, based on what is called the 'Holstmark distribution' between only neighbouring particles and obtained $t_\lambda = n^{1/3} t_c$. However, using an average force of the whole system, the relation $t_\lambda = t_c$ follows. This is identical to the estimates of Kandrup and his collaborators and similar to those of Goodman *et al.*, who suggested the following weak dependence on n: $t_L = t_c / \log(\log n)$. Further studies of this problem based on the geometrical approach include those of El-Zant (1997) and di Bari and Cipriani (1988).

Regarding the question of classification of the gravitational n-body problem, already Gurzadyan and Savvidy (1986) have shown that the gravitational n-body system does not satisfy the requirements for it to be classified as an Anosov C-system. However, as Kandrup showed, in 1990, the probability of a positive curvature *decreases* with *increasing* n. El-Zant (1997) pointed out that conventional methods used for quantifying phase-space mixing, like the computation of the KS entropy (a suitably defined average of positive Liapunov exponents), are not adequate for such systems. We have seen in Chapter 6 that positive KS entropy is the defining property of mixing systems, and thus it is a sufficient condition for a Hamiltonian system to approach a statistical equilibrium by means of a 'hard' enough chaos. This property is guaranteed, however, only if the phase space of the system is compact, as indeed it is in most 'conventional' Hamiltonian systems, but not in the gravitational n-body system in which singularities (collisions and escapes) are possible. Thus, if no softening is assumed in the Newtonian potential, the situation is more complicated – e.g., no final state is guaranteed. This led El-Zant to propose that local geometrical methods be used instead of the global ones. He also proposed that a less stringent condition for randomness seems to be supplied by the mean (Ricci) curvature employed in his work. He performed high precision n-body simulations and calculated the mean curvature on the trajectories, and was led to the conclusion that the question of evolutionary timescales, which is obviously cardinal if one wants to apply these results to realistic systems, can be estimated by finding the absolute value of the mean curvature. He also found that significant softening of the potential tends to drive the mean curvature to be positive, and this has consequences for the interpretation of large n-body simulations. For a more recent detailed study of softened gravitational systems in this context, see El-Zant (2002).

Sophisticated numerical simulations of collisionless systems, combined with theoretical considerations of the kind mentioned above, have contributed to our understanding of the stability and evolution of such systems. There is no doubt that chaos plays a dominant role in the processes driving their evolution on relatively short timescales. It is within this context that the process of *violent relaxation*, proposed by Lynden-Bell in 1967 in the context of collapsing collisionless gravitational n-body systems, is naturally interpreted. The process, which is driven by time fluctuations of the gravitational potential, has a timescale of the order of the crossing time t_c (hence the name 'violent'), and it actually produces a 'quasi-equilibrium', because it acts in the same way on all masses and thus does not cause energy equipartition and thereby separation between massive and light stars. This is in contrast to collisional relaxation, which leads to a true 'thermodynamic' equilibrium with energy equipartition and thus mass separation in the cluster.

It is clear that more work should be done in this direction, since it seems that Hamiltonian chaos (with its various subtleties) may have an important role in the behaviour of gravitational n-body systems. The complexity of this problem, relative to numerous other problems (some of which are mentioned in this book) in which chaos was shown to be instrumental, impedes the achievement of clear-cut and elegant results. The presence of various different timescales, such as those resulting from exponential divergence of trajectories and possibly also collective instabilities, discreteness noise etc., complicate the physical interpretation of the results, that is, definite determination of what is the implication of the apparent chaos on the evolution of a gravitational n-body system. In particular, in the true unsoftened problem, with its singularities, the configuration space is 'incomplete' and some relevant mathematical theorems may not hold. In contrast, it can be shown that infinite spherical gravitational systems with a smooth density distribution give zero curvature, in agreement with the statement that collisionless spherical systems are integrable. Still, the quest for unravelling the possible influence of chaotic behaviour on the physical characteristics of realistic n-body systems (different galaxy and cluster types), and the relevant timescales of the instabilities remain, in my view, a primary challenge in galactic and cluster dynamics.

10

Irregularly variable astronomical point sources

> Our whole knowledge of the world hangs on this very slender thread:
> the re-gu-la-ri-ty of our experiences.
> Luigi Pirandello, *The Pleasure of Honesty.*

The classical astronomical sources of radiation, stars and star-like objects, are spatially unresolvable. This fact does not exclude, however, the possibility of time-variability, and indeed a variety of point sources have been found to possess such *intrinsic* variability, that is, one that remains in the light curve after atmospheric and other local effects are properly eliminated. Different classes of objects exhibit a wide range of variability timescales, often depending also on the spectral range.

When an astronomical source emits a time-variable signal, the natural first step in the data analysis is to search for periodicity. The identification of well-defined periods provides extremely valuable information, which can be used in understanding the relevant physical processes and therefore in constructing viable physical models of the astronomical source. This is obvious if we consider as an example the simplest periodic physical system of them all, the harmonic oscillator. Its period immediately reveals the ratio of the inertia to the restoring force and since every sufficiently small oscillation is to a good approximation harmonic (i.e., linear), the number of physical systems modelled with the help of this paradigm and its generalisations (multidimensional linear systems) has been very large. In astronomy the most prominent examples of this kind are pulsating stars. The famous period–luminosity relations of the classical Cepheids and other pulsating variables have not only been instrumental in the development of stellar pulsation theory, they have also played an important role in establishing the cosmic distance scale. Numerous other astronomical systems exhibit intrinsic periodic variations as well, or have periodic intensity variations resulting from geometrical effects (e.g., eclipses and other periodic phenomena related to the Kepler motion in close binary stars).

There exist, however, variable astronomical sources whose light curves are definitely not periodic nor even quasiperiodic. Some of them, even though they belong to a class in which most sources are regular, appear to be aperiodic. This is also the case in pulsating stars and the natural tendency, until quite recently, has largely been to ignore such pulsators (after classifying them as 'irregular') and this is quite understandable. It is difficult to expect that a professional astronomer should devote precious telescope time to mere monitoring of an irregular pulsating star about which, it seems, nothing can be said, except that its light curve is irregular. In some classes of variables sources, like quasars for example, irregular variability is the rule rather than the exception. This property has generally been attributed to stochasticity, brought about by a very large number of degrees of freedom inherent in the physical processes thought to be operative at the source.

We have seen in Chapter 5 of this book that modern findings on irregular time series analysis indicate that aperiodic signals may be the result of deterministic chaotic processes. When conventional techniques fail to detect periodicity or even quasiperiodicity one may still look for low-dimensional behaviour. This kind of analysis requires, as we have seen in Chapter 5, a very large amount of high-quality data. Astronomical time series, resulting from the photometric monitoring of a particular irregular source, typically contain a limited number of unevenly spaced data points which are often contaminated by noise. Thus the task of a nonlinear analysis of such a signal is often very difficult, if not impossible. Fortunately, photometry is usually technically much simpler than other types of astronomical observation and amateur astronomers favour variable sources. Also, some of the irregular sources are interesting enough so that data from repeated observations of them are available. Still, it appears that a meaningful search for low-dimensional behaviour in astronomical time series must rely on some definite theoretical method to complete and analyse the raw data. Tools of this kind have recently been developed and successfully applied to pulsating stars.

Assume that such an analysis of an aperiodic signal, emitted by some variable astronomical source, *definitely* indicates the presence of low-dimensional and fractal behaviour. As we have hinted above and will discuss in more detail later on, such a conclusion cannot usually be reached easily. What lesson can the theorist learn from this? A definite observational period certainly appears more valuable than a fractal dimension derived from the data set. Indeed, with the present state of affairs, we are unable to propose any complete and well founded general principles. Some suggestive basic observations can, however, be made. Low-dimensionality presumably hints at some simple underlying physical mechanism, so that there is no need to invoke a very large number of degrees of freedom or random processes in order to model the system. For example, in the case of pulsating stars often a very small number of interacting modes and possible resonances between them can be identified as the mechanism behind aperiodic variability. This observation is far from trivial and it largely escaped the attention of astrophysicists until quite recently. When the basic physical mechanisms are understood one can proceed to full modelling and check if the predicted behaviour has the same nonlinear characteristics as the observed data.

In this chapter we shall discuss several astrophysical problems in which this kind of approach has been fruitful, at least to some extent. We shall start by describing several explorations of chaos and low-dimensionality in observational data, focusing mainly on irregular stellar pulsators, for which the results appear to be the most reliable and convincing. Data analyses performed on irregularly variable sources, thought to be accreting objects, have failed so far to provide definite detection of low-dimensional behaviour. However, these studies have supplied some important insights, among them caveats to the analysis itself, and we shall discuss these briefly as well. The theoretical part of this chapter will mostly deal with stellar pulsation, starting with a few specific models of chaotic stellar pulsators and including also a discussion of new and general approaches to nonlinear pulsation. We shall conclude with a few additional, perhaps less definite but certainly promising if not definite, ideas on accreting X-ray binaries.

10.1 Observed aperiodic astronomical signals

The output of photometric monitoring of astronomical systems is naturally cast in the form of time series. We have devoted a full chapter of this book to the description of ideas and techniques that can be applied to the analysis of such series. Some of the difficulties arising during the analysis of astronomical time series, such as uneven sampling and limited coverage, have already been mentioned. The presence of observational errors poses an additional non-trivial challenge to the data analysis. We may expect that in most cases astronomical time series will be contaminated by instrument errors and may also be corrupted by signals resulting from irrelevant sources. Thus a sufficient signal-to-noise strength may often be unavailable over the timescale of interest.

These difficulties are by no means exclusive to irregular aperiodic signals. Astronomers analysing periodic and quasiperiodic data have always been well aware of such problems, and a large amount of work has been done in an effort to devise effective techniques to 'cleanse' the data. As the description of these findings is certainly beyond the scope of this book, we refer the reader to the review by Scargle (1997) on the new methods for studying periodic and aperiodic astronomical signals. Scargle himself and his collaborators have contributed substantially to this subject in a series of papers, the references to which, along with other works, can be found in the above review paper.

We turn now to the description of several attempts to analyse aperiodic astronomical time series, which have quite recently appeared in the literature. As stated above we shall discuss two different types of variable astronomical objects: pulsating stars, for which there exist some reliable results, and accreting systems (X-ray binaries and AGN), for which the situation is less satisfactory. Some of the pitfalls of a straightforward application of the Grassberger–Procaccia (see Chapter 3) method on the reconstructed state space in the context of accreting X-ray binaries were described by Norris and Matilsky (1989). They and Scargle (1990), in a more general work, reached the conclusion that an adequate predictive analysis (like a model or mapping) is most probably indispensable in the studies of irregular astronomical series, which suffer from insufficient amounts of data and a relatively poor signal-to-noise ratio. Further general caveats and in particular those relevant to studies of AGN variability were given by Vio *et al.* (1992). It is recommended that these considerations be taken into account whenever one tries to find low-dimensional behaviour in astronomical data.

Pulsating stars

We start with what appears to be the first convincing detection of low-dimensional dynamics and chaos in an astronomical signal – the analysis of the light curve of the pulsating variable R Scuti. This star is a Population II Cepheid (of the RV Tau type) and has been known to pulsate irregularly, displaying variations in the depth of luminosity minima and large modulations in the amplitude. The typical 'period' (actually the characteristic timescale of light variations) is ~ 150 days, while the amplitude modulation timescale appears to be larger by a factor of ~ 20. Motivated by the irregular pulsation found in numerical hydrodynamical calculations of W Virginis (a related type of Cepheid) models, which were shown to be a manifestation of low-dimensional chaos (see the review by Buchler 1993), Buchler and co-workers (1995) performed a detailed analysis of the observational light curve of R Scuti (R Sct) and found low-dimensional chaotic behaviour. Another star of this type, AC Herculis, was analysed as well in subsequent work by the same researchers. Very recently Buchler, Kolláth and Cadmus (2004) found low-dimensional chaos also in the light curves of a number of semi-regular variables.

In what follows we shall focus on the work done on the light curve of the irregular variable R Sct and only briefly mention the results on other stars, which were achieved through similar analysis. A rather large amount of data on R Scuti's (and of the other stars as well) light curve has been collected over the years, mainly by amateur astronomers. In particular, AAVSO (American Association of Variable Star Observers) compiled data resulting from 31 years of observations of R Sct.

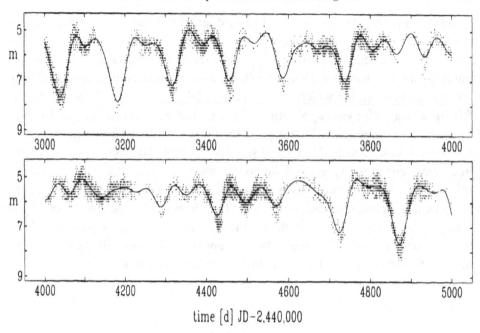

Fig. 10.1. Typical observed light curve segments of R Scuti. The raw data points are indicated by the dots and the smoothed filtered signal by the continuous line. Reprinted, with permission, from Buchler *et al.* (1996).

The large pulsational amplitude of this star has been particularly advantageous, reducing the relative importance of observational errors.

Before analysing these data sets, Buchler and his collaborators performed a number of preparatory steps in an effort to separate and eliminate the 'noise' from the signal. In this category they included observational errors, other external random contributions and also any possible high-dimensional but low-amplitude jitter, resulting perhaps from internal processes in the star, like e.g., turbulence. The raw magnitude data were binned, smoothed and interpolated with a cubic spline, and finally the high-frequency 'noise' was eliminated using a filter in the Fourier space (low pass). The raw data and the smoothed light curve are displayed in Figure 10.1. All such data preparation procedures are quite delicate and Buchler *et al.* have also experimented with other smoothing and filtering techniques, guided by Abarbanel *et al.* (1993). A detailed discussion of these, as well as further technical issues (see below) of their work, is given in Buchler *et al.* (1996).

After preparing the data, Buchler and co-workers performed an embedding procedure, based on the time-delay vector idea (see Chapter 5), in order to reconstruct the dynamics of the hypothetical underlying system (the global flow). They sought to find the dimension of the attractor by using the calculated values of the Liapunov exponents, i.e., the Liapunov dimension D_L (see Chapter 3). In spite of the fact that the R Sct data extend over many years of observations, the data set is still probably too small to estimate the fractal dimension directly from it. We do not know a priori what is the expected dimension, but even if it is as low as 4, for example, the rough criterion given in Chapter 3 requires at least 10^4 data points. Remembering that the original data set was also quite noisy, the number 10^5 appears to be a more conservative estimate of the number of data points needed. Thus, to facilitate the calculation, Buchler *et al.* resorted to finding a *discrete global mapping* in the reconstructed pseudo-state space, with the help of which any number of additional points could be generated. We have already remarked that a predictive theoretical tool or model is necessary for a reliable analysis of this kind and such mapping can answer, at least in principle, this need. We sketch here the procedure, as used by Buchler *et al.*, by means of which a mapping of this kind can be derived from the data. The technique, termed *the global flow reconstruction method*, is described in detail in a paper fully devoted to this purpose (Serre, Kolláth & Buchler, 1996).

Assume that the desired mapping has the general form

$$\mathbf{X}_{i+1} = \mathbf{F}(\mathbf{X}_i) \tag{10.1}$$

where $\mathbf{X}$ is the state vector in the reconstructed space (compare with the second expression in equation 5.18). The explicit form of such mapping is of course unknown, although its *existence* and equivalence to the original dynamics is guaranteed by the embedding theorems. A reasonable possibility is to try to represent the mapping function $\mathbf{F}$ by an expansion in an orthonormal set of functions (see Abarbanel *et al.*, 1993) and Buchler and co-workers indeed used this idea. The details of the original dynamical system in this problem are unknown, but the information that it is a pulsating star of a rather well defined type can be useful in the decisions that have to be made in the choice of an appropriate mapping (10.1). Stellar pulsation of Cepheids and related objects is usually only *weakly* nonlinear. This naturally led Buchler *et al.* to represent the mapping function by a finite polynomial expansion, that is, to set

$$\mathbf{F}(\mathbf{X}) = \sum_{\mathbf{k}} \mathbf{C}_{\mathbf{k}} P_{\mathbf{k}}(\mathbf{X}) \tag{10.2}$$

where $P_{\mathbf{k}}(\mathbf{X})$ are polynomials of order up to p, say, of the vector components X^α (with $\alpha = 1, 2, \ldots, d$) of $\mathbf{X}$. The subscript $\mathbf{k}$ is a shorthand for the combinations of integers $(k_1, k_2, \ldots)$ with $\sum_j k_j \le p$ and $\mathbf{C}_{\mathbf{k}}$ are constant vectors. The polynomials can be written explicitly as

$$P_{\mathbf{k}}(\mathbf{X}) = \sum_{\alpha} A_{\mathbf{k}}^{\alpha} \left(X^{\alpha_1}\right)^{k_1} \left(X^{\alpha_2}\right)^{k_2} \left(X^{\alpha_3}\right)^{k_3} \ldots$$

where $\alpha \equiv (\alpha_1, \alpha_2, \ldots)$ denote the vector component indices and the polynomial powers $\mathbf{k}$ have been defined before. The coefficients

$$A_{\mathbf{k}}^{\alpha} \equiv A_{k_1, k_2, \ldots}^{\alpha_1, \alpha_2, \ldots}$$

should be chosen so as to satisfy an orthogonality relation and the polynomials can be constructed using the data set. Buchler *et al.* chose a *singular value decomposition* algorithm to construct the polynomials and found this technique to be superior to other available approaches.

We shall skip here most of the technical details, related to the evaluation of the global map, and refer the interested reader to Serre *et al.* (1996). In that work the derivation of the map is explained in detail and it also includes a thorough validity test of the full procedure, by applying it to the Rössler oscillator, a well-known prototype of chaotic dynamical systems (see Section 4.2). The key finding was that the mapping in the reconstructed state space, derived by the method proposed by Buchler and his collaborators from *limited* data sets, chosen from numerical solutions of the Rössler ODEs, reproduced very well (qualitatively as well as quantitatively) the *full* dynamics of the Rössler system. They also found that the method is robust and works well even when a rather large amount of noise is added to the signal. Serre *et al.* also applied the method to the results of full numerical hydrodynamical calculations of a stellar pulsation model (W Virginis). The numerical data in this case appeared to be qualitatively similar to the Rössler data and this was the reason that the technique was so thoroughly tested on the Rössler oscillator itself. Without entering into any details of these numerical calculations (see below) we only point out that the global map derived from a subset of data gave satisfactory reproduction of the full data set it this case as well. In the analysis of the observational signals mentioned above, Buchler *et al.* experimented with various choices of the polynomial order p, the time delay τ, and embedding dimension d for the 'synthetic' data, generated with the help of the mapping (which was itself derived from a significantly smaller 'real' data set of R Sct). They found that the dynamics which most probably lie behind the R Sct signal can be effectively embedded in $d = 4$ dimensions, and that the Liapunov dimension of the attractor is $D_{\mathrm{L}} \approx 3.1$. The results of the analysis performed later on the AC Her data indicated that for that data set, D_{L} is approximately between the values of 2 and 2.5. These results indicate that the dimension of the dynamical system responsible for the pulsation in R Sct and AC Her may be as low as 4 and 3 respectively. In both cases the largest Liapunov exponent was found to be positive, indicating that the signal is indeed chaotic. By showing that the R Sct data set is incompatible with quasiperiodic behaviour and that other standard tools of linear analysis fail as well, Buchler *et al.* have further strengthened their conclusion that their discovery of low-dimensional chaotic behaviour in the observational signals of what appear to be highly dimensional systems (pulsating stars), is an exciting *reality*.

In subsequent, very recent work, Buchler, Kolláth and Cadmus (2004) presented analyses of the photometric observations of the light curves of five large-amplitude, irregularly pulsating stars R UMi, RS Cyg, V CVn, UX Dra, and SX Her. They eliminated multiperiodicity and stochasticity as possible causes of the irregular behaviour, leaving low-dimensional chaos as the only viable alternative. They then used the global flow reconstruction technique in an attempt to extract quantitative information from the light curves, and to uncover common physical features in this class of irregular variable stars that straddle (in the H-R diagram) the RV Tau to the Mira variables. Similarly, as in the case of R Sct discussed above, it was found that the dynamics of R UMi, RS Cyg, V CVn (and probably of UX Dra as well) take place in a four-dimensional dynamical phase space, suggesting that two vibrational modes are involved in the pulsation. The physical nature of the irregular light curves may be identified as resulting from a resonant mechanism with the first overtone mode having a frequency close to twice that of the fundamental one. This is similar to the case of R Scuti. If the fundamental mode is self-excited while the resonant one is by itself stable, we may expect chaotic behaviour as a result of the Shilnikov scenario (see Chapter 4 and later in this chapter).

Accreting systems

Since the launch of the first astronomical X-ray satellites it has been established that many compact galactic X-ray sources exhibit frequently irregular temporal variability. These sources have usually been identified as accreting low mass binaries, in which the accreting star is compact. Among the most famous examples we can mention Her X-1, Sco X-1 (the accreting object in these sources is a neutron star) and Cyg X-1 (thought to contain an accreting black hole). The X-ray light curves of these objects and other sources of this kind have been found to exhibit various degrees of irregularity, which has, at first, generally been attributed to the complicated nature of the hydrodynamical and/or magneohydrodynamical processes occurring in the accretion flows.

Her X-1 was the first object whose X-ray variability was tested for low-dimensional behaviour by applying a direct phase space reconstruction method (see Chapter 5). Voges, Atmanspacher and Scheingraber (1987) found that the X-ray signal from Her X-1 exhibits low-dimensional behaviour with a fractal dimension of ~ 2.3 during its 'on' phase (when the source is thought to be unobscured by other components of the system). A significantly higher dimension (between 7 and 9) was suggested by the data during the phase when the X-ray source is behind the accretion disc, and no low-dimensional behaviour at all could be inferred during eclipses.

Following this work a similar analysis was done for the Cyg X-1 data by Lochner, Swank and Szymkowiak (1989). They found that the data do not exhibit a truly low-dimensional behaviour and, at best, an attractor of dimension $\sim 9-10$ may be inferred. They also pointed out that noise inherent in the signal may significantly affect the results of such a straightforward analysis. This last issue was addressed in detail by Norris & Matilsky (1989), who conducted a series of tests on the Her X-1 data for the purpose of determining the reliability of the attractor dimension determination by Voges *et al.*

Norris and Matilsky calculated the correlation dimension of the data using the pair correlation function $C(r)$ in the manner explained in Chapter 5. They analysed several observations of Her X-1 and reached the conclusion that the signal-to-noise level in the EXOSAT and in the HEAO data sets is too low for a reliable conclusion on the existence of a low-dimensional attractor. Her X-1 contains within it a pulsar, which obviously contributes a periodic component to the spectrum and simulations with varying admixtures of random noise to such a periodic signal showed that this prescription can mimic the existence of a fractal dimension in the data. Motivated by these findings Norris and Matilsky proceeded to subtract the pulsed component from the Her X-1 data sets, in which Voges *et al.* found a low-dimensional attractor, and got a signal with just a characteristic noise signature. They then proceeded to argue convincingly that astronomical signals by themselves cannot, in general, provide clear-cut proof of low-dimensional behaviour, because they typically contain an insufficient quantity of data and have a relatively small signal-to-noise ratio. In order to circumvent this problem, Norris and Matilsky proposed that additional predictive tools should be used in the data analysis, and that these could be provided by a model or a mapping (as, for example, the one used in the case of pulsating stars and described above).

When analysing the X-ray light curve of Sco X-1, Scargle *et al.* (1993) indeed introduced a model with the help of which the light curve could be reproduced quite well. The model was based on a system called the *dripping handrail*, a coupled map lattice (see Chapter 7). We shall describe it in some detail in the last section of this chapter. Here we shall just summarise the most important findings on the light curves of Sco X-1 and other sources having similar characteristics (notably the AM Her systems, thought to be binaries containing an accreting magnetic white dwarf). It has been found that these light curves exhibit a broad peak in their power spectrum (the defining property of what has been called in the astronomical literature 'quasiperiodic oscillations' or QPO) and very low frequency noise, i.e., stochastic fluctuations on long timescales. Another interesting feature of many light curves of these sources (and of AGN as well, see below) is the self-similarity (or scaling) of the light curve itself. Scargle *et al.* used wavelet analysis and found that the scaling exponent of the Sco X-1 light curve is approximately 0.6. This means that if $X(t)$, say, is the signal then $X(\lambda t) = \lambda^\alpha X(t)$, where α is the scaling exponent and λ is an arbitrary constant. In practice, the value of λ is obviously limited by the finite time interval between successive data points and the total duration of the signal.

Scargle *et al.* stressed in the above work, and this conclusion is repeated in subsequent studies by the same authors, that low-dimensional behaviour was not found in these cases. Still, the data are clearly aperiodic, and some of their important characteristics can be found by using the appropriate tools of nonlinear analysis and by comparing them to the model (see below in the last section of this chapter).

Quasars, and more generally AGN, constitute an additional important class of sources, whose variability has for quite a long time been recognized as having an unpredictable and irregular nature. A number of studies, published in the 1980s and 1990s, were conducted for the purpose of determining the nature of the apparently random variability of such sources. It is quite clear that the identification of stochasticity (like shot noise or white noise) in the signal, or, in contrast, the presence of low-dimensional chaos, may have important implications for the physical models (or 'scenarios', as they are called) devised to explain the behaviour of these systems. Although the amount of data, collected over years of observation of some systems, is rather large, the variability studies have not provided major progress. This seems to be the result of some basic difficulties, which are inherent in the data analysis process itself, when performed on systems of this sort. Vio *et al.* (1992) addressed this issue in detail, critically examining the classical as well as more modern approaches to data analysis. We shall not discuss in detail all the important findings of this paper, but rather stress only several points related to the characterisation of observational signals by means of approaches developed in the context of nonlinear dynamics.

One important issue of this kind is the question of whether and how it is possible to find *intermittency* in a signal. We have already encountered this concept when discussing, in Chapter 4, the different routes to chaos. Intermittency was characterised there qualitatively by the existence of periods of enhanced irregular activity in an otherwise regular signal. A quantitative diagnosis of intermittency can be based on the determination of the *multifractal* (see Chapter 3) properties of the signal itself. Considering a time series $x(t)$ one can look at the distribution of the increments $\Delta x(t) \equiv x(t + \Delta t) - x(t)$ for a given time difference Δt and regard $(\Delta x)^2$ as the positive-definite measure of the power in the fluctuations (or activity) in the signal. Looking at the points in the signal for which $(\Delta x)^2$ is larger than a given threshold, Δ_0 say, it is possible to calculate the generalised fractal dimensions $D^{(q)}$ (see 3.91) of the set composed of these points. As we already know, the values of $D^{(q)}$ for increasing q, refer to the distribution of the higher moments of the signal. A signal for which all $D^{(q)}$ are equal is composed of a true fractal (a monofractal) and thus there is no intermittency in the signal. In contrast, a multifractal distribution of the data points in the signal indicates intermittency, because computing $D^{(q)}$ for increasing q is essentially equivalent to computing $D^{(0)}$ for increasing values of the 'activity' threshold Δ_0 defining the data set. Vio *et al.* performed a multifractal analysis of the signals of a BL Lac object, PKS 2155-304 and the quasar 3C 345 and found that in the X-ray light curve of the former object, displaying a continuous 'white noise' activity superposed on some long-term trend, all generalised dimensions are almost equal. In contrast, the light curve of 3C 345 shows clear evidence of multifractality with a significant decrease of $D^{(q)}$ with increasing q and thus intermittency, which indicates a nonlinear process. The uneven sampling and observational errors complicate the analysis and we refer the reader to the original paper for technical details.

In addition to suggesting that multifractal analysis should be used as a diagnostic tool in AGN lightcurves, Vio *et al.* demonstrated that the straightforward application of phase space reconstruction methods to the data may lead to conceptual errors. They showed that the saturation of the correlation dimension D_C (see Chapter 5) does *not* always indicate that the analysed signal has a low-dimensional attractor. By analysing a *stochastic* signal having a power-law power spectrum Vio *et al.* obtained correlation dimension saturation, a characteristic property of a signal resulting from a true chaotic system. They thus concluded that as the derivation of the system's dynamics from a time series may be more involved than just a simple embedding procedure, it is necessary to have some a priori information on the system under study. This further strengthens the conclusion mentioned above, that appropriate modelling is needed before reaching a reliable conclusion that a system is indeed low-dimensional. In a subsequent study Provenzale *et al.* (1994) analysed in detail the light curve of the quasar 3C 345 (mentioned above in the context of intermittency) and found that while direct phase space reconstruction does suggest the presence of an underlying low-dimensional attractor, more sophisticated tools, like phase randomisation and signal differentiation (see the paper for details), show that the above result is in fact generated by the power-law shape of the spectrum, and by the long memory in the signal and *not* by low-dimensional deterministic chaos. They concluded that the signal from 3C 345 is consistent with the output of a nonlinear stochastic process having a well-defined intermittent nature.

We conclude this section on data analysis with a general remark, based on the work of Provenzale *et al.* (1994). Low-dimensional chaotic dynamics and strange attractors are extremely important new concepts and they have been shown to be relevant to various physical systems studied in the laboratory. Determination of whether or not these notions are relevant to the observations of astrophysical (and geophysical) systems, which by their nature are not controlled experiments, still remains, in most cases, a nontrivial challenge. It seems that only very few, if any, presumed discoveries of low-dimensional attractors in observational data of natural systems, have survived refined scrutiny. Still, the discovery of genuine low-dimensional behaviour in astrophysical systems is of great interest, and it seems that the analysis of signals from some irregular pulsating stars by Buchler and his collaborators is among the most promising in this context. This is the reason for devoting most of the following theoretical part of this chapter to pulsating stars.

10.2 Nonlinear stellar pulsators – some specific models

Among the different attempts to analyse aperiodic observational astronomical time series discussed in the previous section, irregular stellar pulsators seem to be the best candidates for low-dimensional chaotic behaviour. It is thus only natural to look for physical mechanisms in models of pulsating stars that are capable of producing chaotic variability. Indeed, a number of such theoretical models exist in the literature, some of which had been proposed well before the above mentioned detailed data analysis appeared. Baker's one-zone model was very successful in identifying the physics of pulsational instability of stars in general, and consequently the first attempts to understand chaos in stellar pulsation were based on similar simplistic models.

We have already mentioned in Chapter 2 a simple one-zone *dynamically* driven stellar oscillator and such models of the outer layers of long-period variables will be briefly discussed later on. However, the model that we wish to describe first is based on a *thermally* driven one-zone stellar oscillator. The dynamical system resulting from this model is very similar mathematically to the Moore and Spiegel oscillator, which was proposed in the context of thermal convection and discussed in Chapter 2 also. The possibility that the MS oscillator may be relevant to irregular stellar pulsation had not escaped the attention of those authors and, together with Baker, they explicitly proposed its application to stellar pulsation (Baker, Moore & Spiegel, 1966). We discuss here in some detail a related system, resulting from one-zone modelling of an extended ionisation zone in a star, which was proposed by Buchler and Regev in 1982, when deterministic chaos in dynamical systems had just started gaining widespread attention.

The dynamical equation of such a zone, placed on top of a stellar 'core', was given in Chapter 2 in Equation (2.5), and we rewrite it here for convenience:

$$\ddot{R} = -\frac{GM}{R^2} + 4\pi R^2 \frac{P(\rho, s)}{m} \equiv f(R, s) \tag{10.3}$$

where the symbols here have the same meaning as in Equation (2.5), but the specific entropy s of the zone is now considered as a time-dependent function, to be determined by an additional differential equation.

The required equation for $s(t)$ is obviously derived from the heat equation and in the one-zone model it has the form

$$\dot{s} = -\frac{1}{T} \frac{L - L_{\mathrm{c}}}{m} \equiv \epsilon h(R, s) \tag{10.4}$$

where T is the temperature of the zone, L its luminosity and L_{c} the (constant) luminosity of the core. The terms f and g on the right-hand side of Equations (10.3) and (10.4) can be expressed as well-defined functions of $R(t)$ and $s(t)$ by using

(i) mass conservation, i.e., $\rho(R) = 3m/4\pi(R^3 - R_{\mathrm{c}}^3)$, where R_{c} is the constant core radius,
(ii) equation of state, giving $P(\rho, s)$ and $T(\rho, s)$,
(iii) a radiative transfer equation for L.

Buchler and Regev approximated the latter by the following expression

$$L = \frac{4\sigma(4\pi R^2)^2}{3\kappa(\rho, T)} \frac{T^4}{m'} \tag{10.5}$$

which is a straightforward application of the diffusion approximation to the radiative transfer through the zone. Here σ is the radiation constant, κ the opacity and m' is some fraction of m (the mass of the zone).

Scaling all the physical variables by their typical values makes the variables s and R, as well as the functions $f(R, s)$ and $h(r, s)$, nondimensional and brings out the value of the nondimensional constant ϵ in terms of the units used in the scaling and physical constants. Obviously, ϵ reflects the ratio of the dynamical to the thermal timescales of the system, and this quantity was naturally chosen to be the control parameter for the problem.

Buchler and Regev decided to choose the conditions in the zone to be close to the second ionisation of helium and calculated the equation of state accordingly. The details of the model, including the opacity law, can be found in Buchler & Regev (1982). Figure 2.3 in this book, which is the result of a detailed calculation, shows that because the equation of state includes the possibility of ionisation, the curve $f(R, s)$ for a fixed s exhibits an S shape and for a definite range of the entropy variable s it cuts the axis ($f = 0$) thrice. As explained in Chapter 2, the three hydrostatic equilibria that exist for these fixed values of the entropy, correspond to a fully recombined, fully ionised and partially ionised gas. In the middle (partially ionised) solution for the equilibrium radius, the adiabatic exponent drops below $4/3$ and therefore this hydrostatic equilibrium is *dynamically* unstable (the other two solutions are dynamically stable). Numerically integrating the dynamical system (10.3)–(10.4) Buchler and Regev found that for (a quite narrow) range of ϵ the solution consists of irregular oscillations. Outside this range the solution was found to be periodic (for small values of ϵ the radius and the luminosity of the model star exhibit regular *relaxation* oscillations). The mechanism for this behaviour can be identified as resulting from thermal instability of both (dynamically stable) hydrostatic equilibria, Thus, the outer zone is dynamically driven towards one of these stable hydrostatic equilibria, and oscillations on a dynamical timescale, τ_{dyn}, are present. However since both equilibria are thermally unstable the model cannot relax and is driven (on a thermal timescale, τ_{th}) away from each of the hydrostatic equilibria towards the other one. If $\epsilon \sim \tau_{dyn}/\tau_{th} \ll 1$ the solution consists of relaxation oscillations. When ϵ is in an appropriate range, the relaxation oscillations caused by the thermal evolution, from one dynamically stable equilibrium towards the other one (via the middle dynamically unstable equilibrium), become irregular, because the two relevant timescales become very close to each other. This situation is shown in Figure 10.2. The general behaviour of the solutions is mathematically similar to the original MS oscillator although the physics behind the two-model systems is quite different.

Auvergne and Baglin (1985) studied further a similar ionisation-zone model and derived, by using suitable approximations, an analytical form for the model dynamical system. They found a third-order differential equation for the nondimensional

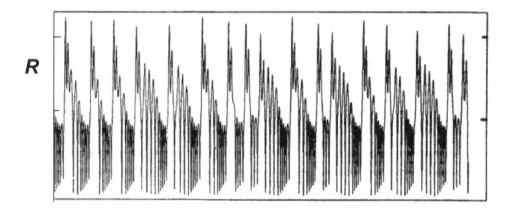

Fig. 10.2. Irregular oscillations occurring for an appropriate value of ϵ (here 0.15) in the model of Buchler and Regev (1982). The radius of the outer zone is shown as a function of time. Nondimensional units are used with a typical short oscillation lasting ~ 1 year. Reprinted, with permission, from Buchler & Regev (1982).

amplitude which, as they have shown, is related to the MS system (2.27). The results of their numerical integrations are very similar to those of Buchler and Regev and confirm the appearance of irregular oscillations, in an appropriate narrow range of the control parameter.

Additional simple models of irregular stellar pulsators have appeared in the literature since the early 1980s but we shall not discuss all of them here. We would like to mention briefly just a few cases, which we find interesting.

The work of Perdang and Blacher (1982) demonstrated that nonlinear coupling between *adiabatic* radial linear modes can lead to chaotic oscillations, that is, dissipative effects are not a necessary condition for chaotic behaviour of stellar oscillations. This paper is just one in a series of works by Perdang and his collaborators, invoking a Hamiltonian system approach to irregular stellar oscillations.

An additional one-zone model was introduced in 1988 by Tanaka and Takeuti. This model gives rise to a nonlinear nonadiabatic oscillator, but it differs from the above mentioned models, based on the MS oscillator, in that it does not involve dynamical instability. Models of this type were discussed in detail by Saitou, Takeuti and Tanaka (1989), where they showed that these systems are similar (although not identical) to the Rössler oscillator.

In a more recent work, Icke, Frank and Heske (1992) proposed to model the irregular pulsation of some Mira-type variables by an outer stellar layer driven by the pulsation of an underlying zone of instability. Their model consists of a driven nonlinear oscillator and they analyse it within the Hamiltonian system formalism.

The ability of very simple models, which are usually referred to as toy models, of stellar pulsators to exhibit chaotic behaviour, and the mounting evidence that the observed light curves of at least some irregular stars are low-dimensional (in the sense explained above) is all rather speculative still. It is not easy to convince the sceptics that the behaviour of a simple model with just a few degrees of freedom is useful for the understanding of a considerably complex object such as an irregularly pulsating star. Indeed, the traditional theoretical approach to the modelling of stellar pulsation has been based on full scale numerical hydrodynamics computations. This approach is clearly the one that can yield a detailed and accurate description of the pulsations, but its shortcomings are obvious. Each model requires a heavy computational effort, and it is usually difficult to extract from the results the underlying systematics.

We give here just one example of such a calculation and have chosen a case having obvious relevance to the subject of this book. The W Vir stars have already been mentioned before in this chapter, where it was described how low-dimensional behaviour was detected in their observational light curves. Before this analysis was done, Buchler and Kovács, had numerically studied sequences of hydrodynamical models appropriate for these stars. They found (for reference see Buchler, 1993) that several sequences of models undergo successive period-doubling bifurcations (see Chapter 4), when the control parameter (the effective temperature) is gradually lowered. In some other sequences (having a higher luminosity) the transition to chaos occurred through a saddle-node (tangent) bifurcation, which appeared after just a few period-doublings. These results are quite remarkable, especially in view of the fact that at least the period doubling route to chaos in the appropriate models seems to be rather robust. It occurred in many model sequences, was confirmed by other hydrodynamical numerical codes (Aikawa, 1987), and survived even the inclusion of time-dependent convection in the codes. Buchler and his collaborators have suggested that the period-doubling bifurcation is associated, in this case, with the excitation of vibrational overtones, brought about by resonances.

Simple models, like the ones mentioned above, and full numerical hydrodynamical modelling, may be considered two conceptual opposites. Somewhere between these two extremes lies what has usually been called in this context the 'dynamical-system approach'. This approach, which will be reviewed in the next section, is still far from complete and applicable to all complex phenomena, but we have gone a long way since the short abstract of Baker, Moore and Spiegel was published in 1966. The procedures outlined in the next section allow viewing of some simple models as asymptotically correct limits of the full equations, or as model equations, in the sense discussed in Chapters 3 and 7 of this book. We have seen there that complicated dynamical systems can be effectively reduced in various circumstances (e.g., close to an instability threshold) and simple, generic equations, whose behaviour can be studied in detail, can be derived for them. Just how this can be accomplished for pulsating stars is discussed in the next section.

10.3 A dynamical-system approach to stellar pulsation

Consider a spherically symmetric star, whose structure variables are expressed as functions of the coordinates m and t, that is, the mass within the relevant radial point and the time. This is usual practice in stellar-structure theory and the explicit equations can be found in standard texts. For our purposes it will be advantageous to make the notation compact by grouping all the relevant functions, like $r(m, t)$, $\rho(m, t)$, $T(m, t)$ etc., in one general vector function $\mathbf{U}(m, t)$, say. The dynamical system describing the *radial* motion of the star can then be written as

$$\partial_t \mathbf{U} = \mathbf{F}(\mathbf{U}, \partial_m) \qquad (10.6)$$

where we have explicitly written that the vector functional $\mathbf{F}$ of $\mathbf{U}$ may also depend on the latter's spatial derivatives. Assuming that $\mathbf{U}_0(m)$ is an equilibrium solution of the system, that is, it describes a static star, we may define $\mathbf{u} \equiv \mathbf{U} - \mathbf{U}_0$ to be a deviation from the equilibrium solution. The time evolution of this perturbation will be governed by an equation which can be written in the form

$$\partial_t \mathbf{u} = \mathcal{L}\mathbf{u} + \mathcal{N}(\mathbf{u}) \tag{10.7}$$

conveniently separating the linear and nonlinear parts of the functional by means of an appropriate expansion around $\mathbf{U}_0$. This equation has the form of the general dynamical system discussed in Chapter 7. It is obviously possible to identify suitable control parameters, whose variation may cause change in the stability of the equilibrium solution, but we do not write it here explicitly as in equation (7.3).

One way to proceed is by performing a dimensional reduction of the system (10.7) in the spirit of the methods demonstrated in Chapter 7 for PDEs, and in Chapter 3 for a large number of coupled ODEs. The first step is to look at the linear problem (neglecting the nonlinear part) and using, as usual, the time dependence $\propto \exp st$ in the linear solution. In Chapter 7 we have also assumed an exponential spatial dependence of the linear solution, but this simplification cannot work here, because the spatial structure of the linear eigenfunctions is now somewhat more complicated. Thus we use here for the linear problem an as yet unspecified spatial dependence $\Phi(m)$ and get the ODE eigenvalue problem

$$\mathcal{L}\Phi = s\Phi \tag{10.8}$$

Clearly, s may be complex and we write, using the notation of Chapter 7, $s = \sigma + i\omega$. Labelling the eigenvalues and eigenfunctions of the linear problem by the index k, say, we may write any solution of the linear problem as a sum of the eigenmodes (each having its own constant amplitude).

We now attempt to apply to the full nonlinear problem (10.7) what we have called in Chapter 7 a 'projection and slave mode elimination technique' (see Section 7.3). The crucial assumption is that we are near criticality, that is, the control parameter(s) are tuned in such a way that $\sigma_k \leq 0$ for all modes but for only *very few*, p say, modes $|\sigma_k|$ are small. These modes are the slow modes and for all other modes (the fast ones) $\sigma_k \ll 0$. This corresponds to a case in which most of the eigenvalues are well separated from the ones near marginality in the general situation depicted in Figure 3.14.

Thus the solution for the full *nonlinear* problem can be formally written as

$$\mathbf{u}(m,t) = \sum_{k=1}^{p} A_k(t)\Phi_k^{\mathrm{s}}(m) + \sum_{l=1}^{\infty} B_l(t)\Phi_l^{\mathrm{f}}(m) \tag{10.9}$$

where the superscripts s and f in the eigenfunctions refer to the slow and fast modes respectively. In writing the solution in this way we obviously assume completeness, that is, that every solution can be decomposed and written in the above modal way.

The substitution of the expansion (10.9) into the original nonlinear equation (10.7) can now yield a set of *amplitude equations* for the $A_k(t)$ and $B_l(t)$. Since the amplitudes' time dependence contains within them the linear contribution $\exp \sigma t$, the fast modes (with large and negative σ) evolve quickly and after this transient passes, the system can be characterised by only the slow (or principal) modes, that is, only the first finite sum in (10.9) is actually relevant. The fast modes can be eliminated because after a very short time their amplitudes nonlinearly saturate at a constant value, and the equations corresponding to them are algebraic. The fast modes are said to be *slave* since their value quickly adjusts to the slow evolution of the principal modes.

The amplitude equations thus become a set of p nonlinear ODEs for the, generally complex, amplitudes of the slow modes, having the form

$$\frac{\mathrm{d}A_k}{\mathrm{d}t} = s_k A_k + g_k(A_1, A_2, \ldots, A_p) \tag{10.10}$$

for $k = 1, 2, \ldots, p$, where the g_k are nonlinear functions. This set of equations thus reflects the dimensional reduction of the original PDE problem.

As we have said above, a similar procedure for the derivation of amplitude equations was demonstrated in Chapter 7 for the Swift–Hohenberg model equation. We recall also that amplitude equations can be derived more formally by performing a singular perturbation analysis (this procedure was demonstrated in Section 3.4 for a general ODE dynamical system near a Hopf bifurcation). For such perturbative approaches to be applicable we need a small parameter and it can be the ratio of the dynamical to thermal timescales, for example. Thus the amplitude equations are in the slow time, that is, they describe the modulation of the fast pulsation.

The description of the dimensional reduction given above is rather heuristic and is based on the examples and the discussion given in the first part of the book. As remarked there, more rigorous treatments for generic systems are available in the literature. The idea of a centre manifold plays an essential role in the implementation of the dimensional reduction, and we refer the reader to the articles of Buchler (1993) and Spiegel (1993) for more general and complete discussion and references. The important point is the fact that the functions g_k in (10.10) are composed of generic expressions, the normal forms. Which normal forms appear in a given theoretical model depends on the nature of the bifurcation at marginality and this is determined by the *linear problem*. The numerical values of the *constants* appearing in the normal forms depend obviously on the specific parameters of the original system, but the qualitative behaviour is set by the normal form, i.e., by the bifurcation type.

The amplitude equations formalism has been applied to particular stellar models by various authors and the results have been compared to full hydrodynamical studies (for references see Buchler, 1993). It seems, however, that finding the normal forms and studying them for *generic* types of behaviour near criticality is particularly useful for the understanding of the mechanisms behind the different types of behaviour. Buchler and his collaborators published a number of studies of this type and we refer the interested reader to the work of Goupil and Buchler (1994), where a full exposition of the amplitude equations formalism for general (nonradial and nonadiabatic) stellar pulsation is given. This article, as well as Buchler (1993), also contains an extensive list of references.

The form of the amplitude equations for different classes of systems depends only on the number of dominant (i.e., slow, see above) modes that are excited, and on the nature of possible resonances between them. Generally, it is assumed that the nonlinearity is weak, that is, only the nonlinear term of lowest order in the amplitudes is kept. The amplitude equations themselves are then derived using multi-time perturbation techniques (like the method of multiple scales, see Chapter 3). The simplest applications of this formalism are obviously for the cases of just a single or a small number of nonresonant modes. In the case of a single critical mode in a pulsating star, the relevant amplitude equation turns out to be

$$\frac{\mathrm{d}A}{\mathrm{d}t} = (\sigma + i\omega)A + Q|A|^2 A \tag{10.11}$$

where the complex constant Q depends, as we already know, on the properties of the particular stellar model. Writing $A \equiv x + iy$ and $Q = \alpha + i\beta$, it is easy to see that the real and imaginary parts of the above amplitude equation yield the normal form characteristic of the Hopf bifurcation, as given in the summary of Section 3.3. Thus, and this fact is perhaps quite trivial, stellar pulsational instability is nothing more than a Hopf bifurcation. If $\alpha = \mathrm{Re}\, Q$ is negative (and this is the usual situation in pulsating stars), the nonlinear saturation of the amplitude gives rise to a limit cycle, that is, a regular periodic pulsation.

The amplitude equations for double-mode pulsators, without resonances, and for pulsators in which integer and half-integer resonances occur, are somewhat more involved. Such equations have been derived for various cases and the coefficients compared with the results of detailed numerical hydrodynamic calculations (see Buchler 1993 for references). These studies of Buchler and his collaborators have already supplied what can be considered as a basic understanding of the nature of stellar pulsation. For example, the 2:1 resonance between the fundamental mode and the second overtone has been found to be responsible for the Hertzsprung progression in classical Cepheids (see, e.g., Simon & Schmidt, 1976). This explanation had been conjectured long before the amplitude equation formalism helped to show it explicitly. Other resonances have been identified as being responsible for a number of other phenomena, known to occur in pulsating stars as well, but it should be said that when the amplitude equations are truncated to lowest nonlinear terms, they are obviously unable to explain all of the interesting phenomena.

A comprehensive review on the status of nonlinear pulsation theory, including approaches based on dynamical system theory, can be found in Buchler (1998).

Amplitude equations derived by multi-time perturbation techniques, like the ones discussed above, are not adequate, at least not a priori, for cases in which the dynamical and thermal timescales become comparable. That this situation should be of interest for *chaotic* pulsations is clearly hinted at by the MS oscillator and its pulsational offspring. Moreover, this seems to be the case also in the above mentioned hydrodynamical W Vir numerical models. A more *general* method, not based on the separation of two timescales, for the derivation of amplitude equations to predict the time dependence of a system governed by PDEs, when that system is close to criticality for the onset of instabilities, has been given by Coullet and Spiegel (1983). This paper's mathematical level is above that of our own book and so we have refrained from discussing it in the first part of this book. Spiegel (1993) reviewed the application of this formalism to stellar pulsation and we are thus able to give here a sketch of such an application, demonstrating how chaos can occur in pulsating stars.

Returning back to (10.7) we follow the procedure outlined above, but write the amplitude equations in vector form

$$\frac{d\mathbf{A}}{dt} = \mathcal{M}\mathbf{A} + \mathbf{g}(\mathbf{A}) \tag{10.12}$$

which is equivalent to (10.10) when the components of the vectors $\mathbf{A}$ and $\mathbf{g}$ are the A_k and g_k of that equation, respectively, and the matrix $\mathcal{M}$ has the eigenvalues s_k. The method of Coullet and Spiegel has been especially devised for polycriticality, that is, for the case in which the real parts of *several* eigenvalues pass through zero almost simultaneously, when suitable parameters are varied. All the other modes are stable and fast (as before).

Wishing to consider the simplest case and guided by the fact that the Baker one-zone model and the BMS (the MS oscillator as applied to stellar pulsation) system was of third order, Spiegel and we here, after him, consider the tricritical case, that is, when there are *three* control parameters, λ, μ, ν say, in the problem, and thus three linear modes may become *slow* together.

It turns out that in the asymptotic limit, as we approach tricriticality, the following generic amplitude equations emerge

$$\frac{dA_1}{dt} = A_2$$

$$\frac{dA_2}{dt} = A_3$$

$$\frac{dA_3}{dt} = \nu A_1 + \mu A_2 + \lambda A_3 + A_1^2 \tag{10.13}$$

where the parameters λ, μ, ν determine the eigenvalues of the matrix $\mathcal{M}$ of (10.12), s_k ($k = 1, 2, 3$), via the characteristic equation

$$s^3 = \lambda s^2 + \mu s + \nu \tag{10.14}$$

For some more mathematical details involved in the derivation of these amplitude equations (by means of an appropriate perturbational procedure, which *does not* demand the existence of several different timescales) we refer the reader to Spiegel (1993) and the works cited there.

This set is clearly equivalent to the third-order system

$$\partial_t^3 A - \lambda \, \partial_t^2 A - \mu \, \partial_t A - \nu A = A^2 \tag{10.15}$$

where we have put $A \equiv A_1$ and opted for the notation $\partial_t \equiv d/dt$ for simplicity.

An interesting result is obtained if the parameters λ, μ, ν are chosen in such a way as to guarantee that the eigenvalues of the three slow modes can be expressed as

$$s_1 = \sigma_1, \quad s_{2/3} = \sigma_2 \pm i\omega \tag{10.16}$$

with $\sigma_1 > 0$ and $\sigma_2 < 0$. This means that the slow modes consist of a monotonically growing solution and damped oscillations and the former becomes unstable first. Close to marginality σ_1 and $|\sigma_2|$ are small and Shilnikov's theorem (see Chapter 4) implies that in such a case (10.15) will have an infinite number of unstable periodic solutions. Thus the system may exhibit chaotic behaviour as the solution 'wanders' from one unstable periodic orbit to another. These arguments imply that if the stellar pulsation equations have the above properties near criticality, the oscillations can become aperiodic due to low-dimensional deterministic chaos. The same kind of reasoning can apply to other polycritical situations as well.

Concentrating on the system (10.15) in the case when the eigenvalues satisfy (10.16) we find that the origin of its three-dimensional state space, spanned by A, $\dot{A}$ and $\ddot{A}$, or alternatively by A_1, A_2, A_3 of (10.13), is an unstable fixed point. The simplest homoclinic orbit of this system leaves the fixed point exponentially along the direction set by the eigenvector corresponding to the positive real eigenvalue σ_1 and spirals back in (because the other eigenvalues are a complex conjugate pair with a negative real part). More involved homoclinic orbits exist as well and in these the orbit loops several times before returning back to the fixed point. Viewed as a function of time the simplest homoclinic orbit (which is sometimes called 'principal') has the shape of a pulse with an exponential rise and an oscillatory tail. Numerical solutions of (10.15) started at the unstable fixed point typically show trains of pulses, with the spacings between them depending on the parameter values.

We have seen in the last example of Chapter 2 and in Section 7.6 how a PDE can be reduced to an ODE (the associated ODE) whose heteroclinic solutions correspond to localised structures, or defects, in the PDE. The system discussed there was spatially extended and the localised structures were *fronts* (stationary or moving) and we shall return to this problem in the next chapter. Here the spatial structure is factored out at the outset by including it in the pulsational eigenfunctions and we are left with only temporal ODEs (the amplitude equations) whose homoclinic solutions correspond to pulses in *time*. Problems of this type, that is, when the original PDE is reduced to a set of ODEs having homoclinic or heteroclinic solutions, corresponding to localised structures in the PDE (pulses or fronts, respectively) may be treated by employing perturbation methods, similar to the effective particle approach in field theory. Such an approach was suggested by Coullet and Elphick (1987) and we shall discuss it only briefly here. An application of a similar technique to a spatio-temporal problem supporting fronts was discussed in detail in Section 7.5 and has been applied also to spatial patterns in the interstellar medium (see Chapter 11).

The idea is to try and construct a more general solution of (10.15) from the primary homoclinic solutions, which we shall call $H(t)$. Because the differential system we are dealing with is translationally invariant (in time), the phase of the pulse $H(t)$ is arbitrary, and thus $H(t-\tau)$, with τ being an arbitrary constant, is also a solution. Consider now a superposition of pulses with different phases, that is, the sum $\sum_m H(t - \tau_m)$, which looks like a sequence of primary pulses, provided that the pulses do not overlap. Clearly, this is guaranteed if the separations of the pulses ($|\tau_j - \tau_{j-1}|$) are larger than a typical pulse width. The above sum is obviously not an exact solution of the ODE (10.15) because the superposition principle does not apply to nonlinear equations. However, if the pulses are widely separated (in the above sense), we may reasonably try to write it in the form of the following Ansatz

$$A(t) = \sum_m H(t - \tau_m) + \epsilon R(t, \epsilon) \tag{10.17}$$

where the additional term is viewed as a small correction.

The natural choice for the small parameter in this problem is

$$\epsilon = \exp(-\sigma\hat{\tau})$$

where $\hat{\tau}$ is the mean spacing of the pulses and $\sigma = \sigma_1$. To simplify this discussion we assume that the parameters are chosen in such a way that $-\sigma_2 = \xi\sigma_1$, with ξ of order unity, and this means that the exponential growth rate of the solution (on the 'trailing' side of the pulse) is of the order of the decay rate (on the 'leading' side). Different growth rates complicate the treatment but it can be done as well.

Singular perturbation theory (see Chapter 3) can now be used and the solvability conditions (here the condition is that the correction term does not diverge) yield in this case a set of ordinary differential equations for the pulse *phases*, that is, positions on the time axis. These ODEs are quite simple if one considers the influence on a pulse of just its nearest neighbours. This is obviously consistent with the assumption of very large pulse separations. The above procedure is similar to the case discussed in Section 7.6

Thus we get

$$\dot{\tau}_k = \zeta_1 \exp[-\sigma(\tau_{k+1} - \tau_k)] + \zeta_2 \exp[-\xi\sigma(\tau_k - \tau_{k-1})] \cos[\omega(\tau_k - \tau_{k-1}) + \theta] \tag{10.18}$$

where ζ_1, ζ_2 and θ are constants derivable from the pulse solution $H(t)$.

The form of these equations is suitable for our case (a pulse having a monotonous exponential growth and an oscillatory exponential decay with the growth and decay rates of the same order). In other cases, that is, when the eigenvalue spectrum of the slow modes is of different nature, these equations are more complicated and we defer the discussion of some other cases to Chapter 11. The key point here is the (further) reduction of the dynamical system to the (approximate) dynamics of the localised structures, i.e., solutions of the amplitude equations, which are themselves a reduction of the original PDE.

We may examine the nature of a *steady* solution of the set of ODEs (10.18) for all k by substituting $\dot{\tau}_k = 0$, or alternatively a uniformly moving (on the time axis) one by putting $\dot{\tau}_k = \text{const}$. This leads to a pattern map, which provides the position of the $m + 1$ pulse from the mth one. For the example discussed here this map has the form

$$Z_{m+1} = C - K Z_m^{\xi} \cos(\omega \ln Z_m - \Theta) \tag{10.19}$$

with $Z_m \equiv \exp[-\sigma(\tau_m - \tau_{m-1})]$ and where C, K and Θ are constants depending on the constants in (10.18), that is, on the pulse shape. The map (10.19) is complicated enough and is known to have chaotic solutions for suitable parameter values.

Of course, for such a procedure to be quantitatively good for a particular stellar model, the derivation of the eigenfunctions and eigenvalues and of the amplitude equations must be performed numerically. The nature of the stability eigenvalues may be more complicated and this generally results in more involved pattern maps. These maps describe series of pulses, phase-locked to each other, and this is made possible by the fact that the pulse tails (at least one of them) are oscillatory. However special the case described here may be, the results tell us that when there are a few slow modes in the problem, the stellar pulsations may exhibit irregularly spaced pulses and that this behaviour may be low-dimensional. A tricritical situation, like the one described here, gives rise to a third-order system, and we may view the above discussion as a justification of the applicability of the simple models we have discussed in the beginning of this chapter.

It is not obvious that the regular or chaotic patterns of locked pulses are stable and this issue may lead to complications. Other refinements, like the inclusion of next-to-nearest neighbour interactions, may lead to more complicated two-dimensional pattern maps and thus to richer behaviour. In any case, the ideas and techniques described in this section, that is, the derivation of amplitude equations and the examination of the nature of their possible solutions, can be very useful for the understanding of the mechanisms that are responsible for the different types of behaviour found in stellar pulsators and full numerical simulations thereof.

10.4 Some models of accreting systems

We have said in the first section of this chapter that the attempts to find low-dimensional behaviour in the light curves of accreting sources, like close binaries and AGN, have not as yet yielded reliable positive results. This is perhaps not surprising in view of the highly complex accretion flows which are thought to power these sources. Shocks, turbulence and often also hydromagnetic interactions, on a variety of spatial and temporal scales, greatly complicate the dynamics, and it seems that it would be too naive to expect that these processes could finally give rise to a low-dimensional attractor in the signal. But this is also largely true in the case of pulsating stars, although they are in comparison somewhat more simple. Yet, we have seen how the emergence of effective low-dimensional dynamics in the latter systems may be understood with the help of theoretical models. Thus the inherent complexity of accreting sources and the fact that deterministic chaos was not found in them observationally, should not rule out models that stop short of utilising full scale numerical magnetohydrodynamics. In this section we shall describe a few attempts to devise simplistic models of accreting systems, using ideas from nonlinear dynamical system theory. Their relevance to real astrophysical objects is still quite questionable, although some of them seem to identify a number of qualitative aspects of real astrophysical objects. Perhaps the first model of this kind was proposed by Livio and Regev (1985), who were motivated by an early qualitative observation of Celnikier (1977). Celnikier pointed out the similarity between the behaviour of the rapid burster (a famous X-ray burst source) and that obtained from a nonlinear population dynamics model, based on the logistic map. Livio and Regev devised a simple two-zone model of the outer layers of an accreting neutron star, in which a thermonuclear burning shell (the inner zone) is surrounded by an accreted envelope. They assumed a thermal coupling between the zones through the diffusion of radiation, and obtained a dynamical system of the type

$$\frac{d}{dt}T_1 = F_1(T_1, T_2, m)$$

$$\frac{d}{dt}T_2 = F_2(T_1, T_2, m) \tag{10.20}$$

where T_1 and T_2 are the temperatures of the inner and outer zones respectively, and the functions F_1 and F_2 contain quite simple analytical expressions, devised to model the physics of nuclear burning (in F_1 only) and radiation transfer (see the original paper for the explicit form of these functions). Here, m is the mass of the outer zone and if it is assumed to be constant the dynamical system (10.20) is two-dimensional and as such cannot, of course, exhibit chaotic behaviour. In an earlier study the same authors found that this system with m constant admits steady (fixed point) or periodic (limit cycle) solutions, depending on the system parameters. The steady solutions correspond to situations in which the nuclear burning is steady and the periodic solutions are essentially a series of bursts, similar to shell flashes. This type of behaviour was found earlier in very similar two-zone models proposed by Buchler and Perdang, and others, in an effort to qualitatively understand shell flashes in stars evolving out of the main sequence.

In the work described here Livio and Regev added a third degree of freedom to the dynamical system by allowing m to vary. The physical processes responsible for the time variation of m were cast into a third-differential equation, coupled to (10.20) and expressing mass conservation of the outer zone. It was assumed that m can vary because of both accretion of fresh material from the outside and loss of burned material, deposited inwards (onto the neutron star). Wishing to induce rich behaviour of the model, Livio and Regev assumed that the accretion rate depends on the peak luminosity of the *previous* burst, so that the rate of change of m can be written as

$$\frac{d}{dt}m = \lambda(L_B - L_0)(1 - L_B) - F_3(T_1)m \tag{10.21}$$

The second term on the right-hand side of this dimensionless equation is the rate at which the material in the shell is being burnt (and this depends on T_1 and linearly on m) and the first term is a combination of two feedback mechanisms, one negative and the other one positive. The peak luminosity of the last burst, L_B, is expressed in units of some maximal value so that $L_B < 1$. Thus the factor $1 - L_B$ simply models a dilution of the accretion rate due to the burst and the concomitant increase in the radiation pressure, and the factor $L_B - L_0$ models, again in a very simplified manner, the mechanism by which the X-rays may induce, by an evaporative process, an increase in the mass loss from the secondary star (which is close to its Roche limit); λ and L_0 are parameters. The dilution of the accretion rate as well as the induced mass loss have both been proposed before by various authors in this context; Livio and Regev have just parametrised it in what seems to be the simplest way.

Fixing all the parameters save λ, Livio and Regev numerically obtained a series of models for different values of λ. Periodic solutions (a series of bursts) were found for sufficiently small values of λ, followed by a *period-doubling* bifurcation of the Feigenbaum type (see Chapters 1 and 4) on the way to chaotic behaviour. The attractor, when λ is in the chaotic region, was found to have a typical 'band' structure and a map, defined by expressing the temperature T_1 (or alternatively the luminosity) at a burst as a function of the same variable at the previous burst, was also calculated. This map (which can be perceived as a first return map for this system) had a quadratic shape, demonstrating the low-dimensional, deterministic nature of the irregular burst sequence.

The next model that we wish to describe here, has been developed somewhat beyond just a qualitative suggestion as was the previous example. A good summary of the model, its properties and possible applications to accreting astrophysical systems can be found in Young and Scargle (1996). It was first proposed in the context of quasi-periodic oscillations (QPO) of the low mass X-ray binaries, like Sco X-1 (in 1993), and magnetic cataclysmic variables of the AM Her type (in 1994). In these studies Scargle and Young were joined by a number of researchers, but here we shall base our discussion only on the paper of Young and Scargle (1996), and refer the reader to references therein.

The model is based on what is known as the *dripping handrail* (DHR), a coupled-map lattice model, which we have mentioned in Chapter 7. The abstract DHR model has been adapted to the problem of accretion flow by representing within it several dynamical effects, that are thought to take place at the inner edge of an accretion disc. Assuming that matter is fed onto the handrail (the inner edge of the disc) from the outside (the body of the disc) it was allowed to stay on the handrail, supported centrifugally and maybe also radiatively, with its density being smoothed out in the tangential direction (by diffusive processes). Only after exceeding a certain critical density, matter could drip from the rail (fall onto the star) because of some plasma or fluid instability. We recall from Chapter 7 that in the original DHR model a 'future-coupled totalistic' kernel was used, with the local dynamics given by the linear circle map (7.100). After adopting this coupled-map lattice Young and Scargle simplified it somewhat. The dynamical variable was chosen to be the (dimensionless) mass density ρ, and consequently the slope parameter in the map (7.100) was set as $s = 1$, to ensure mass conservation. In addition, the authors opted for a simpler form of the spatial coupling term and chose linear Laplacian coupling, representing a linear diffusion process along the handrail. The

CML describing this version of the dripping-handrail model thus came out to be

$$\rho_i^{n+1} = \left[\rho_k^n + D\left(\rho_{k+1}^n - 2\rho_k^n + \rho_{k-1}^n \right) + \Omega \right] \quad (\text{mod } 1) \tag{10.22}$$

where the superscripts denote the time step, the subscripts refer to the spatial lattice (the location on the rail), and the dimensionless parameters D and Ω measure the mass diffusion and accretion rate respectively. As is apparent from (10.22), the handrail is assumed to drip when the density exceeds the critical value $\rho_{\text{crit}} = 1$.

The behaviour of the DHR model defined by (10.22), as found by numerical iterations, critically depends on the ratio D/Ω and this can quite easily be understood qualitatively. If $D \gg \Omega$ the system is dominated by diffusion and thus the threshold density is reached at all sites simultaneously (for the same n). The dynamics are then essentially one-dimensional (in space) and proceed for all cells in the same way, filling them until the threshold is reached, after which they are all emptied. The initial conditions do not affect this evolution because the information about them is quickly lost by the diffusion. In the other extreme ($D \ll \Omega$) accretion dominates and different cells become essentially decoupled. Each cell independently fills and dumps with the phase, that is, the timestep at which the threshold is reached for each cell, depending on the initial conditions. In both cases the evolution is periodic with a fundamental period $1/\Omega$ determined by the rate at which the cells are filled by accretion. In the accretion dominated regime higher harmonics and spatial structures can also appear, while in the opposite limit diffusion damps them out.

More interesting behaviour is expected in the intermediate cases, that is, when D is of the order of Ω. Indeed, the results of numerical calculations in these cases have revealed complex characteristics of the signal, which were classified by the authors as very low frequency noise (VLFN) and quasiperiodic oscillations (QPO). Defining the relevant parameter, measuring the importance of diffusion relative to accretion, as $\theta \equiv \arctan(D/\Omega)$, Young and Scargle found that when the value of $\theta \approx 1$ is approached, the power at low frequency begins to increase steadily, while that at the fundamental frequency decreases. Right before the onset of the VLFN-type behaviour, a broad band in the spectrum, characteristic of QPO, appears as well. These features have been identified before in spectra of some real accreting systems, and the basic physical ingredients of the DHR model – diffusion and accretion with threshold – seem to be sufficient for the explanation of this type of behaviour in a simple and unified framework, at least in principle.

From the calculation of the Liapunov exponents of the signal generated by the DHR coupled map lattice, Young and Scargle concluded that chaos (i.e., a *persistent* positive value of the largest Liapunov exponent) cannot be guaranteed in this case. However, in all simulations the largest Liapunov exponent was found to have a nearly constant positive value for a significant amount of iterations, that is, chaotic behaviour could be confirmed during a long transient, whose duration was found to increase with the number of sites. Thus, although the system is not chaotic and it asymptotically approaches regular behaviour, it is endowed with a property that can be called *transient chaos*. Young and Scargle thus concluded that the observed behaviour of a variety of accretion systems might be associated with the transient behaviour similar to the DHR system. Any temporal and/or spatial inhomogeneities, which are bound to occur in real accreting systems, may tend to 'restart' the system before the chaotic transient has had time to decay to the regular asymptotic state. It is perhaps reasonable, therefore, to expect that observations of accreting systems should yield chaotically variable light curves (an expectation that has not yet been supported by analyses of observational data).

We conclude this section with the description of one example from a variety of models, which were devised to elucidate the optical 'flickering' of cataclysmic variables (CV). This rapid and apparently random variability is seen in every CV and the amount of energy contained in it is, in general, a significant fraction of the total luminosity of the system. The power spectrum of the flickering exhibits a power-law behaviour and some researchers have seen it as the CV analogue of the aperiodic variability of other accreting systems (see above). In a number of quite recent studies it has been proposed to model the CV flickering with the help of a special kind of CML, a *cellular automaton*, a mathematical device known for its application to a number of dynamical systems. We have not explicitly discussed cellular automata in the first part of the book and for our purposes here, it is sufficient to view it as a coupled-map lattice in which the dynamical variable itself (in addition to the time- and space-independent variables) is allowed to take discrete values only. A cellular automaton is thus essentially a set of rules that fix the (discrete) value of a dynamical variable at a given time-step and lattice site, in terms of the values of this variable at the same and neighbouring sites, but at a previous time-step. The book by Wolfram (1994) is recommended for readers interested in the theory and applications of cellular automata, among which are numerical simulations of fluid equations (Frisch *et al.* 1986) and chemical turbulence (Winfree *et al.* 1985).

Yonehara, Mineshige and Welsh (1997) devised a cellular automaton to model the accretion flow through a thin disc by dividing the disc plane into a lattice in cylindrical coordinates (r_i, ϕ_j) with $i = 1, 2, \ldots, N_r$ and $j = 1, 2, \ldots, N_\phi$. The radial index i was chosen to increase inward, that is $r_i > r_{i+1}$, and each cell of the lattice was set to have the same extent in the radial $(r_i - r_{i+1} = \Delta r)$ and angular $(\phi_{j+1} - \phi_j = \Delta\phi)$ directions, and was assigned an initial mass $M_{i,j}$. The evolution of $M_{i,j}$ was set to proceed according to the following set of rules.

(i) Choose one cell randomly at the *outermost* ring ($i = 1$) and add a fixed mass m to it.
(ii) Choose one cell randomly at each ring and let a given small mass $m' \ll m$ be moved from this cell into the adjacent inner cell (with the same j and i, larger by 1).

(iii) Check all cells of the disc for their new mass $M_{i,j}$ and if a cell is found in which $M_{i,j} > M_{\mathrm{crit}}(r)$, i.e., a given maximal mass for a cell in a given ring, let

$$M_{i,j} \;\to\; M_{i,j} - 3m$$
$$M_{i+1,j\pm 1} \;\to\; M_{i+1,j\pm 1} + m$$
$$M_{i+1,j} \;\to\; M_{i+1,j} + m \qquad\qquad (10.23)$$

These rules represent the physical processes of external feeding of the disc (i), viscous mass diffusion through the disc (ii), and an accretion instability giving rise to 'avalanches' (iii). Starting from some initial configuration the above steps are repeated many times to let initial transients decay. Each iteration represents a time-step of a free fall through one cell (Δr), and thus a desired evolution time can be translated into the number of iteration steps.

The cellular automaton (10.23) was in general found to evolve towards a characteristic state, whose qualitative behaviour resembled other well known physical systems. This state can be characterised by persistent general growth (mass accumulation), interrupted by a variety of sudden fast decays (avalanches). In most cases the avalanches are localised and quite small and terminate rapidly, but occasionally a local avalanche may trigger avalanches in the adjacent cells as well, sometimes giving rise to very prominent events. The probability of large avalanches is small because they occur only if prior to such an event the masses of a number of radially adjacent cells are all near the critical value. This generic behaviour has been found to occur in a variety of open dissipative systems, and the state in which it occurs was called by Bak, Tang and Weisenfeld (1988) *self-organised critical-ity*. Such states generally do not have positive Liapunov exponents and are thus not truly chaotic, but they clearly exhibit scaling behaviour. The most famous of these systems is the 'sand-pile' model, which was studied by Bak and co-workers. The results of an actual integration of the cellular automaton (10.23) can be found in that paper and they demonstrate the properties described above, which remind some people also of the behaviour of stock markets.

Yonehara *et al.* found that the behaviour of the cellular automaton (10.23) is quite robust, that is, similar characteristics are obtained even if the avalanche rules are quantitatively modified. For the purposes of applying the model to astrophysical accreting systems Yonehara *et al.* had to somehow model the radiation processes involved in the various accretion events. We refer the reader to the original article and references therein for the details of this modelling. The significant result is that the light curves and their power spectra, resulting from a rather simple cellular automaton model of the accretion, omitting the complicated details the hydrody-namics involved, can still qualitatively reproduce some of the basic properties of the flickering. The most important of these is probably the relative insensitivity of the power-law shape of the fluctuation power spectra to wavelength. This feature is typical in observations and is difficult to explain by models invoking blobs rotating

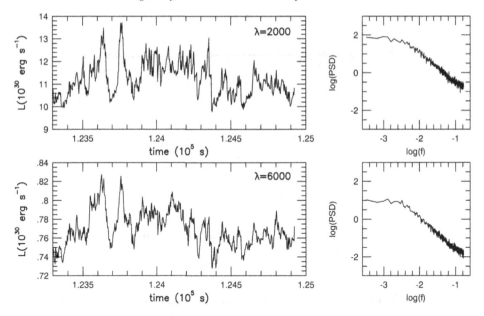

Fig. 10.3. Light curves and power spectral densities produced by the cellular automaton model of Yonehara *et al.* (1997) for the flickering of cataclysmic variables. The upper panels correspond to 2000 Å emission, while the lower ones to emission at 6000 Å. Reprinted, with permission, from Yonehara *et al.* (1997).

at a fixed radius as the origin of the flickering. In Figure 10.3 the flickering light curves at two different wavelengths are shown along with the power spectral density (PSD) as calculated by Yonehara *et al.* for one particular parameter choice. For details we refer the reader to the original paper.

The three models given in this section are examples of utilising ideas from dynamical system theory in the construction of very simple models of the complex processes occurring in accreting astrophysical systems. It is obviously too early to claim that models of this kind can provide detailed understanding of the complicated variability of gas accretion processes in real astrophysical systems. Still, these attempts (and some similar ones, which have not been mentioned here) may provide some basic understanding of the underlying physical processes (whether chaotic or not), and help to devise detailed numerical models in the future.

11

Complex spatial patterns in astrophysics

> The aim of science is always to reduce complexity
> to simplicity.
> William James, *The Principle of Psychology.*

The great majority of astronomical systems are at enormous distances from Earth, but the extent of some of them is large enough as to be spatially resolvable in observations. Surveys in the optical as well as in other spectral ranges have revealed an inherently complex spatial distribution of matter (stars and gaseous nebulae) in the Galaxy and beyond. In this chapter we shall discuss some possible applications of nonlinear dynamics and pattern theory to the study of two (a priori unrelated) topics in this context: the complexity of the interstellar medium (ISM), and the properties of the large scale distribution of matter in the universe.

Interstellar clouds are invariably spatially complex and they pose a rather non-trivial challenge to the theorist, whose primary goal is the understanding of the physical processes shaping them. It is plausible that fluid turbulence plays an important role in this respect. This enigmatic problem is still largely unsolved and we shall not discuss here its possible applications to the ISM complexity (some selected aspects of turbulence will be addressed, however, in the next chapter). After briefly reviewing some observational features of the interstellar medium, which are probably relevant to our discussion, we shall mention some theoretical approaches, among them a pattern and complexity forming toy-model and a couple of global models, motivated by mathematical methods of statistical mechanics.

Gas clouds may appear very prominent and their spatial complexity catches the eye (and the telescope), but the vast majority of observable matter composing the universe is locked in stars, whose spatial distribution exhibits hierarchical clustering on an enormous range of scales – from stellar clusters in a galaxy up to the largest superclusters of galaxies. The readers of this book certainly realise by now that what the astronomers call 'hierarchical spatial clustering' is analogous, at least qualitatively, to the notion of a fractal distribution. Indeed, the suggestion that the universe itself is endowed with fractal matter distribution was instantly advocated by Mandelbrot, when he proposed to apply fractal geometry to the modelling of many physical systems (see Mandelbrot, 1982). The question whether the distribution of matter continues to be a simple fractal, or a multifractal, beyond the scale of the largest galaxy clusters, has evoked a fair degree of controversy, and we shall try to give here some of the key arguments that have been put forward in attempts to clarify this issue and its implications to cosmology.

11.1 Spatial structures in the interstellar medium

Do you see yonder cloud that's almost in shape of a camel?
William Shakespeare, *Hamlet*.

Modern multi-wavelength observations have by now produced detailed maps of various galactic cloud systems and have thus provided valuable information on the spatial distribution of some relevant physical parameters (temperature, velocity fields, chemical composition etc.), in addition to mass density. Interstellar clouds of many sorts have been identified and quite extensively studied observationally, with various tools of statistical analysis being applied to the results. This topic is vast and complex and we shall not even attempt to properly review it here. Only a few of the observational findings and attempts to quantify these results will be given. Our bias is naturally towards those works in which fractals are claimed to be found in the ISM data, but we shall cite some of the opposing views as well. On the theoretical side, again, we shall single out only some of the studies that have been done, in an effort to understand the physical source of the alleged fractal shapes. In particular, we shall give a rather detailed description of a simplistic mathematical toy model of an extended gaseous medium endowed with a local thermal bistability and heat diffusion. It turns out that these physical ingredients, together with some assumed spatial forcing, are already sufficient for producing cloud complexity, not unlike the one described above for interstellar clouds. We shall also examine the effect of the inclusion of fluid advection on these results. It would certainly be naive to expect that a model of this kind is sufficient to serve as the explanation of the ISM complexity and our purpose in discussing it will be, as frequently in the book, largely pedagogical. The model is used mainly as an illustration of what can be done when a modern dynamical-system approach is applied to a problem of this kind. We shall conclude by briefly describing two more ambitious attempts to obtain the ISM scaling laws and fractal dimension using some novel (in astrophysics) theoretical approaches – statistical field theory and the Langevin equation, both of which have been extensively used in condensed-matter physics.

Interpretations of ISM complexity observations

The classical picture of interstellar clouds as separate, well defined entities appears today to be, at best, an oversimplification. If viewed over a sufficient dynamical range the structure is usually hierarchical and appears scale-free. The latter is manifested, e.g., by power-law distributions of cloud sizes and masses (see Elmegreen & Falgarone, 1996 and references therein). It appears that the remarkable relations between the mass, radius and velocity dispersion of the various ISM regions were first noticed by Larson (1981) and confirmed by many other independent studies (see Scalo, 1987). The value of the power-law exponent of the size distribution in a set of self-similar objects (a fractal) is obviously related to the fractal dimension D of this set. If one applies the definition (3.90) to a distribution of this kind (in a similar way as was done in Chapter 3 with respect to the Cantor set, for example) it is quite easy to see that the distribution should behave like

$$N(L) \propto L^{-D} \tag{11.1}$$

where $N(L)$ is the number of objects whose size is *larger* than L. If the mass distribution of some objects is that of a simple fractal it can be expected that, since

$$M \propto L^D \tag{11.2}$$

the mass function $n(M)$, that is, the number of objects (per unit mass interval) whose mass is between M and $M + dM$, should behave like

$$n(M)dM \propto M^{-2}dM \tag{11.3}$$

This and similar global predictions were tested empirically in a number of works and the fractal dimension has been estimated, but before discussing these results we would like to mention some earlier attempts to find the fractal dimension from ISM observational data in a different way.

Bazell and Désert (1988) were the first to apply tools appropriate for the analysis of fractals to the study of the 'infrared cirrus', discovered by *IRAS*. These cloud complexes appear to have a wispy structure which is generally associated with 21 cm H I emission clouds, and can sometimes also be associated with high-latitude molecular clouds. To facilitate the analysis Bazell and Désert examined selected 100 μm images (after removing foreground emission) at a certain contour level, obtaining a set of two-dimensional objects with well defined boundaries. The area of such a cloud and its perimeter could then be measured by appropriate pixel counting procedures.

We have seen in Chapter 3 that the simplest way to express fractal dimension is by using the scaling relation, which gives rise to the definition of the box-counting dimension (3.90) (see also above). Consider the perimeter of the cloud as it is projected on the sky. It is quite easy to show that if D_p is the fractal dimension of such a two-dimensional *perimeter* of an object in the set, the relation between the projected area of the object $\mathcal{A}$ and its perimeter ℓ should read

$$\mathcal{A} \propto \ell^{2/D_p} \tag{11.4}$$

(see Hentschel & Procaccia, 1984). Thus the fractal dimension of the projected cloud boundary can be derived from a plot of the logarithm of the areas versus the logarithm of the perimeters, for a set of clouds. If (11.4) holds, the plot should be linear with the fractal dimension being easily computable from its slope.

Basing their analysis on the above notions, Bazell and Desért were able to establish the self-similarity and thus fractal nature of the interstellar clouds and cloud complexes they studied, at a specific brightness level. They found the fractal dimension of the cloud perimeters to be $D_p = 1.26 \pm 0.04$ and this result was independent of the brightness level and the region in their sample. It is not obvious what the physical significance of these findings is and we shall refrain from drawing any far reaching conclusions. It will be sufficient for us to just conclude that the physical mechanism(s) responsible for the shaping of interstellar clouds have to admit spatial complexity of this kind.

Several subsequent studies of interstellar clouds traced by the 100 μm emission or the 21 cm H I line gave a rather significant scatter $1.2 < D_p < 1.55$ in the fractal dimension of the cloud perimeters (see Vogelaar, Wakker & Schwarz, 1991). Still, since these studies gave essentially consistent results and in spite of the fact that the analyses were done over scales spanning only several decades, it appears that the fractal nature of interstellar cloud boundaries has been reasonably established, similarly to results found previously for atmospheric clouds.

In a series of works published in the 1990s Falgarone and her collaborators engaged in observational analysis of the complexity of the dense interstellar medium, that is, of molecular clouds. Falgarone, Phillips and Walker (1991) applied the above-mentioned perimeter–area relation to their observational data of CO lines at high resolution. They studied two molecular-cloud complexes and found the fractal dimension of $D_p = 1.36 \pm 0.02$ for the projected cloud boundaries. The fact that this result was independent of the cloud size should have theoretical significance since the largest structures (~ 100 pc) in this study are clearly self-gravitating, while the smallest ones (~ 0.1 pc) are not.

We now return to the work of Elmegreen and Falgarone (1996), in which the fractal dimension of a number of molecular-cloud complexes was estimated by compiling data from various previous studies, and by fitting them to the appropriate scaling laws. It turns out that size distribution data and fits to the scaling relation (11.1) give an average fractal dimension of $D = 2.3 \pm 0.3$. We stress again that a perimeter of an object, as described above, is actually the perimeter of this object's two-dimensional projection and therefore we can expect the relation $D = D_p + 1$ to hold if isotropy is assumed. Thus this result is quite consistent with the perimeter studies mentioned above.

Regarding the mass function, it turns out that the empirical mass–size relation actually contains a rather significant scatter in the value of the exponent and if one writes

$$M \propto L^{\kappa} \tag{11.5}$$

in place of 11.2, the values of κ range from 2.38 ± 0.09 and up to 3.68 ± 0.71 for the various cloud complexes. These results were found by Elmegreen and Falgarone by considering the slope of the descending parts of the distributions, and it is not clear whether or not they include some systematic errors. It should also be pointed out that earlier studies (see Scalo, 1987) gave significantly smaller κ values (between 1.5 and 2).

Combining (11.5) with the relation (11.1) and the fact that $n(M) = dN/dM$ gives

$$n(M) \propto L^{-(D+\kappa)} \propto M^{-\alpha_M} \tag{11.6}$$

where $\alpha_M \equiv 1 + D/\kappa$. If $\kappa = D$, as it should be in a mathematical fractal (see 11.2), we get $\alpha_M = 2$ and relation (11.3) follows, independently of D. Empirically, however, the above results for κ and D, which include a significant scatter for κ and probably do not extend over a large enough scale span, are consistent, on average, with α_M in the range of 1.5–2. Elmegreen and Falgarone concluded that the size and mass distributions of interstellar clouds are consistent with a fractal of dimension $D \sim 2.3$, valid in the size range of $\sim 0.01-100$ pc. This possibility may also have important implications for the structure of the intercloud medium. Elmegreen (1997) dealt with this issue and concluded that that this medium is actually structured by the same mechanisms that shape the clouds themselves (in his view, turbulence).

In more recent work Elmegreen (2002) computer generated fractal distributions of a particular kind (such that the density distribution is log-normal). He then clipped these fractals at various intensity levels and computed the size and mass-distribution functions of the clipped peaks (and all the smaller peaks). The distribution functions obtained this way (by clipping) should be analogous to the cloud-mass functions which are determined from maps of the ISM using various intensity thresholds for the definition of clouds. The model mass functions were found to be power laws with exponents in the range from -1.6 to -2.4, depending on the clipping level. This is consistent with the assumption that the complex, scale-free ISM can be regarded as a fractal (over a range of scales), or more strictly speaking a multifractal, despite the fact that the *density* distribution function (in a log–log plot) may appear to be a Gaussian, that is, seemingly indicating a preferred scale.

However, even the sheer notion that interstellar clouds exhibit fractal structure has been challenged in a number of papers. For example, Blitz and Williams (1997) found that the 'clumpy' model (having a preferred length scale) fits well with the observational data of the Taurus molecular-cloud complex and hence molecular clouds are not fractal (see however Stutzki *et al.*, 1998 and the argument given in Elmegreen, 2002). In any case, the question of the validity of modelling a real physical system with a mathematical fractal is always non-trivial. This is particularly so when the object is astronomical and the observational limitations are quite severe. Still, it seems that we now have ample evidence for the existence of fractal structures in the interstellar cloud complexes of various kinds, at least in some representative cases and obviously for some range of scales. A possible connection, reconciling the above 'clumpy' model of the ISM and the multifractal one, can perhaps arise when clouds are viewed as isolated peaks in the fractal distribution, as pointed out by Elmegreen (2002) (see above), who also discussed several additional important aspects of this controversy.

The appearance of a fractal distribution, having a particular dimension, in a given physical system should probably be the signature of the physical process shaping the system. As we saw in Chapter 3, some simple algorithms create fractals with a definite dimension (e.g., the Cantor set or Koch's curve). Even in such idealised mathematical problems, if a given process produces a fractal of definite dimension, it is conceivable that the process is not unique and thus the fractal dimension alone cannot unequivocally identify the formation process. Falgarone *et al.* (1991) noticed the remarkable similarity of their fractal dimension estimates to the ones obtained in laboratory turbulent flow experiments. Consequently, gas dynamical models involving turbulence have been put forward by various researchers to explain the origin of the ISM structure. Fluid-dynamical turbulence still belongs to one of the most poorly understood physical processes and its modelling is mainly numerical (with obvious disadvantages due to limitations in the dynamical range). We shall not deal here with the turbulent ISM models but refer the interested reader to the comprehensive list of works referenced in Elmegreen's (2002) paper.

Complexity from thermal instability

The idea that thermal instability plays an important role in shaping the ISM and perhaps, combined with some other processes, actually generates its complex density structures, has been rather influential since its introduction in the 1960s. We have already mentioned some basic aspects of this problem in Section 2.5, when it was used as one of the astrophysical examples in which methods of dynamical systems and pattern theory, described in the first part of this book, could be used. Immediately after it had been introduced to astrophysics by Field (1965), its applications to the ISM started to appear in the literature. Although this astrophysical application has since then become widespread, thermal instability may possibly also be important in other astrophysical environments (e.g., solar corona, intergalactic gas).

A comprehensive review and discussion of the rather large body of research done on the thermal instability in the ISM is beyond the scope of this book. In a recent work Vázquez-Semadeni, Gazol and Scalo (2000) have investigated numerically, in detail the role of the thermal instability in shaping the ISM. They concluded that it is probably *not* the dominant process and they even questioned the validity of the so-called multiphase model of the ISM. Most of the relevant references on this (still somewhat controversial) issue may be found in this work and we shall not discuss it further here. Instead, in accordance with the spirit of this book, we shall elaborate on the above-mentioned didactic example, even if it cannot be considered as contributing to a realistic model of the ISM. We simply want to demonstrate what can be done, in the context of thermal bistability, using an approach based on dynamical systems and pattern theory. After all, as hinted above, thermal instability leading to the formation of a multi-phase (in this case two-phase) structure in extended media may be relevant to other astrophysical systems as well. For example, in a very recent work, Maller and Bullock (2004) base their ideas on how to resolve one outstanding problem in galaxy-formation theory on the possibility that a significant fraction of the available mass becomes locked in gas clouds, which remain external to a forming galaxy. Such clouds are assumed to be formed as a result of the separation of the primordial medium into two phases, because of thermal bistability.

We start our discussion here with a simple one-dimensional model, whose physical ingredients are clearly only the most essential ones. As explained in Chapter 2, these may reasonably include one-dimensional Lagrangian hydrodynamics supplemented by a heat equation, which incorporates a bistable cooling function and diffusive heat transport. This allows the derivation of an equation like (2.33) as the single model PDE (in the Lagrangian variable σ and the time t) for the temperature-like dependent variable $Z \equiv T^\alpha$ (where α is the exponent arising from the dependence of the thermal conductivity coefficient on temperature), capturing all the above essential physics of the one-dimensional extended medium. We rewrite this equation here for convenience

$$\partial_t Z = Z^\lambda \big[G(Z; \eta) + \partial_\sigma^2 Z \big] \tag{11.7}$$

remembering that $\lambda \equiv 1 - \alpha^{-1}$.

The non-dimensional cooling function $G(Z; \eta)$, appearing in this equation (see definition (2.34)), should model the essential features of the heat balance (and thus incorporate bistability), and as such it has to depend on the parameter η, the non-dimensional value of the fixed pressure. For details see the discussion in Section 2.5.

The basic notion of localised structures (fronts in this case) was introduced and briefly discussed in Chapter 2, following the derivation of the above heat equation. In Chapter 7 we have elaborated on ways to find localised structures (defects) in some generic nonlinear PDEs, and on methods to calculate the interactions between the defects, thus enabling the formulation and understanding of their dynamics. Some essential properties of Equation (11.7) with a bistable cooling function G are similar to that of the Landau equation (7.54), one of the generic 'reaction–diffusion' equations. The latter has been given as one of the generic examples in Chapter 7. In particular, in Section 7.5, some of the advantages of perceiving this equation as a gradient system, for the purpose of deriving the one-dimensional dynamics of its defects (localised structures) was given in detail. It has been shown there how the basic inverse cascade (bottom-up), that is, the growth of large 'clouds' from small perturbations, is the natural outcome of the interactions between the boundaries separating the two different phases, and how spatial forcing may induce a spatially chaotic pattern.

Let us assume a polynomial (specifically, cubic) shape for the cooling function, so as to capture just the essentials of bistability, taking advantage of the function's simplicity. If we write, for example

$$G(Z; \eta) = (Z_0 - Z)^3 - \Delta^2(Z_0 - Z) - \beta \log(\eta/\eta_Z) \tag{11.8}$$

and substitute this expression in the Lagrangian heat equation (11.7), that model equation becomes explicitly very similar to the Landau equation (7.54). Thus it can be expected that it will also have the same generic properties. The parameters Z_0, Δ and β control the shape of the cooling function, or equivalently the spread and the relative depth of the minima of an appropriate potential function (located at the stable zeros of G), that is, the thermally stable states (see the discussion in Section 2.5). It should be remembered that the variable Z is defined as a suitable power of the temperature, and λ is a parameter (both are related to the thermal conductivity power-law dependence on T), and η is the non-dimensional, uniform pressure. The term η_Z is the Zeldovich–Pikelner value of the pressure, that is, the value for which the minima are of equal depth. These matters were discussed in detail in Section 2.5 and we have repeated the definitions here just for convenience.

Elphick, Regev and Spiegel (1991b) investigated this kind of equation (with fluid motions neglected altogether, and consequently with σ being replaced by a spatial variable x and with a slightly different definition of Z, giving rise to $\lambda = 1 -$ see below) and showed that the solution, if started from some random initial condition, generally leads to an inverse cascade with increasingly large clouds predominating, arriving finally at a state with just one or no surviving fronts. The evolution of the front positions follows an equation very similar to (7.71), which was derived in Chapter 7 for the Landau equation. If a spatially periodic excitation source is added to the system, the solution approaches, for appropriate choices of the parameters, a steady state of spatial chaos. The chaotic pattern map, giving the locations of the fronts – kinks and antikinks ('cloud' boundaries) – in the final steady configuration, is essentially identical to the one found for the Landau equation (see Equation 7.77). The methods and calculations are described in detail in Section 7.5 of this book and in the paper cited above. Figure 7.3 summarises the evolution of a one-dimensional system of this kind in the free and forced cases, and is very similar to the results in the astrophysical application. Elphick, Regev and Shaviv (1992) showed that these findings also carry over to the case where fluid motion is included, since in one dimension, the resulting equation is of the same type. The only difference is in the value of the power $\lambda > 0$ and in the fact that a Lagrangian variable σ replaces the spatial coordinate x.

The extension of this kind of study to a multi-dimensional case is certainly called for, since in this case the dynamics of defects resulting from problems of the Landau equation type, whose localised structures are domain walls (curved fronts), differs substantially from the one-dimensional case. The effects of curvature are paramount and the dynamics of domain walls in the purely reactive–diffusive case is dominated by the interplay between curvature and the nature of the bistability (see the eikonal equation (7.84) and the discussion in Section 7.6). In addition, the fluid dynamical aspects of the astrophysical problem can no longer be simplified by using Lagrangian hydrodynamics.

The full multi-dimensional problem of a perfect gas medium of uniform chemical composition, whose dynamics is dominated by the pressure force alone, and which is heated (and cooled) by given thermal processes and is also capable of heat conduction, is described by the fluid dynamical equations, including the energy equation. Two thermodynamic variables (the density and temperature fields, say) are sufficient to completely describe the thermodynamic state of the gas (the other ones follow from an equation of state) and they are complemented by the velocity $\mathbf{v}(\mathbf{r}, t)$ field. We have

$$\frac{\partial \rho}{\partial t} = \nabla \cdot (\rho \mathbf{v}) \tag{11.9}$$

$$\rho \frac{D\mathbf{v}}{Dt} = -\nabla p \tag{11.10}$$

$$\rho T \frac{Ds}{Dt} = -\rho L(T, p) + \nabla \cdot (\kappa \nabla T) \tag{11.11}$$

where D/Dt is the Lagrangian derivative ($\equiv \partial/\partial t + \mathbf{v} \cdot \nabla$) and p and s are the pressure and specific entropy fields respectively, expressible by the density and temperature $\rho(\mathbf{r}, t)$, $T(\mathbf{r}, t)$ through the perfect gas law. Also, as before, L is the cooling function and κ the thermal conductivity.

If the cooling function L has the appropriate structure to cause bistability (see above and in Section 2.5), we are faced with the generalisation of the one-dimensional problem discussed above to two or three dimensions. However, the full set of the governing equations can no longer be generally reduced to a single thermal equation, as was the case in one dimension, even if one assumes (as we did in that case, see Section 2.5) that the pressure is constant. Such reduction can still be possible if, in addition, the velocity is assumed to be zero. The former assumption, i.e., a uniform constant pressure, is reasonable if the dynamical time (the sound crossing time) is much shorter than other relevant time scales, which is generally very well satisfied in the astrophysical cases of interest (in the general literature it is referred to as the short wavelength limit – see Aranson, Meerson & Sasorov, 1993). The pressure then equilibrates to a constant value, imposed on the boundaries, before any significant changes can occur. The latter assumption ($\mathbf{v} = 0$) is, however, never strictly correct, as density variations give rise to fluid motion so as not to violate mass conservation. As mentioned above, in the one-dimensional case a single equation could still be derived using a Lagrangian variable. In the two-dimensional case, a single equation can fully describe the system only if the motions in the plane are completely suppressed and the velocity is confined to only the third *perpendicular* direction (e.g., by magnetic fields).

Aharonson, Regev and Shaviv (1994) studied the evolution of such a two-dimensional system, that is, a medium under the assumption of uniform pressure and without fluid motion in the plane. The thermal equation can then be cast into the form

$$\rho c_p \frac{\partial T}{\partial t} = -\rho L(T; p) + \nabla \cdot (\kappa \nabla T) \tag{11.12}$$

As is well known and as we have seen in detail in our discussion of this problem in Chapter 2, it is the sign of $(\partial L/\partial T)_p$ that determines the linear stability of an equilibrium state (the solution of $L = 0$). We are interested, as was the case in the one-dimensional study, in a situation in which three uniform equilibrium solutions for a given uniform p (perceived as a parameter of the problem) exist, two of them linearly stable and one (the intermediate one) unstable.

Non-dimensionalisation of the above heat equation, by using typical values as the physical units of the variables, and the assumption that the heat conductivity varies as a certain power of the temperature (α, say) gives rise to the following single-model equation

$$\partial_t Z = Z[G(Z; p) + \nabla^2 Z] \tag{11.13}$$

for the field $Z(\mathbf{r}, t)$, defined as a suitable power of the temperature ($Z = T^{1+\alpha}$ in this case). The term $G(Z; p)$ is the appropriate functional derived from the cooling function (with p perceived as a parameter). As in the one-dimensional problem one may choose the simplest non-trivial generic form for the function G to model the bistable behaviour (see Equation 11.8). Aharonson *et al.* (1994) did just that and, with their choice of the Zeldovich–Pikelner value as the pressure unit, $G(Z; p)$ was actually given by (11.8) with the non-dimensional pressure parameter redefined as $p \equiv \eta/\eta_Z$, that is, the critical pressure being $p = 1$ in these units.

Note that Equation (11.13) is not exactly the two-dimensional extension of (11.7) or (2.33), because the power λ in those equations is generally not equal to 1. Indeed, this two-dimensional problem with no fluid motion is actually the extension of an earlier one-dimensional study (Elphick *et al.*, 1991), which neglected fluid motion as well. Still, the value of the (positive) power λ should not change the qualitative behavior of the solutions, because the essence of defect dynamics is determined generically by stability properties of the zeros and by the general shape of $G(Z)$. The multiplying factor, Z to some positive power, i.e., a positive monotonic function of Z, does not change that in a significant way (see below).

The *one-dimensional* properties and dynamics of fronts separating the two uniform stable phases ($Z = Z_0 \pm \Delta$) were discussed above and in Sections 2.5 and 7.6 of this book, and in the papers of Elphick *et al.*, where the fact that the model equation is a gradient system was crucial to the treatment. Although the two-dimensional case is inherently more complex, the model equation (11.13) remains a gradient system. Specifically, it can be written as

$$\partial_t Z = -Z \frac{\delta \mathcal{F}}{\delta Z} \tag{11.14}$$

where δ denotes functional derivative, and the functional $\mathcal{F}$ is given by

$$\mathcal{F}[Z] = \int \left[\frac{1}{2} (\nabla Z)^2 + V(Z; p) \right] \tag{11.15}$$

with the potential

$$V(Z; p) \equiv \frac{1}{4}(Z - Z_0)^4 - \frac{1}{2}\Delta^2(Z - Z_0)^2 + \beta(Z - Z_0)\log p + V_0(p) \tag{11.16}$$

where $V_0(p)$ is an arbitrary function of the parameter p. The fact that we are dealing here with a two-dimensional system is explicitly apparent in the form of the function $Z(\mathbf{r}, t)$, where the spatial variable $\mathbf{r}$ is a two-dimensional vector as is the gradient operator in (11.15).

This formulation of the problem (as a gradient system) is useful in the multi-dimensional case as well, particularly for finding some general principles of the front dynamics. Elphick *et al.* (1991) have used it to show that in the two- (or more) dimensional case the only stable stationary solutions (i.e., those minimising the Liapunov functional $\mathcal{F}$) are the homogeneous states $Z_0 \pm \Delta$. This is so even for the critical pressure, $p = 1$ (in the units chosen). Recall that in the one-dimensional case, the single-front solution was a third stable stationary state. Such a solution is thus lost when the dimension of the system is more than just one. This result has been rigorously derived before for a generic system of this kind (7.79)–(7.80) by Rubinstein, Sternberg and Keller (1989), who have moreover shown that, in general, the curves delineating fronts (or domain walls) have the tendency to shorten. That is, 'clouds' (encompassed by a closed domain wall) always 'evaporate' even though the pressure is critical. The topic is discussed in detail in Section 7.6 of this book, in the context of the Landau equation, but the conclusions are applicable to our case as well, as the relevant derivation is essentially the same. If $p \neq 1$ the symmetry between the two stable phases is broken and the evolution can be driven by this asymmetry – one of stable states becomes metastable only as the minima of the potential V (see 11.16) are no longer of equal depth.

Thus, the overall pattern evolution is generally driven by the tendency of curve shortening (curvature or 'surface tension' effect), asymmetry of potential minima (metastability or 'pressure' effect) and also by front interaction (if two fronts become close). Clearly, the third effect is appreciable only if the separation between the fronts is very small, as the interaction strength between two fronts decays exponentially with distance (see the discussion of the one-dimensional case in Section 7.5). In contrast, the curvature effect leads to front velocity proportional to the curvature (see the eikonal equation (7.84) and the discussion following it in Section 7.6).

To get some quantitative estimate for the above we consider a circular cloud of radius $R(t)$, whose evolution as a separate entity (curved-front self-interaction) can be simply represented (see Section 7.6) by the equation

$$\frac{dR}{dt} \propto -\frac{1}{R} \tag{11.17}$$

(front velocity is proportional to the curvature). The solution of this equation $R_0^2 - R^2 \propto t$, where R_0 is the initial radius, allows one to estimate what is the maximal distance between two circular clouds so that the mutual attraction of their boundaries locally overcomes their tendency to shrink. In such a case they will touch and merge before the self-interaction due to curvature has enough time to induce circularisation and shrinking. If D is such maximal separation, it is easy to see that the following relation between D and R_s (the radius of the smaller, and therefore the one that shrinks faster, circular cloud), should hold: $D = a \log R_s + b$, where a and b are constants dependent on the parameters of the function G above.

This said, we have to admit that the multi-dimensional evolution of a cloudy pattern in a medium governed by thermal bistabilty and diffusion is certainly too complicated to be fully grasped by a qualitative or a semi-analytical perturbative approach, like the one applied to the one-dimensional problem. Therefore Aharonson, Regev and Shaviv (1994) approached the problem numerically, solving (using a standard finite-difference technique) the PDE (11.13) in a rectangular domain with Neumann boundary conditions ($\partial_n Z = 0$, that is, vanishing normal derivative). The initial conditions were set to consist of the hot ('intercloud') phase near the boundaries and a random value around Z_0 elsewhere (binned in squares of 5×5 grid points). The grid spacing was chosen so as to be one characteristic length (the Field length, see Section 2.5) which is of the order of the front width, because the objective was the treatment of an extended cloudy medium and not of a resolved single front (cloud boundary).

The results of a variety of test calculations supported the expected findings, which could be predicted from the analytical estimates given above (like the shrinking rate of a single cloud, the fact that the critical distance for merging, D, goes like $\log R_s$ etc.). But the main and significant result was obtained when the evolution of the filling factor (the area fraction occupied by the dense phase) and of the fractal dimension of the cloud boundaries were computed in fully-fledged numerical calculations. The evolution of the system, initialised as described above, was followed for approximately 1000 cooling times. During this time the complexity resulting from the random initial configuration gradually simplified, with small structures merging and forming larger ones. Ultimately, one large cloud with a complicated boundary emerged and this boundary slowly shortened and became smoother, on the way to the cloud's gradual shrinking and ultimate disappearance. The area fraction occupied by the dense cloud phase started (after a transient of a few hundred cooling times) to decay linearly in time. After a similarly lasting transient, the fractal dimension of the boundary settled on the value of approximately 1 (within the sampling error of the 'measurement' based on the box counting dimension definition). This result can clearly be perceived as the two-dimensional analogue of the inverse cascade found for the one-dimensional case and described above, and in more detail in Section 7.5 of this book.

Aharonson *et al.* next turned to finding the effect of a spatio-temporal forcing on this type of evolution. Such forcing should be natural in the astrophysical setting, since the spatial distribution of the heating sources (e.g., stars in the ISM) is generally not uniform and their intensity may also vary in time. In addition, local heat sinks may also be present (because of increased radiative losses, say). While these processes may be extremely complicated, it is tempting to model them with a spatially random and temporally periodic (i.e., one Fourier component of the time variability) simple perturbation added to the cooling function. In the work described here the authors added to the cooling function $ZG(Z)$ (see Equations (11.13) and (11.8)) a perturbation of the following form (expressed here in Cartesian coordinates)

$$\delta(x, y, t) = \epsilon f(x, y) \sin[\omega t + \phi(x, y)] \tag{11.18}$$

where the spatial functions f and ϕ are generated by a random process, ω is some typical frequency (see below), and the parameter $\epsilon < 1$ reflects the perturbation strength.

Choosing the perturbation length scale to be of the order of 10 Field lengths and thus assuming the spatial functions to be constant in squares of 10×10 grid points and equal to a random number between -1 and 1, fixing ω so that it corresponds to a timescale of 10 cooling times and setting ϵ at the value of 0.1, Aharonson *et al.* obtained the following significant result. After an initial transient, similar to the one that was present in the unperturbed evolution, both the dense cloud area fraction and the fractal dimension of the cloud boundary settled on a value that remained approximately constant for the whole calculation (over 10^3 cooling times). It is particularly interesting to note that the fractal-dimension value settled on $D = 1.2 \pm 0.1$. Thus the perturbation induced a persistent cloud structure with complex boundaries, characterised by a fractal dimension not too far from the 'observational' values of ISM clouds. The evolution of the fractal dimension as a function of time in one representative calculation of this sort is given in Figure 11.1.

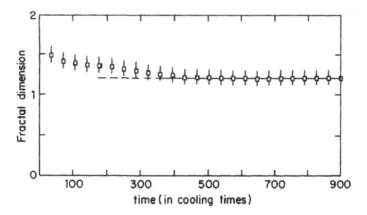

Fig. 11.1. Fractal dimension of the boundary between the two stable phases (calculated by the box-counting method) as a function of time for the system evolution including spatio-temporal forcing in the cooling function. For details see text. Reprinted, with permission, from Aharonson, Regev & Shaviv (1994).

Aharonson *et al.* experimented with other choices of parameters in the perturbing function as well and summarised their findings in the following conclusions. A random initial condition develops rather quickly (in ~ 10 cooling times) into a single large cloud surrounded by the tenuous intercloud phase (the one imposed on the boundary). If no forcing is present, the cloud boundary gradually becomes smoother and shorter, on the way to complete cloud disappearance. The dominant mechanism in this stage (which lasts a very long time, as compared to say $\sim 10^3$ cooling times) is curvature-driven motion. Front interactions remain important only during the first transient stages, when many small clouds are close to each other. Thus, although two-dimensional effects dominate here, the final result is qualitatively similar to the one-dimensional case. If a spatio-temporal excitation due to local heat sources and sinks is applied, cloud boundaries may remain rather complicated for extremely long times, owing their shape to the perturbation. A $\sim 10\%$ perturbation is sufficient to maintain complex cloud boundaries without, at the same time, significantly disturbing the bulk of the two stable phases. Perturbations having significantly smaller amplitudes than that cannot overpower for a long time the tendency described above for the unperturbed state, while too-large perturbation amplitudes completely destroy the order and do not allow for clouds of significant size to form. The spatial scale of the perturbation (~ 10 Field lengths in the calculation described above) affects somewhat the fractal dimension of the cloud boundaries. For very large scales the fractal dimension naturally approaches the value of 1, while for small scales it may increase above the value of 1.2, found in the described calculation. One should not forget, however, that scales too closely approaching the Field length cannot be meaningful in this context, because this is the front width scale.

A typical length-scale estimate for ISM conditions gives a value of around a few parsecs for the Field length, while the cooling time is of the order of 10^6 years. Thus the size of the system simulated in the calculation performed by Aharonson *et al.* and described above is of the kiloparsec order of magnitude, with the perturbation length scale of the order of tens of parsecs, and a time scale of tens of millions of years. These scales seem to be a reasonable choice for a spatio-temporal modulation of the cooling function due to a fairly massive star population. Thus the value found for the fractal dimension of the cloud boundaries corresponds to a model with reasonably realistic parameters. Most of the above calculations were performed under isobaric conditions with the critical pressure. Experiments with pressure variations around the critical value (e.g., a pressure pulse) were found to have only a transient effect. In general, a system like the one described in this paragraph would have its clouds disappearing in $\sim 10^8 - 10^9$ yrs in the unperturbed case. With a perturbation of the type described above the clouds persisted significantly longer and their boundary was found to be fractal.

When fluid motions in all directions are allowed in a multi-dimensional problem of the sort described above, the full set of equations (11.9)–(11.11) cannot be reduced any more to just a single model equation. Thus the numerical approach becomes more involved, even if the assumption of a constant pressure is kept. Shaviv and Regev (1994) performed a series of numerical calculations of the full 2-dimensional problem, relaxing also the isobaric assumption. They found results that resembled, in their overall tendency, the above-described purely thermal–diffusive case. However, the advective motions, which are necessary for mass conservation, obviously affected some important quantitative characteristics of the pattern evolution. Using methods from non-equilibrium statistical mechanics and pattern theory, Shaviv and Regev were able to assess analytically some of the important features of the pattern evolution. The numerical calculations supported these findings. The conclusions on two such aspects of the patterns, dynamical exponents and cloud size distribution, will be summarised here. All the details of the calculations and results can be found in the original paper.

In order to understand qualitatively their results, Shaviv and Regev considered the following simplified problem. Imagine a *spherically symmetric* configuration developing in accordance with equations (11.9)–(11.11), where the cooling function has the bistability property, as before. Spherical symmetry allows one to define a Lagrangian mass variable and reduce the problem to a single equation, as in the planar one-dimensional case. There is however an important difference – the spherically symmetric case is endowed with front curvature and therefore its effects can be examined. Indeed, in a D-dimensional space the Lagrangian mass coordinate, m say, satisfies $dm = \alpha r^{D-1} dr$, where α is a geometrical factor, containing also the density of the inner (to r) phase, and r is the spatial spherical coordinate. One may reasonably assume that a spherical front of this sort satisfies an eikonal equation, but in the *mass* space, that is

$$u_{\perp m} + K = c_m \qquad (11.19)$$

where $u_{\perp m} \equiv dm/dt$ is the spherical front propagation velocity in mass space, K its curvature and c_m is the velocity the front would have if it were flat, that is, the one resulting from just the stability asymmetry (pressure not equal to its critical value) of the phases. This equation is essentially the same as the original eikonal equation for a purely reactive diffusive medium (7.84), but here the velocities are in mass space and, in addition, the velocity units were normalised so as to eliminate the constant in front of the curvature term. Note that the above definitions of m and $u_{\perp m}$ yield

$$u_{\perp m} dt = \alpha r^{D-1} dr$$

Let the critical pressure be denoted by p_c (previously we have used it as the pressure unit and thus its value was simply 1) for clarity. The stability asymmetry (the difference in depth of the two potential minima) and therefore c_m are obviously dependent on the value of $p - p_c$. The larger $|p - p_c|$ is, the faster the front should move in the direction of the less stable (metastable) phase (see the discussion of this issue for straight fronts in Elphick, Regev & Shaviv, 1992) and obviously $c_m = 0$ for $p = p_c$. Using these considerations in the two opposite extreme cases we can deduce the following from equation (11.19).

(i) Near the critical pressure, that is, for $c_m \sim 0$, i.e., its being negligible with respect to $K = (D - 1)/r$ (curvature of a D-dimensional 'sphere') one gets

$$u_{\perp m} = -K \quad \rightarrow \quad \alpha r^{D-1} dr = -\frac{D-1}{r} dt \qquad (11.20)$$

with the solution

$$r^{D+1} - r_0^{D+1} = \alpha^{-1}(1 - D^2)t$$

where r_0 is the cloud's radius at $t = 0$. This means, as expected, that a spherical cloud will shrink because of curvature-driven motion. To prepare for a generalisation of these results, we now define a typical length scale of the growing domain, ℓ by $\ell^{D+1} \equiv r_0^{D+1} - r^{D+1}$ and obtain that ℓ grows in time, thus

$$\ell(t) \propto t^{\frac{1}{D+1}}$$

(ii) Far from the critical pressure, that is for the curvature term being negligible with respect to c_m, we have

$$u_{\perp m} = c_m \quad \rightarrow \quad \alpha r^{D-1} dr = c_m dt \qquad (11.21)$$

with the solution

$$r^D - r_0^D = \alpha^{-1} c_m D t$$

The cloud will thus grow or decay according to the sign of c_m and defining as above the length scale of the growing domain by $\ell^D \equiv |r_0^D - r^D|$ we get

$$\ell(t) \propto t^{\frac{1}{D}}$$

One can now conjecture that these time–growth laws of the length scale, found for a spherical cloud, are the same as the correlation length growth in a general cloudy system, viewed as a disordered system. The only rationale behind this conjecture is dimensional – ℓ is the only relevant length scale in the system. In a spherical cloud it is the radius and in a disordered system it should be the correlation length, defined with the help of the correlation function. In Chapter 3 we have defined the two-point correlation function of a point set (see (3.92)). It can be shown that it is possible to generalise this definition to any continuous system characterised by a field $T(\mathbf{x})$, say, so that the correlation function is

$$\xi(\mathbf{r}) \equiv \langle T(\mathbf{x}+\mathbf{r})T(\mathbf{x}) \rangle - \langle T(\mathbf{x}+\mathbf{r}) \rangle \langle T(\mathbf{x}) \rangle \qquad (11.22)$$

where $\langle\rangle$ indicates an ensemble average. For large $r \equiv |\mathbf{r}|$ one expects an exponential decay of the correlation function

$$\xi \propto e^{-r/\lambda_0} \qquad (11.23)$$

where λ_0 is the correlation length (sometimes denoted by ξ). If the field is time dependent, all the functions in the previous two formulae can be regarded as time dependent and in particular, the correlation length itself. If the correlation length grows as a power of t, $\lambda_0 \propto t^z$, say, z is referred to as the *dynamical exponent* of the correlation length.

Shaviv and Regev (1994) went on and tested their conjecture numerically confirming the values of the dynamical exponents suggested above. Choosing a randomly generated system with variation on small scales and pressure not too far from critical, the evolution was initially curvature driven, that is, the correlation length grew as $\propto t^{1/(1+D)}$. After some critical time there was a cross-over to a pressure driven evolution with the correlation length increasing as $\propto t^{1/D}$. A typical curvature-driven evolution sequence of a two-dimensional cloudy medium, started from a randomly prepared (on a small scale) configuration, is depicted in Figure 11.2.

In this work no spatio-temporal forcing of any kind was introduced, as the main purpose was to understand the development of a multi-dimensional cloudy medium in thermally bistable conditions, with the effects of fluid motion (resulting from mass conservation) included. Thus the effects of curvature and pressure in driving the pattern dynamics were the only ones to be considered. Typically, curvature evolution dominates in the early stages, because large curvatures are found in an initially very complex (starting from a random) distribution, in which very small structures abound. Pressure (if different from the critical value) effects should take over later on, because the emerging large structures (after the small structures have been wiped out) are no longer endowed with large curvature. If a suitable spatio-temporal perturbation is introduced it may, of course, interfere with this evolution, in a way that should probably be similar to the case studied by Aharonson *et al.* (see above). Shaviv and Regev did not study the effects of forcing but rather concentrated on the characterisation of the complexity, which they found in some of their models persisted for quite long time periods. They were able to obtain relationships between various statistical measures of the distribution, among them the fractal dimension, and their development in time (for details see the original paper). In concluding this discussion it should be remarked that the models described here may be valuable for the understanding of the physical processes included in

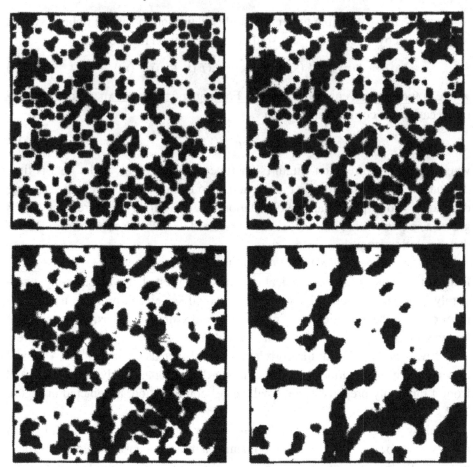

Fig. 11.2. The evolution of a 2-dimensional thermally bistable medium in the 'curvature-driven' ($p = p_c$) phase. The dark regions correspond to one of the stable phase (dense, cool 'clouds') while the bright regions represent the other stable phase (tenuous, hot 'intercloud' medium). The upper-left panel is chronologically the earliest while the lower-right one is the latest. The correlation length in this case grows with time as $\propto t^{1/3}$. For details see text. Reprinted with permission, from Shaviv & Regev (1994), © American Physical Society.

them, but cannot provide, of course, a complete description of media in which high velocity dynamical processes (explosions, shocks etc.) are important. Fluid motion, when included, was the result of only thermal effects and small pressure fluctuations.

Global statistical models for fractal ISM

Our last topic in this section on spatial structures in the ISM will be a brief discussion of some of the ideas that have been advanced in an effort to explain the formation of a fractal ISM structure.

A number of suggestions on how scale-free gas distributions of the ISM can be produced, invoking strongly turbulent motions of a self-gravitating gas (sometimes with dynamically important magnetic fields as well) have been advanced over the years. Since turbulence is a very difficult and still rather poorly understood topic (despite the high expectations that chaos theory will provide a breakthrough in its understanding, see next chapter), most of those studies have relied mainly on large-scale numerical calculations. The morphology of laboratory turbulent flows often reveals fractal structures not unlike the ones observed in the ISM (we shall touch upon this briefly in the next chapter). There are, however, some important differences between the laboratory and ISM turbulence. The value of the Reynolds number is perhaps the most obvious of these, but the effects of compressibility and long range forces (gravitational, electromagnetic) should be important as well. The huge Reynolds numbers encountered in astrophysical flows makes numerical simulations problematic and the small scales must somehow be modelled. It is possible, or even quite probable, that fluid turbulence is instrumental in shaping the ISM. However, the status of our analytical understanding of this enigmatic phenomenon, and consequently the need to rely on numerical experiments to model it, make this topic rather inappropriate for this book. We shall therefore not discuss turbulent ISM models here but mention turbulence, per se, only in a very limited and rudimentary context in the next chapter.

We would like to discuss here instead two alternative global approaches to ISM complexity. Although these studies do not exactly use methods of dynamical systems or pattern theory, like the ones introduced in the first part of this book, they seem to offer new (at least in astrophysics) ideas, with the help of which novel analytical tools may be provided. These tools can then be used in addition to (or even instead of) full-scale numerical simulations. This possibility arises because the general premise is to refrain from identifying, at the outset, hydrodynamical or magnetohydrodynamical turbulence as the dominant ISM complexity generator. This is the reason for including these approaches, albeit rather briefly, in our following discussion.

As the first work of this kind we choose the field theoretical approach introduced by de Vega, Sánchez and Combes (1996). Their motivation lies in the fact that in spite of the diversity of physical processes operating in the ISM – radiative heating and cooling, supersonic motions and shock dissipation, turbulence, and so on – the cloud mass size and internal velocity dispersion size relationships are observed to have a scale-free power law nature. The former relationship was already mentioned above (11.5), and the latter is given, e.g., in Scalo (1987) as $\Delta v \propto L^q$ with $0.3 \leq q \leq 0.6$. These relationships seem to be valid over approximately four decades of scale and it is reasonable to idealise the situation by invoking single universal scaling exponents, valid over all scales, for the distribution functions of at least a particular kind of ISM clouds. De Vega and collaborators focused their attention on the cold ISM and decided to treat it globally by employing methods of statistical field theory.

These non-trivial methods have been widely used in the treatment of critical phenomena (which are basically not very different from bifurcations). As this topic is clearly beyond the scope of this book we have not included statistical field theory in the first part. The interested reader is referred to the book by Parisi (1988), for example. Here we shall describe just the highlights resulting from this approach, quoting them from the paper mentioned above and discussing them briefly.

De Vega *et al.* considered a system of non-relativistic particles which interact with each other through gravity and are in a thermodynamic equilibrium, characterised by temperature T. It can be shown that this system is exactly equivalent to the field theory of a single scalar field with exponential interaction and de Vega *et al.* studied such a field theory, defined with the help of an appropriate grand canonical partition function. They have shown that mass fluctuations (ΔM) and velocity dispersion (Δv) in this model scale with distance (L) as the following power laws

$$\Delta M \propto L^d \quad \text{and} \quad \Delta v \propto L^q \tag{11.24}$$

where $d \equiv 1/\nu$, $q \equiv (1 - \nu)/2\nu$, and ν is the correlation-length critical exponent for the field, which is one of the central numerical parameters of the theory. In order to find the value of ν within a particular model it is necessary to determine the long distance critical behaviour of the field. A perturbative approach gives $\nu = 0.5$, while the renormalisation group approach (a single scalar field) indicates that this behaviour is governed by the well known exact Ising model result, which gives $\nu = 0.631\ldots$ (for details see de Vega *et al.*, 1996 and references therein).

Using these results one gets for the fractal dimension (denoted here by d) values of 2 or $1.585\ldots$, and for the velocity dispersion exponent $q = 0.5$ or $q = 0.293\ldots$. These values are certainly within the observational results summarised by Scalo (1987), but d is somewhat lower than the values of κ deduced by Elmegreen and Falgarone (1996) (11.5), although the average overall value of the fractal dimension advocated by them is quite close ($D = 2.3$). Keeping in mind the fact that the observational values of the scaling exponents are still rather crude and that it may very well be that they are scale-dependent (i.e., we are actually not dealing with a purely fractal distribution), the results of the field theoretical approach are quite encouraging.

It is well known in the theory of critical phenomena that the *equilibrium* scaling laws can be found without recourse to study of the system's dynamics. It is not clear, of course, whether the ISM can be considered as an equilibrium system, in this sense, or if this is only a good approximation in some range of spatial scales. The ISM dynamics consists of fluid flows (often turbulent ones), among other things, and as such they may be relevant to the ISM physics in the above sense. If so then they should be included in the theory, but as remarked above laboratory results for incompressible flows with modest Reynolds numbers (from which reliable information can be extracted) may have no relevance to the ISM altogether. Given our present understanding of turbulence, it is difficult to get theoretical results (they are usually assumed rather then derived from the theory). In this respect, any approach that circumvents the need to deal with turbulence is certainly welcome.

Recent work by Casuso and Beckman (2002) can be considered as an another alternative global approach. They employed the stochastic Langevin equation in an effort to study the evolution of the mass function of ISM clouds. The motivation for this approach is based, as they explain, on their view of the interpretation of observational findings (some of which were mentioned above). The wish was to incorporate the two competing views of the ISM: a totally scale-free one versus one with some preferred scales. The latter is based on observational indications that the mass function turns over and decays at some prescribed (quite low) mass value. This could also be seen in the data analysis of Elmegreen and Falgarone (1996), but they preferred to interpret it as an artificial cutoff created by observational limitations. Casuso and Beckman (2002) chose to incorporate these 'bumps' (for additional observational data on this, see the references quoted in their paper), as they call them, in the otherwise scale-free, power-law behaviour as an indication that two competing physical processes leading to fractal and at the same time 'clumpy' behaviour may actually coexist in the ISM. To achieve the goal of modelling these two types of behaviour simultaneously, Casuso and Beckman employed the Langevin equation for the evolution of the velocities of the elementary objects (actually small gas clumps), incorporating into it a friction-like dissipation term, a fluctuating forcing term and, in addition, a term representing a magnetic force.

The small (sizes below $\sim$1pc) gas clumps were thus treated as 'particles' executing Brownian motion and being influenced by frictional and magnetic forces. The appropriate Langevin equation for the velocity u of these particles assumed thus the usual form

$$\frac{du}{dt} = -\beta u + A(t) \tag{11.25}$$

where $A(t)$ is a stochastic, time-dependent function, and the dissipation coefficient β includes viscous (in the Stokes limit) and magnetic terms, i.e.,

$$\beta \equiv \frac{6\pi b\eta + qB}{m} \tag{11.26}$$

Here b is the particle size, m its mass and q its charge, η is the viscosity coefficient and B the magnetic field.

One of the possible, plausible physical points of view is to consider ISM clouds as aggregates of the gas clumps, which were considered as 'particles' and whose motion is governed by Equation (11.25). Employing a rather standard (in statistical physics) procedure it is possible to obtain the probability distribution $P(x, t)$ that a clump arrives at a cloud surface after a path length x, assumed to be the same as the distance between clouds, and time t. An additional assumption is that the motions of the gas clumps are spatially limited (by gravitation, say), and this fact may be approximately accounted for by formally confining the motion to within a box. This modifies the above mentioned probability distribution to a double Gaussian. Using a mass–size distribution of the type (11.5) one can convert the result into the following asymptotic (valid for large t) mass function

$$n(M) \propto \left\{ \exp(-M^{2p}/\sigma) + \exp\left[-(M_C^p - M^p)^2/\sigma\right] \right\} M^{p-1} \qquad (11.27)$$

where $p \equiv 1/\kappa$ comes from the observationally determined mass-size relation (see (11.5) and the discussion earlier in this section), σ is the appropriate Gaussian width, derived from the 'diffusion coefficient' (which contains the parameters of the Langevin equation) and M_C is the mass corresponding to the length scale of the confining box (see above).

The above mass function contains thus, in addition to an underlying power-law, representing a scale-less fractal distribution, local changes of gradient (i.e., bumping) at certain masses as well. Casuso and Beckman (2002) have pointed out that most data sets indeed clearly show such a duality: a general trend following an approximate power-law, but with several local peaks at specific masses. They fitted their theoretical mass-function formula to several samples of observational data and so obtained appropriate numerical values for the parameters p, M_C and σ. As it turns out, different data sets gave rather different values of these parameters and Casuso and Beckman, in discussing their findings, acknowledged that the results seem as yet incomplete, because the model is probably too simplistic. The main advantage of this approach seems to lie however in its qualitative ability to reconcile the two opposing views of the ISM complexity (scale-free fractal versus an array of clouds with a preferred scale) within a single framework.

It would probably be fair to sum up this section on ISM complexity by stating that, so far it has not been possible to achieve a consensus on a single well-defined theory capable of providing a satisfactory explanation of the observational findings. The findings themselves, still suffer from significant observational errors, some of them probably systematic, and there exists a considerable scatter in even the basic quantifiers of ISM complexity. We hope that the approaches summarised here may provide some basis for theoretical progress towards understanding the processes shaping the observed fascinating ISM complexity.

11.2 The large-scale structure of the Universe

But somewhere, beyond Space and Time
Is wetter water, slimier slime!
Rupert Brooke, *Heaven.*

Virtually all viable cosmological models are derived from Einstein's general theory of relativity and are based on the *cosmological principle.* In its simplest form this principle asserts that the universe is *homogeneous* and *isotropic.* However, direct observations of the *luminous* matter in the universe around us have always revealed a substantial degree of clumpiness. Optical catalogues of galaxy positions, compiled over the years, have become progressively better in resolution and depth (some of them have also provided distance information, inferred from measuring the redshift). Straightforward examination of the pictures reveals galaxy groups, clusters, superclusters, voids, filaments and walls (see, e.g., Figure 1 in Wu, Lahav & Rees, 1999). Attempts to quantify the observed distributions, which have accompanied the compilations of the data, reflect the desire to rationalise what one sees and thus give meaning to the identification of the various structures. Similar quantitative approaches have also been applied to the results of numerical gravitational N-body simulations in an expanding universe, which have generally produced structures similar to those found in observations.

Since the luminous matter constitutes (to the best of our knowledge) but a few percent of the mass in the universe, optically observable galaxies can be viewed as being just tracers. These galaxies thus move in the underlying dark matter gravitational potential, probably accumulating in its troughs (where they are also preferably formed). Thus, the inclusion of various possibilities for the dark matter properties in the computer simulations, and comparison of the numerical results with the observed distribution may provide clues on the dark matter and the cosmological parameters. These issues are clearly crucial to cosmology, i.e., to the question of which cosmological model, if any, faithfully describes our universe as we see it. It is therefore difficult to overestimate the importance of a reliable and meaningful quantification of the galaxy distribution data. This vast issue is, by itself, well beyond the scope of this book and we shall address it only in the context of our subject here, and that relates to the question of whether the large-scale distribution of matter in the universe is fractal or homogeneous. Observational surveys of the galaxy distribution leave little doubt that some relevant statistical measures, notably the correlation dimension (see Chapter 3), of the luminous matter distribution in the universe are scale-dependent. It also appears that the distribution of luminous matter on scales less than about $50h^{-1}$ Mpc (h is the Hubble constant in units of 100 km/s/Mpc), is indeed fractal.

It is the smoothness of the galaxy distribution (or lack thereof) on the largest scales that is still the key point in the debate between the proponents of a 'fractal universe' and the more traditional cosmologists. The cosmic microwave background (CMB) radiation, for which several years of data (since the launch of COBE) are now available, can provide a measure of the spatial fluctuations in the universe on the grandest scales – of the order of $\sim 1000h^{-1}$ Mpc. Fairly recent results indicate that the correlation dimension of the mass distribution at these scales is not different from the value of 3 by more that a few tens of 10^{-6}, indicating that the universe is quite smooth on the largest scale.

The question how (and on what scale) does the transition from a fractal distribution to a smooth one occur is still quite open. It should also not be forgotten that various attempts to evaluate the mass distribution in the universe are based on the observation of different entities, e.g., luminous matter (galaxies) versus the CMB (the universe at a very early epoch, when galaxies were still absent). Indeed, some very recent optical surveys aim at an impressively deep coverage (up to $\sim 300-400h^{-1}$ Mpc) but the grandest scales can be probed, at the present time, only by other means. Modern observations in the radio, microwave and X-ray bands have enabled us to take a deeper 'look', well beyond the existing and upcoming deep optical surveys. Some recent radio surveys and X-ray background (XRB) data can probably probe scales of up to $\sim 500h^{-1}$ Mpc, and thus bridge the gap between the optical surveys and the largest CMB scale. Other probing tools, like the Lyman α forest, may also be valuable in this respect.

In this section we shall try to summarise the current situation regarding the fractal distribution of matter in the universe. For a fairly recent comprehensive review we refer the interested reader to Wu, Lahav & Rees (1999), where an extensive list of references can also be found.

Galaxy distribution and clustering

The studies of the spatial distribution of galaxies, incorporating a variety of data sets, have usually been based on the analysis of the correlation function (see Peebles, 1980) or the correlation integral (see Paladin & Vulpiani, 1987). These two notions have already been discussed in considerable detail in Chapter 3 of this book, but we would like to add here that the actual analysis in most studies has been aimed at the derivation of $D^{(2)}$, which is equivalent to the correlation dimension of the distribution. The idea has been to look for a scaling behaviour, that is, power-law relations of the type $C(r) \propto r^{D^{(2)}}$ for the correlation integral, or $\xi(r) \propto r^{-\gamma}$ (with γ a constant) for the correlation function. The actual procedure employed in the case of finite sets (as galaxy samples) for obtaining the two-point correlation integral can be based on (3.93) and (3.94), while the one appropriate for the correlation function approach has usually been based on (3.92) (for details see Peebles, 1980). For a small enough r, i.e., when $1 + \xi(r) \approx \xi(r)$, one expects that the relation $D^{(2)} = 3 - \gamma$, between the powers scaling the correlation integral and function, actually holds.

A lucid and comprehensive summary of the statistical descriptors of this type and their application to galaxy distribution can be found in the second section of Murante *et al.* (1997). Following their work we remark that in actual galaxy samples one cannot expect to get a mathematically perfect scaling behaviour. Thus the value for $D^{(2)}$ or γ, as obtained from plots of the correlation integral or function versus distance (called r above, and ϵ in the definitions in Chapter 3), actually depends on the particular spatial scale of the domain used. This behaviour also occurs for higher-order correlations and there seems to be no doubt that scaling behaviour in real samples can be identified only over limited ranges of spatial scales and the powers actually vary with r.

Another source of complication is the apparent ambiguity in interpreting the observed galaxy distribution, arising from the fact that some authors look for a scaling behaviour (by power-law fitting) in the correlation function $\xi(r)$, while others prefer to quantify this behaviour for $C(r)$ (that is, actually for $1 + \xi$). Although, as mentioned above, these two quantities approach each other for $r \rightarrow 0$, discrepancies may be found, because r can be too large (this is particularly noticeable in dense sets).

Early studies (especially of angular position catalogues) were based on fitting a power law to the correlation function, and have yielded the value of $\gamma \approx 1.8$ for scales up to $\sim 5-10h^{-1}$ Mpc (Peebles, 1980). The precise value of γ seems to depend also on the morphological types and possibly also on the luminosities of the galaxies in the sample. Such a technique was applied later to the CfA catalogue (which includes also redshift, that is, distance information) and $\xi(r)$ was found to behave as a power law, essentially consistent with $\gamma = 1.8$, up to scales of $\sim 10-15h^{-1}$ Mpc (Davis & Peebles, 1983). For larger scales this study found a breakdown from a power-law behaviour with $\xi(r)$ becoming zero (and eventually negative). Davis and Peebles interpreted the point $r_0 \simeq 5h^{-1}$ Mpc, where $\xi(r_0) = 1$ as the correlation length of the sample, that is, a scale above which the distribution is essentially homogeneous. Coleman, Pietronero and Sanders (1988) challenged this result, pointing out that it is caused possibly by the wrong *assumption* that the distribution is actually homogeneous for scales close to, but smaller than, the size of the sample. Their analysis seemed to indicate that the absence of scale in the

CfA catalogue actually continues all the way up the size of the sample, and should be interpreted as reflecting a fractal structure with $D^{(2)} \approx 1.2$ up to $20h^{-1}$ Mpc and possibly even above that. This last conclusion has since then been at the heart of an ongoing controversy, hinted at above.

A subsequent study, published by Guzzo *et al.* (1991) and based on the analysis of the Perseus–Pisces cluster, also gave $D^{(2)} \approx 1.2$ (but only on scales of $1.0-3.5h^{-1}$ Mpc). This kind of scaling has also been confirmed by the output of several numerical simulations (see, e.g., Murante *et al.* 1996 and references therein). At larger scales, however, Guzzo *et al.* (1991) reported $D^{(2)} \approx 2.2$ ($r \sim 3.5-20h^{-1}$ Mpc), and similar values $D^{(2)} \approx 2.25$ (for $r \sim 1.0-10h^{-1}$ Mpc) were also found by Martinez and Coles (1994) for the QDOT redshift sample. This last work found a still larger value ≈ 2.77 for larger scales ($10-50h^{-1}$ Mpc) in the same sample, and a number of more recent, deeper studies have indicated that there is a gradual growth of the dimension with scale (see Wu, Lahav & Rees 1999) up to values consistent with homogeneity already on scales of $300-400h^{-1}$ Mpc (the ESO slice project, see Scaramella *et al.*, 1998). This behaviour has naturally been regarded as an observational 'proof' of the approach to homogeneity, that is, a space filling dimension of 3, at the largest scales. Still, some have challenged such high values (i.e., significantly larger than 2) of the correlation dimension as being the result of overestimates in the data analysis.

It seems therefore that it would be fair to conclude that the two opposite claims, i.e., the one invoking a universal fractal distribution up to very large scales, and the other, suggesting a relatively small correlation length with a homogeneous distribution already at moderate scales, have both to be abandoned. Observational data indicate a clear increase of $D^{(2)}$ with scale up to several hundred h^{-1} Mpc (see Wu, Lahav & Rees, 1999), even if the exact value of its maximum is not precisely determined. What remains still quite controversial is the nature of the 'transition to homogeneity'. Here there seems to be even some confusion regarding the meaning of the relevant statistical measures (see Gaite *et al.*, 1999; Pietronero *et al.*, 2002). These issues are the subject of the next subsection and in the remainder of this one we shall describe some theoretical efforts aimed at trying to propose and understand the physical mechanism which may give rise to a galaxy distribution, whose correlation dimension increases with scale, as it is deduced from the observational studies mentioned above. In one such theoretical work, Murante *et al.* (1997) suggested that the fractal scaling exponents of the galaxy distribution, which change with scale (see above), may actually be the result of what they call the 'singularity picture'. This notion refers to the observational and theoretical (mainly from computer simulations) finding that the large-scale structure contains prominent mass concentrations of various kinds. These generally lead to a number of very strong, local condensations, whose limiting behaviour suggests that they evolve towards the formation of singularities (see, e.g., White, 1996 and references therein). The creation of tight local clusters seen in the simulated (and real) galaxy distributions suggests that it may be reasonable to regard these singularities as resulting from an (approximately) spherical collapse. That process is known to lead to the formation of singular densities in finite time with a density profile $n(R) \propto R^{-\alpha}$, where R is the distance from the singularity centre and α is a constant depending on the physical characteristics of the spherical collapse (usually $1.5 \le \alpha \le 3$). An isothermal gas sphere with pressure, for example, corresponds, as is well known, to $\alpha = 2$. It is possible to show that the generalised dimensions (see the definition in Equation 3.91) resulting from a singular distribution of the kind described above satisfy $D^{(q)} \to 3 - \alpha$ for $q \to \infty$, while for any finite q the value of the dimension $D^{(q)}$ is somewhat larger than that.

Murante *et al.* (1997) randomly superimposed a few tens of singularities of the above kind and generated from them point distributions (by a suitable sampling procedure). They then computed the generalised dimensions of these point sets as a function of scale. Several cases, corresponding to different values of α, were explored. A qualitative resemblance to the actual galaxy distribution data was found for sufficiently strong singularities ($\alpha \ge 1.5$), with the scaling behaviour being quite good for generalised dimensions $q \ge 3$ and at sufficiently small scales.

The singularity picture, that is, simply assuming the existence of a random array of singular density condensations, can thus be considered as an alternative to other processes, which have been invoked in order to explain structure formation with fractal characteristics. Among the latter we would like to mention the model of nonlinear gravitational clustering proposed by Provenzale *et al.* (1992). They introduced a simple but dynamically motivated family of multifractal cascades, which are particularly well suited for a bottom-up scenario of gravitational evolution. We should note, however, that just a simple-minded singularity picture model (as described above) can only be applied at best to the smallest scales. If the structure on these scales is indeed singularity driven, and on the largest scales we approach homogeneity (see below in the next subsection), that is, a space-filling distribution of dimension 3, the scaling at intermediate scales remains to be addressed. The specific version of the multifractal cascade model mentioned above is also not adequate for reproducing the scaling behaviour at intermediate scales. Provenzale *et al.* (1992) claim, however, that if one chooses equal probabilities for over- and underdense fluctuations (as is appropriate for linear density fluctuations, which may be preferred at the intermediate-scale regime), the multifractal cascade model gives a scaling with $D^{(q)} \to 2$ for $q \to \infty$. This is similar to the observational finding by Guzzo *et al.* (1991) who obtained $D^{(2)} \approx 2.2$ (for $r \sim 3.5-20h^{-1}$ Mpc) as mentioned before. Alternatively, this intermediate scale value may have a 'topological' origin, that is, it may be the result of domain walls, pancakes, or cosmic strings (see Provenzale *et al.* 1992 for more details on this suggestion and references).

Provenzale, Spiegel and Thieberger (1997) suggested that additional statistical measures of the galaxy distribution may be useful for a meaningful characterisation and perhaps deeper understanding of the scaling behaviour at intermediate scales. Recognising that the singularity picture presented above may work well only for the smallest scales, they suggested looking at higher moments of the galactic point set, that is, a generalisation of the two-point correlation integral (see the definition (3.93)). This generalisation is analogous to the introduction of the generalised dimensions (3.91), and there is obviously a relationship between the generalised correlation integral and the corresponding generalised dimension for any q.

The qth moment of a finite point set is defined as

$$C_q(r) = \left\{ \frac{1}{M} \sum_i^M \left[\frac{1}{N} \sum_j^M \Theta(r - |\mathbf{x}_i - \mathbf{x}_j|) \right]^{q-1} \right\}^{1/(q-1)} \tag{11.28}$$

and clearly reduces to the two-point correlation integral (3.93) for $q = 2$ and at the limit of an infinite point set. Following Provenzale *et al.* (1997) we point out that it can be expected that C_q goes to zero (with $r \to 0$) as $\propto r^{D^{(q)}}$, and therefore obeys, for small enough r, the relationship

$$\log C_q = D^{(q)} \log(r/r_0) \tag{11.29}$$

where r_0 is some characteristic separation. If this last relationship is viewed actually as just one term in an appropriate expansion in powers of $\log r$, going further gives

$$\log C_q = D^{(q)} \log(r/r_0) + \log \Lambda + O\left(\frac{1}{\log(r/r_0)}\right) \tag{11.30}$$

where the nature of Λ can be seen from keeping just the first two terms of the expansion (11.30) and thus obtaining the explicit relationship

$$C_q(r) = \Lambda r^{D^{(q)}} \tag{11.31}$$

Now, Λ does not strictly need to be constant and may depend on $\log r$, provided $\log \Lambda$ is of the order of unity (i.e., $[\log(r/r_0)]^0$). Generalising the definition of Mandelbrot (1982), who perceived Λ as a prefactor and called it *lacunarity*, appropriate for a simple monofractal, Provenzale *et al.* (1997) suggested the term *lacunarity function* for the more general case. The lacunarity function is periodic in $\log(r/r_0)$ if a scaling relation, $C_q(r) = AC_q(Br)$, with A and B constants, exists. For details on the reasoning leading to this conclusion and other examples of lacunarity functions see Provenzale *et al.* (1997), who performed an analysis of the

CfA catalogue data and obtained the lacunarity function for the galaxy distribution contained in that catalogue. The logarithmic amplitude that was found was quite small but nevertheless significant, reflecting the fact that, as with other fractals, the apparent fluctuations around the relation (11.29) are mainly a manifestation of the lacunarity function.

The discovery of the lacunarity function in the galaxy distribution may have important astrophysical consequences if these results are confirmed and quantitatively strengthened by studying newer and richer galactic data sets. By its inclusion in the analysis of detailed galactic distribution data sets, we may hope to better understand the processes that produce clusters and to see, perhaps, an analogy between the cascades forming galaxies, clusters of galaxies etc. and the energy spectrum of turbulence. In the latter case the fractal characteristic of the phenomena and the oscillatory nature of the associated lacunarity are of great interest and are the subject of intensive research. It is certainly conceivable that this kind of analogy may provide valuable insights and help improve our understanding of structure formation in the universe. We shall mention below two such specific ideas, originating in some recent studies in the context of condensed-matter and statistical-field theory.

The question of homogeneity on the largest scales and the transition to it

As we have already seen, the growth of the correlation dimension with scale is regarded by the astrophysical community, at least at the time of writing this chapter, to be a rather robust observational finding. Values as high as $D^{(2)} = 2.9$, and even perhaps a little more than that, are claimed when scales of $300-400h^{-1}$ Mpc are probed (Scaramella *et al.* 1998).

The grandest scales ($\sim 1000h^{-1}$ Mpc) of the universe are naturally probed by the cosmic microwave background (CMB), thought to be a direct relic of the Big Bang itself. The COBE satellite, measuring this radiation, has detected only tiny deviations from a homogeneous distribution of the physical properties of the universe on these largest scales. The gap between the direct optical surveys on the largest scale (like the ESO slice project) and the CMB scale is however still unexplored directly. Various objects, which can be seen at enormous distances, may serve as tracers of the galaxy distribution at these scales, but this clearly introduces some complications because it is not clear a priori if the clustering of such tracer objects to that of the galaxies is similar, and also how does this distribution (and possibly also the brightness) evolve with time? Among the possible tracers of galaxy clustering at high redshifts one may consider radio galaxies, quasars, some very high redshift galaxies, Lyman α clouds and the X-ray background. Wu, Lahav and Rees (1999) review and discuss all of these and critically examine what can be learned from the distribution of these objects. It seems that the case for homogeneity (with very small fluctuations thereof) and for fractal clustering (reflecting very large density fluctuations) at relatively small scales, with the fractal dimension increasing with scale, is quite convincing. Some controversy still remains on the subject of *transition* to homogeneity and the proper statistical tools to explore it.

Gaite, Domínguez and Pérez-Mercader (1999) have stressed, in this context, the differences between several statistical concepts, which are sometimes confused in the astronomical literature. The two-point correlation function of a distribution, which we denote by $\xi(r)$, is the basic concept and we have dealt with it in Chapter 3 (see (3.92)) and also above in the context of ISM complexity. The important distinction in this respect is between the correlation length λ_0, as it is defined in statistical physics of critical phenomena, that is, the scale at which there is a crossover between a power law and an exponential decay of the correlation function $\xi(r)$ with distance (i.e., $\xi \propto e^{-r/\lambda_0}$); and the length scale r_0, at which $\xi = 1$. The latter has sometimes been called 'correlation length' as well, hence the confusion. Clearly, at scales of $r \gg \lambda_0$ the fluctuations are small and the system is structureless, that is, homogeneous. For $r \ll r_0$ (but above some lower cutoff reflecting the mean separation of two points in the set) the fluctuations are very large and exhibit structure. This can be perceived as the fractal regime. In between the fractal and homogeneous regimes there is the critical regime $r_0 \ll r \ll \lambda_0$. Here the fluctuations are small, but they are correlated and thus give rise to extended structures. Since the correlation function still obeys a power-law behaviour, the fluctuations over the background exhibit a self-similar, fractal structure. Gaite and collaborators propose to call r_0 the *nonlinearity scale* and thus properly distinguish between the different regimes. For example, this would naturally account for the fact that galaxy clustering is observed even on scales much larger than r_0 (when the fluctuations are small). A meaningful definition of the transition to homogeneity thus seems to be the scale at which a *crossover* between a power law and exponential correlation function decay occurs (see Pietronero, Gabrielli & Sylos Labini, 2002 for further discussion and their view on this issue), that is, the correlation length λ_0.

A novel, very interesting and promising idea to reconcile the controversy regarding the fractal dimension of the galaxy distribution has recently been proposed by Eckmann *et al.* (2003). It is based on a mathematical theorem asserting that it is impossible to faithfully determine the dimension of a set, in which the positions of its members are determined by observations from a single vantage point, if the true dimension of this set is more than 2. This difficulty arises not because of projection effects, but rather is caused by galaxies 'hiding' behind other galaxies. Thus it may be that the dimension of the full set of galaxies is actually 3, as various indirect methods indicate, but direct observations should reveal significantly lower values.

Indeed whether the distribution of galaxies has a fractal dimension that gradually increases with scale, or there is actually a crossover between two different types of behaviour, or perhaps indeed the large-scale distribution of galaxies is homogeneous but only *appears* to be fractal, with a dimension close to the value of 2, still appears to remain open. Further and improved observations and analytical tools are needed in order to clarify the issue.

We shall conclude this discussion by mentioning two additional theoretical studies, which attempt to model the behaviour of the overall distribution of galaxies in the universe, using global approaches, which have mainly been applied in condensed-matter physics. Studies of this kind are quite interesting and may perhaps help, as we have already stated above, in our understanding of the large-scale structure and in particular in resolving the above mentioned controversies.

The first of these is the attempt to explain the distribution with the help of a nonequilibrium reaction–diffusion model (see Chapter 7 for the discussion of models of this kind). The scheme, known in the literature as the 'forest-fire' model, was proposed by Bak, Chen and Tang (1990) for the purpose of capturing some of the essential features of turbulent systems, in which energy is injected at the largest scale and dissipated at a small length scale. It is an essentially geometrical model consisting of a three-dimensional cubic lattice, on whose empty sites 'trees' (representing energy) are grown randomly at a rate p, say. During one discrete time unit these trees can 'burn down' (and thus leave the site for new trees to grow) but they also ignite trees at the neighbouring sites. The burning represents dissipation, while the ignition of neighbours – diffusion. The discreteness of this simple reaction–diffusion system makes it a cellular automaton, a special case of coupled map lattice (see Chapter 7), and its dynamics can be quite easily followed numerically. It turns out that after a transient such a system enters a statistically stationary state, with a rather complex distribution of fires. Interestingly, this distribution, while not being self-similar over all scales (and therefore not a true fractal), can be attributed with an apparent fractal dimension that varies with scale. It becomes homogeneous beyond a correlation length (λ_0, as it is defined in statistical physics – see above), which depends on the injection rate parameter p. At scales below λ_0 the following approximate relation, between the apparent fractal dimension D and the length scale at which it is computed r, was numerically found to be

$$D(r) \sim 3\frac{\log r}{\log \lambda_0} \qquad (11.32)$$

where both r and λ_0 are expressed in units of the model lattice spacing.

Bak & Chen (2001) proposed to apply their 'forest-fire' model to the distribution of luminous matter in the universe. Using the data compiled by Sylos Labini, Montuori and Pietronero (1998), Bak and Chen were able to fit three different galactic data sets to the 'forest-fire' model prediction (11.32), thus obtaining a value of the order of 300 Mpc for the correlation length λ_0. The lower cutoff scale, which actually represents the model lattice spacing, that is, the length unit, can also be estimated from these data, albeit with a low precision. Bak and Chen found it to be of order of the mean intergalactic spacing. The picture emerging from the 'forest-fire' model is thus of a non-fractal (when viewed over the entire range of scales) and non-homogeneous galaxy distribution. The scale-dependent dimension implies that at small distances the distribution is point-like and it becomes filamentary (or string-like), that is, of dimension 1, on scales of the order of Mpc. When viewed at larger scales the distribution gradually becomes two-dimensional (wall-like) and the dimension continues to grow with scale. Finally, above the correlation length, the universe becomes uniform (of dimension 3).

The simplicity of the 'forest-fire' model makes it quite appealing and its application to the problem of large-scale structure can serve as yet another example of using a generic dynamical system to try and understand a seemingly very complex astrophysical system. The combined effect of turbulent fluid dynamical processes, including collisions, mergers, explosions, shocks etc. may perhaps be quite well approximated, on large scale, by a reaction–diffusion system. However, at least one of the model's important predictions, the crossover to uniformity at ~300 Mpc must await the availability of deeper galactic distribution data before it can be observationally verified. It is also instructive to compare the 'forest-fire' model findings with more conventional interpretations of the large-scale structure, i.e., the analysis based on the behaviour of the correlation function itself. Bak and Chen discuss this issue and conclude, among other things, that the effect of the evolution of the universe can be phenomenologically expressed by the increase of the correlation length λ_0 with time, describing a universe whose density decreases in time.

The absence of the gravitational force from the 'forest-fire' model ingredients is quite intriguing because the conventional wisdom seems to have been that it is precisely this force that actually controls the evolution of the large-scale structure of the universe. Still, the results are quite remarkable and plausible if the gravitational force, on large scales, is indeed not as instrumental to structure formation as it may appear e.g., because of the, essentially unknown, distribution of dark matter and its role. In this context it should be mentioned that Seiden and Schulman (1986) proposed some time ago that a similar reaction–diffusion mechanism may shape galaxies.

An alternative global approach to the problem of structure formation in the universe (and in the ISM as well – see the previous section) is the field-theory approach to the thermodynamics of self-gravitating systems introduced by de Vega, Sanchez and Combes (1998). Using the theory of critical phenomena and finite-size scaling they showed that gravity may provide a dynamical mechanism creating a fractal structure. Assuming that the galaxy distribution (within some range of scales) is in a quasi-isothermal equilibrium and in a critical state, they derived the critical exponent v, which gives the fractal dimension $D = 1/v$. As in their earlier work on the ISM patterns (discussed in the previous section), here too, the value of the fractal dimension depends on the assumption of what governs the long range behaviour. The Ising model fixed point gives $D = 1.585$, while that of mean field theory $D = 2$. Although there are no free parameters in this kind of approach, the fact that it predicts one fixed dimension (and in a rather limited range of scales) makes it perhaps useful in interpreting the galaxy distribution but only in a relatively small scale range. In summary, it is very clear that further work (both observational and theoretical) is needed on this still quite controversial subject. New data from deep surveys will soon become available, and new methods of statistical analysis together with fresh theoretical ideas (maybe along the line of the works sketched here) are probably necessary in the ongoing quest to properly quantify and physically understand the large-scale structure of the universe.

12

Topics in astrophysical fluid dynamics

Dond'escono quei vortici
Di foco pien d'orror?
Lorenzo da Ponte, *Don Giovanni.*

Where the hell do all these
vortices come from?
Ed Spiegel, *Astrophysics Seminar.*

Developments in the theory of nonlinear conservative dynamical systems and
Hamiltonian chaos have often been motivated, as we have seen, by problems in
celestial mechanics. Fluid dynamics has, likewise, motivated much of the dissipative dynamical systems and pattern theory. However, while celestial mechanics is
certainly a part of astronomy (a discussion of chaotic dynamics in planetary, stellar
and galactic *n*-body systems can be found in Chapter 9 of this book), fluid dynamics can be regarded as an essentially separate discipline of the physical sciences
and applied mathematics. Its uses and applications are widespread, ranging from
practical engineering problems to abstract mathematical investigations, and its importance to astrophysics stems from the fact that most of the observable cosmos is
made up of hot plasma, whose physical conditions are very often such that a fluid
(or sometimes magneto-fluid) dynamical description is appropriate.

As is well known, and we shall shortly discuss explicitly, fluid dynamics has as
its basis the description of matter as a continuum. Various assumptions then give
rise to appropriate sets of the basic equations and these are usually nonlinear PDEs.
The nonlinear PDEs of fluid (or magneto-fluid) dynamics are complicated dynamical systems and they have so far defied (except for the simplest cases) a rigorous
and complete mathematical understanding. The problem of fluid turbulence has
been particularly notorious in this respect. The rapid developments in the theory of
nonlinear dynamical systems, following the works published in the early 1980s on
the subject of temporal chaos in maps and ODEs, and on spatio-temporal complexity and patterns in PDEs, which appeared later on, had naturally been regarded as

381

promising directions towards a possible breakthrough in the understanding of fluid turbulence. It is still difficult to judge how close we have come to such a breakthrough, and even whether the direction is a correct one. It would be fair to say, however, that the so-called *dynamical-system approach* to fluid dynamics in general, and to turbulence in particular has already yielded some fruitful insights and many of the methods described in the first part of this book (especially in Chapters 4, 5 and 7) have been successfully implemented in the study of the behaviour of some particular fluid flows and problems.

For obvious reasons, our exposition here on the subject of fluid dynamics and its application to astrophysical systems will be very far from complete. The literature on fluid dynamics is vast and we choose to single out just two works and recommend them as references: the book by Thompson (1971) and the classical text of Landau and Lifshitz (1987). The former is a very comprehensive (including a detailed discussion of compressible flows) and didactic textbook. The latter reminds one of a wine of exceptional quality – it is best used only after achieving some maturity, and the appreciation for it increases with age (of both the book and the reader).

Several entire books devoted to the subject of fluid dynamics from an astrophysical perspective have been published quite recently, establishing perhaps the discipline of *astrophysical fluid dynamics* (AFD). In the books by Shu (1992), Shore (1992) and Battaner (1996) the reader can find traditional expositions on fluid dynamics, oriented towards astrophysical applications. The related discipline of GFD (geophysical fluid dynamics) has by now become firmly established and has produced a rather extensive literature on the dynamics of rotating stratified fluids. Simplistic fluid models, capable, however, of exhibiting complex behaviour, have often been used in GFD and the understanding gained from their study has significantly contributed to research on the Earth's atmosphere and oceans. The books by Pedlosky (1987) and Salmon (1998) contain excellent accounts on this subject. The Les Houches course on AFD, edited by Zahn and Zinn-Justin (1993), is similar in its nature to the approaches used in GFD and is probably the closest in spirit to the material given in this chapter.

As stated above, we shall not even be able to mention here many important topics of AFD. The choice of those that we have included, because we find them well-suited to this book, is subjective but not altogether arbitrary. Their characterising property can perhaps be described as ones in which a nonlinear dynamical system approach has been successfully applied to fluid dynamical problems of astrophysical interest. However, this obviously can neither have a precise meaning nor even be objectively defined. In particular, we shall say very little on the problem of turbulence (which is clearly within the nonlinear realm and is certainly relevant to astrophysics). We refer the reader interested in the 'state of the art' of turbulence research to two recent works. The comprehensive book by Pope (2001) contains a good summary of the classical approaches to turbulence theory and practical implementation thereof, while a summary of dynamical system approaches to turbulence can be found in the book by Bohr *et al.* (1998). Both books can also serve as very good sources for references. Still, we hope that the material given in this chapter will shed some new light on a number of interesting fluid dynamical phenomena of astrophysical interest, and perhaps even stimulate new directions in their research.

This last and somewhat special chapter of the book will start with two sections, containing what may be perceived as a fluid-dynamics primer. Most of the basic concepts and equations (also including properties of special flows), needed for the rest of this chapter, will be introduced here. Most of this material is quite elementary and can be found in standard texts on fluid dynamics, but we have decided to include it nevertheless; the reasons are two-fold. First, we would like this chapter to be as self-contained as possible. Second, some of the rudiments of fluid dynamics are presented here in a way that is quite different from the ones used in traditional texts on astrophysical fluid dynamics. These formulations are, in our view, very useful not only for the material given here, but also for future AFD work.

We shall discuss next some aspects of hydrodynamical stability, and focus, in particular, on the problem of the onset of instability in shear flows. The failure of linear stability analysis to correctly predict what is observed in the laboratory in such flows has caused a number of suggestions on the nature of these phenomena. We shall discuss some recent ideas in this context and examine their possible relevance to one, still outstanding, astrophysical problem.

This will be followed by a section introducing some of the basics needed for modelling laboratory thermal convection and a rather detailed discussion of the famous Lorentz model, the paradigm of dissipative chaos. The much more difficult and largely unsolved problem of stellar convection will be mentioned only briefly.

General Hamiltonian formulations of fluid dynamics, an exciting and elegant topic on its own, and some particular examples will be introduced in the last section, but no explicit applications to astrophysics will be mentioned. The related subject of chaotic mixing, the so called 'Lagrangian chaos', which will be discussed next, has however astrophysical relevance. For example, some recent ideas relating to magnetohydrodynamical (MHD) dynamos, which are thought to be present in various astrophysical systems, make use of it. We shall conclude with a short and rather basic discussion of this complicated subject.

12.1 Basic concepts and equations

At the heart of fluid dynamics lies the notion of a continuum. The fluid is considered to be composed of fluid 'elements' or 'particles' (we shall use these terms interchangeably), keeping their identity as they move and obeying the laws of particle dynamics. These fluid elements must be macroscopic, that is, large enough to contain a very large number of the microscopic constituents of matter (atoms, molecules). Only then do the velocity, density and other macroscopic physical properties of such a fluid element have sense (as suitably defined averages). On the other hand, these elements must be sufficiently small, so that their properties are attributed to a point in space, and the set comprising them is dense.

A sufficient condition for the validity of a continuous description of matter, based on the above mentioned notion of fluid elements or particles, thus appears to be the fulfilment of the strong inequality $l \ll L$, where l is the mean separation of the *microscopic* particles in the system and L is a physical length scale of interest in the system. Thus L is a scale over which the *macroscopic* properties of the system are expected to vary appreciably. If this is the case, the size of a fluid element (whose volume is δV say), $\lambda \equiv \delta V^{1/3}$, can be chosen so that $l \ll \lambda \ll L$. The physical properties of matter can then be regarded as continuous functions of space (and time) if coarse graining over the spatial scale λ is assumed. Some subtle issues related, for example, to a possible exchange of microscopic particles between fluid elements, the meaning of a 'dense set' of fluid elements and so on, can be treated precisely if kinetic theory is used for fundamental derivations, or alternatively, if an axiomatic approach is followed. For our purposes here, however, the *assumption* that a continuum description is possible and faithful, at least for a reasonably long time, will be sufficient.

The characterisation of fluid flow – kinematics

Because the dynamics of the fluid elements is treated like that of particles, the natural kinematical approach is the *Lagrangian* one, which is based on labelling the elements and following (in time) the motion of each of them. A useful way to label the elements is by their positions at some initial time, say. A function $\mathbf{x}(\mathbf{X}, t)$ can thus be defined as the position of a fluid element, which was initially at $\mathbf{X}$, at any later time t. It is possible to express this symbolically as $\mathbf{x} = \boldsymbol{\Phi}_t(\mathbf{X})$ with $\mathbf{X} = \boldsymbol{\Phi}_{t=0}(\mathbf{X})$ and regard the motion of the dense set of all fluid elements as a continuous (in time) differentiable mapping, denoted by $\boldsymbol{\Phi}$, of space onto itself and this is equivalent to a *flow* in the dynamical system sense (see Chapter 3).

The Jacobian of the transformation (which is itself time-dependent, in general), is an important entity in such a flow and we shall denote it by

$$J \equiv \frac{\partial(\mathbf{x})}{\partial(\mathbf{X})} \equiv \det \mathcal{J} \qquad (12.1)$$

where the matrix

$$\mathcal{J}_{ij} \equiv \frac{\partial x_i}{\partial X_j} = \frac{\partial}{\partial X_j}[\Phi_t(\mathbf{X})]_i \tag{12.2}$$

is called the *displacement gradient* matrix or tensor. This is also equivalent to the definition (3.7), which was formulated for dynamical systems flows in Chapter 3. If $J \neq 0$, the invertibility of the the transformation Φ is guaranteed, that is, we have for example $\mathbf{X} = \Phi_t^{-1}(\mathbf{x})$.

Any function G (scalar, vector, tensor), expressing some physical property of the fluid and thus defined on the space occupied by it and time, can be expressed in the Lagrangian (or material) viewpoint as $G(\mathbf{X}, t)$, that is, by giving the value of the property in question of a fluid element, defined by the label $\mathbf{X}$, at time t. The alternative viewpoint, known as *Eulerian*, regards G as a function of space and time $G(\mathbf{x}, t)$, that is, a *field*, giving the physical property of the element that happens to be at the position $\mathbf{x}$ at time t. To convert from the Lagrangian to the Eulerian representation (and vice versa) one has to make use of the mapping Φ_t and/or its inverse.

In particular, it is important to distinguish between the Lagrangian (or 'material', or 'substantial') time derivative and the Eulerian one. The former, which we denote by D/Dt, is a rate of change evaluated while *following* the motion of an element, i.e. with $\mathbf{X}$ fixed, while the latter, written simply as ∂_t or $\partial/\partial t$, is the rate of change evaluated at a fixed position in space $\mathbf{x}$. Symbolically

$$\frac{D}{Dt}G \equiv \left(\frac{\partial G}{\partial t}\right)_{\mathbf{x}}; \quad \partial_t G \equiv \left(\frac{\partial G}{\partial t}\right)_{\mathbf{x}} \tag{12.3}$$

and the relationship between the two derivatives can be established using the chain rule. Since any function $G(\mathbf{x}, t)$ can be written as $G[\mathbf{x}(\mathbf{X}, t), t]$, thus

$$\frac{D}{Dt}G = \partial_t G + \mathbf{v} \cdot \nabla G \tag{12.4}$$

where $\mathbf{v}$ is the velocity of the fluid element which is at point $\mathbf{x}$ at time t, i.e., the appropriate value of the Eulerian velocity field (see below). The material derivative is thus explicitly shown in (12.4) to be the sum of a *local* time change (the Eulerian derivative) and a term reflecting the change due to the fluid particle's motion, that is, an *advective* change.

Because the natural definition of the velocity is that of a particle, that is, the Lagrangian one, a word of explanation is perhaps needed to clarify the conversion between this velocity and the Eulerian one. Let the Lagrangian velocity be denoted for a moment by $\dot{\mathbf{x}}$, that is,

$$\dot{\mathbf{x}}(\mathbf{X}, t) \equiv \frac{D}{Dt}\mathbf{x}(\mathbf{X}, t)$$

The Eulerian velocity field can then be formally expressed as

$$\mathbf{v}(\mathbf{x}, t) = \mathbf{v}\left[\mathbf{x}(\mathbf{X}, t), t\right] \equiv \dot{\mathbf{x}}\left[\boldsymbol{\Phi}_t^{-1}(\mathbf{x}), t\right] \tag{12.5}$$

This said, we shall from now on call both the Eulerian and Lagrangian velocities
$\mathbf{v}$, and use them interchangeably.

Before proceeding to the discussion of the dynamics of fluid flows, we shall introduce a number of additional useful, purely kinematical concepts. The first of
these is that of a *streamline*. This is, at any particular instant t, a curve that has
the same direction as the velocity vector $\mathbf{v}(\mathbf{x}, t)$ at each point. Mathematically, a
streamline can be parametrically expressed as $\mathbf{x}(s)$ from the solution of the following differential equations (given here in Cartesian coordinates)

$$\frac{1}{v_x}\frac{\mathrm{d}x}{\mathrm{d}s} = \frac{1}{v_y}\frac{\mathrm{d}y}{\mathrm{d}s} = \frac{1}{v_z}\frac{\mathrm{d}z}{\mathrm{d}s} \tag{12.6}$$

at a particular time. Clearly, streamlines are fixed in space in *steady flows*, that is,
if the flow functions do not depend explicitly on time. In such cases the streamlines are actually fluid particles' paths. On the other hand, in unsteady flows fluid
particles' paths do not coincide, in general, with streamlines.

The next important concept is that of a *material invariant*. This is any function
$F(\mathbf{x}, t)$, which satisfies

$$\frac{D}{Dt}F = 0 \tag{12.7}$$

indicating that the physical property defined by F is conserved along the motion
of the fluid particles. Note that this property depends not only on the form of F
but also on the flow, as can be easily seen from the explicit form of the material
derivative. As an example, consider a flow in which the density of the fluid element
does not change. This is expressed by $D\rho/Dt = 0$ and this relation defines, as we
shall see below, an incompressible flow, which can take place also in fluids whose
density may be non-uniform.

Two additional fields, related to the velocity field and embodying its properties,
also constitute valuable concepts. These are the scalar field $\xi \equiv \nabla \cdot \mathbf{v}$ and the (more
important) vector field $\boldsymbol{\omega} \equiv \nabla \times \mathbf{v}$, the *vorticity*. These concepts are frequently
useful in theoretical and practical considerations related to specific flow classes.
Their physical meaning becomes quite obvious when one considers the velocity
gradient tensor (or matrix), $\partial v_i/\partial x_j$. Splitting this tensor into its symmetrical and
antisymmetric parts we write

$$\frac{\partial v_i}{\partial x_j} = \mathcal{D}_{ij} + \Omega_{ij} \tag{12.8}$$

where

$$\mathcal{D}_{ij} = \frac{1}{2}\left(\frac{\partial v_i}{\partial x_j} + \frac{\partial v_j}{\partial x_i}\right) \tag{12.9}$$

is the rate-of-deformation (or simply 'deformation') tensor ($\mathcal{D}$) which is symmetric, while

$$\Omega_{ij} = \frac{1}{2}\left(\frac{\partial v_i}{\partial x_j} - \frac{\partial v_j}{\partial x_i}\right) \tag{12.10}$$

is the antisymmetric part, and is called the spin tensor (Ω). The latter is clearly related to the vorticity $\boldsymbol{\omega}$ by the relation $\Omega_{ij} = \frac{1}{2}\epsilon_{ikj}\omega_k$ (ϵ is the antisymmetric Levi–Civita tensor), while the former is related to ξ because $\nabla \cdot \mathbf{v} = \mathrm{Tr}\,\mathcal{D}$. It is possible to show (see e.g., Thompson, 1971) that a material infinitesimal line element $\mathbf{q} \equiv \delta\mathbf{r}$ evolves in a given flow according to

$$\frac{D\mathbf{q}}{Dt} = \mathcal{D}\mathbf{q} + \frac{1}{2}\boldsymbol{\omega} \times \mathbf{q} \tag{12.11}$$

Thus $\mathcal{D}$ deforms the element and minus its trace gives the relative compression rate $D(\ln\rho)/Dt$, while $\boldsymbol{\omega}$ rotates the element at an angular velocity equal to half the vorticity's value.

Conservation laws and the dynamical equations of fluid flow

Before introducing the fluid dynamical equations we wish to remark that the macroscopic property of a material continuum, which is usually used to define it as a *fluid*, is a dynamical concept. It is related to the fluid's *inability to sustain shearing forces* (see below). While a solid, for example, can be deformed by such forces and achieve equilibrium, fluids may be in equilibrium (state of no relative motion between their different parts) only if the resultant shear force, applied on any fluid element, vanishes.

Turning now to the discussion of the dynamical equations, we note that the Eulerian description is usually the more convenient one (except in special circumstances, e.g., one-dimensional flows) although the Lagrangian one appears to be more natural physically. In what follows we shall generally give the equations in the Eulerian formulation but the Lagrangian form of some equations will be mentioned as well. To facilitate the formulation of the equations, which express essentially conservation laws, we assign each fluid particle a definite velocity (the above mentioned average of the macroscopic motions within the element) $\mathbf{v}$, density (the element's mass divided by its volume) ρ and an additional thermodynamic function. The latter may be, e.g., the pressure P within the element, its temperature T, specific internal energy ε and so on. An equation of state, with the help of which any thermodynamic function can be expressed by any other pair of such functions (if the chemical composition is given), is assumed known as well. The use of thermodynamics presupposes that within each fluid element *thermodynamic equilibrium* prevails.

As said before, the fundamental equations of fluid dynamics reflect the usual physical conservation laws and we shall discuss them now in turn. We do not intend to give here the detailed derivations of all the equations and results, and refer the reader to the books mentioned above.

To formulate *mass conservation* we observe that the Jacobian determinant J, as defined in (12.1), gives the ratio of infinitesimal volume elements as they are evolved by the flow. This immediately gives a Lagrangian expression for mass conservation

$$\rho(\mathbf{X}, t) = \rho(\mathbf{X}, 0)J^{-1} \tag{12.12}$$

It is then only a technical (albeit quite tedious) matter to show, using the definition (12.1) and relations (12.4)–(12.5), that $D \ln J / Dt = \nabla \cdot \mathbf{v}$. From the last two equations there follows immediately an expression for the material derivative of the fluid density field $\rho(\mathbf{x}, t)$

$$\frac{D\rho}{Dt} = -\rho \nabla \cdot \mathbf{v} \tag{12.13}$$

and this gives the Eulerian continuity equation

$$\partial_t \rho + \nabla \cdot (\rho \mathbf{v}) = 0 \tag{12.14}$$

which can also be derived heuristically by considering the mass budget of any fixed fluid volume and the Gauss theorem.

The *momentum-conservation equation*, resulting from Newton's second law (conservation of momentum), can be most naturally written for a fluid particle

$$\frac{D}{Dt}\mathbf{v} = \mathbf{f} \tag{12.15}$$

where $\mathbf{f}$ is the total force (per unit mass), acting on the fluid particle.

We remark here at the outset on the distinction between an equilibrium state of the fluid (*hydrostatic* equilibrium) and a state of *steady* flow. In the former the fluid is at rest, that is $\mathbf{v} = 0$ everywhere. This typically happens because the forces acting on any fluid element balance (the resultant force is zero) and no initial motion is induced in the fluid externally. A steady flow is one in which the Eulerian fields explicitly do not depend on time. Thus we have $\partial_t \mathbf{v} = 0$ in (12.15) but the velocity does not have to be zero.

The resultant force $\mathbf{f}$ in equation (12.15) includes, in general, external (i.e., long range) forces, which we shall call *body* forces and denote by $\mathbf{b}$, and internal forces, exerted on a fluid element by its immediate ambiance and acting on it through its surface. In continuum mechanics the latter type of force is called *stress*. The stress is clearly a tensorial quantity, because its force components (per unit area) depend on the orientation of the area element across which they act. The stress tensor σ_{ij} denotes the jth component of force per unit area whose element is oriented in the direction of the ith unit vector. The force of stress, which acts so as to deform a fluid element, can be further subdivided into the pressure force (acting on each fluid volume element in a direction normal to its surface) and shear (acting tangentially to this surface), thus

$$\sigma_{ij} = -P\delta_{ij} + \sigma'_{ij} \tag{12.16}$$

where σ' is the shear stress tensor and the minus sign in front of the pressure is the result of a convention in the definition of the stress.

Fluids, in which the stress is solely due to the pressure, are called *ideal*. The momentum equation for an ideal fluid can thus be written, in the Eulerian form, as

$$\partial_t \mathbf{v} + \mathbf{v} \cdot \nabla \mathbf{v} = -\frac{1}{\rho}\nabla P + \mathbf{b} \tag{12.17}$$

and is called the *Euler* equation.

Obviously this is only a convenient limit (see below for its applicability). In any real physical fluid shear stress forces are also present, because microscopic particles may be exchanged between adjacent fluid elements. Thus, in the general case the Euler equation is replaced by

$$\partial_t v_i + v_k \partial_k v_i = -\frac{1}{\rho}\partial_k \sigma_{ik} + b_i \tag{12.18}$$

written here in Cartesian components. The stress tensor σ_{ik} includes *both* the pressure and the shear stress parts as given in (12.16), and we have used a shorthand notation (e.g., $\partial_k \equiv \partial/\partial x_k$) and the summation convention (on repeated indices).

The exchange of microscopic particles between fluid elements must give rise to macroscopic momentum and heat exchange between these elements, phenomena that can usually be described in a satisfactory manner by macroscopic laws, incorporating appropriate transport coefficients. The internal friction due to momentum exchange goes under the name of *viscosity* and is characterised by positive viscosity coefficients η (shear viscosity) and ζ (the *second* or *bulk* viscosity). These two coefficients, as it turns out, suffice for obtaining (under some reasonable assumptions that go under the name 'Newtonian fluid') a constitutive relationship, linking the shear stress with the rate of deformation. Heat exchange can be similarly treated by linking the heat flux with the temperature gradient using one additional transport coefficient (see below). The momentum equation including viscous stresses, expressed by using the above-mentioned prescription is called the *Navier–Stokes* equation and reads

$$\rho\left(\partial_t \mathbf{v} + \mathbf{v} \cdot \nabla \mathbf{v}\right) = -\nabla P + \rho \mathbf{b} + \eta \nabla^2 \mathbf{v} + \left(\zeta + \frac{1}{3}\eta\right)\nabla(\nabla \cdot \mathbf{v}) \qquad (12.19)$$

This form is only valid if the viscosity coefficients do not change with position and otherwise the equation is more complicated (see Thompson, 1971; or Landau & Lifshitz, 1987; for the assumptions leading to the Navier–Stokes equation, the general form thereof, and a discussion of the physical meaning and significance of bulk viscosity).

The *energy conservation* principle gives rise to an additional equation, coupled to the mass and momentum conservation equations. Energy conservation can be naturally expressed by the first law of thermodynamics – $d\varepsilon = \delta q + (P/\rho^2)d\rho$, where ε is the specific *internal energy* and δq is the amount of heat absorbed by a unit mass. Applying this relation to a fluid particle we get that along its motion

$$\frac{D\varepsilon}{Dt} = \frac{P}{\rho^2}\frac{D\rho}{Dt} + \frac{Dq}{Dt}$$

This equation may be used together with the mass conservation equation and the equation of motion to obtain different forms of the fluid dynamical energy equation. We shall give here, without proof (see the books referenced above for derivations), only one form of the energy equations which will be used later in this chapter.

Using the mass conservation equation and assuming that the internal energy can be written as $\varepsilon = C_V T$, where T is the temperature field and C_V is the (constant) specific heat per unit mass at constant volume (this is obviously true for a perfect gas), the energy equation assumes the form

$$\rho C_V(\partial_t T + \mathbf{v} \cdot \nabla T) = -P\nabla \cdot \mathbf{v} + Q \qquad (12.20)$$

where Q, the rate of heat absorption per unit volume, includes contributions due to various physical processes, which may be important or negligible depending on the type of flow considered. To bring this out we can formally write $Q = Q_{\text{cond}} + Q_{\text{vis}} + Q_{\text{rad}} + Q'$, i.e., separate the contributions due to heat conduction, viscous dissipation, radiative processes and all other heat sources (or sinks) (Q'). In general, all these contributions depend on some of the fields, which have already appeared in the continuity and momentum equations, possibly their gradients, and also on transport coefficients. In particular, we shall discuss in this book examples where thermal conduction is important and therefore write out now explicitly the conductive term (in the case of a constant heat conduction coefficient k)

$$Q_{\text{cond}} = k\nabla^2 T \qquad (12.21)$$

but leave all other (e.g., viscous dissipation, radiative losses and gains, Ohmic dissipation in electrically conducting fluids, nuclear energy generation rate, etc.) terms unspecified.

The above hydrodynamical equations (one vector and two scalar PDEs) can suffice, in principle, for the determination of the velocity field and two additional thermodynamic scalar fields (T and ρ, say). Other thermodynamic functions (e.g., P) appearing in the equations are supplied, as pointed out above, by the equation of state. Additional constitutive relations, giving the transport coefficients in terms of the thermodynamic variables, are also necessary and have to be known or assumed. All this given, the fluid problem may be well-posed, at least in principle, if suitable *initial* and *boundary* conditions are supplied. The solution of the fluid dynamical equations in such a setup is referred to as a hydrodynamical *flow*.

We shall discuss special flows and their properties in the next section, but before that we would like to introduce an additional useful dynamical relation involving the vorticity. It can be regarded as an alternative form of the equation of motion (momentum conservation), and can be brought out when the vector identities

$$(\mathbf{v} \cdot \nabla)\mathbf{v} = \nabla\left(\frac{1}{2}v^2\right) - \mathbf{v} \times (\nabla \times \mathbf{v})$$
$$\nabla^2 \mathbf{v} = \nabla(\nabla \cdot \mathbf{v}) - \nabla \times (\nabla \times \mathbf{v}) \tag{12.22}$$

are substituted into the momentum equation (12.19). Assuming also that the body force is derivable from a potential, $\mathbf{b} = -\nabla\Phi$, and using the thermodynamic identity (due to Gibbs)

$$dP = \rho dw - \rho T ds \tag{12.23}$$

where s is the specific entropy and w the heat function (enthalpy), we get

$$\partial_t \mathbf{v} + \boldsymbol{\omega} \times \mathbf{v} = -\nabla\mathcal{B} + T\nabla s - \nu\nabla \times \boldsymbol{\omega} + \left(\mu + \frac{4}{3}\nu\right)\nabla(\nabla \cdot \mathbf{v}) \tag{12.24}$$

Here μ and ν are the bulk and shear *kinematic* viscosity coefficients, i.e., ζ/ρ and η/ρ respectively, and $\mathcal{B}$ is the *Bernoulli function* defined as

$$\mathcal{B}(\mathbf{x}, t) \equiv w + \frac{1}{2}v^2 + \Phi \tag{12.25}$$

This equation will be useful in the next section when some special flows are discussed.

We end this general discussion of the equations of fluid dynamics with one additional basic observation on the the properties of fluid flows, that is, solutions of these equations. If one finds a solution for the equations for given initial and boundary conditions (a formidable task in general), this solution actually describes not

just one flow but a whole family thereof. This can easily be seen if one nondimen-
sionalises the equations, scaling the fluid fields and coordinates by their characteristic values. Taking, for example, the Navier–Stokes equation for a fluid whose
density is a constant, $\rho(\mathbf{x}, t) = \rho_0$ say, for simplicity, we get the nondimensional
equation

$$\partial_t \mathbf{v} + (\mathbf{v} \cdot \nabla \mathbf{v}) = -\nabla P + \frac{1}{Re} \nabla^2 \mathbf{v} \tag{12.26}$$

After choosing typical values of the velocity and the flow's spatial scale, L and U,
as the length and velocity units respectively, we also scale the pressure by $\rho_0 U^2$
and thus obtain (12.26).

The nondimensional quantity $Re \equiv \rho_0 U L / \eta = U L / \nu$, which appears in this
equation, is called the *Reynolds number*. It plays an important role in fluid dynamics and in particular in stability and turbulence theory (see later in this chapter).
Physically, it is an estimate of the ratio of the inertial term to the viscous one in
the momentum equation. Thus, flows in which the Reynolds number is very large
may be considered inviscid, that is, the Euler equation is appropriate for their description (see also below). In the opposite limit ($Re \ll 1$) we get flows that are
dominated by viscosity. Clearly, a solution of (12.26) for a given fixed Reynolds
number represents a family of physical flows, whose spatial scale, velocity, pressure and density may be different, as long as the non-dimensional combination Re
is the same. Obviously, the boundary and initial conditions must also be scaled
accordingly. All flows belonging to the same family in this sense are called *similar*. Several additional nondimensional quantities play distinguished roles in fluid
dynamics and we shall define and use some of these 'numbers' and discuss their
significance later on in this chapter.

12.2 Special flows and their properties

The fluid-dynamical equations are a set of quite complicated nonlinear PDEs. It
is probably worthwhile to attempt to explore some of their analytical properties
and the properties of some particular flows (solutions). Before attempting to solve
a general and complicated problem (a task that can usually only be done numerically) one can, for example, investigate what can be said on the above mentioned
properties in some special cases. In a number of limits (see below) the equations
become simplified and some analysis is therefore possible and may be useful in
general.

In discussing several simplifying and specific physical assumptions here we shall
generally speak of *flows* rather than fluids, that is, of realisations or solutions of
the fluid-dynamical equations. Thus the special properties will be of flows and
not necessarily of the fluids themselves. For example, in the context of the mass

conservation equation, it is clear (see below) that *incompressible flows* can occur also in fluids of variable density, but only fluids whose density is constant and cannot be changed are referred to as *incompressible fluids*. Incompressible fluids support only incompressible flows, but in view of what was said above, the occurrence of an incompressible flow does not guarantee that the fluid in which it takes place is incompressible. Regarding the momentum equation, we have seen that in the limit Re $\gg$ 1 we can speak of an *inviscid flow* and this term is used rather than of an inviscid fluid, because fluids are never really inviscid. They may however support flows which are approximately inviscid, that is, the viscous terms in the equations can be neglected. In the context of energy conservation, special types of flow can be identified as well. For example if $Q = 0$ and thus the only source of heat for a fluid element is its compression by thermal pressure, we speak of an *adiabatic* flow. Related to this is the notion of an *isentropic* flow, defined by the requirement that the specific entropy s is a material invariant. If the entropy is constant throughout the fluid and does not change in time we speak of *homentropic* fluids and flows.

In this section we shall briefly discuss some important, and frequently also useful, properties of such special types of flows. We shall also briefly discuss the basics of the theory of rotating flows, at various limits. As pointed out above, this theory is the cornerstone of GFD and many AFD applications may benefit from it as well.

Incompressible flows

If the density is a material invariant, that is $D\rho/Dt = 0$, the fluid elements are not compressed along their motion and flows endowed with this property are called *incompressible*. From Equation (12.13) incompressibility is formally equivalent to the statement

$$\nabla \cdot \mathbf{v} = 0 \tag{12.27}$$

This does not necessarily require that the fluid, as such, cannot be compressed at all and the density is constant throughout the system. If however ρ is constant in space and time, all flows of this fluid are obviously incompressible. Incompressible flows constitute a subclass for which some significant simplifications can be made, and therefore it is very important to understand under which physical conditions it is a good approximation to consider a flow to be incompressible.

The most important of these conditions can be most easily understood for steady, pressure-driven (with no body forces) isentropic flows. In such flows the fluid element retains, along its motion, its specific entropy that is, a small variation in the density of the element, $\delta\rho$, caused by a pressure variation δP, satisfies $\delta\rho = \delta P/v_s^2$, where $v_s \equiv (\partial P / \partial \rho)_s^{1/2}$ is the adiabatic speed of sound. On the other hand, we may estimate the pressure change along a streamline when the flow is purely pressure driven by comparing the order of magnitude of the advective term and the pressure gradient term in (12.17). This gives $\delta P \sim \rho v^2$. Equating the above two expressions for δP we get $\delta\rho/\rho \sim (v/v_s)^2$ and thus if we want the flow to be approximately incompressible, we must have $v \ll v_s$, that is, a very *subsonic* flow.

If the flow is not steady, the term $\partial_t \mathbf{v}$, rather than being zero, may be of the order of magnitude of the pressure-gradient term, and thus the above condition (highly subsonic motion) is clearly insufficient. The additional condition for incompressibility arising for this case is $L/v_s \ll \tau$, where L and τ are the spatial and temporal scales, respectively, over which the flow changes appreciably. Likewise, an additional condition arises in the case when body forces act on the fluid. It is $bL \ll v_s^2$, where b is a typical value of the body force and this means that the rate of working by the body force on a unit mass of a fluid element is negligible with respect to its internal energy.

All the above conditions are for near-incompressibility. Strict incompressibility is formally achieved only for $v_s \to \infty$, that is, if the propagation of disturbances may be regarded as instantaneous and in this limit all of the above conditions are clearly satisfied. Still, many flows are formally compressible, but incompressibility is a very good approximation, and assuming this greatly simplifies the problem. The equations of mass and momentum conservation for incompressible flows are thus the incompressibility condition (12.27) and

$$\partial_t \mathbf{v} + \mathbf{v} \cdot \nabla \mathbf{v} = -\frac{1}{\rho} \nabla P + \mathbf{b} + \nu \nabla^2 \mathbf{v} \qquad (12.28)$$

Sometimes it is this set that is called the Navier–Stokes equations and not the more general one (12.14) and (12.19). The energy equation in the case of an incompressible flow is also simplified, because the terms containing $\nabla \cdot \mathbf{v}$, both in the viscous dissipation and in the compressional heating expressions, drop out.

An important relation concerning vortex dynamics in incompressible flows emerges when we restrict ourselves to (incompressible) flows of a fluid having a constant density ρ_0. We also assume that the body force $\mathbf{b}$ is derivable from a potential, that is, $\mathbf{b} = -\nabla \Phi$. Substituting $\nabla \cdot \mathbf{v} = 0$ and $T\nabla s = -\nabla(w + P/\rho_0)$ in Equation (12.24) gives

$$\partial_t \mathbf{v} + \boldsymbol{\omega} \times \mathbf{v} = -\nabla \mathcal{B}_d - \nu \nabla \times \boldsymbol{\omega} \qquad (12.29)$$

where $\mathcal{B}_d = P/\rho_0 + v^2/2 + \Phi$ ($\mathcal{B}_d$ is similar to the Bernoulli function, but with P/ρ_0 replacing w). Taking the curl of the last equation we get

$$\partial_t \boldsymbol{\omega} + \nabla \times (\boldsymbol{\omega} \times \mathbf{v}) = -\nabla \times (\nu \nabla \times \boldsymbol{\omega}) \qquad (12.30)$$

Assuming now that ν is a constant, $\nabla \cdot \mathbf{v} = 0$ and using vector identities this transforms into

$$\frac{D\omega}{Dt} = (\omega \cdot \nabla)\mathbf{v} + \nu\nabla^2\omega \qquad (12.31)$$

which gives the evolution of the vorticity of a fluid element in a flow whose velocity field is known.

In the case of *two-dimensional* flows the description of an incompressible flow can be very significantly simplified by introducing the concept of a *stream function*. This is most easily done in the case of Cartesian coordinates, when a two-dimensional flow is equivalent to what is called a *plane parallel* flow, that is, a flow in which one component of the velocity, v_z say, vanishes and the other two (v_x, v_y in Cartesian coordinates) are independent of z. The incompressibility condition (12.27) guarantees then the existence of a well-defined scalar function $\psi(x, y, t)$ so that

$$v_x = \partial_y\psi \qquad \text{and} \qquad v_y = -\partial_x\psi \qquad (12.32)$$

This obviously automatically satisfies the vanishing of the velocity divergence and thus incompressibility.

$\psi(x, y, t)$ is called the *stream function* and has physical meaning and importance which goes beyond its being just a useful mathematical device to reduce the degree of differential equations. An important property of the stream function follows from (12.32) and that is the statement that the variation of ψ *along* a streamline is zero –

$$\mathbf{v} \cdot \nabla\psi = \partial_y\psi\partial_x\psi - \partial_x\psi\partial_y\psi = 0$$

so that the stream function is constant along streamlines. Thus, if the stream function of a steady flow is known, for example, the streamlines can be trivially found. Moreover, the mass flux across any finite curve segment in the $x-y$ plane is given by the difference of the stream function values evaluated at the segment edges (see e.g., Landau & Lifshitz, 1987).

A useful way of viewing the representation (12.32) is the vector equation

$$\mathbf{v} = \nabla \times (\hat{\mathbf{z}}\psi) \qquad (12.33)$$

which demonstrates the simplification; instead of dealing with the velocity (which is a vector field) it is enough to consider only the scalar field ψ. Taking the curl of this equation gives

$$\nabla \times \mathbf{v} = \omega = -\hat{\mathbf{z}}\nabla^2\psi \qquad (12.34)$$

showing that the value of the vorticity, ω, which has now only a z-component, is actually the source of the stream function in the Poisson equation

$$\nabla^2 \psi = -\omega \tag{12.35}$$

Also, equation (12.31) can then be conveniently written as

$$\partial_t \omega + J(\omega, \psi) = \nu \nabla^2 \omega \tag{12.36}$$

where $J(\omega, \psi) = \partial_x \omega \partial_y \psi - \partial_x \psi \partial_y \omega$. Equations (12.35)–(12.36) constitute a set of equations which are sufficient for a description of a two-dimensional flow of a constant density fluid. It is sometimes referred to as the 'vorticity–stream function' formulation of the problem.

We conclude the discussion of incompressible flows by describing one particularly important example: the case of thermal convection in a horizontal fluid layer, placed in a constant vertical gravitational field. This is usually referred to as the Rayleigh–Bénard experiment or flow. Typically, a flow of this kind satisfies most of the conditions for approximate incompressibility (see the discussion at the beginning of this section). However, the buoyancy force on the fluid elements, which clearly depends on density *variations*, cannot be neglected with respect to the other terms in the vertical momentum equation. In fact, it is precisely this force that drives the flow. Still, the fluid equations in such a setting can be significantly simplified with the help of what is known as the *Boussinesq* approximation (some refer to a Boussinesq *fluid*), named after one of the inventors of this approach. The approximation consists of treating the fluid as incompressible (in fact, of constant density) everywhere *except* in the buoyancy force term.

For a rather formal derivation of the Boussinesq approximation and a discussion on under what conditions it can be justified in the case of a compressible fluid layer, we refer the reader to Spiegel & Veronis (1960). Here we shall give the equations following an essentially intuitive approach (see also Drazin & Reid, 1981). The resulting equations describe a fluid layer of an essentially constant density ρ_0 (see below, however), confined between two horizontal planes, $z = 0$ and $z = h$ say, on which the temperatures are held at the fixed values $T(x, y, 0) = T_0$ and $T(x, y, h) = T_1$. Small density variations, which are relevant to the buoyancy term but negligible otherwise, are assumed to arise from thermal expansion alone and this dependence is assumed to be linear

$$\rho - \rho_0 = -\alpha \rho_0 (T - T_0) \tag{12.37}$$

where $\alpha \equiv -(\partial \ln \rho / \partial T)_P$, is the thermal expansion coefficient.

The equation of motion (12.28) can then be shown to take on the form

$$\partial_t \mathbf{v} + \mathbf{v} \cdot \nabla \mathbf{v} = -\nabla \left(\frac{P}{\rho_0} + gz \right) + \alpha g (T - T_0) \hat{\mathbf{z}} + \nu \nabla^2 \mathbf{v} \tag{12.38}$$

where $\nu = \eta/\rho_0$ is the kinematic viscosity coefficient and g the acceleration due to gravity.

In the energy equation (12.20) all the heating terms, save that resulting from thermal conduction, are assumed negligible (actually the compression term is combined with the temperature derivative term and a perfect gas equation of state is used to simplify the result – for details see e.g., Drazin & Reid, 1981). This gives

$$\partial_t T + \mathbf{v} \cdot \nabla T = \kappa \nabla^2 T \tag{12.39}$$

with the coefficient $\kappa \equiv k/(\rho_0 C_P)$, where C_P is the specific heat at constant pressure. The perfect gas equation of state, referred to above and used in the derivation of the last equation, is $P = (C_P - C_V)\rho T$.

Equations (12.37)–(12.39) and the incompressibility condition (12.27) constitute the *Boussinesq equations*. They have been extensively studied for more than 100 years to model thermal convection in fluid layers under a variety of conditions. We shall use these equations later on in this chapter for the derivation of the Lorentz model, and for that we shall need a slightly modified nondimensional version thereof, which we now derive. First, we notice a state of rest $\mathbf{v} = \mathbf{0}$ with a stationary and linear vertical temperature stratification,

$$T(\mathbf{x}, t) = \overline{T}(z) \equiv T_0 - \delta T \frac{z}{h}$$

where $\delta T \equiv T_0 - T_1$, is a solution of the equations. Although the boundary conditions on the velocity have not been specified so far, the zero solution satisfies most reasonable choices. The physical picture corresponding to this solution is, of course, of a thermally conducting static slab of fluid. It is well known that if δT is positive and large enough the fluid starts to move so as to accommodate upward heat transfer by *convection*. We shall deal with convection in this book only in the context of the Lorentz model, but the form of the Boussinesq equations which we shall give here has been extensively used in the studies of convection (see e.g., Chandrasekhar, 1961; Drazin & Reid, 1981; Manneville, 1990; and references therein).

We proceed by defining the temperature *deviation* (from the above linear profile)

$$\theta(\mathbf{x}, t) \equiv T(\mathbf{x}, t) - \overline{T}(z)$$

and substituting this relation in equations (12.38) and (12.39), which after some trivial rearrangement of terms, give

$$\partial_t \mathbf{v} + \mathbf{v} \cdot \nabla \mathbf{v} = -\nabla \left(\frac{P}{\rho_0} + gz + \alpha g \frac{\delta T}{2h} z^2 \right) + \alpha g \theta \hat{\mathbf{z}} + \nu \nabla^2 \mathbf{v} \tag{12.40}$$

and

$$\partial_t \theta + \mathbf{v} \cdot \nabla \theta = \frac{\delta T}{h} \hat{\mathbf{z}} \cdot \mathbf{v} + \kappa \nabla^2 \theta \tag{12.41}$$

Finally, choosing the length unit as $L = h$, the time unit as the heat diffusion timescale $\tau = \kappa/L^2 = \kappa/h^2$ (thus the velocity unit is $U = L/\tau = h^3/\kappa$) and the temperature deviation unit as δT, we get the nondimensional equations

$$\frac{1}{Pr}(\partial_t \mathbf{v} + \mathbf{v} \cdot \nabla \mathbf{v}) = -\nabla \Pi + Ra\,\theta \hat{\mathbf{z}} + \nabla^2 \mathbf{v} \tag{12.42}$$

and

$$\partial_t \theta + \mathbf{v} \cdot \nabla \theta = \hat{\mathbf{z}} \cdot \mathbf{v} + \nabla^2 \theta \tag{12.43}$$

where

$$\Pi \equiv \frac{h^2}{\nu\kappa}\left(\frac{P}{\rho_0} + gz + \alpha g \frac{\delta T}{2h} z^2\right)$$

is a nondimensional effective pressure variable.

Two new nondimensional numbers appear in Equation (12.42) and both have physical significance. First, the *Prandtl number* (Pr) is the ratio of the kinematic viscosity to the thermal diffusion coefficient

$$Pr = \frac{\nu}{\kappa} \tag{12.44}$$

It measures the relative importance of dissipation of mechanical energy of the flow by shear compared to energy dissipation by heat flow. The Prandtl number is of the order of one, e.g., for water at room temperature. The second nondimensional quantity, the *Rayleigh number* (Ra) measures the tendency of a fluid element to rise because of buoyancy induced by thermal expansion, relative to the opposing tendency, brought about by energy dissipation due to viscosity and thermal diffusion. It is defined as the nondimensional ratio

$$Ra = \frac{\alpha g h^3}{\nu\kappa}\delta T$$

Viewed differently, the Rayleigh number is essentially a product of two characteristic timescale ratios. The timescale of the convective motion, which is driven by buoyancy is clearly $\tau_b = \sqrt{h/g_{\text{eff}}}$ where $g_{\text{eff}} = g\delta\rho/\rho_0 = g\alpha\delta T$. The timescales of dissipation by viscosity and heat conduction are $\tau_\nu = h^2/\nu$ and $\tau_\kappa = h^2/\kappa$, respectively. Thus

$$Ra = \frac{\tau_\nu \tau_\kappa}{\tau_b \tau_b}$$

In practice, the Rayleigh number is a nondimensionless measure, for a given fluid layer, of the temperature difference between the bottom and the top, and thus in most Rayleigh–Bénard convection experiments it serves as the control parameter.

Inviscid flows

We now restrict our attention to *inviscid flows*, that is, flows for which the viscous terms in the equations are neglected. This is justified, as we have seen, when the nondimensional ratio Re, the Reynolds number, is sufficiently large – a condition satisfied by most astrophysical flows.

This allows us to omit the viscous terms in the equations of momentum and energy conservation, which simplifies then considerably. In what follows we shall discuss, however, mainly the dynamical equations involving the vorticity, as they are the most appropriate for the derivation of the results we wish to explore here. For example equation (12.24) with the viscous terms omitted (as appropriate for inviscid flows), gives

$$\partial_t \mathbf{v} + \boldsymbol{\omega} \times \mathbf{v} = -\nabla \mathcal{B} + T \nabla s \tag{12.45}$$

where $\mathcal{B}$ is the Bernoulli function defined in (12.25). It should be remembered that it has been assumed here that the body force is derivable from a potential.

We would also like to remark here that a significant simplification of the fluid dynamical equations may be achieved for flows for which the pressure P is a well-defined function of only the density ρ throughout the fluid. Such flows are called *barotropic* and require a given relation of the type $P = P(\rho)$ (not necessarily reflecting a thermodynamic statement but only a property of a particular flow). This is clearly sufficient to circumvent the need for an energy equation for a formal closure of the fluid dynamical PDE system (polytropic stellar models, well known to the students of stellar structure, are a particular example of a trick of this kind). Note also that isentropic flows are obviously barotropic.

For barotropic flows, the relation $P = P(\rho)$ guarantees that $\rho^{-1} dP$ is an exact differential. Thus we may define (up to an additive constant) a function h by the indefinite integral

$$h = \int \frac{1}{\rho} dP \tag{12.46}$$

This allows us to modify equation (12.45) to read

$$\partial_t \mathbf{v} + \boldsymbol{\omega} \times \mathbf{v} = -\nabla \mathcal{B}_b(\mathbf{x}, t) \tag{12.47}$$

with the modified Bernoulli function (useful for barotropic flows) defined by

$$\mathcal{B}_b(\mathbf{x}, t) \equiv h + \frac{1}{2} v^2 + \Phi \tag{12.48}$$

that is with the heat function, w which appears in the general Bernoulli function, replaced by h. We should remark in this context that the function h, defined above, is in general *not* the usual thermodynamic enthalpy function, w (but they coincide for homentropic flows, of course).

Taking the curl of equation (12.45) we get

$$\partial_t \boldsymbol{\omega} + \nabla \times (\boldsymbol{\omega} \times \mathbf{v}) = \nabla T \times \nabla s \tag{12.49}$$

and if we substitute $T \nabla s = \nabla w - \nabla P / \rho$, a consequence of the Gibbs identity (12.23), in Equation (12.45), before taking its curl, an alternative form of (12.49) is obtained

$$\partial_t \boldsymbol{\omega} + \nabla \times (\boldsymbol{\omega} \times \mathbf{v}) = \frac{\nabla \rho \times \nabla P}{\rho^2} \tag{12.50}$$

With the help of vector identities and some algebra the last two equations can provide us with useful relations on the evolution of the vorticity of a fluid element along its motion. From (12.50) it follows that

$$\frac{D\boldsymbol{\omega}}{Dt} = (\boldsymbol{\omega} \cdot \nabla)\mathbf{v} - \boldsymbol{\omega}(\nabla \cdot \mathbf{v}) + \frac{\nabla \rho \times \nabla P}{\rho^2} \tag{12.51}$$

and using the continuity equation and some additional algebra we can get from this the important equation

$$\frac{D}{Dt}\left(\frac{\boldsymbol{\omega}}{\rho}\right) = \left(\frac{\boldsymbol{\omega}}{\rho} \cdot \nabla\right)\mathbf{v} + \frac{\nabla \rho \times \nabla P}{\rho^3} \tag{12.52}$$

In the last two equations the term $\nabla \rho \times \nabla P$ can obviously be replaced by $\rho^2 (\nabla T \times \nabla s)$, because Equation (12.49) can be used in the derivations, instead of (12.50).

Equations (12.51) and (12.52) were derived under quite general conditions and they are valid for most astrophysical flows, which can be typically considered inviscid (the Reynolds number is very large) and are driven by conservative forces. The consequences of these equations provide some rather deep general insights into the nature of such flows. Moreover, when the barotropic assumption is applied to equations (12.50)–(12.52), the last term in these equations vanishes since in barotropic flows the surfaces of constant pressure and constant density coincide. In fact, the iso-surfaces of *all* thermodynamic functions coincide in barotropes. Alternatively, this result can be derived immediately by using from the outset Equation (12.47) instead of (12.45).

We shall now explicitly state, without proof (see e.g., Landau & Lifshitz, 1987; Salmon, 1998), two important relations, known in the literature as theorems, valid for *inviscid* and *barotropic* flows, in which all the body forces are derivable from a potential.

(i) *Kelvin's circulation theorem*: Let $C(t)$ be any closed contour in the fluid, which changes in such a way that each point on it is moving with the local flow velocity. Then $\Gamma \equiv \oint_C \mathbf{v} \cdot d\mathbf{l}$ is a material invariant of the flow, i.e.,

$$\frac{D}{Dt}\Gamma = 0 \tag{12.53}$$

(ii) *Ertel's theorem*: Let F($\mathbf{r}$, t) be any material invariant of the flow. Then the quantity $(\rho^{-1}\boldsymbol{\omega} \cdot \nabla)F$ is also a material invariant of the flow, i.e.,

$$\frac{D}{Dt}\left[\left(\frac{\boldsymbol{\omega}}{\rho} \cdot \nabla\right)F\right] = 0 \tag{12.54}$$

Kelvin's theorem is obviously related to vorticity as the velocity circulation around a closed contour is equal to the vorticity flux through a surface defined by this contour (by Stokes' theorem). It essentially means that the velocity circulation around a closed contour moving with the fluid or the vorticity flux (the 'number' of vortex lines) through the surface defined by this contour, is conserved by the flow. This guarantees that if the flow is *irrotational* (i.e., $\boldsymbol{\omega} = 0$ everywhere) at some instant it remains always irrotational, a fact which also follows directly from Equation (12.51). An irrotational flow is also called a *potential* flow, because the vanishing of $\nabla \times \mathbf{v}$ guarantees the existence of a scalar velocity potential function $\phi(\mathbf{x}, t)$ from which the velocity derives $\mathbf{v} = \nabla\phi$.

If a flow is viscous and/or if it is *baroclinic* (that is, not barotropic) Kelvin's theorem breaks down. Thus viscosity and/or baroclinicity (i.e., the property that isobaric surfaces are, at least somewhere in the flow domain, inclined to the iso-choric ones by a finite angle) can be considered as vorticity sources in the flow. This can also be readily seen when examining Equations (12.31) and (12.51).

Note, however, that even if a flow is barotropic and inviscid, the presence of an obstacle (e.g., a solid body) in some position downstream presents a vorticity source in an otherwise irrotational flow. Kelvin's theorem breaks down very near the obstacle, simply because it is impossible to draw a closed contour, lying entirely in the fluid, around fluid particles residing just 'on top' of the obstacle. Thus if we assume that the flow is inviscid throughout, a mathematical difficulty (a tangential discontinuity in the flow) appears (see Landau & Lifshitz, 1987). This obviously disappears when one considers a real physical fluid in which viscous forces, however small otherwise, become dominant if the shear is large enough (in a tangential discontinuity there is formally an infinite shear). This is then the vorticity source. Even in flows that can be considered inviscid, far enough from an obstacle a small length scale is created in the flow near the obstacle, a *boundary layer*, so that in this layer viscous forces are dominant. In terms of the Reynolds number this clearly means that if the flow in the vicinity of the boundary is to be taken into account the effective Reynolds number is significantly smaller (because there is a smaller length scale). Obstacles are thus, generally speaking, sources of vorticity.

Ertel's theorem is related to Kelvin's theorem but so far it has not been widely used in the astrophysical literature. In GFD applications, however, the concept of *potential vorticity* (the material invariant of Ertel's theorem with F chosen to be the specific entropy) is considered to be among the most basic notions. It plays a central role in the theory of geophysical rotating fluids (see, e.g., Pedlosky, 1987; Salmon, 1998). Rotation is not uncommon in astrophysical systems as well, and it seems that the potential vorticity concept may also be useful in AFD applications.

We conclude this introduction to fluid dynamics with a short discussion of the basic properties of rotating flows. Assume that we are considering a flow viewed in a frame of reference rotating (with respect to an inertial frame) at a constant angular velocity $\mathbf{\Omega} = \Omega\hat{\mathbf{z}}$, say. Following the usual prescription known from classical mechanics, we include centrifugal and Coriolis terms in the equation of motion in the rotating frame. For the sake of simplicity we shall consider here only a constant-density fluid. This approximation is very good in the case of a liquid in a rotating vessel or of liquid masses on the rotating Earth, for example. GFD and experiments on liquids rotated in the laboratory are indeed the primary motivation of the theory of rotating fluids.

Thus the Navier–Stokes equation in the incompressible case (12.28) takes on the form

$$\partial_t \mathbf{u} + \mathbf{u} \cdot \nabla \mathbf{u} = -2\mathbf{\Omega} \times \mathbf{u} - \mathbf{\Omega} \times (\mathbf{\Omega} \times \mathbf{r}) - \frac{1}{\rho}\nabla P + \mathbf{b} + \nu\nabla^2\mathbf{u} \qquad (12.55)$$

where $\mathbf{u}$ is the velocity and $\mathbf{r}$ is the position vector measured relative to the rotating frame. If the body force is derived from a potential ($\mathbf{b} = -\nabla\Phi$), we may define a 'reduced pressure' p by

$$p \equiv P - \rho\Phi - \frac{1}{2}\rho(\mathbf{\Omega} \times \mathbf{r}) \cdot (\mathbf{\Omega} \times \mathbf{r})$$

and the equation of motion becomes

$$\partial_t \mathbf{u} + \mathbf{u} \cdot \nabla \mathbf{u} + 2\mathbf{\Omega} \times \mathbf{u} = -\frac{1}{\rho}\nabla p + \nu\nabla^2\mathbf{u} \qquad (12.56)$$

so that the Coriolis term appears to be the only prominent change from the non-rotating case. Nondimensionalisation of the last equation, that is, replacing $\mathbf{r}, t, \mathbf{u}, \mathbf{\Omega}$ and p by their scaled counterparts $L\mathbf{r}, \Omega^{-1}t, U\mathbf{u}, \Omega\hat{\mathbf{z}}$ and $\rho\Omega U L p$ respectively, finally gives

$$\partial_t \mathbf{u} + Ro\,\mathbf{u} \cdot \nabla \mathbf{u} + 2\hat{\mathbf{z}} \times \mathbf{u} = -\nabla p + E\,\nabla^2\mathbf{u} \qquad (12.57)$$

Two important nondimensional parameters appear in this equation. They are the *Rossby number*

$$Ro \equiv \frac{U}{\Omega L}$$

and the *Ekman number*

$$E \equiv \frac{\nu}{\Omega L^2}$$

where L, U and Ω are the typical length, velocity (in the rotating frame) and rotational angular velocity scales of the flow, which were also used in the nondimensionalisation of the equation, and v is the viscosity.

The Rossby number reflects the ratio between the typical values of the advective acceleration in the flow (U^2/L) and the Coriolis force per unit mass (ΩU), and as such provides an estimate of the importance of the nonlinear terms. The Ekman number is an overall measure of how the typical viscous force compares to the Coriolis force and is essentially the inverse of the Reynolds number appropriate for the rotating flow. In typical GFD applications (and in AFD obviously, as well) $E \ll 1$ because viscosity does not play a significant role in the kind of scales encountered. Even in the laboratory E is generally very small, and this is the reason that boundary layers are often created near rigid walls confining the flow. The Rossby number can be considered small only if the velocities in the rotating frame are small enough and this happens if, for example, the configuration in the rotating frame is close to equilibrium. If $\mathrm{Ro} \ll 1$ the equation of motion is linear and this is obviously a simplification.

The dimensionless form of the vorticity equation for the conditions considered here is (compare to Equation 12.31)

$$\partial_t\boldsymbol{\omega} + Ro(\mathbf{u} \cdot \nabla)\boldsymbol{\omega} = [(Ro\,\boldsymbol{\omega} + 2\hat{\mathbf{z}}) \cdot \nabla]\mathbf{u} + E\,\nabla^2\boldsymbol{\omega} \qquad (12.58)$$

where $\boldsymbol{\omega} = \nabla \times \mathbf{u}$ here is the vorticity in the rotating frame. Note that in essence this equation means that the frame rotation contribution to the vorticity is $2\boldsymbol{\Omega}$ (in dimensional units).

A very important result in the theory of rotating fluids, known as the *Taylor–Proudman theorem* can be very easily derived from Equation (12.57). The theorem is valid for steady ($\partial_t\mathbf{u} = 0$) and inviscid ($E = 0$) flows, in which also Ro is assumed to be zero. The last assumption means, as we have seen, that the departures of the flow from pure rotation are small. The equation of motion thus reduces to

$$2\hat{\mathbf{z}} \times \mathbf{u} = -\nabla p \qquad (12.59)$$

Taking the curl of this expression and using the incompressibility condition gives

$$(\hat{\mathbf{z}} \cdot \nabla\mathbf{u}) = 0$$

This means that the velocity $\mathbf{u}$ must be independent of the coordinate measured along the rotation axis, which we have chosen to be the z-coordinate, so that $\mathbf{u} = \mathbf{u}(x, y)$ in Cartesian coordinates. Thus steady inviscid flows which are dominated by strong rotation $Ro \ll 1$ are essentially two-dimensional. So since every fluid particle in a vertical column has the same velocity, the entire column (called a *Taylor column*) must move as if it were a single fluid element. For example, if a finite size obstacle is present in such a flow at some height and the fluid is therefore forced to flow around it, an entire column of zero velocity will be present above and below the obstacle. These consequences of the Taylor–Proudman theorem have been observed in the laboratory and in geophysical flows.

In GFD one deals with fluids (atmosphere, oceans) whose vertical extent is very much smaller than their horizontal scale. One of the approximations exploiting this fact, that is very frequently used, is the famous *shallow-water theory*, formulated on a rotating sphere. We shall not discuss this theory here (see e.g., Pedlosky, 1987; Salmon, 1998) and only remark that is some aspects this theory is similar to the treatment usually employed in the study of thin accretion disks (see in the next section). If one considers horizontal motions of such a shallow-fluid layer in a rotating (around a vertical axis) frame of reference in conditions that Equation (12.59), at least approximately (i.e., for a duration much longer than the period of rotation), holds, the flow is then termed *geostrophic* (one also talks of geostrophy or the geostrophic approximation). The meaning of geostrophy is that there is a balance (the geostrophic balance) between the pressure and Coriolis terms. This is reflected, for example, by the fact that winds are blowing *along* isobars (as it can be seen in weather maps), contrary perhaps to the 'intuition' of those who study the behaviour of fluids in an inertial frame. This is obviously a trivial consequence of (12.59).

Stratification and moderate relaxation of some other of the rather stringent conditions for exact geostrophy may be treated with the help of the Boussinesq approximation (discussed above, see Equations 12.37–12.39) and with what is called the *quasi-geostrophic* approximation (see the books on GFD mentioned above). The author of this book is quite convinced that some AFD problems, for example accretion discs, have not yet benefited sufficiently from similar approaches.

In this and the preceding sections we have introduced only the basics of fluid dynamics. Although these may perhaps suffice for the understanding of the topics given in the remainder of this chapter, we refer the reader to the books and references mentioned above for detailed discussion of the various topics in fluid dynamics.

12.3 Hydrodynamical stability

Hydrodynamical stability theory is an important part of fluid dynamics. Fluid flows are, as we have seen in the first section of this chapter, solutions of complicated, nonlinear, partial differential equations of motion. It is generally very difficult, and often impossible, to solve these equations and moreover, not every solution can actually exist in Nature and be observable. Those that do must not only obey the equations, but also be stable.

The concept of stability of solutions of various dynamical systems has been introduced and discussed in considerable detail in the first part of this book. In Chapter 3 we dealt with the stability of fixed points and limit cycles in dynamical systems defined by ODEs or maps. Chapter 7 was devoted in its entirety to the basic aspects of instabilities and the resulting pattern formation in extended systems governed by PDE dynamical systems, a class to which the equations of fluid dynamics clearly belong.

The ideas and methods employed in the study of hydrodynamical stability are similar to the ones discussed in those chapters (specifically, see Sections 3.1, 3.2, 7.1 and 7.2). However, in Chapter 3 they were used for relatively simple dynamical systems (maps and ODE) and in Chapter 7 for only some generic single PDE. All these are considerably simpler than the fluid dynamical equations. Indeed, as we shall see shortly, the straightforward investigation of the linear stability of even the simplest hydrodynamical flows may already be quite non-trivial and must be done, in practice, with the help of numerical calculations.

The importance of the study of hydrodynamical stability and its basic problems had been recognized and formulated by Helmholtz, Kelvin, Rayleigh and Reynolds in the 19th century, long before the modern methods of dynamical system theory were introduced. The mathematical methods of hydrodynamic stability (at least in their linear aspect) have a lot in common with stability theory in other physical systems, and also with the theory of some seemingly unrelated physical systems (e.g., basic quantum mechanics) and this is so because of the similar mathematical basis – spectral theory of linear operators (see e.g., Friedman, 1990). It seems, however, that as a result of the fact that fluid dynamics is a rather mature subject, the body of work in this context is the largest and it has influenced developments in other fields as well. Moreover, the important and outstanding problem of fluid turbulence and the transition to it is intimately related to hydrodynamical stability, as unstable flows often evolve into turbulent ones. It is therefore understandable that the study of hydrodynamical stability has always been a very active field and continues to be at the forefront of research. For obvious reasons we shall not be able to even mention here many of the important facets of this field. We refer the reader interested in a detailed study of hydrodynamic stability to the classic book by Chandrasekhar (1961) and the more modern comprehensive book by Drazin and Reid (1981) for a thorough exposition of this fascinating topic.

The traditional starting point of an investigation of hydrodynamic stability has always been *linear stability analysis*. Although this is not, as pointed out before, our central objective here, we shall briefly introduce it first, as it lies at the foundation of the theory. After formulating the problem of linear stability and demonstrating some of its difficulties and problematic results, like the puzzling (that is, contrary to experimental findings) results that some archetypical incompressible shear flows are always linearly stable, we shall proceed to discussing some relatively new ideas that may hint on ways of a possible resolution of this problem. Finally we shall mention some recent work, in which these ideas are used in the context of instabilities that may drive turbulence in some very important astrophysical objects – accretion discs.

It is impossible to conclude these remarks without mentioning the role of numerical computer experiments in the study of hydrodynamical stability. Computational fluid dynamics has come a long way since the advent of the first computing machines, devised mainly for the development of what we now call 'weapons of mass destruction'. High-speed parallel computers have enabled fluid research to reach a stage where it can rival laboratory investigations by faithfully simulating controlled experiments. In spite of these achievements, there is little doubt (at least in this author's mind) that theoretical understanding of the basic processes remains indispensable in the formulation of a viable theory of hydrodynamical stability and of its practical applications.

Linear stability – modal analysis

The mathematical tool that lies at the basis of linear-stability analysis of a hydrodynamical flow is simply the appropriate eigenvalue problem, resulting from adding a small perturbation to a known solution (usually a steady flow) and linearising around it. The essence of this procedure was discussed in Section 7.1 of this book with the eigenvalue problem of linear stability analysis given by Equation (7.4). As hinted above, some nontrivial complications of the procedure outlined in Chapter 7 can be expected when the PDE system consists of the equations of fluid dynamics. For example, the unperturbed solution (usually referred to, in fluid dynamics, as the *basic flow*) may be more general than just the homogeneous (constant, actually zero) solution, which was used in Chapter 7. Consequently, the spatial dependence of the basic perturbation cannot always be assumed to have the form of a plane wave (as in Equation 7.1), at least not in all spatial variables. It is therefore necessary to formulate here the basic problem of linear hydrodynamical stability (including the notation that will serve us later) and accompany it by some examples.

We start by writing the fluid dynamical equations, appropriate for a particular problem, in the following very general form

$$\partial_t \mathbf{u} = \mathcal{H}(\mathbf{u}; a) \tag{12.60}$$

where a is a nondimensional control parameter (involving e.g., the parameters characterising the flow, like the Reynolds number and so on) and $\mathcal{H}$ is the relevant 'hydrodynamical' operator, whose parametric dependence on a is indicated explicitly.

Splitting this operator into linear and strictly nonlinear parts (as in Equation 7.3) will not be useful here, as the unperturbed solution cannot be assumed to be zero (see below). Note that since (12.60) contains all the relevant equations of a fluid dynamical problem, the vector **u** includes all the variables of this problem (i.e., the velocity **v**, the pressure p and so on).

Assuming that the basic flow **U**(**r**) is a steady solution of (12.60), we perturb it by **u**′(**r**, t) (this is what an Eulerian perturbation is usually called in fluid dynamics, δ**u** being reserved for Lagrangian perturbations), that is, substitute **u** = **U**+**u**′, with **u**′ considered to be very small. Linearising the equation around **U** by neglecting terms which are second-order or higher in the perturbation gives rise to the following linear initial value problem for the perturbation

$$\partial_t \mathbf{u}' = \mathcal{L}(\mathbf{u}'; \mathbf{U}, a) \tag{12.61}$$

where $\mathcal{L}$ is a linear operator (with its possible dependence on **U** and the parameter a indicated here explicitly) whose exact form is determined during the linearisation of (12.60). The linear initial value problem must, of course, be supplemented by appropriate boundary conditions on the perturbation, which depend on the specific setup of the original problem.

The assumption that the basic flow is steady makes the linear operator $\mathcal{L}$, and thus the coefficients of the linear problem (12.61), independent of t. Thus, it is plausible that we may separate the variables, so that the general solution of the linear initial-value problem is a linear superposition of *normal modes*, each of the form

$$\mathbf{u}'(\mathbf{r}, t) = \hat{\mathbf{u}}(\mathbf{r})e^{st} \tag{12.62}$$

The eigenvalues s and the corresponding eigenfuctions $\hat{\mathbf{u}}(\mathbf{r})$ can be found in principle from the resulting boundary value problem

$$s\hat{\mathbf{u}} = \mathcal{L}(\hat{\mathbf{u}}) \tag{12.63}$$

where we have dropped the explicit indication of the linear operator's dependence on the basic flow and control parameter(s), for convenience. As the problem involves real functions the eigenvalues are themselves real or come in complex conjugate pairs and if the spatial domain, over which the problem is defined, is bounded, the eigenvalues are discrete. It is only natural to expect that the basic flow can be considered unstable if and only if at least one of the eigenvalues has a *positive* real part, reflecting an exponentially growing mode. This approach to the problem of linear stability is known as *modal* analysis; it dates back to the earliest studies of hydrodynamical stability and is usually well known in the astrophysical community.

For the sake of clarity we shall now give the explicit formulation of the modal approach, whose abstract form has been given above, for incompressible flows in a domain where the fluid density is constant and there are no body forces. The starting point is thus the appropriate Navier–Stokes equation (12.26) and the equation of continuity (12.27) in their nondimensional form

$$\partial_t \mathbf{v} + (\mathbf{v} \cdot \nabla)\mathbf{v} = -\nabla p + \frac{1}{Re}\nabla^2 \mathbf{v}$$
$$\nabla \cdot \mathbf{v} = 0 \tag{12.64}$$

so that the state vector is $\mathbf{u} \equiv (\mathbf{v}, p)$ with $\mathbf{v}$ the Eulerian velocity field, p the pressure and Re is the Reynolds number.

Assume now that the steady flow $(\mathbf{V}, P)$ is the solution of the above equations and some specific boundary conditions. The linear stability analysis of this flow is performed according to the prescription given above and results in a linear system like (12.61). Explicitly we get

$$\partial_t \mathbf{v}' + (\mathbf{v}' \cdot \nabla \mathbf{V}) + (\mathbf{V} \cdot \nabla \mathbf{v}') = -\nabla p' + \frac{1}{Re}\nabla^2 \mathbf{v}'$$
$$\nabla \cdot \mathbf{v}' = 0 \tag{12.65}$$

for the perturbation $(\mathbf{v}', p')$.

The boundary conditions for the perturbation are usually taken to be the same as those of the basic flow, and the stability of the basic flow can be defined on the basis of the solutions of (12.65). The basic flow is considered to be stable (to small perturbations) if all perturbations, which are sufficiently small initially, remain small for all time, and unstable if at least one perturbation, which is small initially, grows so much that it ceases to be small after some time. *Asymptotic* stability is guaranteed if the basic flow is stable and moreover the perturbations actually tend to zero for $t \to \infty$. These heuristic statements can be given a precise meaning analogously to the definitions in Sections 3.1 and 7.1 of this book (see e.g., Drazin & Reid, 1981), or alternatively by using the energy of the perturbation (Joseph, 1976; and see also Schmid & Hennigson, 2000) but we shall not be concerned with these formal matters here.

As mentioned before, the steadiness of the basic flow makes the coefficients of the linear system time independent and allows for the separation of variables. The resulting boundary value problem (12.63) reads

$$s\hat{\mathbf{v}} = -\hat{\mathbf{v}} \cdot \nabla \mathbf{V} - \mathbf{V} \cdot \nabla \hat{\mathbf{v}} - \nabla \hat{p} + \frac{1}{Re}\nabla^2 \hat{\mathbf{v}}$$
$$\nabla \cdot \hat{\mathbf{v}} = 0 \tag{12.66}$$

with the spatial eigenfunctions $(\hat{\mathbf{v}}, \hat{p})$ and eigenvalues s arising from the separation (12.62). Writing $s = \sigma + i\omega$ we deduce what has already been said in the general discussion above, that is, that the basic flow is stable if and only if $\sigma < 0$ for *all* eigenvalues.

This allows one to define the critical Reynolds number Re_c (if it indeed exists). It is found experimentally that increasing Re as a parameter, in a particular flow setting, typically drives the flow towards instability. Thus the definition of the critical Reynolds number is based upon the statement that if $Re \leq Re_c$ then $\sigma \leq 0$ for all eigenvalues but $\sigma > 0$ at least for one mode, for at least one value of Re in any neighbourhood Re_c. Clearly, the situation $Re = Re_c$ is one of marginality, in the sense explained in Chapter 7.

An obvious advantage of the modal approach arises when the set of eigenfunctions of (12.66) is *complete*, that is, any initial perturbation can be expressed as a superposition of these eigenfunctions, and if the eigenfunctions are *orthogonal*. In such a case, if the largest σ (the real part of s) among all the eigenvalues is negative, that is, the basic flow is linearly stable, all modes decay exponentially and therefore the total perturbation decays as well. In such a case it seems quite safe to assume that stability of a given basic flow is guaranteed if it is *linearly* stable and this, in turn, is governed by the normal modes.

Indeed, in the case of some important classes of flows the modal approach to hydrodynamical stability has been quite successful and has given results that are in accord with laboratory experiments or the observed behaviour of geophysical and astrophysical flows. Laboratory investigations of the onset of instability in a fluid layer placed in a vertical gravitational field and heated from below (Rayleigh–Bénard convection) have been continuing now for over 100 years and they essentially confirm theoretical calculations based on the modal approach. This type of flow is a paradigm of instability that is driven by buoyancy forces, whose origin is essentially thermal and it has obvious geophysical and astrophysical applications. We have already introduced the Rayleigh–Bénard problem in the context of explaining the Boussinesq approximation and will discuss its stability later on in this section, in the context of the derivation of the Lorentz model. Other classical astrophysical examples of the success of the modal approach include the Jeans instability, and the Field (thermal) instability.

In contrast to these successes of the modal analysis, this approach has patently failed in some other important cases, notably in flows in which the instabilities are driven by shear. Among these, the most 'notorious' examples of basic flows are the plane Couette flow (having a linear velocity profile between two infinite flat plates moving parallel to each other) and the pipe Poiseuille flow (having a parabolic profile inside a cylinder). Modal analysis, performed for these flows, indicates that they are stable for all values of the Reynolds number, while in careful modern (performed in the past decade or so) laboratory experiments, which have aimed at avoiding any finite disturbances to the laminar flow, instability and transition have been observed at finite values of the Reynolds number. In other instances, e.g., the plane Poiseuille flow and some cases of the Taylor–Couette flow (between rotating cylinders), for which modal analysis gives the onset of instability at a particular Re_c, actual transition is observed in the laboratory at lower Reynolds numbers (when all computed eigenvalues s have negative real parts). Consequently, these instabilities have been called *subcritical*. For recent laboratory results on these flows and their comparison with the predictions of linear stability analysis see Trefethen *et al.* (1993), Grossmann (2000), Schmid & Hennigson (2000).

It is impossible to give here a detailed discussion of the failure of the modal approach in the above mentioned flows. Nevertheless it seems that discussing just one (quite simple) example may be very useful in clarifying the difficulties arising in the application of the modal approach to certain generic shear flows. In addition, in this way we will be able to introduce one of the basic tools of hydrodynamic stability theory of shear flows, the Rayleigh stability equation (which is actually the inviscid limit of the celebrated Orr–Sommerfeld equation). We choose as our example the linear stability of plane Couette *inviscid* flow. Expressed in Cartesian coordinates it is given by

$$\mathbf{V} = V(y)\hat{\mathbf{x}}; \quad P = \text{const} \tag{12.67}$$

and is assumed to be confined to $y_1 \leq y \leq y_2$, that is, to the region between the planes $y = y_1$ and $y = y_2$. It is also assumed that these planes are actually rigid walls ($v_y = 0$ there) and this constitutes the boundary conditions. It is easy to see that the plane parallel flow (12.67) with any $V(y)$ satisfies (12.64) in its inviscid limit, and is therefore a steady solution thereof.

We proceed now to the linear stability analysis of such a plane parallel base flow and thus consider perturbations $\mathbf{v}'(\mathbf{x}, t)$, $p'(\mathbf{x}, t)$, which also satisfy the vertical impenetrability condition at the confining planes. The linearised perturbation equation (12.65) then reads

$$(\partial_t + V\partial_x)\mathbf{v}' + \frac{dV}{dy}v'\hat{\mathbf{x}} = -\nabla p', \quad \nabla \cdot \mathbf{v}' = 0 \tag{12.68}$$

where the perturbation velocity Cartesian components (v'_x, v'_y, v'_z) are denoted by (u', v', w') and with the boundary condition

$$v'(\mathbf{x}, t) = 0 \quad \text{at} \quad y = y_1, y_2$$

For the sake of convenience we shall henceforth drop the prime from the perturbation and denote by prime the vertical derivative (d/dy). Taking the divergence of the linear equation (12.68) and using the incompressibility condition for the velocity perturbation, one can obtain an expression for the pressure perturbation

Laplacian, which in the new notation is

$$\nabla^2 p = -2V' \partial_x v$$

Now taking the Laplacian of the y component of (12.68) and eliminating the pressure term using the above relation, yields an equation for the vertical velocity perturbation v

$$[(\partial_t + V \partial_x)\nabla^2 - V''\partial_x]v = 0 \tag{12.69}$$

This equation can conveniently be supplemented by a vertical vorticity equation. Its variable is the vorticity (actually its vertical component) perturbation and we shall denote it by ω. It is defined as $\omega \equiv \partial_z u' - \partial_x w'$ and the desired equation can be obtained by differentiating the appropriate components of (12.68) with respect to z and x and subtracting them, with the result

$$(\partial_t + V \partial_x)\omega = -V'\partial_z v \tag{12.70}$$

It turns out that these two equations, supplemented by the boundary condition specified above, constitute a system equivalent to the original problem which is usually referred to as the velocity–vorticity formulation. We shall not prove this fact here and only remark that an alternative approach consists of using equations (12.68) for the three components of the velocity and the pressure, and showing later on (with the help of what is known as the Squire transformation) that to each unstable three-dimensional mode there corresponds an even 'more unstable' two-dimensional one. Thus, looking at the flow as two-dimensional, and setting the second vertical component of the velocity perturbation to zero, does not limit the generality of the linear stability analysis, and finally the modal approach to the problem yields the Rayleigh stability equation (for details of this approach see, e.g., Drazin & Reid, 1981). We shall use here the velocity–vorticity formulation, which will be used in the next subsection, and obtain the Rayleigh equation as well.

The modal approach consists of transforming the initial value problem (12.69)–(12.70) into a boundary value problem by using normal modes for the perturbation

$$v(x, y, z, t) = \hat{v}(y) \exp[i(\alpha x + \beta z) - i\alpha c t] \tag{12.71}$$

where α, β and c are constants. This particular (plane-wave) form of the perturbation in the vertical (to the basic flow) directions is justified because the coefficients of the initial value problem are independent of these directions and therefore the dependence on them may be separated in this way. Also, the time dependence here is obviously equivalent to our general formulation (12.62), but here the eigenvalue $c \equiv is/\alpha$ is used, instead of s, for the purpose of obtaining the Rayleigh equation in its customary form. This follows from the substitution of the above wavelike perturbation into the velocity equation (12.69), giving

$$[(V - c)(\mathcal{D}^2 - k^2) - V'']\hat{v} = 0 \qquad (12.72)$$

where $k^2 \equiv \alpha^2 + \beta^2$ and $\mathcal{D} \equiv d/dy$.

The Rayleigh stability equation (12.72), supplemented by the appropriate boundary conditions, which we now set as $\hat{v} = 0$ at $y = \pm 1$ by properly shifting and rescaling the coordinate y (this leaves the equations invariant), specifies the eigenvalue problem in its most concise form.

If α and β are real, so is k and thus the Rayleigh equation is real. Consequently, the eigenvalues c are real or come in complex conjugate pairs. The choice of real k is known as the *temporal problem*, where the spatial structure of the wavelike perturbation remains unchanged when its amplitude is growing or decaying. This problem is simpler than the one in which one allows for the choice of complex k and hence for the growth of amplitude in space. For our purpose here it will suffice to treat the temporal problem, and as it turns out the result we wish to show holds in general. In addition, because the equation depends only on the squares of α and β, we can choose them to be positive without limiting the generality of the discussion.

Without going into too many details and analysing the various cases (see Drazin & Reid, 1981; for a comprehensive discussion of the Rayleigh stability equation) we can easily get the famous inflection point criterion (first derived by Rayleigh himself) as follows. Assume that the basic flow is unstable to a particular mode corresponding to the eigenvalue c. This means that c must be complex and that its imaginary part, c_i say, is positive (because then we have a time dependence of the form $\exp(-i\alpha c t)$ and when $\alpha > 0$ exponential growth occurs if and only if $c_i > 0$). After dividing the Rayleigh equation by $V - c$ (which cannot be zero because c is assumed to be complex) and multiplying it by $\hat{v}*$, we integrate it over the domain $-1 \leq y \leq 1$. Using integration by parts and the boundary conditions we get

$$\int_{-1}^{1} (|\mathcal{D}\hat{v}|^2 + k^2|\hat{v}|^2)dy + \int_{-1}^{1} \frac{V''}{V - c}|\hat{v}|^2 dy = 0 \qquad (12.73)$$

Notice that the first integral is real valued and positive definite and the second one is, in general, complex valued. Taking the imaginary part of the whole expression thus gives

$$\int_{-1}^{1} \frac{c_i|\hat{v}|^2}{|V - c|^2} V'' dy = 0$$

Since c_i is assumed to be positive and the ratio of the squares of the absolute values of functions in the integrand is non-negative, the only possibility for this integral to vanish is if V'' changes sign inside the domain, that is, at some point in $(-1, 1)$ we have $V'' = 0$. This means that *if* the plane parallel flow is unstable, the profile of this basic flow must have an inflection point. This statement specifies thus a *necessary* condition for instability in plane parallel inviscid flows.

It is a simple matter therefore to conclude from the Rayleigh inflection point criterion that the plane parallel linear flow known as a plane Couette flow, whose velocity profile is e.g.,

$$V(y) = y \quad \text{for} \quad -1 \le y \le 1 \tag{12.74}$$

is *stable*. This is so because there is no inflection point of V in the domain ($V'' = 0$ *everywhere* and thus V' does not change sign), which is necessary for instability. This result appears to be in contrast with experimental findings (see e.g., Grossmann, 2000; and references therein) in which instability is found to occur in such flows for finite values of the Reynolds number ($\sim$1500). Formally, the inviscid analysis is the infinite Reynolds number limit of the viscous case, and therefore the flow appears stable even for $Re \to \infty$ as is confirmed by the viscous flow analysis given below.

The analysis for plane parallel *viscous* flows is more complicated than the inviscid case given above, but it can be done in a similar manner yielding the following velocity–vorticity equations:

$$\left[(\partial_t + V \partial_x) \nabla^2 - V'' \partial_x - \frac{1}{Re} \nabla^4 \right] v = 0 \tag{12.75}$$

$$\left[\partial_t + V \partial_x - \frac{1}{Re} \nabla^2 \right] \omega = -V' \partial_z v \tag{12.76}$$

where the boundary conditions in the viscous case include, in addition to the inviscid conditions, also $v' = 0$ at the rigid walls.

The eigenvalue problem for the velocity of the viscous flow, resulting from the modal approach, that is, the substitution of a perturbation like (12.71), reads

$$\left\{ i\alpha \left[(V - c)(\mathcal{D}^2 - k^2) - V'' \right] - \frac{1}{Re} (\mathcal{D}^2 - k^2)^2 \right\} \hat{v} = 0 \tag{12.77}$$

and is known as the Orr–Sommerfeld equation. This equation is significantly more complicated that the Rayleigh equation, which can be easily obtained as its inviscid limit (formally $Re \to \infty$).

Even the simplest flows (like the plane Couette flow treated above) do not lend themselves to straightforward analytical treatments of the Orr–Sommerfeld problem. Numerical approaches must almost always be employed (only in very few cases some sophisticated analytical techniques are feasible). Among numerical calculations of this type we would like to mention the findings that the spectra of plane Couette viscous flow and of the pipe Poiseuille flow do not contain any unstable eigenvalues. In contrast, the plane Poiseuille flow has been found to be unstable with the critical value of the Reynolds being $Re_c \approx 5770$ (see Schmid & Hennigson, 2000; for details and references). The experimental findings for the three flows mentioned above differ substantially from the results of the modal approach to linear stability analysis. The plane Couette flow and the pipe Poiseuille flow, which are theoretically stable for all Re, exhibit instability and loss of laminarity for values of the Reynolds number of about 2000 and 1300 respectively. The plane Poiseuille flow already becomes turbulent in the lab at about $Re \sim 1000$, significantly below the theoretical value of Re_c quoted above. Attempts to carefully monitor the disturbances (initial perturbations) in the experimental setup so as to avoid a 'nonlinear onset', or at least reduce the perturbation size considerably, have been made in some experiments. This indeed increased the Reynolds number of the transition, but never sufficiently so as to reconcile the experimental findings with the results of linear modal stability analysis (see Grossmann, 2000, for a summary of experimental results and references to the original works).

Non-modal onset of instability in shear flows

The above-mentioned failure of the modal approach to predict the laboratory findings about the onset of instabilities in some basic laminar shear flows remains one of the central problems in the stability theory of fluid motion. This is so because in the traditional approaches the transition to turbulent motion always starts with a linear instability. As recently reiterated by Grossmann (2000), there seems to be no escape from the conclusion that the laminar-to-turbulent transition in Couette or Poiseuille shear flows is *not* the consequence of linear instability of the basic laminar flow. Small disturbances cannot be expected to grow exponentially.

There exists a rather large amount of important and innovative work aimed at explaining the onset of instability and the loss of laminarity in shear flows of this kind. The article by Bayly, Orszag and Herbert (1988) reviews mechanisms based on three-dimensional instabilities of two-dimensional nonlinear waves, which are postulated to spontaneously arise in flows of this type. Other ideas have been proposed as well (see the references in Grossmann, 2000), but none of these have as yet been fully accepted as the ultimate solution to the problem. We would like to introduce here, and explain with the help of a simplistic example, an additional, rather recent idea in this context, based on the fact that the relevant operator arising in the linear stability analysis of shear flows is generally *nonnormal*.

A linear operator $\mathcal{L}$ is called *normal* if it commutes with its adjoint. Self-adjoint (Hermitian) and unitary operators obviously belong to this class. Normality, in this sense, of a linear operator guarantees the mutual orthogonality of its eigenfunctions. When this property is absent, the eigenfunctions are not necessarily orthogonal to each other. As we shall shortly demonstrate by a simple example, there may exist perturbations, consisting of a linear superposition of such nonorthogonal *decaying* modes, that can actually *grow* for some time. It was already known to Orr in 1907 that linear disturbances in shear flows can grow for a while, but the context was eigenvalue degeneracy. The general ideas based on nonnormality of the relevant linear operator have been introduced and pursued only since the late 1980s (Farell, 1988; Butler & Farell, 1992). A good summary of these ideas can be found in Trefethen *et al.* (1993), where an analytical approach to nonnormal operators, based on 'pseudo-spectral analysis' is also introduced. The key result of these studies (a systematic detailed discussion can be found in the recent textbook by Schmid & Hennigson, 2000) is that small perturbations of smooth shear flows may grow, because of the non-orthogonality of the eigenfunctions, by huge factors (e.g., $\sim 10^5$-fold!) as a result of a linear mechanism, even though all the modes, by themselves, exponentially decay (the flow is linearly asymptotically stable).

The nonnormality of the linear operator arising in linear stability analysis of shear flows can easily be noticed by a mere inspection of Equation (12.68) or its viscous counterpart. The left-hand sides of those equations contain terms, arising from $(\mathbf{v}' \cdot \nabla)\mathbf{V}$ in (12.65), which represent advection of the laminar basic flow $\mathbf{V}(\mathbf{r})$ by the perturbation $\mathbf{v}'(\mathbf{r}, \mathbf{t})$. Such terms are bound to introduce asymmetry of the matrix arising in the eigenvalue problem, because it is sensitive to the laminar shear profile. Indeed, in the case leading to (12.68), the only surviving term is the one containing dV/dy, because all the other derivatives (of this and the other possible components) of the basic flow vanish. This asymmetry causes the nonnormality of the relevant operator. The actual occurrence of transient growth (and its maximal extent) of a perturbation in a case governed by such an operator (i.e., whose eigenfunctions are not orthogonal) clearly depends on the perturbation's initial condition. Consequently, the only approach in the investigation of such transient growth is to consider the original initial-value problem, like (12.65) instead of the boundary value problem (12.66) resulting from normal mode analysis.

We shall now illustrate the principal differences between these two approaches by examining a simplistic 'toy' model, which is however capable of capturing the essence. The initial-value problem, resulting from linear stability of a viscous plane parallel shear flow, is given (in the velocity–vorticity formulation) in Equations (12.75)–(12.76). We shall 'model', in a sense, this problem by the simple matrix equation

$$\frac{d}{dt}\begin{pmatrix} v \\ \omega \end{pmatrix} = \begin{pmatrix} -1\, Re^{-1} & 0 \\ 1 & -2\, Re^{-1} \end{pmatrix}\begin{pmatrix} v \\ \omega \end{pmatrix} \tag{12.78}$$

complemented by prescribed initial conditions $v(0) = v_0$ and $\omega(0) = \omega_0$.

This system resembles the fluid problem and, in particular, the matrix (which plays the role of the relevant operator) possesses an off-diagonal structure, which makes it nonnormal (that is non-commuting with its adjoint). In an actual flow of this sort the ratio between the terms responsible for nonnormality and the other terms is proportional to the Reynolds number of the basic flow. We model this by making the off-diagonal term equal to 1, while the diagonal terms are of order $1/Re$. Also, the diagonal terms are chosen to be different from each other to prevent eigenvalue degeneracy, which we do not wish to be the cause of the transient growth here. Note that both Trefethen *et al.* (1993) and Grossmann (2000) have used similar 'toy' models to demonstrate the effect of nonnormality, but they chose to make the off-diagonal terms of order Re. This obviously gave equivalent results, but our choice seems to be more didactic at the limit $Re \rightarrow \infty$.

The linear initial value problem (12.78) has the exact solution

$$\begin{pmatrix} v \\ \omega \end{pmatrix} = v_0 \begin{pmatrix} 1 \\ Re \end{pmatrix} \exp[-t/Re]$$

$$+ (\omega_0 - v_0\, Re) \begin{pmatrix} 0 \\ 1 \end{pmatrix} \cdot \exp[-2t/Re] \qquad (12.79)$$

which can easily be found using standard methods for linear ODEs. Since the eigenvalues $\lambda_1 = -1/Re$ and $\lambda_2 = -2/Re$ are both negative, normal mode analysis would indicate that the perturbations decay exponentially in time. We know already, however, that this is a nonnormal system, and it is not difficult to see that because the eigenvectors are not orthogonal, the solution (12.79) may exhibit transient growth for appropriate initial conditions. Indeed, the solution with the initial condition $\omega_0 = 0$ and an arbitrary $v_0 \neq 0$, for example, is

$$v(t) = v_0 \exp[-t/Re]$$

$$\omega(t) = v_0\, Re(\exp[-t/Re] - \exp[-2t/Re]) \qquad (12.80)$$

For times much shorter than any number of the order of Re (our system is nondimensional, of course) we can expand the two exponentials in the solution for $\omega(t)$, given above, to get

$$\omega(t) \sim v_0 t - \frac{3v_0}{Re} t^2 + \cdots$$

showing that this component of the solution actually exhibits algebraic growth (which however ceases and is taken over by the exponential decay after a sufficient time).

The connection between this and nonnormality or its consequence, nonorthogonality of the eigenvectors, is clearly apparent when we write an expression for the angle ϕ between the directions of the normalised eigenvectors,

$$\Lambda_1 = (1 + Re^2)^{-1/2} \begin{pmatrix} 1 \\ Re \end{pmatrix} \quad \text{and} \quad \Lambda_2 = \begin{pmatrix} 0 \\ 1 \end{pmatrix}$$

by forming their scalar product

$$\cos \phi = \mathbf{\Lambda}_1 \cdot \mathbf{\Lambda}_2 = \frac{Re}{\sqrt{1 + Re^2}} \qquad (12.81)$$

For large Reynolds numbers the eigenvectors are almost parallel and if the initial conditions are chosen in the 'right' way, the resultant vector (the solution) may initially be very close to zero, because the modal components may happen to almost cancel each other out. Later on, however, these components (along the eigenvectors), which may initially be not too small, decay exponentially at different rates. Thus the resultant may grow for some time, until the decay of both components finally drives it towards zero.

Note that for $Re \to \infty$ (i.e., in the inviscid limit) the 2×2 matrix in (12.78) degenerates to $\begin{pmatrix} 0 & 0 \\ 1 & 0 \end{pmatrix}$, whose eigenvalues are degenerate and eigenvectors exactly parallel ($\cos \phi = 0$, see Equation 12.81) and, as is well known, the solution of the initial-value problem is just algebraic growth.

This example clearly shows that in the description of the dynamics of perturbations, governed by a linear initial-value problem, eigenvalues are not sufficient. Eigenvectors, specifically if they are not orthogonal, are no less important. Thus modal (spectral) analysis, based on the study of eigenvalues only, may be insufficient in this case, if one is interested to follow the development of the perturbation and not only its asymptotic fate. Indeed, detailed studies of the basic shear-flow types mentioned above show that the bunching of eigendirections does occur, and increases with the Reynolds number (see Grossmann, 2000; and references therein). A systematic approach to transient growth of perturbations and its quantification can be found in the book by Schmid and Hennigson (2000). Finding the *optimal* perturbations, that is, those perturbations (initial conditions of the initial-value problem) that give rise to maximum possible amplification at a given time, is obviously one of the important issues in this context and there exist a number of methods to do that.

The fact that perturbations on some linearly stable basic flows may temporarily grow, sometimes by a very large factor, and that this factor (and also the duration of the transient) increases with Re, leads to an interesting idea. We have seen that in shear flows the modal approach generally fails to correctly predict the loss of stability and the transition to turbulence. The idea is that transition to turbulence may proceed via nonnormal transient growth, causing the perturbations to grow so much that nonlinear effects come into play. It this way linear instability is actually not needed for the flow to become turbulent. It is *bypassed*. A significant missing link in this, so called, bypass transition still remains, however. Even if the flow is such that the perturbations are optimal and give rise to very significant growth of the disturbances, one still has to propose and understand the mechanism(s) which are actually responsible for the transition to turbulence. The search for such mechanisms has been an active field of research in recent years and several ideas have been proposed. It is quite clear that external forcing can be instrumental, if the response of the system is optimal (see Schmid & Hennigson, 2000), but there is also a need for viable internal mechanism(s), because one would like to account for the possibility of spontaneously occurring transitions (without the need for continuous forcing) in experiments or observations on natural fluid systems. Among the ideas that have been put forward we shall only mention the stochastic forcing (see Farrell & Ioannou, 1998; and references therein), which is clearly of the former category and a nonlinear interaction of transiently growing modes (see Grossmann, 2000; and references therein). These ideas look promising, but at the time of the writing of this book it appears that there are still no generally accepted explicit generic mechanisms that could induce some kind of global nonlinear instability by utilising the transiently (linearly) grown disturbances.

The developments in the stability theory of shear flows described above have only recently begun to find their way into the astrophysical literature, in the context of accretion-disc theory. Accretion discs are thought to be present in a variety of astrophysical systems in which gravitational accretion of matter, endowed with non-zero angular momentum, onto a relatively compact object takes place. These systems include objects as different from each other as supermassive black holes at centres of active galaxies, the primary stars (usually white dwarfs or neutron stars) in close binary systems and newly formed stars. The accretion flow can settle in a disc-like configuration around the accreting object (with mass transport inward) only if some kind of angular momentum transport outward is present. The more efficiently the disc is able to cool, the thinner it becomes, i.e., its vertical extension is much smaller than the radial one. A very thin disc is also essentially Keplerian, that is, the angular velocity of matter, as a function of the distance from the central object, satisfies $\Omega(r) \propto r^{-3/2}$.

The literature on the various aspects and problems of accretion disc theory is vast and we recommend Pringle (1981) and Frank, King & Raine (2002) to readers interested in a good overview of the subject. Here we will focus on just one (albeit quite important) aspect of the central problem in accretion disc theory – the identification of the physical mechanism responsible for angular momentum transport. It is quite clear that the microscopic coefficient of viscosity cannot produce any meaningful angular momentum transfer and accretion rate (these are measured indirectly from the accretion luminosity, resulting from the dissipation of the kinetic energy of the gravitationally infalling matter), and thus some kind of anomalous transport must be invoked.

Already in the pioneering works on accretions discs (Shakura & Sunyaev, 1973; Lynden-Bell & Pringle, 1974) it has been suggested that the required transport can be naturally achieved if the flow is turbulent. Moreover, the proposal to circumvent the complexities, and our lack of understanding of turbulence, by introducing the famous α disc model, has been put forward at the outset. Despite its simplicity (it is essentially a parametrisation based on the simplest dimensional argument) the α prescription for the eddy viscosity parameter has been extremely fruitful, giving rise to a large number of successful interpretations of observational results. Accretion discs have usually been treated as one-dimensional (in the radial direction) steady structures, with the axisymmetry and steady state assumptions complemented by straightforward vertical 'averaging'. More sophisticated models have also been put forward, using a more careful asymptotic analysis. For example, alternative (to the α model) global principles, inspired by shallow water theory in GFD, i.e., uniformity of entropy and *potential* vorticity were used by Balmforth *et al.* (1992) and yielded global equilibria of turbulent accretion discs. In another asymptotic study Regev and Gitelman (2002) (see also references therein) found steady meridional flows in thin accretion discs and investigated their properties. It seems, however, that astronomers have remained quite content with the simplest thin α discs.

A serious theoretical problem has however always accompanied accretion disc modelling, the nature of which makes it related to the topic of this section. Keplerian discs are found, in general, to be linearly asymptotically stable to pure hydrodynamical perturbations. The famous Rayleigh circulation criterion (see Drazin & Reid, 1981) for stability of swirling flows

$$\frac{\mathrm{d}}{\mathrm{d}r}[r^2\Omega^2(r)]^2 \geq 0 \tag{12.82}$$

is clearly satisfied for a Keplerian velocity profile. Even though this criterion is strictly valid only for axisymmetric perturbations, and not the most general basic flows, no linear hydrodynamic instability of any kind has ever been explicitly shown to exist. For some researchers this has been reminiscent of the failure of similar stability analyses in shear flows (as explained above), and they have remained confident that these astrophysical swirling flows with literally astronomical values of the Reynolds number somehow find their way to become turbulent. Others (a large majority) approached this matter more cautiously, constantly seeking alternative (to turbulence) mechanisms for angular-momentum transport, or for instabilities of various kinds which might lead to turbulent behaviour.

In 1991 Balbus and Hawley proposed the magneto-rotational instability (MRI) as the physical trigger of accretion-disc turbulence. This instability can operate in magnetised plasmas (ionised fluids), and the Keplerian velocity profile is linearly unstable. The issue of the physical origin of accretion disc turbulence thus appeared for many to have finally been settled. The review article of Balbus and

Hawley (1998) summarises in detail and gives justifications for this view. Moreover, numerical simulations performed by these authors and their collaborators have suggested that, while purely hydrodynamical planar Couette flows do become unstable and turbulent due to finite amplitude disturbances, turbulence disappears because of rotational effects (i.e., Coriolis forces) whenever the angular velocity profile satisfies the Rayleigh criterion (12.82). The claim thus seemed to be that the magneto-rotational instability is both sufficient and necessary for turbulence in accretion discs.

The search for a non-magnetohydrodynamical source for accretion disc turbulence has not been abandoned, however. In addition to the intellectual challenge, and the wish not to accept the notion that accretion-disc theory must be approached almost exclusively by three-dimensional numerical MHD simulations, this search has been performed for some well-defined and explicit reasons. Baroclinic instabilities, destabilisation by vertical stratification and other ideas have been proposed but the recent efforts, based on the idea of transient growth and bypass transition, seem now to be the most popular.

Ioannou and Kakouris (2001) were the first to investigate the possibility of transient nonnormal growth of perturbations in Keplerian accretion disc flows. They considered incompressible perturbations, which for a two-dimensional problem allow a simplified treatment based on the stream-function formulation. They were able to find, using appropriate techniques, the optimal perturbations which rapidly grow (on the timescale of the rotational period), and to show that these transiently growing perturbations carry angular momentum outward while they grow. Although these perturbations have the converse effect (i.e., angular momentum transport inward) during their subsequent decay phase, Ioannou and Kakouris found that, in the statistically steady state that emerges under random forcing, the net angular momentum transport will be predominantly outward. The desired effect was found to occur when the random forcing was of a broad-band (in space and time) white noise. It is, however, still unclear if the conditions prevailing in real accretion discs indeed provide the appropriate forcing so as to sustain the required angular momentum transport and energy dissipation.

Several more recent works have been based on linear analysis within the shearing-box (or sheet) approximation. The essence of this approximation is in its focusing on a small neighbourhood of a point in the accretion disc, the expansion of the basic flow around this point (thus making the shear linear), and the study of perturbations on the basic flow. The original work of Balbus and Hawley, including the numerical calculations, was also done within this framework. In all these works (for a variety of conditions) transient growth has been shown to exist (Chagelishvili *et al.*, 2003; Tevzadze *et al.*, 2003; Yecko, 2004; Umurhan & Regev, 2004). In the above papers the authors see in the linear transient growth a beginning of the flow's transition to turbulence. Such a bypass (of linear instabilities) transition is obviously viable only if indeed the amplified perturbations, or waves excited by them, are somehow tapped by nonlinear processes and develop into a fully fledged turbulent flow. The details of these positive feedback processes, even in the simpler hydrodynamical-shear flows mentioned before, and certainly in accretion-disc flows, still remain to be understood and elucidated.

Umurhan and Regev (2004) have complemented their linear stability analysis by high-resolution numerical simulations to explore the nonlinear regime. The have used spectral methods in *two dimensions*, and found that the transiently growing modes ultimately start decaying, but some nonlinear feedback mechanisms seem to re-excite these modes and the flow exhibits a complex and irregular time behaviour of recurrent transient growth episodes (see Figure 12.1). This activity is bound to ultimately decay in such two-dimensional flows because of viscous dissipation (as is indeed apparent from the $Re = 5000$ curve in the figure). If the Reynolds number is nominally infinite the decay time is determined by the value of the effective Reynolds number (i.e., the numerical resolution). In the simulations of Umurhan and Regev (2004) the activity was maintained for up to one thousand orbital periods. Although encouraging, these results are still incomplete, mainly because of the two-dimensionality. The existing three-dimensional hydrodynamical simulations of Balbus, Hawley and their collaborators have failed to detect hydrodynamical turbulence in rotating shear flows with angular momentum increasing outwards. Are these simulations inadequate as a result of a too crude spatial resolution, i.e.,

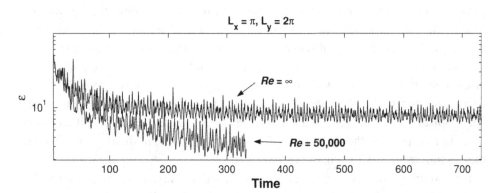

Fig. 12.1. The energy in the disturbance in units of the total energy in the basic shear flow as a function of time (in units of the local orbital period). White-noise initial conditions have been used and the spectral calculation was performed on a rectangular two-dimensional shearing box with periodic boundary conditions. L_x and L_y are the dimensionless values of the box extent in the radial and azimuthal directions respectively. The spikes are caused by recurrent transient growth events and the overall decay by viscosity. For $Re = \infty$ the effective Reynolds number is determined by the 'numerical resolution' (inclusion of 512×256 spectral modes). Reprinted, with permission, from Umurhan & Regev (2004).

a too low effective Reynolds number, or perhaps they indicate that such flows are indeed stable even if the Reynolds number is enormous? It is also possible that we still lack some basic understanding and thus fail to include some relevant physical ingredients in the models. Clearly, further analytical, experimental and numerical work is needed before this issue is finally clarified.

12.4 Thermal convection and the Lorentz model

Thermal convection is a very common fluid-dynamical phenomenon. It is a simple matter to see that the hydrostatic equilibrium of a stratified self-gravitating fluid, or even more simply, a fluid layer in a constant external gravitational field, may be unstable with the instability being driven by buoyancy. The physical mechanism behind this *convective* instability can be readily understood (see below), but some of the phenomena associated with the resulting flow still defy satisfactory mathematical modelling. The situation is particularly severe in highly unstable, thick, compressible fluid layers, which are characteristic of astrophysical applications. In these cases the convection is turbulent and we cannot compute reliably even the heat flux – the stellar convection problem, for example, is still unsolved although it has been central for more than a century. As it turns out, however, the simple mixing-length algorithm has sufficed for many purposes, notably for good numerical modelling of stellar structure and evolution. Still, although our ignorance has perhaps not impeded the progress of stellar astrophysics, better understanding of astrophysical convection could certainly be very helpful. A simple version of the convection problem arises when a fluid slab, lying horizontally in a uniform gravitational field, is heated from below (the Rayleigh–Bénard experiment). The linear stability problem relevant to this case can be directly approached by applying the methods discussed in the previous section) to the Boussinesq equations (12.42)–(12.43). A detailed discussion of this problem can be found for example in Drazin & Reed (1981), but we remind anyone who has attended a basic stellar structure course (or reached page 7 in Landau & Lifshitz, 1987) that if dissipation is neglected (i.e., there is no viscosity and no heat transport) hardly any analysis is needed. Considering a small adiabatic vertical displacement of a little parcel of fluid (maintaining pressure equilibrium with its surroundings), it is easy to obtain the well known Schwarzschild criterion for convective *stability*. In stellar structure books it is usually formulated as subadiabacity of the temperature gradient, but often (e.g., Landau & Lifshitz, 1987) it is spelled out as the condition that the specific entropy increase outward, that is, $ds/dz > 0$ (when g is directed towards the negative z direction).

If dissipation is allowed, the relevant quantity becomes the Rayleigh number and convection starts when Ra is above some critical value, Ra_c (e.g., Drazin & Reid,

1981; and see below). Good agreement has been found between the measured value of Ra_c and the theoretical predictions of linear stability analysis. Although this situation is much more satisfactory than in the case of shear flows, at least as far as the instability onset is concerned, linear theory has very little to say on the spatio-temporal patterns appearing in a convecting fluid under different circumstances. This is so because the perturbations eventually grow to the point of being large enough for nonlinear effects to be dominant.

In practice, both simple 2-dimensional rolls (see below) and more complex hexagonal cells are quite common. The flow in these cells is ordered, but it becomes unstable when Ra_c is increased further, usually leading to time-dependent complex motions and ultimately to turbulence. The precise sequence of events depends critically on a variety of factors, like boundary conditions, the temperature dependence of the viscosity and the Prandtl number. Detailed discussion of convection, even in its simplest setting, is beyond the scope of this book. From the many possible references we single out the books by Drazin and Reid (1981) and Maneville (1990) and recommend them for the interested reader. The latter, being the more modern, is closer in its approach to the spirit of this book and in particular the methods mentioned in Chapter 7. We also strongly recommend the modest-looking, but rather rich in meaning, summary by Spiegel (1995). A review of convection in the stellar context can be found in Spiegel (1971, 1972). In this section we shall discuss only the Lorentz model, one of the paradigms of chaos theory, whose origins lie in the study of thermal convection. This nonlinear ODE system, which exhibits temporal chaos, is formulated for just three temporal functions, and we shall first show how it can arise in the context of thermal convection. After briefly describing the complex behaviour of this model, we shall try to assess what the connection is (if any) between this behaviour and thermal convection.

From the Boussinesq to Lorentz equations

Most chaos books give some kind of 'derivation' of the Lorentz model, from the much more general problem of hydrodynamical convection. The Lorentz equations are usually shown to emerge from a very severe truncation of a Fourier series expansion. This makes their relevance to a quantitative description of convection rather doubtful. Yet, thinking about the convection problem using the 'parcel' approach mentioned above, but in a refined and deeper way (Spiegel, 1995), yields the Lorentz model as well. It is thus not very clear what the value of each of the various 'derivations' is, including the one given here. Nevertheless, we choose to describe below a rather detailed procedure, in which it is at least apparent what approximations and assumptions are used.

We use the Boussinesq approximation, i.e., Equations (12.42–12.43) and assume that the flow is two-dimensional (plane parallel with all the functions independent of the Cartesian coordinate y, say, and with $\mathbf{v} \cdot \hat{\mathbf{y}} = 0$). The nondimensional equations can then be written in the following explicit component form

$$\frac{1}{Pr}(\partial_t u + u\,\partial_x u + w\,\partial_z u) = -\partial_x \Pi + \nabla^2 u$$

$$\frac{1}{Pr}(\partial_t w + u\,\partial_x w + w\,\partial_z w) = -\partial_z \Pi + Ra\,\theta + \nabla^2 w$$

$$(\partial_t \theta + u\,\partial_x \theta + w\,\partial_z \theta) = w + \nabla^2 \theta \qquad (12.83)$$

where, for simplicity, the notation u and w for the x and z velocity components, respectively, has been used. Also $\nabla^2 = \partial_x^2 + \partial_z^2$ here.

Remembering now that in this approximation the velocity field is divergence free, we can make use of the stream function $\psi(x, z, t)$, with a choice appropriate for our geometry (compare to 12.32)

$$w = \partial_x \psi \quad \text{and} \quad u = -\partial_z \psi \qquad (12.84)$$

The substitution of these relations in Equations (12.83) gives rise to a set of three rather complicated equations for the functions ψ, θ and Π. However, if one takes the x derivative of the second equation and subtracts from it the z derivative of the first one, the pressure term drops out and we are left, after some simplifications resulting from straightforward algebra, with the following two equations for the stream function $\psi(x, z, t)$ and temperature deviation $\theta(x, z, t)$

$$\partial_t(\nabla^2\psi) = \partial_z\psi\,\partial_x(\nabla^2\psi) - \partial_x\psi\,\partial_z(\nabla^2\psi) + Pr\,Ra\,\partial_x\theta + Pr\,\nabla^4\psi \qquad (12.85)$$

$$\partial_t\theta = \partial_z\psi\,\partial_x\theta - \partial_x\psi\,\partial_z\theta + \partial_x\psi + \nabla^2\theta \qquad (12.86)$$

The choice of the boundary conditions is also dictated by simplicity (and some physical reasoning) and it is assumed that the temperature variation, vertical velocity $w = \partial_x\psi$, and shear $\partial_z u = -\partial_z^2\psi$ at the top and bottom limiting horizontal surfaces, $z = 0$ and $z = 1$, all vanish. This is supplemented by periodic boundary conditions in the lateral (x) direction, for convenience.

The equations are still too complicated to be solved analytically. We may however exploit a fact that was already known to Rayleigh, namely that the amplitude of a spatial Fourier component of the stream function, whose form is $\propto \sin(\pi a x)\sin(\pi z)$, accompanied by a corresponding temperature deviation $\propto \cos(\pi a x)\sin(\pi z)$ (both clearly satisfy the boundary conditions), will grow if Ra is sufficiently large. This can be seen by direct substitution of these functions into the equations, but to be more systematic we try the double Fourier expansions

$$\psi(x, z, t) = \sum_{m=1}^{\infty} \sum_{n=1}^{\infty} \psi_{mn}(t) \sin(m\pi ax) \sin(n\pi z)$$

$$\theta(x, z, t) = \sum_{m=0}^{\infty} \sum_{n=1}^{\infty} \theta_{mn}(t) \cos(m\pi ax) \sin(n\pi z) \qquad (12.87)$$

which satisfy the boundary conditions. The substitution of these expansions in the Equations (12.85)–(12.86) gives rise to an *infinite* set of ODEs for the mode amplitudes $\psi_{mn}(t)$ and $\theta_{mn}(t)$, and therefore does not, in itself, constitute any progress towards solving the problem. As we have seen in Chapter 7, however, methods for approximate but systematic treatment of such infinite-dimensional ODEs do exist, and one may reduce the dimension drastically and capture the evolution of just the few relevant modes in a small set of amplitude equations. The problem of Raleigh–Bénard convection is a well studied example of the application of such analytical methods, but as stated before we will not discuss this approach to convection here (see Maneville, 1990; and references therein).

We shall follow here Lorentz (1963), who performed an extreme (and essentially ad hoc) truncation leading to the now famous highly non-trivial three-dimensional ODE system. The Ansatz

$$\psi(x, z, t) = \psi_{11}(t) \sin(\pi ax) \sin(\pi z)$$

$$\theta(x, z, t) = \theta_{11} \cos(\pi ax) + \theta_{02}(t) \sin(\pi z) \qquad (12.88)$$

can be shown to satisfy Equations (12.85)–(12.85), provided that

$$\dot{\psi}_{11} = \frac{Pr\,Ra}{\alpha} \theta_{11} - Pr\,\alpha\psi_{11} \qquad (12.89)$$

$$\dot{\theta}_{11} = a\pi\,\psi_{11} - \alpha\theta_{11} - a\pi^2\psi_{11}\theta_{02} \qquad (12.90)$$

and

$$\dot{\theta}_{02} = \frac{1}{2}a\pi^2\psi_{11} - 4\pi^2\theta_{02} \qquad (12.91)$$

where $\alpha \equiv \pi^2(1 + a^2)$.

The particular form of the spatial part of the Ansatz chosen above is not totally arbitrary. It is apparent that it describes a velocity field similar to the convective rolls observed near the onset of instability. Indeed, the velocity components

$$u = -\partial_z\psi \propto \cos(\pi z) \sin(\pi ax)$$

$$w = \partial_x\psi \propto \sin(\pi z) \cos(\pi ax)$$

are appropriate for rolls whose axis is parallel to the $\hat{\mathbf{y}}$ direction. The temperature deviation has two parts. The first gives the temperature difference between the upward and downward moving portions of a convective cell, and the second gives the deviation from the vertical linear temperature profile at the centre of a cell.

The set of the three amplitude equations (12.89)–(12.91) was found by equating coefficients of the different combinations of spatial-mode products. Because of the nonlinearity, the truncation of the Fourier series as above is not quite sufficient. A term containing $\sin(\pi z)\cos(2\pi z)$ appears and this can be split into a sum of $\sin(\pi z)$ and $\sin(3\pi z)$ terms. To effect the truncation and enable closure, leading to the above set of just three amplitude equations, one has to simply drop the relatively rapidly varying $\sin(3\pi z)$ term.

Scaling now the independent and dependent variables as follows: $t' \equiv \alpha t$

$$X(t) \equiv \frac{a\pi^2}{\alpha\sqrt{2}}\psi_{11}(t); \quad Y(t) \equiv \frac{r\pi}{\sqrt{2}}\theta_{11}(t) \text{ and } Z(t) \equiv r\pi\theta_{02}(t)$$

where the reduced Rayleigh number r is defined as

$$r \equiv \frac{\pi^2 a^2}{\alpha^3}Ra = \frac{a^2}{\pi^4(1+a^2)^3}Ra$$

we obtain finally the Lorentz system, as it is known in the chaos theory literature, in its canonical form

$$\begin{aligned} \dot{X} &= \sigma(Y - X) \\ \dot{Y} &= -XZ + rX - Y \\ \dot{Z} &= XY - bZ \end{aligned} \tag{12.92}$$

The parameters of this system are $\sigma \equiv Pr$, $b \equiv 4/(1+a^2) > 0$ (a geometric factor) and r, as defined above, the reduced Rayleigh number.

The Lorentz system has played a paramount role in the development of chaos theory. The geometrical form of its strange attractor is very well known today from popular and semi-popular publications (its buildup from a trajectory is even available as a screen saver for personal computers) and its mathematical behaviour has been extensively studied. We shall discuss it briefly below, but before that we would like to make some final short comments on the relation between this model and the fluid-dynamical convection problem. Lorentz's original paper was published in 1963 in the *Journal of Atmospheric Science*, and it seems that when talking of existing and known flow patterns which 'vary in an irregularly, seemingly haphazard manner, and .... do not appear to repeat their previous history' he had in mind some experiments on thermally driven motions in a rotating annulus, crudely modelling perhaps the Earth's atmosphere. His analysis was done, however, for a two-dimensional Boussinesq fluid layer, such as the one discussed here.

We have also seen that a chaotic model, similar but not identical to the Lorentz model, can be obtained by simplifying the convection problem to the study of the motion of a rising and oscillating fluid element (the Moore and Spiegel oscillator discussed in considerable detail in Chapter 2 of this book). Moreover, as already remarked above, the Lorentz model can arise, almost by inspection, from considerations pertaining to developed convection in an astrophysical context (Spiegel, 1995). It is thus interesting to know whether convection (either in the Rayleigh–Bénard or other physical settings) behaves similarly to the Lorentz model. This seems improbable, at least for the Rayleigh–Bénard case, because of the drastic truncation. Similarly, it is difficult to see how the essence of the enormously complicated thick convective layers of astrophysical fluids can be captured by the general arguments of the type mentioned above. It is perhaps for this reason that Lorentz's paper, as well as the work of Moore and Spiegel, were largely ignored for at least a decade after their publication. Only in the late 1970s and early 1980s, after deterministic chaos appeared at the centre of the scientific stage, was it realised that real convection may exhibit chaotic behaviour, similar in many ways to simple generic chaotic systems.

It has been found that truncations, similar to the one performed above but with a larger number of modes, have led to systems whose behaviour has similar features to the Lorentz system, including the presence of strange attractors. It is very clear, however, that an explanation of the behaviour of even a system as simple as the Rayleigh–Bénard experiment is qualitatively beyond the capabilities of the Lorentz model. For example the flow loses spatial two-dimensionality and becomes turbulent most probably via the quasiperiodic Ruelle–Takens–Newhouse scenario (see Chapter 4). This differs from the transition to chaos in the Lorentz system (see below) and, in any case, fluid turbulence is more complex than just temporal chaos. In addition, it has been shown with the help of numerical experiments that two dimensions are not sufficient for a Rayleigh–Bénard system to exhibit turbulent behaviour. It is worth mentioning, however, that a number of mathematical investigations of the Navier–Stokes equations in general, and the Rayleigh–Bénard system in particular, indicate that both two- and three-dimensional flows can be completely described by a *finite* number of modes (see the last chapter of Lichtenberg & Lieberman, 1992; for references).

Behaviour of the Lorentz system

The Lorentz system (12.92) has been studied extensively since the publication of the discoveries (see the early chapters of this book) leading to the enormous interest in deterministic chaos. Sparrow (1982) devoted a whole book to the behaviour of this system, a fact which is certainly remarkable in view of the 'innocent' look of the Lorentz equations.

We shall now employ some of the methods introduced in the first part of this book to briefly analyse the Lorentz system. It will be helpful to remember that the state variables X, Y, Z have specific physical interpretation in the convection problem. Variable X is related to the amplitude of the streamfunction and as such reflects the intensity of the convective motion in the rolls. Variable Y reflects the maximal temperature difference between the rising and falling fluid, and Z measures the distortion of the temperature vertical profile from a linear one.

By a mere inspection of the system it is easy to see that it is invariant under the transformation $X \rightarrow -X, Y \rightarrow -Y, Z \rightarrow Z$. It is also useful to note that, in an actual physical setup, the parameters σ and b are essentially determined by the properties of the fluid and the geometry, while r can quite easily be changed in an experimental setup (by adjusting the bottom–top temperature difference). Thus r appears to be the natural control parameter for the problem. In his original paper Lorentz chose $\sigma = 10, b = 8/3$ (that is, $a = 1/\sqrt{2}$) and we shall refer here to this choice as the 'canonical values'. He then integrated the equations numerically for various values of the control parameter r.

Below we summarise the properties of the Lorentz system (the first three items belong to 'local geometrical properties' in the nomenclature of Chapter 3).

- *Phase volume:* Calculating the divergence of the phase-space velocity-vector field we find

$$D = \frac{\partial \dot{X}}{\partial X} + \frac{\partial \dot{Y}}{\partial Y} + \frac{\partial \dot{Z}}{\partial Z} = -(b + \sigma + 1)$$

Thus we have $D < 0$ for the canonical choice, i.e., phase volume shrinks and therefore the system possesses an attractor whose dimension is less than 3. The rate of contraction for these parameter values is quite large. As $D \approx -13.7$, the volume contracts by a factor of $e^D \sim 10^{-6}$ in unit time. Moreover, it can be shown that the solutions are all bounded for positive times, and even very large initial values of X, Y, or Z decay by the motion and remain bounded for all times.
- *Fixed points:* Three fixed points $\dot{X} = \dot{Y} = \dot{Z} = 0$ are readily found.

  (i) $\mathbf{X}_0$ with $X_0 = Y_0 = Z_0 = 0$, the origin, corresponding to the state of no convection, that is, pure heat conduction with no fluid motion.
  (ii) The pair $\mathbf{X}_{1,2}$ with $X_{1,2} = Y_{1,2} = \pm\sqrt{b(r-1)}$, $Z_{1,2} = r - 1$, corresponding to two symmetric, convective steady states *provided* $r > 1$, that is, these fixed points exist only when the Rayleigh number is above its critical value, which is $Ra_c \approx 657.5$ (for the canonical value $a^2 = 1/2$).

- *Linear stability of fixed points:* If one performs a linear stability analysis of the fixed points (see Chapter 3) it is easy to see the following.

(i) The fixed point at the origin X_0 is a stable attracting point for $r < 1$. For this value of the parameter it is also the only fixed point. If $r > 1$ this point loses stability (one eigenvalue acquires a positive real part) and the two additional fixed points appear. The stability of the origin depends only on the value of the Rayleigh number.

(ii) The fixed points X_1 and X_2, which appear for $r > 1$, are attracting (stable) as long as $1 < r < r_1$, with

$$r_1 = \frac{\sigma(\sigma + b + 3)}{\sigma - b - 1}$$

and lose stability at $r = r_1$. Note that this condition can be satisfied only if $\sigma > (b + 1)$, and the actual value of r_1 depends on σ and b. With the canonical choice we have $r_1 = 24.74 \ldots$. This loss of stability of the pair of fixed points means that the convective states lose their ordered rolls structure.

- *Nonlinear aspects of the system's behaviour – chaos:* The description here is based on extensive numerical and analytical studies of the Lorentz system (see, e.g., Sparrow's book) and we use concepts introduced in Chapter 4 of this book.

  (i) At $r = 1$, when the origin loses stability and two stable spiral points appear, the system undergoes a bifurcation. It is important to note that the two spiral points initially have separate basins of attraction, and *almost* all (that is, except a set of measure zero) trajectories converge to one of these points. The rarely occurring trajectories that do not do so still stay in the vicinity of the origin.

  (ii) Near $r = 13.926$, the origin becomes a homoclinic point and the basins of attraction of the two stable spiral points are no longer clearly distinct. This means that trajectories can cross backward and forward, between the neighbourhoods of the two points, before settling down.

  (iii) The bifurcation point $r = r_1$ is an *inverted* Hopf bifurcation, namely, as the fixed points lose their stability they do *not* develop into stable limit cycles. For $r > r_1$ there is no simple attractor, however, as we have seen above, the phase volume shrinks. In his numerical study Lorentz (1963) followed the system's trajectory for $r = 28 > r_1$, and the initial condition $X(0) = Z(0) = 0, Y(0) = 1$. After some transient oscillations the motion was found to become erratic. During this phase, when different integrations were started at the same prescribed phase point the results differed after some time! This result obviously reflects SIC and the divergence of nearby trajectories on a strange attractor, but for Lorentz it was surprising that a difference in just the computer truncation error could produce such trajectory difference.

  (iv) The motion on the Lorentz attractor (as it is now called) can be characterised by oscillations, i.e., spiralling around one of the fixed points, and after some time a rather abrupt 'jump' of the trajectory to the neighbourhood of the second fixed point. This is followed by a spiralling motion around that point for some arbitrary time until a jump back (to the first point's neighbourhood) occurs, and so on. This combination of spiralling (out), along the fixed point's unstable manifold and later returning to that point's vicinity along its stable manifold, gives rise to the creation of complexity by means of the 'stretching and folding' action by the phase flow. The motion on the Lorentz's attractor has been proven to have a positive Liapunov exponent and to be mixing and ergodic (Bunimovich & Sinai, 1980). We have defined the last two notions in Chapter 6 for Hamiltonian systems, but dissipative systems may possess these properties as well.

(v) Other predictive criteria (see Chapter 4) indicate chaotic behaviour as well. The Fourier spectrum of $X(t)$, for example, is continuous and wide (see e.g., Lichtenberg & Lieberman, 1992), Poincaré sections display typical (to chaos) behaviour and a mapping $Z_{n+1} = F(Z_n)$, where Z_m indicates the values of the variable Z at successive *maxima*, has everywhere a slope whose magnitude is greater than unity and is therefore chaotic. Remarkably, this return map was found and published in 1963 by Lorentz himself, long before the occurrence of chaos and its significance in simple mappings was elucidated in the work of Feigenbaum and others.

• *Special properties*: It is perhaps interesting to note some idiosyncratic properties of the Lorentz attractor, which are not necessarily present in other chaotic systems.

(i) While the strange attractor appears as the only possibility emerging at $r = r_1$, when r is *increased* from below (the fixed points are destroyed at this point), when one *decreases* r down from above $r_1 \approx 24.74$, the strange attractor does not disappear until $r_2 \approx 24.04$. The system thus exhibits hysteresis, with three attractors (two stable fixed points and a strange attractor) coexisting in the narrow range $r_2 < r < r_1$ in this case.

(ii) The sequence of events leading to chaotic behaviour, which occurs when r is increased towards r_1, does not include periodic regimes or a nice period-doubling sequence (which are among the generic routes to chaos, as explained in Chapter 4). For very large values of r, however, periodic limit cycles have been found to appear with the strange attractor ceasing to exist. The periodic behaviour is then superseded by the return of chaotic behaviour, for somewhat larger values of r. Thus, we may say that *intermittency* is observed for large r in the Lorentz system. The first two periodic windows are for the r values approximately in the range 145–148 and 210–234.

In Figure 12.2 a typical trajectory on the Lorentz attractor is shown. This is a familiar picture, which has become very famous in the popular literature and is now one of the chaos icons in the context of what has been termed the 'butterfly effect' in weather unpredictability (i.e., SIC, see Chapter 4).

To sum up, we reiterate that the behaviour of fluid systems, and even the particular setup for which the Lorentz system has been 'derived', cannot be faithfully modelled by the system's behaviour, certainly not in the chaotic regime. The most problematic step in the derivation is, as we have seen, the severe truncation of the Fourier expansion. We have already remarked that models, in which significantly more modes are retained, do seem to ultimately give rise to motion resembling trajectories on the Lorentz attractor, and moreover studies of the Navier–Stokes

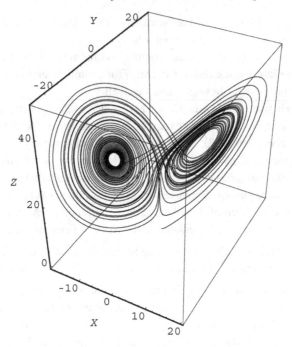

Fig. 12.2. A chaotic trajectory on the Lorentz attractor computed for $r = 28$ and the canonical values of σ and b

system indicate that a *finite* number of modes are sufficient for a complete description of the solutions. It appears, however, that we are still far from an understanding of hydrodynamical turbulence, of the type arising in Rayleigh–Bénard convection, for example. Temporal chaos, of the type occurring in the Lorentz system, even when complemented by new techniques of pattern theory (see Chapter 7), are still insufficient to accomplish this understanding. More than 25 years after the discoveries that brought deterministic chaos to centre stage in research, and made it to be regarded as a promising tool towards the cracking of the turbulence enigma, it is quite disappointing to note that this long-lasting problem still remains rather far from being understood in a satisfactory manner.

12.5 Hamiltonian formulations and related topics

In this book we have reviewed some of the extensive knowledge that is available on chaotic behaviour of ODE systems. In particular, Chapter 6 is devoted in its entirety to Hamiltonian systems, which allow elegant mathematical formulations of their properties and behaviour. Astrophysical applications of Hamiltonian dynamical system theory are reviewed in Chapter 9 in the context of planetary, stellar and galactic dynamics. We have seen that the Hamiltonian formulation can often be exploited in gaining deep understanding of the system (for example, the connection between symmetry properties and conservation laws), and also sometimes enables the use of effective, practical and computational tools (e.g., symplectic integrators).

A priori, it seems that there is little connection between the complicated PDEs of fluid dynamics discussed in this chapter, describing the motion of generally dissipative continuous media, and the elegant mathematical structures encountered in Hamiltonian dynamics. We have already introduced, however, the concept of a 'fluid Lagrangian' (9.11) when discussing the role of Hamiltonian chaos in tidal capture binaries (Chapter 9). It turns out that a Hamiltonian formulation of fluid dynamics is indeed possible, at least for special types of flows.

In the beginning of this chapter we pointed out that a hydrodynamical flow, that is, a solution of the fluid dynamical equations may be viewed, analogously to ODE flows, as a continuous transformation of space onto itself. This was naturally apparent in the Lagrangian description of fluid motion, in which one uses the initial positions of the fluid elements as their Lagrangian labels. Thus the motion of a fluid particle can be described by the function $\mathbf{x}(\mathbf{X}, t)$, giving the position, at time t, of the fluid particle labelled by $\mathbf{X}$ (its position at time $t = 0$). As an analogy to discrete particle dynamics (see Chapter 6) we may form the Lagrangian appropriate for fluid particle dynamics by summing (that is, integrating) over the continuous distribution of the particles

$$L = \int d^3\mathbf{X}\, \rho_0 \left[\frac{1}{2}\dot{\mathbf{x}} \cdot \dot{\mathbf{x}} - \varepsilon - \Phi(\mathbf{x}) \right] \tag{12.93}$$

where $\rho_0 = \rho(\mathbf{X}, 0)$ is the initial (at $t = 0$) density of the fluid particle whose initial position is $\mathbf{X}$, that is, $\rho_0 d^3\mathbf{X}$ is the mass of the differential fluid element centred on position $\mathbf{X}$ at time $t = 0$.

The integrand (the Lagrangian density) consists, as usual, of the kinetic energy minus the potential energy (both per unit mass) of the element. The total 'potential energy' is taken, in the fluid case, to also include the thermodynamic internal energy and not only the part arising from body forces (the assumption that body forces are derivable from a potential Φ is made, as usual).

Because we use here the Langrangian formulation of fluid dynamics (that is, the one in which one follows fluid particles) in all the terms of (12.93) that are functions of position, the particle position function $\mathbf{x}(\mathbf{X}, t)$ is used. Thus, the velocity is

$$\dot{\mathbf{x}} = \left(\frac{\partial}{\partial t} \mathbf{x}(\mathbf{X}, t) \right)_{\mathbf{X}} \tag{12.94}$$

where we have opted for the 'overdot' notation for the Lagrangian time derivative of the position function, instead of D/Dt (compare with 12.3), for convenience.

Also, in the case of *barotropic* flow, which we are assuming here, the internal energy is a function of specific volume (the inverse of the density) alone, which may change as the fluid element is moving, that is, we have

$$\varepsilon = \varepsilon \left(\rho_0^{-1} \frac{\partial (\mathbf{x})}{\partial (\mathbf{X})} \right) \tag{12.95}$$

Here the relation (12.1) has been used in the Lagrangian mass conservation equation (12.12). Finally, the potential energy Φ depends on $\mathbf{x}(\mathbf{X}, t)$. Thus the action is actually expressed as a functional of $\mathbf{x}$, and its derivatives. The Hamiltonian variational principle is achieved in this notation by postulating the stationarity of the action $\int L \mathrm{d}t$ with respect to arbitrary variations $\delta \mathbf{x}$ in the fluid particles' locations.

A direct calculation (we omit it here for the sake of economy, see Salmon, 1998) then gives the components of the equation of motion

$$\frac{\mathrm{d}^2 x_j}{\mathrm{d}t^2} = -\frac{1}{\rho} \partial_j P - \partial_j \Phi \tag{12.96}$$

where P has been *defined* as $P = -\mathrm{d}\varepsilon/\mathrm{d}(1/\rho)$ and is thus identical the thermodynamic pressure.

This equation is obviously identical to the Euler equation and has been derived from a Hamiltonian variational principle, formulated for barotropic ideal (that is inviscid) fluids with the body forces being derivable from a potential. A variational principle for a non-barotropic fluid can be formulated as well, but the specific energy must then be a function of an additional thermodynamic function (e.g., specific entropy). Although an analogous momentum equation can be derived from the variational principle in this case as well, issues relating to thermodynamic equilibrium and to the need for an additional equation (e.g., for the entropy evolution) have to be taken care of.

Among the important advantages of a Hamiltonian formulation is, as we have mentioned above, the correspondence between symmetry properties of the Lagrangian and the conservation laws of the resulting dynamical equation. Explicit time and position invariance corresponds, as is well known, to energy and momentum conservation. It is worth mentioning that in fluid dynamics particle-relabelling symmetry (the Lagrangian does not change if one chooses to vary the particle labels) corresponds to the conservation law embodied in Ertel's theorem (12.54). The proof of this assertion is somewhat lengthy and we refer the reader to Salmon (1998) for the details of the derivation of this exciting result.

Hamilton's variational principle using the Lagrangian (12.93) corresponds, as we have seen, to the Lagrangian formulation (or notation) in fluid dynamics. The Eulerian notation (that is, considering the relevant functions as position- and time-dependent fields) is however the more common one in a variety of fluid-dynamical applications. For example, in the tidal-capture problem discussed in Chapter 9 the fluid Lagrangian (9.11) was written in the Eulerian notation. Unfortunately, the various Eulerian forms of Hamilton's principle for fluid dynamics are quite cumbersome and the derivations are non-trivial.

The idea behind the transformation to Eulerian form is quite simple, however. The Lagrangian form of Hamilton's principle given above was the requirement that the action defined with the Lagrangian (12.93) should be stationary with respect to arbitrary variations $\delta\mathbf{x}$. A simple exchange between the dependent and independent variables, that is between $\mathbf{x}$ and $\mathbf{X}$ is needed, in order to write the Lagrangian density as a function of $\mathbf{X}$ and its derivatives, and the action is then found by integrating on $d^3\mathbf{x}$. The Hamiltonian principle in its Eulerian form will thus follow from the requirement that this action be stationary for arbitrary variations $\delta\mathbf{X}$.

The complications that arise are from the requirement to express the velocity $\mathbf{v} \equiv \dot{\mathbf{x}}$ as derivatives of $\mathbf{X}$. While we have obviously

$$(\partial_t + \mathbf{v} \cdot \nabla)\mathbf{X} = \frac{D\mathbf{X}}{Dt} = 0 \tag{12.97}$$

the natural way to exploit this relation in the transformation to be effected is to append these three equations as constraints. Thus the action written in the Eulerian notation, that is

$$L = \int d^3\mathbf{x}\, \rho_0 \frac{\partial(\mathbf{X})}{\partial(\mathbf{x})} \left[\frac{1}{2}\mathbf{v} \cdot \mathbf{v} - \varepsilon - \Phi(\mathbf{x}) - \boldsymbol{\Theta} \cdot \frac{D\mathbf{X}}{Dt} \right] \tag{12.98}$$

Here we have

$$\varepsilon = \varepsilon\left(\rho_0 \frac{\partial(\mathbf{X})}{\partial(\mathbf{x})} \right) \tag{12.99}$$

that is, again we assume a barotropic flow and $\boldsymbol{\Theta}$ is a vector of Lagrange multipliers.

The form (12.98) of the Lagrangian makes the computation of the variation in the general case quite involved (see Salmon, 1998). It is, however, simplified considerably for flows that are vorticity free (that is, potential flows). In such a case a velocity potential exists, so that $\mathbf{v} = \nabla\phi$, and as it turns out the above-mentioned constraints can be expressed in the Lagrangian density with the help of the velocity potential. The fluid Lagrangian then takes on the form

$$L = -\int d^3\mathbf{x}\, \rho \left[\partial_t\phi + \frac{1}{2}\nabla\phi \cdot \nabla\phi + \varepsilon(\rho^{-1}) + \Phi(\mathbf{x}) \right] \tag{12.100}$$

where we have explicitly written the density function ρ in place of $\rho_0\partial(\mathbf{X})/\partial(\mathbf{x})$. For details of the procedure for bringing the fluid Lagrangian of a barotropic potential flow to the form (12.100), see Salmon (1998) and Bateman (1944).

We shall now consider a particular example of a Hamiltonian dynamical system that is based on the fluid-dynamical equations. This system does not have direct relevance to the general topic of Hamiltonian formulation of fluid dynamics, which we have discussed above, but is a rather special case of flow for which a traditional fluid-dynamical approach leads to a Hamiltonian system. The system arises when one considers a two-dimensional incompressible flow, a situation in which the vorticity has only one component and it is simply related to the stream function ψ. Using Equations (12.33) and (12.34), which are applicable to a z-independent (in Cartesian coordinates) flow, and denoting the z component of the vorticity by the scalar ω, we know (it was shown explicitly in Section 2 of this chapter) that this scalar field is actually the source of the stream function in a Poisson equation

$$\nabla^2 \psi = -\omega \tag{12.101}$$

Assume now that the conditions are such that the vorticity in a flow of this kind is essentially zero almost everywhere, that is, except for a finite number of regions of small spatial extension. Such a situation, in which the flow possesses localised regions of high vorticity, is not uncommon; GFD and AFD flows, especially when they are rotating, often exhibit strong and tight vortices. Hurricanes and tornadoes in the Earth's atmosphere, vortical motion in the ocean, and the famous vortices on the Jovian planets are being continuously studied, observationally and theoretically. Vortices probably abound also on rotating astrophysical objects, like rotating stars and accretion discs. The literature on these subjects is truly vast and we shall just mention a few rather recent contributions in which the existence of coherent vortices on accretion discs and their effects have been investigated (Abramowicz *et al.*, 1992; Bracco *et. al.*, 1999; Li *et al.*, 2001; Umurhan & Regev, 2004).

We shall consider here a two-dimensional incompressible flow in which the vorticity field is approximated by a number of discrete singular points, outside of which the vorticity is zero. These point vortices (actually lines, parallel to the z-axis, say) are thus singular sources of the stream function in the Poisson equation (12.101), analogously to the case of line charges in potential (e.g., electrical or gravitational) theory. Let there be n such point vortices, each of a given strength Γ_i. Then we can write

$$\omega(\mathbf{x}) = \sum_{i=1}^{n} \Gamma_i \delta(\mathbf{x} - \mathbf{x}_i)$$

where $\mathbf{x} \equiv x\hat{\mathbf{x}} + y\hat{\mathbf{y}}$ is the position vector in two dimensions, $\mathbf{x}_i$ is the location of the ith point vortex, and δ is Dirac's delta function.

It is not difficult to see that the solution of the Poisson equation for this kind of source is

$$\psi(\mathbf{x}) = -\frac{1}{2\pi} \sum_{i=1}^{n} \Gamma_i \ln|\mathbf{x} - \mathbf{x}_i| \tag{12.102}$$

as is the case for a collection of line charges in electrostatics. Note that any function $\Psi(\mathbf{x})$, whose Laplacian vanishes, may be added to $\psi(\mathbf{x})$ and the sum will also be a solution. Uniqueness is achieved by setting the boundary conditions, as usual.

The Cartesian velocity components (in Eulerian notation) are obviously given by the appropriate spatial derivatives of ψ as in Equation (12.32). We will, however, be interested here in the motion of the point vortices themselves, and this can best be expressed in the Lagrangian notation, giving the motion of a fluid particle which is at point (x, y) at time t, say. The instantaneous velocity components of this particle (understood in the Lagrangian sense), are simply $\dot{x} = v_x$ and $\dot{y} = v_y$ and thus the equations of motions of this (and for that matter any other) fluid particle are

$$\dot{x} = \frac{\partial \psi(x, y, t)}{\partial y}$$

$$\dot{y} = -\frac{\partial \psi(x, y, t)}{\partial x} \tag{12.103}$$

Using Equation (12.102) for ψ we can now express the two velocity components as

$$\dot{x} = \partial_y \psi(x, y) = -\frac{1}{2\pi} \sum_{i=1}^{n} \Gamma_i \frac{(y - y_i)}{|\mathbf{x} - \mathbf{x}_i|^2}$$

$$\dot{y} = -\partial_x \psi(x, y) = \frac{1}{2\pi} \sum_{i=1}^{n} \Gamma_i \frac{(x - x_i)}{|\mathbf{x} - \mathbf{x}_i|^2} \tag{12.104}$$

Equations (12.104) describe the motion of fluid particles in the flow whose stream function is (12.102), that is, under the influence of a 'field' of the collection of point vortices. In particular, it is interesting to examine the motion of the fluid particle located on a particular vortex (the kth one, say), which is actually the motion of the point vortex itself. This is easily done by looking at (12.104) for $\mathbf{x} = \mathbf{x}_k$. To avoid singularity, the term $i = k$ in the sum must be excluded (this is a diverging self-interaction term of the familiar type). It is then possible to write the equations of motion of the kth vortex in the form

$$\Gamma_k \dot{x}_k = \partial_{y_k} H; \qquad \Gamma_k \dot{y}_k = -\partial_{x_k} H \tag{12.105}$$

where

$$H(\mathbf{x}_i, \mathbf{x}_j) \equiv -\frac{1}{4\pi} \sum_{i,j} \Gamma_i \Gamma_j \ln |\mathbf{x}_i - \mathbf{x}_j| \tag{12.106}$$

where the terms $i = j$ are obviously excluded from the double summation.

A short reflection on (12.105) for $1 \leq k \leq n$ is enough to identify here a bona fide Hamiltonian system, describing n 'particles', whose conjugate phase coordinates and momenta are given by

$$q_k = \sqrt{|\Gamma_k|}x_k \quad \text{and} \quad p_k = \text{sgn}(\Gamma_k)\sqrt{|\Gamma_k|}y_k$$

respectively, and the Hamiltonian function is simply (12.106). Note that in the definitions of the coordinates and momenta, we have explicitly allowed for the strength parameter to take on negative values, reflecting the sense of the vortex motion. The sign function introduced in p_k takes care of this possibility in a proper way.

Numerical studies of an n vortex system defined by (12.105) have revealed that for $n > 3$ this system is generically non-integrable, displaying the typical characteristics of Hamiltonian chaos. Little is known, however, on the limit $n \gg 1$, which might be naively interpreted as approximately representing a flow with a continuous vorticity distribution. For a review on this problem see Aref (1983).

It is conceivable that a problem of this kind may be formulated for an astrophysically interesting case, including perhaps a background flow (introduced, for example, through the gauge function Ψ), to gain some preliminary basic understanding on the role and effect of vortices in flows that are rich in them.

We now wish to introduce and briefly discuss another fluid-dynamical topic, which is not directly connected to Hamiltonian dynamics, but some of its aspects are formally related and analogous to Hamiltonian systems. Already at the beginning of this chapter it was explained that a hydrodynamical flow may be viewed as a differentiable mapping of space onto itself. This can be expressed for any time t by $\mathbf{x} = \mathbf{\Phi}_t(\mathbf{X})$ with $\mathbf{X} = \mathbf{\Phi}_{t=0}(\mathbf{X})$, and in this way a vector function $\mathbf{x}(t)$ is obtained for every initial condition $\mathbf{x}(0) = \mathbf{X}$.

A function of this kind is also obtained, for every initial condition, as a solution of any well posed ODE dynamical system, as for the one defined by the simple looking Equation (12.107), say. The main difference between the two cases is in the nature of the equations for which the flow is a solution. In general, an ODE flow $\mathbf{x}(t)$ is the solution for an ODE

$$\dot{\mathbf{x}} = \mathbf{F}(\mathbf{x}, t) \tag{12.107}$$

where $\mathbf{F}$ is some prescribed well behaved function and an initial condition of the type $\mathbf{x}(0) = \mathbf{X}$ is given. In contrast, the equations generating hydrodynamical flows look, in general, much more complex than that and it is very difficult to solve them in general, but the resulting flows can be expressed in the Lagrangian representation of fluid dynamics in the same way as ODE flows. An ODE flow is a mapping of *phase* space onto itself, while in a hydrodynamical flow the relevant space is the Euclidean *physical* space itself. However, this fact should not make much difference in the geometrical properties of the flows, and if a particular hydrodynamical flow is *known* the possibility of analysing it with the help of tools developed for ODE flows (and described in the first part of this book) obviously exists.

A number of physically interesting hydrodynamical flow characteristics, that are important in a variety of applications, can be qualitatively and quantitatively studied using the above mentioned methods. The nature of these studies remains however purely kinematical because, as hinted above, the methods are usually applied to *known* fluid flows. Solving for them is another matter. In describing the mathematical tools devised for the study of ODE systems, both dissipative and Hamiltonian (in Chapters 3,4 and 6), we have often used geometrical characterisation of the motion in phase space. In particular, we have seen that complex geometrical structures arise in chaotic phase flows (see Chapter 4). In dissipative systems, we have explored how strange attractors of fractal dimension may arise as a result of shrinking and repetitive stretching and folding of volumes in phase space by the flow. In Hamiltonian flows volume is conserved, but complexity may still arise as a result of the breakup of KAM tori (see Chapter 6) on which regular periodic or quasiperiodic Hamiltonian motion proceeds. Chaotic regions may start appearing in phase space, interwoven with regions of regular motion, and ultimately (when a suitable control parameter passes a critical value) chaos becomes widespread.

All chaotic ODE flows invariably greatly distort simple regions of phase space (like, e.g., balls) into highly complex structures and in doing so essentially mix, in a sense, phase space. This notion was given a somewhat more precise meaning in Chapter 6 in the context of Hamiltonian systems, but the stretching and folding, which is present in dissipative systems as well, must give rise to distortion and mixing also.

If we encounter a hydrodynamical flow (be it a volume-preserving one or not) which is analogous, in the sense explained above to a chaotic ODE flow, it is to be expected that it will mix the fluid elements. The kinematics of mixing in hydrodynamical flows and in particular in the chaotic ones has been investigated for a variety of applications. A large number of published articles have been devoted to this and even an entire book, Ottino (1989). Here we shall just give the reader a taste of what is involved, and point out two applications (to mixing of fluid species and MHD dynamo theory) that may have astrophysical consequences.

We shall start, as usual, with a simple example: an incompressible two-dimensional flow. When looking at the system (12.103), describing this flow, one is immediately struck by its Hamiltonian structure (which we have already discussed in the context of point vortex motion). If we consider $x(t)$ as a coordinate of a particle, $y(t)$ as its conjugate momentum, and postulate that the stream function $\psi(x, y, t)$ is the Hamiltonian, then (12.103) are Hamilton's equations of motion for that particle. Thus the physical $x-y$ plane in which our two-dimensional incompressible fluid flow takes place is the phase plane of a Hamiltonian dynamical system. The Hamiltonian flow induced by this system is therefore nothing more than the actual fluid flow (in the fluid-dynamical Lagrangian notation). This is especially appealing if observed experimentally, when fluid particles originating from different parts of the system are physically labelled by dyes of different colour. In these experiments one observes the mixing of the different parts of an incompressible fluid, but the image provides a concrete glimpse into the abstract phase flow of a Hamiltonian system.

If the stream function does not contain time explicitly, the full phase space of the Hamiltonian system (12.103) is two-dimensional, and chaotic solutions are impossible. Thus, the streamlines of the corresponding steady fluid flow are actually the invariant tori, i.e., energy surfaces (here, curves) on which the trajectories of the Hamiltonian system must lie. If, however, the fluid system is experimentally prepared in such a way that $\partial_t \psi \neq 0$, an autonomous dynamical system containing (12.103) also has to include an expression for $\dot{\psi}$, and is thus three-dimensional (losing its exact Hamiltonian analogy, though). We have already seen at least one example of such a case – the forced Duffing system (2.15). If the dissipation term is suppressed ($\delta = 0$) it is a driven Hamiltonian system; however, the resulting three-dimensional autonomous system, is obviously not Hamiltonian (phase space is odd-dimensional). If the driving term has a period, τ say, a stroboscopic picture of the fluid flow (with time intervals τ) represents the Poincaré surface of section of the dynamical-system phase space.

In practice, the analogy between the surface of section of a driven Hamiltonian, one degree of freedom dynamical system, and a stroboscopic picture of an incompressible, two-dimensional unsteady flow, can only be achieved if we have a way to know explicitly the stream function $\psi(x, y, t)$. This is not an easy task but some beautiful examples exist in the literature. For example, Chaiken and his collaborators performed an experiment on a flow having a very small Reynolds number (it is then called *Stokes flow*), taking place between two rotating eccentric cylinders. In Stokes flows the advective term in the Navier–Stokes equation (12.19) is neglected with respect to the viscous term. If the flow is steady, the Navier–Stokes equation for such a flow is

$$0 = -\nabla P + \nu \nabla^2 \mathbf{v} \tag{12.108}$$

where ν is the kinematic viscosity ($\nu = \eta/\rho$) and is assumed constant. This equation can be reduced to the biharmonic equation for the stream function if one takes its curl. Because the flow is two-dimensional (taking place in the x–y plane, say), the vorticity can be expressed as the Laplacian of the stream function as shown in Equation (12.34). Thus we get

$$\nabla^4 \psi = 0 \tag{12.109}$$

In two dimensions, closed-form solutions of this equation are known in the geometry described above, when one of the cylinders is rotating at a fixed angular velocity and the other one stationary. The streamlines seen in such an experiment faithfully represent the orbits of the Hamiltonian system, whose time-independent Hamiltonian is equal to the stream function, satisfying (12.109) for the proper boundary conditions.

If the cylinders are now rotated in an alternate and repetitive way, each for a time τ, with the other one being stationary during that time, a stroboscobic picture of the resulting unsteady flow reproduces well the surface-of-section of the corresponding driven Hamiltonian system. The degree of chaos (see Chapter 6) depends in this case on the rotation rates of the cylinders and on the degree of their eccentricity. The correspondence between computer-generated surfaces of section and the stroboscopic pictures is remarkable. The figures are reproduced in Tabor (1989), where the reference to the original article can also be found.

The chaos appearing in the mixing pattern of such types of flow is referred to as *chaotic advection* or *Lagrangian chaos*, a term reflecting its connection to the motion of fluid particles. It is not related a priori to the phenomenon of true fluid turbulence, in which the velocity field **v** and the other fluid Eulerian fields exhibit seemingly random fluctuations as a result of flow instabilities (see below, however). Indeed, the velocity of a fluid exhibiting Lagrangian chaos can be found by taking partial space derivatives of ψ, which is a perfectly well behaved time-dependent function. More examples of Lagrangian chaos can be found in Ottino (1989) (see also Moon, 1992).

It seems rather unlikely that the abstract surface-of-section corresponding to an actual astrophysical system, modelled by a periodically driven two-dimensional Hamiltonian system, can be realised in the lab by stroboscopically photographing the mixing of a fluid in some contrived container; but the possibility alone is very exciting and challenging. We shall conclude this last chapter of the book by mentioning some examples, in which techniques used in the analysis of ODE flows (Hamiltonian as well as dissipative), have proven themselves useful in the study of mixing and distortion of passive (or active) physical fields in hydrodynamical and magnetohydrodynamical flows.

A Rayleigh–Bénard-like convection problem, but containing a fluid mixture of two species, has been experimentally investigated in the context of mixing (of a passive scalar, in this case) in a number of studies. Knobloch and his collaborators considered the corresponding theoretical problem, using methods from Hamiltonian dynamical system theory. As we have seen in the discussion of the Rayleigh–Bénard problem earlier in this chapter, the stream-function formulation is viable in this case. We recall that as long as the gradually increased Rayleigh number is small enough, the flow remains steady (non-convecting at first and supporting a steady roll pattern later). Ultimately, the steady roll pattern (whose nature depends on the boundary conditions) loses its stability and the problem becomes time-dependent. The problem, when the control parameter is close to its critical value, can be studied semi-analytically using perturbation methods. The streamfunction for a steady flow (or equivalently, in the case of plane waves, when it is expressed in the co-moving frame) is a time-independent integrable Hamiltonian, in the sense expressed in Equation (12.103) and discussed after the equation. The streamlines of the flow (lines of constant streamfunction) look simple, but they reveal the existence of regions containing trapped (inside a fixed region, or a region co-moving with the wave) particles, which are separated from regions containing untrapped particles by a streamline – actually a heteroclinic orbit of the Hamiltonian system. When the control parameter is above the critical value and the streamfunction becomes time-dependent (actually, a small periodic forcing term is added to it), this structure represents a surface of section of the driven Hamiltonian system and a typical structure is revealed. The separatrix, corresponding to a heteroclinic orbit in the steady case, thickens into a layer containing chaotic orbits. The trapped and untrapped particle regions, which are now smaller, are thus separated by diffusing and mixing particles. For additional details on this problem and quantitative estimates characterising the mixing see Lichtenberg & Lieberman (1992) and the original paper by Knobloch and Weiss (1989) and references therein.

As we have seen in the introductory sections of this chapter, hydrodynamical flows invariably transport various physical properties of the fluid by advection. If s is some scalar property, per unit mass, of the fluid, the advection term in a flow, characterised by the velocity field $\mathbf{v}$, is the second term of the material derivative (see Equation 12.4). Vector properties are obviously advected as well, but the appropriate term is a little more complicated, because it has to include not only the fact that the vector is being carried by the fluid, but also that it is being turned and stretched by the motion. In (12.11) we have seen how these effects can be written for an infinitesimal material line element, but for a general vector field $\mathbf{F}$, say, the advective contribution to its evolution (in an incompressible flow) is $(\mathbf{v} \cdot \nabla)\mathbf{F} - (\mathbf{v} \cdot \nabla)\mathbf{F}$ (see for example the case when $\mathbf{F}$ is the vorticity in an incompressible flow, i.e., Equation 12.31 with $\nu = 0$, see below).

The expressions for the full kinematical evolution of scalar and vector fields in fluid flows also contain the local change, and therefore one talks of suitably defined material time derivatives

$$\frac{Ds}{Dt} \equiv \frac{\partial s}{\partial t} + \mathbf{v} \cdot \nabla s$$

for transport of a scalar field s, and

$$\frac{\mathcal{D}\mathbf{F}}{\mathcal{D}t} \equiv \frac{\partial \mathbf{F}}{\partial t} + (\mathbf{v} \cdot \nabla)\mathbf{F} - (\mathbf{F} \cdot \nabla)\mathbf{u}$$

for transport of a vector field **F**, by an incompressible flow. The second of these is the *Lie derivative*, a concept in differential geometry, which may be familiar to the reader from its use in General Relativity, for example. These material derivatives are used in the evolution equations, which in general are statements equating them to possible diffusive processes that proceed at the microscopic level. For example, in (12.39) the scalar field is the temperature and we have on the right-hand side a diffusion term resulting from heat conduction

$$\frac{DT}{Dt} = \kappa \nabla^2 T \tag{12.110}$$

On the other hand (12.31), expressing the evolution of the vorticity *vector* in an incompressible flow, can be expressed as

$$\frac{\mathcal{D}\omega}{\mathcal{D}t} = \nu \nabla^2 \omega \tag{12.111}$$

where the diffusive right-hand-side term is the contribution due to viscosity.

When the hydrodynamical velocity field (the flow) is given, the transport problem, like the one given in Equation (12.110) or (12.111), is purely kinematical. If the transported fields back-react on the flow, their transport equations become coupled to the hydrodynamical equations, and the resulting dynamical problem becomes significantly more complicated. But there is still considerable interest in the simpler kinematical problem, especially the one consisting of studying the effect of a fluid flow on an immersed vector field through (12.111), for prescribed **v**(**x**, *t*).

There exist two physically important vector fields, that seem to evolve kinematically in an analogous way (in a prescribed flow). The first of these, the vorticity $\omega(\mathbf{x}, t)$, has already been mentioned, and if the flow does not admit dissipation (is inviscid, in this case) and is incompressible we get from (12.111)

$$\frac{\mathcal{D}\omega}{\mathcal{D}t} = 0 \tag{12.112}$$

For compressible fluids a similar expression exists for potential vorticity, but only if the flow is barotropic (see Equation 12.52). Since the early studies of G.I. Taylor, it has been observed that vorticity is typically enhanced and concentrated in turbulent flows. Therefore, studies of (12.112) for velocity fields that produce such vorticity concentrations have been conducted, and in particular, more recently, for chaotic flows.

Balmforth *et al.* (1993) studied some mathematical properties of the advection of vector fields by chaotic flows, governed by an equation like (12.112), but for a general vector field **V**. In addition to being relevant for the vorticity (or potential vorticity) evolution problem in prescribed flows – the *kinematic* turbulence problem, as these authors called it – it is also applicable to magnetohydrodynamics. The vector field then is the magnetic field $\mathbf{B}(\mathbf{x}, t)$ itself or $\mathbf{B}/\rho$ in compressible flows. In the perfectly conducting limit the magnetohydrodynamical (MHD) equation describing the evolution of the magnetic field is simply

$$\frac{D}{Dt}\mathbf{B} - (\mathbf{B} \cdot \nabla)\mathbf{v} \equiv \frac{\mathcal{D}\mathbf{B}}{\mathcal{D}t} = 0 \qquad (12.113)$$

for an incompressible flow and with $\mathbf{B}$ replaced by $\mathbf{B}/\rho$ otherwise. This well known result, from which the Alfvén notion of 'frozen-in' magnetic-field lines or flux is obtained, is a direct consequence of the pre-Maxwell system of MHD equations.

As we have not discussed MHD equations in this book, the interested reader is referred to virtually any book on magnetohydrodynamics, but preferably to Shu (1992), where the subject is presented in a physically comprehensive yet succinct way. The analogy between vorticity and the magnetic-field advection thus seems complete, but while there exists a direct relation between the former and the flow field $\boldsymbol{\omega} = \nabla \times \mathbf{v}$, no analogous relation exists for $\mathbf{B}$. This should not cause, however, any difference, as long as only kinematics is concerned. In the case of $\mathbf{B}$ the problem is referred to as the *kinematic dynamo* and its importance to MHD in general, and to geophysical and astrophysical applications in particular, is obvious. Balmforth *et al.* (1993) studied the kinematic dynamo problem for a particular chaotic flow, generated by a suitably chosen ODE and also for a chaotic mapping, and defined an integral transfer operator, which gives the large scale evolution of the field. This work, which included only a couple of examples, has been superseded by the book of Childress and Gilbert (1995), in which the subject of the fast MHD dynamo has been described and treated in detail.

It is certainly beyond the scope of this (already rather expansive) book to go into the, frequently nontrivial, details of this problem. We would like, however, to encourage readers who are interested in the dynamo problem to study this important work, and we believe that our book and in particular its first part provides sufficient mathematical background for that. Childress and Gilbert (1995) explain in detail, using simple examples at first, how fast dynamos, that is, processes by which the magnetic field grows exponentially, can occur in chaotic flows. The idea that stretching, twisting and folding of magnetic-field lines, as a result of chaotic flows, is responsible for the dynamo action is examined and appropriate mathematical methods, both analytical and numerical, are introduced and explained. Finally, the difficult *dynamical* problem is introduced and discussed briefly. The magnetic-dynamo problem is probably one of the most difficult applications of methods from nonlinear dynamics, chaos and complexity to astrophysics. It is only appropriate to end this book expressing the hope that significant progress in the understanding of this problem is possible and imminent, and that it will be achieved through the work of one (or more) of the readers of this book.

References

Some books are to be tasted,
others to be swallowed,
and some few to be chewed and digested.
Francis Bacon, *Of Studies*.

Aarseth, S. (1998), in *Impact of Modern Dynamics in Astronomy* (IAU Coll 172), ed. J. Henrard & S. Ferraz-Mello, New York, Springer.

Abarbanel, H.D.I., Brown, R., Sidorovich, J.J. & Tsimring, L.S. (1993), *Reviews of Modern Physics*, **65**, 1331.

Abraham, R.H. & Marsden, J.E. (1978), *Foundations of Mechanics*, Reading MA, Benjamin.

Abraham, R.H. & Shaw, C.D. (1992), *Dynamics, the Geometry of Behaviour*, New York, Addison-Wesley.

Abramowicz, M.A., Lanza, A., Spiegel, E.A. & Szuszkiewicz, E. (1992), *Nature*, **356**, 41.

Aharonson, V., Regev, O. & Shaviv, N. (1994), *Astrophysical Journal*, **426**, 621.

Aikawa, T. (1987), *Astrophysics & Space Science*, **139**, 281.

Applegate, J., Douglas, M.R., Gursel, Y., Sussman, G.J. & Wisdom, J. (1986), *Astronomical Journal*, **92**, 176.

Aranson, I., Meerson, B. & Sasorov, P.V. (1993), *Physical Review*, **E 47**, 4337.

Aref, H. (1983), *Annual Review of Fluid Mechanics*, **15**, 345.

Arnold, V.I. (1978), *Mathematical Methods of Classical Mechanics*, New York, Springer.

Auvergne, M. & Baglin, A. (1985), *Astronomy & Astrophysics*, **142**, 388.

Bak, P. & Chen, K. (2001), *Physical Review Letters*, **86**, 4215.

Bak, P., Chen, K. & Tang, C. (1990), *Physics Letters*, **A147**, 297.

Bak, P., Tang, C. & Weisenfeld, K. (1988), *Physical Review*, **A38**, 364.

Baker, N., Moore, D. & Spiegel, E.A. (1966), *Astronomical Journal*, **71**, 845.

(1971), *Quarterly Journal of Mechanics & Applied Mathematics*, **24**, 389.

Balbus, S.A. & Hawley, J.F. (1998), *Reviews of Modern Physics*, **70**, 1.

Balmforth, N.J., Meachem, S.P., Spiegel, E.A. & Young, W.R. (1992), *Annals of the New York Academy of Sciences*, **675**, 53.

Balmforth, N.J., Cvitanović, P., Ierley, G.R., Spiegel, E.A. & Vattay, G. (1993), *Annals of the New York Academy of Sciences*, **706**, 148.

Barnsley, M. (1988), *Fractals Everywhere*, San Diego, Academic Press.

Bateman, H. (1994), *Partial Differential Equations of Mathematical Physics*, New York, Dover.

Battaner, E. (1996), *Astrophysical Fluid Dynamics*, Cambridge, Cambridge University Press.

Bayly, B.J., Orszag, S.A. & Herbert, T. (1988), *Annual Review of Fluid Mechanics*, **20**, 359.

Bazell, D. & Désert, F.X. (1988), *Astrophysical Journal*, **333**, 353.

Bender, C.M. & Orszag, S.A. (1999), *Advanced Mathematical Methods for Scientists and Engineers*, New York, Springer.

Bergé, P., Pomeau, Y. & Vidal, C. (1986), *Order within Chaos*, New York, Wiley.

Binney, J. (1982), *Monthly Notices of the Royal Astronomical Society*, **201**, 1.

Binney, J. & Tremaine, S. (1987), *Galactic Dynamics*, Princeton, Princeton University Press.

Blitz, L. & Williams, J.P. (1997), *Astrophysical Journal*, **488**, L145.

Boas, M. (1983), *Mathematical Methods in the Physical Sciences*, 2nd edn., New York, Wiley.

Bohr, T., Jensen, M.H., Paladin, G. & Vulpiani, M. (1998), *Dynamical System Approach to Turbulence*, Cambridge, Cambridge University Press.

Bracco, A., Chavanis, P.H., Provenzale, A. & Spiegel, E.A. (1999), *Physics of Fluids*, **11**, 2280.

Brouwer, D. & Clemence, G.M. (1961), *Methods of Celestial Mechanics*, San Diego, Academic Press.

Buchler, J.R. (1993), *Astrophysics & Space Science*, **210**, 9.

(1998), in *A Half Century of Stellar Pulsation Interpretation: A Tribute to Arthur N. Cox*, eds. P.A Bradley and J.A. Guzik, ASP Conference Series **135**, 220.

Buchler, J.R. & Regev, O. (1982), *Astrophysical Journal*, **263**, 312.

Buchler, J.R., Serre, T., Kolláth, Z. & Mattei, J. (1995), *Physical Review Letters*, **73**, 842.

Buchler, J.R., Serre, T., Kolláth, Z. & Mattei, J. (1996), *Astrophysical Journal*, **462**, 489.

Buchler, J.R., Kolláth, Z. & Cadmus Jr., R.R. (2004), *Astrophysical Journal*, **613**, 532.

Bunimovich, L.A. & Sinai, Ya.G. (1980), *Communications in Mathematical Physics*, **78**, 247.

Butler, K. & Farell, B. (1992), *Physics of Fluids A*, **4(8)**, 1637.

Casuso, E. & Beckman, J. (2002), *Publications Astronomical Society Japan*, **54**, 405.

Celnikier, L.M. (1977), *Astronomy & Astrophysics*, **60**, 421.

Chagelishvili, G.D., Zahn, J.-P., Tevzadze, A.G. & Lominadze, J.G. (2003), *Astronomy & Astrophysics*, **402**, 401.

Chandrasekhar, S. (1961), *Hydrodynamic and Hydromagnetic Stability*, Oxford, Oxford University Press.

Childress, S. & Gilbert, A.D. (1995), *Stretch, Twist and Fold: The Fast Dynamo*, New York, Springer.

Coddington, E.A. & N. Levinson, N. (1955), *Theory of Ordinary Differential Equations*, New York, McGraw-Hill.

Coleman, P.H., Pietronero, L., & Sanders, R.H. (1988), *Astronomy & Astrophysics*, **200**, L32.

Contopoulos, G. (1987), *Annals of the New York Academy of Sciences*, **497**, 1.

(2002), *Order and Chaos in Dynamical Astronomy*, New York, Springer.

Coullet, P. & Elphick, C. (1987), *Physics Letters*, **A121**, 233.

Coullet, P. & Spiegel, E.A. (1983), *SIAM Journal of Applied Mathematics*, **43**, 775.

Coullet, P. & Tresser, C. (1978), *Journal de Physique*, **39**, C5.

Cross, M.C. & Hohenberg, P.C. (1993), *Reviews of Modern Physics*, **65**, 851.

Crutchfield, J.P. & Kaneko, K. (1987), in *Directions in Chaos, Volume 1*, ed. Hao Bai-lin, Singapore, World Scientific, p. 272.

Danby, J.M.A. (1988), *Fundamentals of Celestial Mechanics*, 2nd edn., Richmond VA, Willmann-Bell.

Davis, M. & Peebles, P.J.E. (1983), *Astrophysical Journal*, **267**, 465.

de Vega, H.J., Sánchez, N. & Combes, F. (1996), *Physical Review*, **D54**, 6008.

de Vega, H.J., Sánchez, N. & Combes, F. (1998), *Astrophysical Journal*, **500**, 8.

di Bari, H. & Cipriani, P. (1998), *Planetary and Space Science*, **46**, 1543.

Drazin, P.G. (1992), *Nonlinear Systems*, Cambridge, Cambridge University Press.

Drazin, P.G. & Reid, W.H. (1981), *Hydrodynamic Stability*, Cambridge, Cambridge University Press.

Duncan, M.J. & Lissauer, J.J. (1998), *Icarus*, **134**, 303.

Duncan, M.J. & Quinn, T. (1993), *Annual Review of Astronomy & Astrophysics*, **31**, 265.

Eckmann, J.-P., Järvenpää, E., Järvenpää, M. & Procaccia, I. (2003), in *Simplicity Behind Complexity (Euroattractor 2002)*, W. Klonowski, ed., Lengerich, Pabst Science Publishers.

Elmegreen, B.G. (1997), *Astrophysical Journal*, **477**, 196.

(2002), *Astrophysical Journal*, **564**, 773.

Elmegreen, B.G. & Falgarone, E. (1996), *Astrophysical Journal*, **471**, 816.

Elphick, C., Meron, E. & Spiegel, E.A. (1988), *Physical Review Letters*, **61**, 496.

Elphick, C., Ierley, G.I., Regev, O. & Spiegel, E.A. (1991a), *Physical Review*, **A44**, 1110.

Elphick, C., Regev, O. & Spiegel, E.A. (1991b), *Monthly Notices of the Royal Astronomical Society*, **250**, 617.

Elphick, C., Regev, O. & Shaviv, N. (1992), *Astrophysical Journal*, **392**, 106.

El-Zant, A.A. (1997), *Astronomy & Astrophysics*, **326**, 113.

(2002), *Monthly Notices of the Royal Astronomical Society*, **331**, 23.

Falgarone, E., Phillips, C.K. & Walker, C.K. (1991), *Astrophysical Journal*, **378**, 186.

Farell, B. (1988), *Physics of Fluids*, **31**, 2039.

Farrell, B.F., & Ioannou, P.J. (1998), *Theoretical & Computational Fluid Dynamics*, **11**, 215.

Feigenbaum, M.J. (1978), *Journal of Statistical Physics*, **19**, 25.

Fernandez, J.A. (1994), in *Asteroids, Comets, Meteors (IAU Symp. 160)*, eds. A. Milani, M. Di Martino & A. Cellino, Dordrecht, Kluwer, p. 223.

Ferraz-Mello, S. (1999), *Celestial Mechanics and Dynamical Astronomy*, **64**, 43.

Field, G.B. (1965), *Astrophysical Journal*, **142**, 531.

Frank, J., King, A.R. & Raine, D.J. (2002), *Accretion Power in Astrophysics*, 2nd edn., Cambridge, Cambridge University Press.

Friedman, B. (1990), *Principles and Techniques of Applied Mathematics*, New York, Dover.

Frisch, U., Hasslacher, B. & Pomeau, Y. (1986), *Physical Review Letters*, **56**, 1505.

Gaite, J., Domínguez, A. & Pérez-Mercader, J. (1999), *Astrophysical Journal*, **522**, L5.

Gingold, R.A. & Monaghan, J.J. (1980), *Monthly Notices of the Royal Astronomical Society*, **191**, 897.

Goldreich, P. & Peale, S.J. (1966), *Astronomical Journal*, **71**, 425.

(1968), *Annual Reviews of Astronomy & Astrophysics*, **6**, 287.

Goldstein, H. (1980), *Classical Mechanics*, 2nd edn., San Francisco, Addison-Wesley.

Goodman, J., Heggie, D.C. & Hut, P. (1993), *Astrophysical Journal*, **415**, 715.

Goodman, J. & Schwarzschild, M. (1981), *Astrophysical Journal*, **245**, 1087.

Goupil, M.J. & Buchler, J.R. (1994), *Astronomy & Astrophysics*, **291**, 481.

Grassberger, P. & Procaccia, I. (1983), *Physica*, **9D**, 189.

Gringrod, P. (1991), *Patterns and Waves*, Oxford, Oxford University Press.

Grossmann, S. (2000), *Reviews of Modern Physics*, **72**, 603.

Guckenheimer, J. & Holmes, P. (1983), *Nonlinear Oscillations, Dynamical Systems and Bifurcations of Vector Fields*, New York, Springer.

Gurzadayan, V.G. & Savvidy, G.K. (1986), *Astronomy & Astrophysics*, **160**, 230.

Gutzwiller, M.C. (1990), *Chaos in Classical and Quantum Mechanics*, New York, Springer.

(1998), *Reviews of Modern Physics*, **70**, 589.

Guzzo, L., Iovino, A., Chincarini, G., Giovanelli, R. & Haynes, M.P. (1991), *Astrophysical Journal*, **382**, L5.

Habib, S., Kandrup, H.E. & Mahon, M.E. (1997), *Astrophysical Journal*, **480**, 155.

Hasan, H., Pfenniger, D. & Norman, C.A. (1993), *Astrophysical Journal*, **409**, 91.

Hénon M. (1969), *Quarterly of Applied Mathematics*, **27**, 291.

(1976), *Communications in Mathematical Physics*, **50**, 69.

(1997), *Generating Families in the Restricted Three-body Problem*, New York, Springer.

Hénon M. & Heiles, C. (1964), *Astronomical Journal*, **69**, 73.

Hentschel, H.G.E. & Procaccia, I. (1984), *Physical Review*, **A29**, 1461.

Holman, M., Touma, J. & Tremaine, S. (1997), *Nature*, **386**, 254.

Icke, V., Frank, A. & Heske, A. (1992), *Astronomy & Astrophysics*, **258**, 341.

Ioannou, P.J. & Kakouris, A. (2001), *Astrophysical Journal*, **550**, 931.

Iooss, G. & Joseph, D.D. (1980), *Elementary Stability and Bifurcation Theory*, New York, Springer.

Joseph, D.D. (1976), *Stability of Fluid Motions*, Berlin, Springer.

Kandrup, H.E., Mahon, M.E. & Smith Jr., H. (1994), *Astrophysical Journal*, **428**, 458.

Kaneko, K. (1989), *Physica*, **D94**, 1.

Kaplan, J.L. & Yorke, J.A. (1979), *Communications in Mathematical Physics*, **67**, 93.

Kevorkian, J. & Cole, J.D. (1981), *Perturbation Methods in Applied Mathematics*, New York, Springer.

Klahr, H.H. & Bodenheimer, P. (2003), *Astronomy & Astrophysics*, **582**, 869.

Knobloch, E. & Weiss, J.B. (1989), *Physical Review*, **A40**, 2579.

Kumar, P. & Goodman, J. (1996), *Astrophysical Journal*, **466**, 946.

Kuramoto, Y. (1984), *Chemical Oscillations, Waves and Turbulence*, New York, Springer.

Lamb Jr., G.L. (1980), *Elements of Soliton Theory*, New York, Wiley.

Landau, L.D. & Lifshitz, E.M. (1987), *Fluid Mechanics, 2nd edn.*, London, Pergamon Press.

Larson, R.B. (1981), *Monthly Notices of the Royal Astronomical Society*, **194**, 809.

Laskar, J. (1989), *Nature*, **338**, 237.

(1990), *Icarus*, **88**, 266.

(1991), in *Chaos, Resonance and Collective Phenomena in the Solar System (IAU Symp. 152)*, ed. S. Ferraz-Mello, Kluwer, Dordrecht, p. 1.

(1993), *Physica*, **D67**, 257.

(1994), *Astronomy & Astrophysics*, 287, L9.

(1996), *Celestial Mechanics & Dynamical Astronomy*, **64**, 115.

Laskar, J. & Robutel, P. (1993), *Nature*, **361**, 608.

Laskar, J., Joutel, F. & Robutel, P. (1993), *Nature*, **361**, 615.

Lecar, M., Franklin, F.A., Holman, M.J. & Murray, N.W. (2001), *Annual Review of Astronomy & Astrophysics*, **39**, 581.

Lepp, S., McCray, R., Shull, J.M., Woods, D.T., & Kallman, T. (1985), *Astrophysical Journal*, **288**, 58.

Li, H., Colgate, S.E., Wendroff, B. & Liska, L. (2001), *Astrophysical Journal*, **551**, 874.

Libchaber, A. & Maurer, J. (1980), *Journal de Physique*, **41**, 13.

Lichtenberg, A.J. & Lieberman, M.A. (1992), *Regular and Chaotic Dynamics*, 2nd edn., New York, Springer.

Lissauer, J.J. (1995), *Icarus*, **114**, 217.

(1999), *Reviews of Modern Physics*, **71**, 835.

Livio, M. & Regev, O. (1985), *Astronomy & Astrophysics*, **148**, 133.

Lochner, J.C., Swank, J.H. & Szymkowiak, A.E. (1989), *Astrophysical Journal*, **337**, 823.

Lorentz, E.N. (1963), *Journal of Atmospheric Science*, **20**, 130.

Lynden-Bell, D. (1967), *Monthly Notices of the Royal Astronomical Society*, **136**, 101.

Lynden-Bell, D. & Pringle, J.E. (1974), *Monthly Notices of the Royal Astronomical Society*, **168**, 603.

Maller, A.H. & Bullock, J.S. (2004), *Monthly Notices of the Royal Astronomical Society*, **355**, 694.

Mandelbrot, B. (1982), *The Fractal Geometry of Nature*, San Francisco, Freeman.

Manneville, P. (1990), *Dissipative Structures and Weak Turbulence*, San Diego, Academic Press.

Mardling, R.A. (1995), *Astrophysical Journal*, **450**, 722 & 732.

Mardling, R.A. & Aarseth, S.J. (2001), *Monthly Notices of the Royal Astronomical Society*, **321**, 398.

Martinez, V.J. & Coles, P. (1994), *Astrophysical Journal*, **437**, 550.

May, R.M. (1976), *Nature*, **261**, 459.

Meron, E. (1992), *Physics Reports*, **218**, 1.

Merritt, D. & Fridman, T. (1996), *Astrophysical Journal*, **60**, 136.

Merritt, D. & Valluri, M. (1996), *Astrophysical Journal*, **471**, 82.

Milani, A., Nobili, A.M., Fox, K. & Carpino, M. (1986), *Nature*, **319**, 386.

Miller, R.H. (1964), *Astrophysical Journal*, **140**, 250.

Moon, F.C. (1992), *Chaotic and Fractal Dynamics*, New York, Wiley.

Moons, M. (1997), *Celestial Mechanics & Dynamical Astronomy*, **65**, 175.

Moore, D.W. & Spiegel, E.A. (1966), *Astrophysical Journal*, **143**, 871.

Murante, G., Provenzale, A., Borgani, A., Campos, A. & Yeppes, G. (1996), *Astroparticle Physics*, **5**, 53.

Murante, G., Provenzale, A., Spiegel, E.A. & Thieberger, R. (1997), *Monthly Notices of the Royal Astronomical Society*, **291**, 585.

Murray, N. & Holman, M. (1999), *Science*, **283**, 1877.

Murray, N., Holman, M. & Potter, M. (1998), *Astrophysical Journal*, **116**, 2583.

Nayfeh, A.H. (1973), *Perturbation Methods*, New York, Wiley.

Nayfeh, A.H. & Balachandran, B. (1995), *Applied Nonlinear Dynamics*, New York, Wiley.

Norris, J. & Matilsky, T. (1989), *Astrophysical Journal*, **346**, 912.

Oppenheim, A.V. & Schafer, R.W. (1975), *Digital Signal Processing*, New Jersey, Prentice-Hall.

Ott, E. (1993), *Chaos in Dynamical Systems*, Cambridge, Cambridge University Press.

Ottino, J.M. (1989), *The Kinematics of Mixing: Stretching, Chaos and Transport*, Cambridge, Cambridge University Press.

Paladin, G. & Vulpiani, A. (1987), *Physics Reports*, **156**, 147.

Parisi, G. (1988), *Statistical Field Theory*, Philadelphia, Perseus Publishing.

Pedlosky, J. (1987), *Geophysical Fluid Dynamics*, New York, Springer.

Peebles, P.J.E. (1980), *The Large Scale of the Universe*, Princeton, Princeton University Press.

Perdang, J. & Blacher, S. (1982), *Astronomy & Astrophysics*, **112**, 35.

Pfenniger, D. (1986), *Astronomy & Astrophysics*, **165**, 74.

Pietronero, L., Gabrielli, A. & Sylos Labini, F. (2002), *Physica*, **A306**, 395.

Pismen, L. (1989), *Physical Review*, **A39**, 12.

Pope, S.B. (2001), *Turbulent Flows*, Cambridge, Cambridge University Press.

Pringle, J.E. (1981), *Annual Review of Astronomy & Astrophysics*, **19**, 137.

Provenzale, A., Galeotti, P., Murante, G. & Villone, B. (1992), *Astrophysical Journal*, **401**, 455.

Provenzale, A., Vio, R. & Christiani, S. (1994), *Astrophysical Journal*, **428**, 591.

Provenzale, A., Spiegel, E.A. & Thieberger, R. (1997), *Chaos*, **7**, 82.

Regev, O. & Gitelman, L. (2002), *Astronomy & Astrophysics*, **396**, 623.

Rubinstein, J., Sternberg, P. & Keller, J.B. (1989), *SIAM Journal of Applied Mathematics*, **49**, 116.

Ruelle, D. (1989), *Chaotic Evolution and Strange Attractors: The Statistical Analysis of Time Series for Deterministic Nonlinear Systems*, Cambridge, Cambridge University Press.

Saitou, M., Takeuti, M. & Tanaka, Y. (1989), *Publ. Astronomical Society of Japan*, **41**, 297.

Salmon, R. (1998), *Lectures on Geophysical Fluid Dynamics*, Oxford, Oxford University Press.

Scalo, J. (1987), in *Interstellar Processes*, ed. D.J. Hollenbach & H.A. Thornson Jr., Dordrecht, Reidel, p. 543.

Scaramella, R. *et al.* (1998), *Astronomy & Astrophysics*, **334**, 404.

Scargle, J.D. (1990), *Astrophysical Journal*, **359**, 469.

(1997), in *Astronomical Time Series*, eds. D. Maoz, A. Sternberg, & Leibowitz, E.M., Dordrecht, Kluwer, p.1.

Scargle, J.D., Steiman-Cameron, Young, K., *et al.* (1993), *Astrophysical Journal*, **411**, L91.

Schmid, P.J. & Hennigson, D.S. (2000), *Stability and Transition in Shear Flows*, New York, Springer.

Schwarzschild, M. (1979), *Astrophysical Journal*, **232**, 236.

Seiden, P.E. & Schulman, L.S. (1986), *Science*, **233**, 425.

Sellwood, J.A. (1987), *Annual Review of Astronomy & Astrophysics*, **25**, 151.

Serre, T., Kolláth, Z. & Buchler, J.R. (1996), *Astronomy & Astrophysics*, **311**, 833.

Shakura, N.I. & Sunyaev, R.A. (1973), *Astronomy & Astrophysics*, **24**, 337.

Shaviv, N.J. & Regev, O. (1994), *Physical Review*, **E50**, 2048.

Shore, S.N. (1992), *An Introduction to Astrophysical Fluid Dynamics*, San Diego, Academic Press.

Shu, F.H. (1992), *The Physics of Astrophysics II – Gas Dynamics*, Berkeley, University Science Books.

Simon, N.R. & Schmidt, E.G. (1976), *Astrophysical Journal*, **205**, 162.

Sparrow, C. (1982), *The Lorentz Equations, Chaos and Strange Attractors*, New York, Springer.

Spiegel, E.A. (1971), *Annual Review of Astronomy & Astrophysics*, **10**, 261.

(1972), *Annual Review of Astronomy & Astrophysics*, **11**, 323.

(1993), *Astrophysics & Space Science*, **210**, 33.

(1995), in *Physical Processes in Astrophysics*, eds. E.L. Schatzman, J.-L.. Masnou & I.W. Roxburgh, New York, Springer, p.10.

Spiegel, E.A. & Veronis, G. (1960), *Astrophysical Journal*, **131**, 442.

Stutzki, J., Bensch, F., Heithausen, A., Ossenkopf, V. & Zielinsky, M. (1998), *Astronomy & Astrophysics*, **336**, 697.

Sussman, G.J. & Wisdom, J. (1988), *Science*, **241**, 433.

Sussman, G.J. & Wisdom, J. (1992), *Science*, **257**, 56.

Sylos Labini, F., Montuori, M. & Pietronero, L. (1998), *Physics Reports*, **293**, 61.

Tabor, M. (1989), *Chaos and Integrability in Nonlinear Dynamics*, New York, Wiley.

Tevzadze, A.G., Chagelishvili, G.D., Zahn, J.-P., Chanishvili, R.G., & Lominadze, J.G. (2003), *Astronomy & Astrophysics*, **407**, 779.

Thompson, P.A., (1971), *Compressible Fluid Dynamics*, New York, McGraw-Hill.

Trefethen, L.N., Trefethen, A.E., Reddy, S.C. & Driscoll, T.A. (1993), *Science*, **261**, 578.

Umurhan, O.M. & Regev, O. (2004), *Astronomy & Astrophysics*, **427**, 855.

Valtonen, M. & Mikkola, S. (1991), *Annual Review of Astronomy & Astrophysics*, **29**, 9.

Vázquez-Semadeni, M., Gazol, A. & Scalo, J. (2000), *Astrophysical Journal*, **540**, 271.

Vio, R., Christiani, S., Lessi, O. & Provenzale, A. (1992), *Astrophysical Journal*, **391**, 518.

Vogelaar, M.G.R, Wakker, B.P. & Schwarz, U.J. (1991), in *Fragmentation of Molecular Clouds and Star Formation*, eds. F. Falgarone, F. Boulanger & G. Duvert, Dordrecht, Kluwer.

Voges, W., Atmanspacher, H. & Scheingraber, H. (1987), *Astrophysical Journal*, **320**, 794.

Walgraef, D. (1997), *Spatio-temporal Pattern Formation*, New york, Springer.

Weigend, A.S. & Gershenfeld, N.A. (1994), *Time Series Prediction*, San-Francisco, Addison-Wesley.

White, S.D.M. (1996), in *Gravitational Dynamics*, eds. O. Lahav, E. Terlevich & R. Terlevich, Cambridge, Cambridge University Press.

Whitham, G.B. (1974), *Linear and Nonlinear Waves*, New York, Wiley.

Winfree, A.T., Winfree, E.M. & Seifert, H. (1985), *Physica*, **D17**, 109.

Wisdom, J. (1982), *Astronomical Journal*, **87**, 577.

(1987), *Astronomical Journal*, **94**, 1350.

Wisdom, J., Peale, S.J. & Mignard, F. (1984), *Icarus*, **58**, 137.

Wolf, A., Swift, J.B., Swinncy, H.L. & Vastiano, J.A. (1985), *Physica*, **16D**, 285.

Wolfram, C. (1994), *Cellular Automata and Complexity*, Philadelphia, Perseus Books.

Wu, K.K.S., Lahav, O. & Rees, M.J. (1999), *Nature*, **397**, 225.

Yecko, P. (2004), *Astronomy & Astrophysics*, **425**, 385.

Yonehara, A., Mineshige, S. & Welsh, W.F. (1997), *Astrophysical Journal*, **486**, 388.

Young, K. & Scargle, J.D. (1996), *Astrophysical Journal*, **468**, 617.

Zahn, J.-P. & Zinn-Justin, J. eds. (1993), *Astrophysical Fluid Dynamics (Les-Houches – Session XLVII)*, Amsterdam, North-Holland.

Zeldovich, Ya.B. & Pikelner, S.B. (1969), *Soviet Physics JETP*, **29**, 170.

Index

451

Printed in the United States
By Bookmasters